Optimal Control of GREENHOUSE CULTIVATION

Optimal Control of GREENHOUSE CULTIVATION

Gerrit van Straten • Gerard van Willigenburg
Eldert van Henten • Rachel van Ooteghem

CRC Press
Taylor & Francis Group
Boca Raton London New York

CRC Press is an imprint of the
Taylor & Francis Group, an **informa** business

CRC Press
Taylor & Francis Group
6000 Broken Sound Parkway NW, Suite 300
Boca Raton, FL 33487-2742

Printed in the United States of America on acid-free paper
10 9 8 7 6 5 4 3 2 1

International Standard Book Number: 978-1-4200-5961-8 (Hardback)

Library of Congress Cataloging-in-Publication Data

Optimal control of greenhouse cultivation / Gerrit van Straten ... [et al.].
p. cm.
Includes bibliographical references and index.
ISBN 978-1-4200-5961-8 (alk. paper)
1. Greenhouses--Environmental engineering. 2. Greenhouses--Climate. 3. Greenhouses--Heating and ventilation. I. Straten, G. van (Gerrit van), 1946-

SB416.O68 2010
631.5'83--dc22 2010036526

Visit the Taylor & Francis Web site at
http://www.taylorandfrancis.com

and the CRC Press Web site at
http://www.crcpress.com

Contents

Preface

MOTIVATION AND GOAL

With the advancement of more and more sophisticated greenhouses all over the world, automatic control of greenhouse climate has become imperative. Computerized control systems on the market today in countries with a well-developed greenhouse industry offer the grower the opportunity to manipulate the indoor environment according to his wishes. The task of the computer system is to realize the climate schedule desired by the grower and to provide information and feedback about the system's behavior. Automatic control in countries with an upcoming greenhouse industry is developing along the same lines. The focus of commercially available control systems is on control of the climate. The gradual piling up of ad hoc solutions ultimately has led to untransparent systems with hundreds of user settings. How the settings should be chosen is left to the grower. While the importance of these systems for the success of the greenhouse industry is acknowledged, on leaning back, one may ask: Are these systems really solving the true problem of the grower? Will these systems, at the end of the day, yield the largest profit to the grower? Will they lead to the desired sustainability so badly needed in the industry? The answers to these questions have motivated us and other researchers from various parts of the world to rethink the system and to look for different approaches. They form the motivation for writing this book.

This book is about optimal control of greenhouse cultivation. We use the word *cultivation* rather than *climate* to express that it is not so much the indoor climate we are interested in, but rather the cultivation of horticultural crops. Climate control is an instrument to reach that goal, but without considering the crop the system is not complete. Secondly, we use the word *optimal* to express that our aim is to achieve the result in which the grower is interested: maximum profit, within environmental, legal, and societal constraints. When we say optimal, we really mean optimal in an economic sense, and not in the sense of the rather loose use of the word to indicate some measure of best technical performance. Thinking in terms of economic optimality immediately leads to a different view on climate control. Standard setpoint control design is trying to suppress the effect of disturbances. In the case of greenhouses, the major disturbance is the weather. At the same time, it is a resource, as, after all, the greenhouse is a solar collector. So, instead of suppressing the effect of the weather, economic optimal control will try to exploit the opportunities offered by the weather as much as possible.

If the economic problem can be solved by optimal control methods, will the grower then become superfluous? Not at all. The role of the grower as entrepreneur will be as important as always. However, as we hope to demonstrate in the book, the way business information—for instance, about expected prices of sold products or about developments on the energy market—can be conveyed to the daily operation of the greenhouse is completely transparent in the optimal control setting. This is in contrast to the current situation, in which a grower has to translate his economic expectations into hundreds of settings for the greenhouse climate computer, virtually without any help from the system. The grower also remains important in judging risks and setting constraints to the freedom of operation of the optimal controller, if necessary, but in contrast to the current situation, the optimal control framework offers an opportunity to show the effects of his actions.

SCIENCE AND MODELS

Truly optimal control of greenhouse cultivation cannot be reached without using scientific information. The best way to encapsulate, communicate, and implement scientific knowledge is via mechanistic models that describe the dynamics of the climate and the crop. While solutions that formalize expert knowledge in some form may provide useful practical solutions, they will never be able to offer real optimality. The same is true for black-box methods that are based on data for a single particular solution. Therefore, in this book, the emphasis will be on the use of science-based mechanistic dynamic models to achieve our goal. These models must obviously be cast in a form that is suitable for the methodology of optimal control, and for this reason, our methods are based on the state–space representation of the relevant dynamic models.

TARGET GROUP, PHILOSOPHY, AND CONTENTS

The target group we had in mind are researchers working in the area of greenhouse control, as well as engineers employed by system providers. The book may also be of interest to practitioners who are advising growers, or to scientifically trained growers. The book begins with an introductory chapter that briefly points to the growing importance of the greenhouse industry in producing food and flowers worldwide. The greenhouse–crop system is briefly described. This part elucidates the central role of fast physiological crop processes, i.e., photosynthesis and evapotranspiration, in the process, as these affect both the greenhouse climate and crop growth and development. They are therefore the pivot in any integrated greenhouse–crop model. Finally, it is shown that the greenhouse cultivation control problem has a hierarchical structure and that, in general, much less feedback information is available for the crop than for the greenhouse. It sets the scene for what will appear to be a key issue in the book: in which way is information best exchanged between the various hierarchical layers?

Chapter 2 provides a tutorial introduction to optimal control, on the basis of a very simple but illustrative example. Having a model, a goal function, and an optimization method, the control inputs that are required to achieve the best goal function value are computed. To keep things clear, the example avoids uncertainties at this stage. Hence, the solutions obtained are open-loop solutions—the controls are computed in advance and are applied as computed. The purpose is to give the reader a feel for what optimal control is about. In passing, we note here that optimal control is really optimal steering. Finding the open-loop solution is also known in the literature as "dynamic optimization." Throughout the book we will use these terms rather loosely.

In Chapter 3, the theory of open-loop optimal control is given. The necessary conditions for optimality that play a central role in solving the problem are discussed, and important notions like the Hamiltonian and the adjoint variables or costates are introduced. Also, direct and indirect solution methods are summarized and illustrated on the basis of examples. The discussion is kept as concise as possible. The reader who is interested in more in-depth treatment is referred to a number of excellent books on the topics. The main idea here is to demystify the idea that optimal control is really something for the brave. The fact that tools are readily available now should convince the practitioner of the feasibility of the method, we hope.

Up until Chapter 3, uncertainties have not played a significant role. In the greenhouse, however, these uncertainties are dominantly present. For instance, the weather is variable and partly unpredictable, and, hence, our models will not be perfect. Some form of feedback is necessary, thus leading to a closed-loop system. This is the topic dealt with in Chapter 4. An explanation is given of why "standard" linear quadratic feedback is not going to work in greenhouse climate control. The solution is found in time-scale decomposition, which separates the slow crop biomass response from the fast greenhouse climate response. This ultimately leads to a two-step solution. First a slow optimal control problem is solved. This is, in fact, dynamic optimization, and it leads to optimal state trajectories and optimal costate trajectories for the crop states. Next, the online control is realized with a receding horizon optimal controller, which uses the *same economic cost criterion*, augmented with a term to value the long-term development of the crop. It appears that the costates of the slow variables, i.e., the crop biomass, are the pivot variables in this approach. The receding horizon controller is a model-predictive controller that updates its initial state on the basis of observed information, thus providing the feedback that is of paramount importance. However, unlike standard model predictive controller (MPC), the criterion is still economic. The treatise in Chapter 4 can be seen as the core of the methodology developed in the book.

Chapter 5 places the developed optimal control method in the frame of hierarchical control, and it provides an extensive overview of the literature on model-based or model-inspired greenhouse climate control in general. Methods that are not model-based, such as expert systems, are mentioned but not discussed in detail, as they cannot provide optimality. As said before, the key differences between various hierarchical solutions offered in the literature lie in the way information is transferred between the layers. In the hierarchical setup, first a dynamic optimization of the crop system is used, usually considering the greenhouse dynamics as pseudo-static. The focus is on the crop. Then, there are two different lines. One uses the state variables as setpoints for lower level controllers. The controller can be anything in this setup, and various solutions offered in the literature are reviewed. The other is the optimal control solution in Chapter 4, which conveys the slow costates and uses at the lower level again an (economically) optimal controller. Our goal in Chapter 5 is to place the numerous control studies reported in the literature into the perspective of hierarchical optimal control. It is clear that an essential step in achieving optimality is the dynamic optimization over a season on the basis of crop models. If this step would become part of the control system, already a big leap would be made toward true optimality, even if our proposed second step of trying to achieve optimality at the online level as well is not adopted.

The next three chapters discuss examples of various realistic applications of the fully integrated control with optimality at all levels. Chapter 6 deals with lettuce, which is a single-harvest crop. It has, among other topics, an interesting discussion on the economic interpretation of the costates. Chapter 7 describes a real experiment with a continuous-harvest crop, the tomato. A big-leaf, big-fruit model is presented as an approximation of the crop behavior. Calibration of crop and greenhouse models is briefly discussed. Ultimately, a number of interesting outcomes for various periods of the year are discussed at some length. In Chapter 8, the problem is solved for a solar greenhouse to illustrate the applicability of the approach to more complicated systems. It also presents unique comprehensive models for the crop photosynthesis and evapotranspiration and for the greenhouse physics. It also shows that optimal control studies can be used to study the effect of specific pieces of equipment on the overall performance of the system, thus linking control to design. Our purpose in presenting these examples is to show the feasibility of the approach and to point out to the reader a number of specific points that require attention in practical applications.

Finally, in Chapter 9, we sketch a number of exciting developments in the greenhouse industry. Most of these are inspired by the strongly felt need to make the industry more sustainable. Our expectation is that these innovations will call upon more sophisticated control systems than are available today. We argue why the optimal cultivation control methodology presented in this book offers an excellent starting point for the development of the systems of the future. We also discuss a number of societal and organizational bottlenecks that may preclude the adoption of more advanced technologies, together with a number of technical points that need further attention. This chapter is mainly intended as a source of inspiration of further work by scientists and also to give practitioners backing if they want to convince their superiors to move in new directions.

THE AUTHORS' CONTEXT

In agriculture, there are many applications of systems and control theory, but the most eye-catching ones are robotics and climate control. As people interested in control issues, and being at Wageningen University and Research Centre, with its roots in agriculture, it is only natural that we have to work in these directions. In addition, being in The Netherlands, which hosts an important greenhouse industry and possesses a world reputation in advanced greenhouse applications, there was really no way not to become involved in greenhouse climate control. The history of greenhouse work in Wageningen goes back to the 1970s, with pioneers like Alexander Udink ten Cate and Gerard Bot, who started modeling and control in The Netherlands and also had a leading role internationally. In the 1990s, under the inspiring lead of Hugo Challa, who unfortunately died way too early, a working group was established aiming at bringing greenhouse control to a higher scientific level. A large number of Ph.D. students worked in this frame, among them Eldert van Henten, one of the authors of this book, who in The Netherlands pioneered the field covered by this book in the late 1980s and early 1990s. Under the supervision of Jan Bontsema, he introduced the two–time scale approach that is an important theme in this book. He is responsible for Chapter 6 about lettuce application. Further Ph.D. studies were guided by, among others, Gerard van Willigenburg, one of the strongest advocates of the optimal control methodology. He is the principal author of Chapters 2–4, and was strongly urging to perform the practical test of the tomato example in Chapter 7, which resulted in the Ph.D. dissertation of Frank Tap. The current Chapter 7 is a full remake of the work of Tap, for which Gerrit van Straten is responsible. Another student of Gerrit and Gerard was Rachel van Ooteghem, who has written Chapter 8 about the solar greenhouse. Gerrit van Straten is the principal author of the other chapters and the main coordinator of the whole project. Chapter 9 is the result of a number of formal and informal discussions among us, as authors, and with scientists and representatives of internationally operating greenhouse climate computer vendors.

FINALLY

Our main motivation to write the book has not been altered over the years. While the greenhouse industry has a high potential to make a significant contribution to the needs of mankind, its survival will critically depend upon the ability to reach sustainability. Developments toward this goal will lead to more and more complex solutions. Even though this may not yet be visible in upcoming nations, it will be the worldwide trend. It is our strong conviction that well developed modern greenhouse crop cultivation cannot be based upon standard basic control solutions. Far more advanced solutions are needed to reach a profitable and sustainable greenhouse production system without complicating the life of the grower. We hope to have demonstrated in this book that the optimal control framework is, indeed, a most powerful, science-based, and feasible solution to achieve this goal. We hope that the reader will find inspiration in this book, and we will be glad to receive any feedback from our audience.

Gerrit van Straten
Gerard van Willigenburg
Eldert van Henten
Rachel van Ooteghem
Wageningen, The Netherlands

Acknowledgments

With this all Dutch authorship, we are fully aware of the "Dutch" bias in our work. It is evident from Chapter 5 that worldwide many scientists have done very significant work in the area, both on greenhouse climate control in general as well as in bringing optimality toward greenhouse cultivation. We are grateful to have met several of them as visitors or cooperators in a number of projects. In particular, we would like to mention the stimulus received from visiting scholars Marc Tchamitchian, Zaid Chalabi, Qichang Yang, and Mao Hanping. And above all, the stimulating and very instrumental discussions with Ido Seginer and, later, other members of his group, in particular Ilya Ioslovich and Rafi Linker, must be greatly acknowledged. Ido Seginer has certainly been a pioneer in advocating dynamic optimization in greenhouse cultivation for almost half a century. We also would like to acknowledge the vivid exchange of ideas we had, again with Ido and our Belgian partners, in the frame of the European NICOLET project, led by Fokke Buwalda, and later in the frame of the European Watergy project, instigated by Martin Buchholz from the TU Berlin in Germany, and in cooperation with scientists from the experimental station "Las Palmarillas" of Cajamar in Almería, Spain. In view of the multitude of opportunities to exchange ideas with researchers from all over the world, the reader may rightly conclude that this is much less a Dutch project than it may seem.

The writing of the book took quite a bit longer than originally envisaged. The first ideas go back to the IFAC2004 in Prague, where Gerrit van Straten met Frank Lewis, who immediately showed a very instrumental and stimulating enthusiasm. When we started writing, we had quite a different book in mind, and as often in a project like this, there were several deadlocks. In fact, we have been writing nearly two books over time, but it was only after setting our ambitions at more realistic grounds that we could make further progress. We would like to thank Sigurd Skogestad at NTNU in Trondheim, Norway, and Jerónimo Pérez Parra, former director of Cajamar Las Palmarillas in El Ejido, Almería, Spain, for their hospitality during two short sabbatical visits of the first author. These periods have been very important in bringing this long-drawn-out project to an end.

Acknowledgments

Authors

Prof. G. (Gerrit) van Straten (1946) holds an M.Sc. in chemical engineering from Eindhoven University and a Ph.D. from Twente University, The Netherlands, on research related to photosynthesis and algal blooms in surface waters. From 1979 to 1980, he was a visiting scholar and project leader at the International Institute for Applied Systems Analysis (IIASA) in Laxenburg, Austria. In 1990, he accepted a position as full professor at Wageningen University, where he currently leads the Systems and Control Group. The group develops and applies systems and control methodology to study the behavior of dynamical systems in the bio- and agro-sciences and to realize automated systems in the agro, environmental, and food industries. Apart from being author and editor of several books and proceedings, he published over 175 international scientific papers advocating the use of systems modeling and control in a wide spectrum of applications in agriculture, food processing, and environmental technology, with special emphasis on greenhouse cultivation control. He received the IFAC Outstanding Contribution Award and served as Chair of the IFAC Technical Committee on Control in Agriculture (2003–2008). He is also Associated Editor for Control Engineering Practice (CEP) and Editor-in-Chief of Computers and Electronics in Agriculture (COMPAG).

Dr. L. G. (Gerard) van Willigenburg (1958), assistant professor, received his M.Sc. in electrical engineering from Delft University of Technology (1983) and a Ph.D. degree from Delft University of Technology (1991) on digital optimal control of a rigid manipulator. His research interests include digital optimal control, reduced-order control, adaptive control, and model predictive control (receding horizon control). The application areas are indoor climate control (greenhouses and animal housings), robot control, automatic guidance of agricultural field machines, and the control of processes in the food industry (e.g., sterilization and drying).

Prof. E. J. (Eldert) van Henten (1963) received his M.Sc. degree in 1987 with honors and his Ph.D. degree in 1994 in agricultural and environmental sciences at Wageningen University, The Netherlands, with a Ph.D. dissertation entitled "Greenhouse Climate Management: An Optimal Control Approach."

He is working as a senior scientist in the Business Unit Glass of Plant Research International, and as of September 2005 he is also holding a post as full professor of agricultural engineering at the Farm Technology Group of Wageningen University. His research interests are biorobotics, robot motion planning, optimal robot design and modeling and (optimal) control of biological systems with greenhouse crop production as the main application field.

Dr. R. J. C. (Rachel) van Ooteghem, M.Sc. (1969), received her B.Sc. in control systems engineering from the Polytechnic Heerlen, The Netherlands, and her M.Sc. in control systems engineering with honors from the Polytechnic Arnhem & Nijmegen/University of Hertfordshire (1998). In 2007, she received her Ph.D. from Wageningen University on her thesis "Optimal Control of a Solar Greenhouse." Her research is directed to reduction of energy demand in greenhouses and implementation issues of greenhouse climate control for economically optimal crop production.

Notation Conventions

Notation of Greenhouse and Crop Principal Variables

Symbol	Description	Unit	Symbol Variables	Description	Unit
Φ	mass flow rate	kg s^{-1}	φ	specific mass flow rate	kg m^{-2}[gh] s^{-1}
Q	heat exchange rate	W (J s^{-1})	q	specific heat exchange	W m^{-2}[gh]
F	volumetric flow rate	m^3 s^{-1}	j	heat or mass flux (generic)	units m^{-2}[gh]
T	temperature	K	T^C	temperature	°C
C	concentration	kg m^{-3}	p	pressure	Pa (N m^{-2})
			RH	relative humidity	(–)
			v	velocity	m s^{-1}
A	surface area	m^2	V	volume	m^3
W	biomass (dry weight)	kg[dw] m^{-2}[gh]	B	biomass (fresh weight)	kg[fw] m^{-2}[gh]
D	development stage	–	S	integral value	(depends on subscript)

Subscripts

a	greenhouse air below screen	o	outdoor
al	artificial lighting equipment	out	going out of the system
aq	aquifer	p	pipe system (unspecified)
as	greenhouse air above screen	q	with respect to heat
B	assimilate buffer	r	roof (cover)
boil	boiler	ri	roof indoor side
c	crop	ro	roof outdoor side
CO_2	carbon dioxide	roots	roots
e	equipment (utility)	s	upper soil layer
F	fruit	sc	screen
g	greenhouse compartment	sk	sky
he	heat exchanger	ss	lower soil layer (subsoil)
hp	heat pump	stem	stem
H_2O	water	T	with respect to temperature
in	going into the system	u	upper heating net
l	lower heating net	uc	upper cooling net
L	leaf	v	with respect to ventilation
		V	vegetative

Superscripts

Ap	aperture	high	upper bound
cond	conduction	low	lower bound
cons	condensation	lsd	leeward side
dem	demand	lwv	longwave radiation
fw	fresh weight	max	maximum

Superscripts *(continued)*

min	minimum	vp	valve position
ppm	ppm	vent	ventilation
rad	global radiation	wb	wet bulb
ref	reference	wsd	windward side
sat	saturation	%	percentage
sp	setpoint		

Note: *With concentrations*: the first subscript refers to the substance, e.g., CO_2, H_2O, unless it is clear from the context what is meant. Then there is a comma, followed by an indicator for the location/compartment, e.g., a (air), s (soil), and c (crop).

Flows and transport terms are denoted by capital letters if they are in units mass or energy per time unit (e.g., Φ, Q) and in lower case if they refer to flows per unit greenhouse area (e.g., φ, q). The first subscript indicates the substance, e.g., CO_2. It followed by a comma and then source and destination separated by an underscore, e.g., Φ_{CO_2,a_c}.

Superscripts are used to denote specific attributes. If there is no ambiguity, they are left out.

Parameters are often chapter specific. They are defined at first appearance and are also summarized in tables per chapter.

Generic Systems Notation

Variable	Symbol	Vector Dim
State	$\mathbf{x}$	n_x
Control input	$\mathbf{u}$	n_u
External input (disturbance)	$\mathbf{d}$	n_d
System output (noise-free)	$\mathbf{y}$	n_y
Measured (observed) output	$\mathbf{y}^{obs}$	n_y
Auxiliary variables of interest (computable)	$\mathbf{z}$	n_z
Parameters	$\mathbf{p}$	n_p
Costate (adjoint variable)	$\boldsymbol{\lambda}$	n_x
Cost function	J	
Hamiltonian	H	
Terminal cost	Φ	
Penalty	P	
Running cost	L	
Unit price	c	

System Notation Superscripts and Overscripts

Lower bound	x^{min}	Steady state	x^{ss}
Upper bound	x^{max}	Estimated/forecasted	$\hat{x},\hat{d}$
Optimal	x^*	Averaged or filtered/slow subproblem	$\bar{x},\bar{d}$
Slow	x^s	Nominal	d^{nom}
Fast	x^f	Observed	y^{obs}

1 Introduction and Problem Statement

1.1 GREENHOUSE-CROP CULTIVATION—BENEFITS AND CHALLENGES

Of all agricultural production activities, the greenhouse industry is worldwide the fastest growing sector. There are two major reasons for this. First, the greenhouse separates the crop from the environment, thus providing some way of shelter from the direct influence of the external weather conditions. This enables the production of crops that otherwise could not be produced at that specific location. Second, the greenhouse enclosure permits the manipulation of the crop environment. This asset allows the grower to steer the cultivation in a desirable direction. It leads to higher crop yield, prolonged production period, better quality, and less use of protective chemicals. The added value per unit surface area in greenhouse crops is much higher than that in open-field cultivation. The downside of this intensification is that with current designs, greenhouse production has a higher demand per unit area for resources. In moderate climate zones, energy is needed, whereas in (semi) arid zones, the cooling and availability of water is of major concern. On the other hand, in view of the growing concern about sustainability, one has to realize that, after all, the greenhouse is a solar collector, and this will ultimately be another factor that will contribute to the growth of the greenhouse sector in the future.

Over time, greenhouses have evolved from very simple structures with little or no options for control to very advanced, modern industrial structures, with various ways to manipulate the environmental conditions experienced by the plant. A greenhouse in this book means a structure that covers the crop and that has at least one device that can be manipulated to modify the internal environmental conditions. This latter condition excludes simple plastic shields over crops on the open soil. Slightly paraphrasing Hanan (1998), it will also be assumed that the greenhouse is intended to grow crops that have an economic value. A factor that is common to all greenhouses is that solar energy is allowed to enter the structure to provide energy for photosynthesis. We will not deal so much with the plant factory, where the light is coming exclusively from lamps, although the core methodology is applicable there as well. An important implication of having the sun's radiation as input is that it introduces a degree of unpredictability that has to be dealt with. Because the sun is a resource, the task is not so much to suppress its effect on the internal climate but, rather, to exploit it.

In general, it can be assumed that the goal of the grower is to make a profit. The economy of greenhouse cultivation is determined by a number of factors, determined by decisions that the grower has to make. These can be listed as follows:

- Investments in greenhouse type and infrastructure. These are guided by arguments related to the target crops and available designs on the market, which differ in terms of type of substrate, expected resource use, flexibility of operation, degree of automation, and expected performance. Once the structure is chosen, it is relatively difficult to modify it.
- Strategic choice on the kind of crops to be grown and on the initiation of a new batch of crops. Once the decision has been made, it cannot be changed, but another decision is possible for the next batch.

- Operational costs, which exist of more or less fixed capital costs, that is, interest on investment loans and depreciation costs, and variable resource costs, being labor costs, costs of logistics and materials, and energy costs. There is normally a positive correlation between total variable costs and amount of resources used, but the actual amount to be paid can be influenced by contracting.
- Income from selling the crop and other potential revenues against marketing costs. There is in general a positive correlation between crop yield and income, but the actual realization of these relationships depends among other things on product quality, time of delivery, branding, contracting, and market prices.

The use of materials and energy as well as crop yield and quality can be influenced by operating the adjustable components of the greenhouse, such as heating input, window opening, screening, and CO_2 dosage. Hence, it can be expected that the way these controls are operated influences the final economic result. The final result will also be determined by the actual realization of the weather, which is beyond the control of the grower and which will always lead to year-by-year variability in income. No controller can prevent this, but what we can expect from a control strategy is that, ideally, under the given circumstances, the control exploits the opportunities and contributes in the best possible way to the net profit of the grower. To go for this ideal is the main philosophy in this book.

1.2 AUTOMATIC CONTROL

To fully exploit the enhanced possibilities for crop and resource management in greenhouses, it is indispensable to perform the adjustment of the control variables in an automatic way. This is because it is almost impossible for a human being to understand and manipulate systems with more than two dependent processes without additional aid. Changing, for instance, the opening of a window with the purpose of reducing the relative humidity, also will have an effect on temperature and will therefore call for additional measures. Moreover, if the opening of the windows had to be done manually, as in the early days, the labor costs would be unaffordable in our current time. Hence, the introduction of automatic controllers and computer-controlled greenhouses in the second half of the twentieth century was a major step forward to economically attractive crop production. Even the most basic automatic control will enhance the capacities of the greenhouse industry in emerging greenhouse areas all over the world. In addition, the availability of automatic control systems opens up new avenues for optimization of greenhouse-crop cultivation, as will be explored in this book.

1.3 ELEMENTARY DESCRIPTION OF THE GREENHOUSE-CROP SYSTEM

By way of example, Figures 1.1 and 1.2 show two types of differing greenhouses. One is the parral type, as is frequently used in warmer areas, for instance, in the Mediterranean area, Latin America, and several parts of China. The control options are to open the side ventilators, using a roller bar, and the roof ventilators by changing the opening angle. The openings are generally covered by insect screens to prevent insect intrusion. The other is a Venlo type of greenhouse, as in use in moderate climate zones, for instance, in The Netherlands and other Western European countries, North America, and New Zealand. Here, there is a heating system to supply heat, the windows can be opened to provide ventilation, a screen can be used to prevent heat loss during the night (not shown), and there can be CO_2 dosage. There can be short-term and long-term heat buffers, together with heat exchangers and heat pumps, which are left out here for simplicity. Also, the irrigation and fertigation system is left out. It is assumed that these water-sided systems are operated in such a way that water and nutrient supply are not limiting crop growth and development.

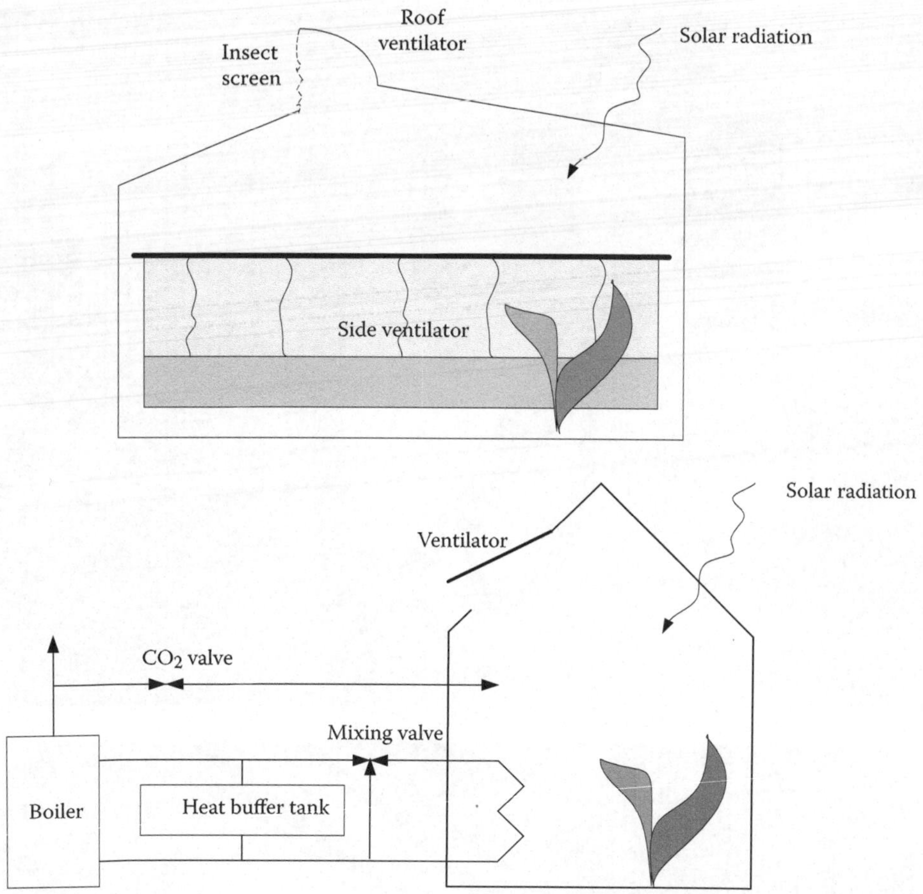

FIGURE 1.1 Schematic examples of greenhouse layouts. Top: parral greenhouse. Bottom: Venlo-type greenhouse.

Radiation from the sun is used for photosynthesis and also acts as a heat source to the greenhouse. Surplus moisture content, generated by crop evapotranspiration, is ventilated to the outside air. Heat losses occur via the greenhouse cover and by ventilation. Ventilation also exchanges CO_2 with the outside air.

The crop experiences the local environmental variables temperature, moisture content, and CO_2 concentration. Crop photosynthesis, evapotranspiration, growth, and development depend on these variables. On the other hand, the crop influences these variables itself via photosynthesis, respiration, and evapotranspiration. The greenhouse climate also depends on the external weather conditions. Apart from solar radiation, the most important external disturbances are outside air temperature, moisture content, and wind speed. Wind speed influences the heat exchange coefficient of the wall and cover and also affects the ventilation rate through openings in the cover.

In the Venlo design, the environmental conditions of the crop can be further manipulated by supplying heat and CO_2. Heating is often also added to the parral greenhouses. Depending on the greenhouse layout, there can be many more control processes that influence the climate and the crop, such as shading, cooling, and supplementary lighting. Other control methods used by the grower are manipulations with the crop itself, like spacing, pruning, removing leaves, and harvesting. These manipulations are not done in an automatic fashion.

Despite the differences between greenhouse nurseries, the essential behavior can be described by generic energy and mass balances. The accumulation of energy, mass, and biomass in greenhouse

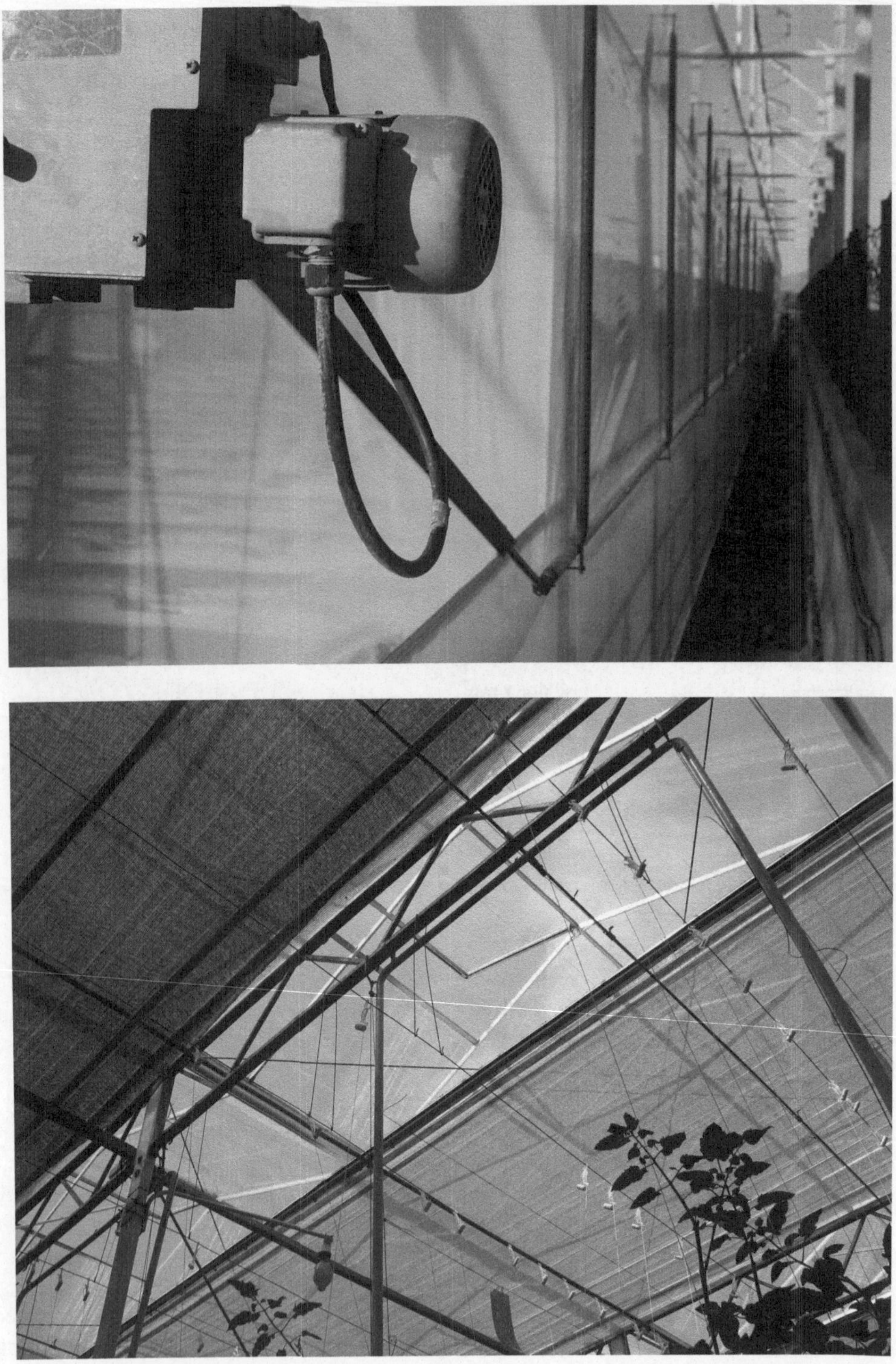

FIGURE 1.2 Details of actuator structures. Top: roll-up side ventilator in a parral greenhouse. Bottom: window opening construction in a modern greenhouse. Photographs by G. van Straten.

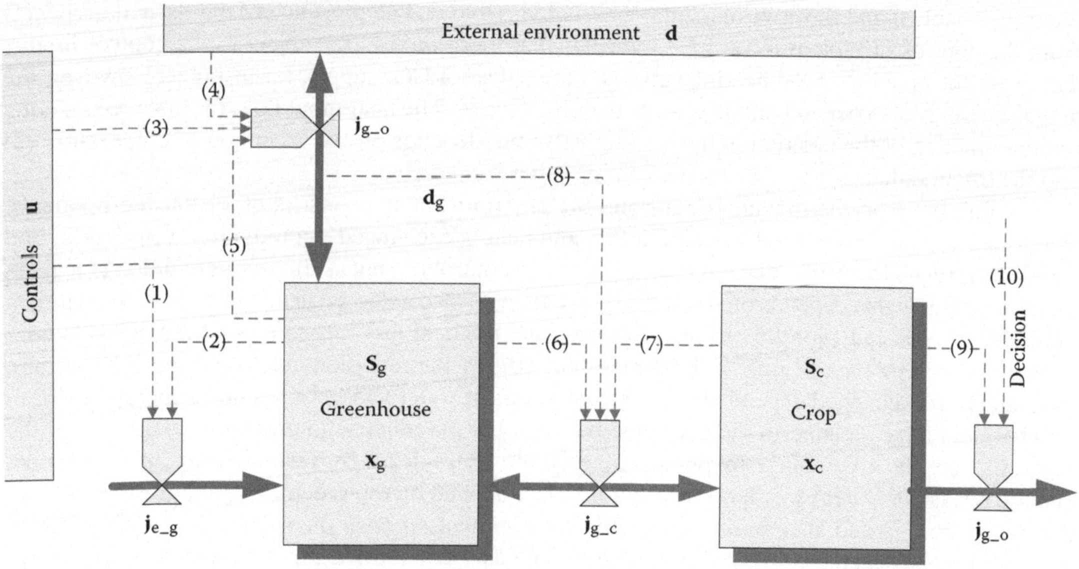

FIGURE 1.3 Elementary energy and mass fluxes and influencing factors in greenhouse-crop systems.

and crop and the physical flows between them are visualized in the scheme in of Figure 1.3 abstract form.

In Figure 1.3, the subscripts g, c, o, and e are used to denote the greenhouse compartment (g), the crop compartment (c), the environment outside the greenhouse (o), and the equipment or utilities that supply the resources (e), respectively. The shadowed rectangles denote the greenhouse and crop compartments that have storage capacity. The stored energies and masses in greenhouse and crop per unit projected greenhouse area are formally denoted here by the vectors $\mathbf{S}_g$ and $\mathbf{S}_c$, respectively. The solid arrows denote flows of energy, water, or carbonaceous material. It is customary to express them per unit projected greenhouse area so that flows become fluxes,* denoted by the vector $\mathbf{j}$. The subscripts and the arrow direction denote in which direction the fluxes are counted positive; for example, $\mathbf{j}_{e\_g}$ represents fluxes of heat and CO_2 toward the greenhouse from the resource utility equipment. Fluxes can be negative, for instance, in the case of withdrawal of energy by pad-and-fan cooling.

The masses and energies $\mathbf{S}_g$ and $\mathbf{S}_c$ depend on the size of the system and are therefore extensive variables. They can easily be coupled to intensive variables such as concentrations or temperatures. The intensive variables are indicated formally in the scheme by the vectors $\mathbf{x}_g$ and $\mathbf{x}_c$ for greenhouse and crop, respectively. Although mass and energy balances are most easily set up in terms of extensive variables, it is often convenient to work with intensive variables, not only because they are more directly related to variables that are measured but also because the fluxes depend directly on these intensive variables. For instance, the flux of carbon dioxide from the greenhouse to the crop depends on the CO_2 concentration as well as—depending on the model—on the leaf area index, an intensive variable that is in a rather complicated way related to the extensive variable crop biomass. Similarly, the heat exchanged by ventilation depends on the temperatures of the greenhouse air and the ambient temperature as well as on the latent heat difference determined by the humidity ratio, which is an intensive variable.

The dashed arrows represent by which factors the fluxes are influenced. They can therefore be seen as information flows. There are two types of influential factors, commonly called inputs: the

* The term "flux" is used here merely as shorthand for "flows per unit greenhouse projected area" and should be sharply distinguished from a flux through an associated contact area.

control variable **u** and the environmental external variable **d**. The presence of the dashed arrow (1) from the control variable **u** toward $\mathbf{j}_{e\_g}$ indicates that these fluxes are subject to the control inputs. These are the opening of the heating valve and the valve for CO_2 supply, for instance. However, the actual flux may also depend on the state of the greenhouse. The heat input flux, for instance, is not a unique function of the position of the heating valve but depends on the greenhouse temperature (2) and the direct radiation received by the heating pipes.

The flux between the greenhouse air and the environment $\mathbf{j}_{g\_o}$ consists of various components. Water and CO_2 are exchanged via ventilation, and heat is exchanged via radiation, ventilation, and transport through the walls. The window opening is a control (3), but as the ventilation flux at a given window opening also depends on the wind speed, there is also a dashed arrow from the environment (4). Similarly, the radiation flux through screens, and the heat loss through thermal screens is controlled not only by the opening of the screens but also by the radiation itself (4). Clearly, moisture, CO_2, and heat exchanged depend on the concentrations of water vapor, CO_2, and temperature (5).

The main fluxes related to the exchange between the greenhouse internal environment and the crop ($\mathbf{j}_{g\_c}$) are the CO_2 uptake by photosynthesis, the CO_2 release by various forms of respiration, and the release of water by evapotranspiration. They depend on the greenhouse states (6) as well as the crop states (7) and also, indirectly, on the environment, *in casu* the solar radiation (8). This is expressed in the scheme by $\mathbf{d}_g$, which can be viewed as a direct throughput; that is, $\mathbf{d}_g$ is an instantaneous function of **d**. By screening or artificial lighting, $\mathbf{d}_g$ can be manipulated and hence have a direct influence on the greenhouse-crop fluxes. Otherwise, these fluxes cannot be manipulated directly, except by measures not related to the greenhouse climate, such as watering and application of growth stimulating or suppressing means. As the crop is harvested, there is a flux of mass from the crop compartment to the environment ($\mathbf{j}_{c\_o}$), depending on the crop state itself (9). Also picking leaves, removing surplus buds, and so forth belong to this group. All of these are generally based on discrete actions. A dashed line marked "decisions" is used in the scheme to indicate these non-automatic control influences (10). The resulting fluxes obviously depend on the state of the crop (9). The measures indicated as "decisions" are not in the scope of the greenhouse controller, but they do affect its operation because they influence the state of the crop. There is no fundamental reason why they cannot be included in the control, but in this book they are not considered further.

The previous description provides the basis for modeling of greenhouse and crops, which plays a central role in design and control of greenhouse cultivation. It should be noted that the controllable fluxes have constraints that are determined not only by the installed capacity but also by the environmental conditions and the system variables. The ventilation flow is an example because at maximum window opening the actual flow still depends on wind speed and greenhouse temperature. Similarly, the maximum heat flow from heating pipe to greenhouse is not fully defined by maximum valve opening but also depends on the temperature difference between boiler temperature and greenhouse air temperature. Hence, controllable fluxes have time-varying constraints, and in modeling one has to be prepared to cope with this additional complication.

1.4 MEASUREMENTS AND INSTRUMENTATION

An important component of modern greenhouses is the instrumentation. Most physical variables relevant in a greenhouse can be measured by automatic sensors. This holds for wet and dry bulb temperature, CO_2 concentration, and relative humidity. The absolute moisture content can be computed from these data. Inside radiation can also be measured, although it is somewhat less common. The most important disturbances can be measured with sensors as well, that is, outdoor temperature, outdoor CO_2 concentration, outdoor relative humidity, wind speed, and diffuse and direct solar radiation. All these data are sampled data, that is, samples are taken and stored electronically at regular interval or, sometimes, only at times when something is changing. Also, in principle, the control inputs are known, although it must be said that these important data are not always recorded. Overall, the measurements provide quite a good input–output picture of the physical part of the greenhouse-crop system. Sensor information about the state of the crop is less easy to obtain and is not standard in the current greenhouse industry, but there are some developments, such as continuous measurement of crop weight on measurement gullies, observations of crop evapotranspiration with lysimeters, and some ways of automatic measurement of photosynthesis, for example, by fluorescence.

An issue of considerable practical interest in installing sensors and using sensor information is that the spatial distribution within the greenhouse is usually not homogeneous. Developments in wireless sensor technology make it possible to deploy a large array of sensors, especially temperature sensors, which gradually allow to see ever increasing detail of the distribution and its dynamics. The existence of spatial distributions is a factor to account for, but it does not preclude the use of optimal control. Therefore, in order not to complicate the treatment of control principles more than necessary, in this book we will pretend that the greenhouse is homogeneous, unless otherwise indicated.

1.5 DECOMPOSITION, FLUXES, AND INFORMATION FLOWS

Formally, mass and energy balances for the greenhouse-crop system have the following general form:

$$\dot{\mathbf{S}}_{\mathrm{g}} = \mathbf{j}_{\mathrm{e\_g}} + \mathbf{j}_{\mathrm{o\_g}} - \mathbf{j}_{\mathrm{g\_c}} + \mathbf{j}_{\mathrm{g\_g}} \tag{1.1}$$

$$\dot{\mathbf{S}}_{\mathrm{c}} = \mathbf{j}_{\mathrm{g\_c}} - \mathbf{j}_{\mathrm{c\_o}} + \mathbf{j}_{\mathrm{c\_c}} \tag{1.2}$$

The additional terms $\mathbf{j}_{\mathrm{g\_g}}$ and $\mathbf{j}_{\mathrm{c\_c}}$ have been introduced here to allow for exchange of mass and energy between various components of the state vector within a compartment: in the greenhouse, for instance, the conversion of vapor into condensed water in the greenhouse or the exchange of heat between greenhouse air and soil, and in the crop, for instance, the conversion from assimilates into structural matter.

It is clear from Equations 1.1 and 1.2 that only the term $\mathbf{j}_{\mathrm{g\_c}}$ appears in both. It underlines what is already obvious from Figure 1.3, namely, that the exchange of energy and matter between greenhouse and crop plays a central role. In practice, $\mathbf{j}_{\mathrm{g\_c}}$ encompasses photosynthesis, respiration, and evapotranspiration.

As the fluxes depend on the intensive variables, such as temperature and concentration, rather than the extensive variables, it is more convenient to set up models of the system in terms of intensive variables. The relation between energy and mass extensive quantities and intensive variables can be expressed formally as

$$\mathbf{S} = \boldsymbol{K}\mathbf{x} \tag{1.3}$$

where matrix $\boldsymbol{K}$ stands for capacities, typically volume for concentrations, and heat capacity for temperature, expressed per unit greenhouse area. Provided that the number of differential equations in Equations 1.1 and 1.2 was sufficient to describe the system, the number of independent intensive variables for which a differential equation is required must be equal to the number of independent extensive variables. Hence, the matrix $\boldsymbol{K}$ is square. It has the principal capacities on the diagonal, but occasionally off-diagonal elements occur; for instance, the extensive variable latent heat is coupled to the intensive variables temperature and moisture content. Written out, we have

$$S_i = k_{i1}x_1 + k_{i2}x_2 + \cdots + k_{in}x_n = \sum_{j=1}^{n} k_{ij}x_j \tag{1.4}$$

so that index ij links the ith extensive variable to the jth intensive variable and where most of $k_{ij}, i \neq j$ are zero.

In principle, the capacities may be functions of the intensive variables. An example is the temperature dependency of density of the air. So, formally, we have

$$\dot{\mathbf{S}} = \dot{\boldsymbol{K}}\mathbf{x} + \boldsymbol{K}\dot{\mathbf{x}} \tag{1.5}$$

Let

$$\dot{k}_{ij} = \sum_m \frac{\mathrm{d}\,k_{ij}}{\mathrm{d}\,x_m}\dot{x}_m \tag{1.6}$$

Then it can be seen that, overall, it is possible to write

$$\dot{\mathbf{S}} = \boldsymbol{M}\dot{\mathbf{x}} \tag{1.7}$$

where $\boldsymbol{M}$ is a complicated capacity term that depends not only on the principal capacities $\boldsymbol{K}$ but also on the working point $\mathbf{x}$ and the sensitivities $\mathrm{d}k_{ij}/\mathrm{d}x_m$ appearing in Equation 1.6. Making the reasonable assumption that the greenhouse capacities do not depend on the crop-intensive variables and vice versa, we may formally write

$$\dot{\mathbf{x}}_{\mathrm{g}} = \boldsymbol{M}_{\mathrm{g}}^{-1}\left(\mathbf{j}_{\mathrm{e\_g}} + \mathbf{j}_{\mathrm{o\_g}} - \mathbf{j}_{\mathrm{g\_c}} + \mathbf{j}_{\mathrm{g\_g}}\right) \tag{1.8}$$

$$\dot{\mathbf{x}}_{\mathrm{c}} = \boldsymbol{M}_{\mathrm{c}}^{-1}\left(\mathbf{j}_{\mathrm{g\_c}} - \mathbf{j}_{\mathrm{c\_o}} + \mathbf{j}_{\mathrm{c\_c}}\right) \tag{1.9}$$

which together with constitutive relations that link the fluxes to the intensive variables yield a model expressed in intensive variables. Because of the dependencies of the expanded capacities $\boldsymbol{M}$ on the actual working point, these equations become nonlinear, even if the fluxes are linear. On the other hand, the dependencies of the capacities on the intensive variables are rather weak over the operating range encountered in greenhouses, and hence the contribution is generally small.

Equations 1.8 and 1.9 show that the central role of the evapotranspiration and the net crop photosynthesis is preserved when the equations are written as differential equations for the intensive variables of greenhouse and crop. In addition, they show that in steady state it suffices to equate the fluxes, and the complication resulting from state-dependent capacities vanishes. A steady state for the crop may seem less relevant, but in crops that continue to deliver fruits, such as tomato, there could be a steady state, and Equation 1.9 together with a policy to maintain the number of fruits constant then yields a harvest control law.

Writing out the flux terms using the information in Figure 1.3 gives

$$\dot{\mathbf{x}}_{\mathrm{g}} = \boldsymbol{M}_{\mathrm{g}}^{-1}\left(\mathbf{j}_{\mathrm{e\_g}}\left(\mathbf{x}_{\mathrm{g}}, \mathbf{u}, \mathbf{d}\right) + \mathbf{j}_{\mathrm{o\_g}}\left(\mathbf{x}_{\mathrm{g}}, \mathbf{u}, \mathbf{d}\right) - \mathbf{j}_{\mathrm{g\_c}}\left(\mathbf{x}_{\mathrm{g}}, \mathbf{x}_{\mathrm{c}}, \mathbf{d}_{\mathrm{g}}(\mathbf{u})\right) + \mathbf{j}_{\mathrm{g\_g}}\left(\mathbf{x}_{\mathrm{g}}, \mathbf{d}_{\mathrm{g}}(\mathbf{u})\right)\right) \tag{1.10}$$

$$\dot{\mathbf{x}}_{\mathrm{c}} = \boldsymbol{M}_{\mathrm{c}}^{-1}\left(\mathbf{j}_{\mathrm{g\_c}}\left(\mathbf{x}_{\mathrm{c}}, \mathbf{x}_{\mathrm{g}}, \mathbf{d}_{\mathrm{g}}(\mathbf{u})\right) - \mathbf{j}_{\mathrm{c\_o}}(\mathbf{x}_{\mathrm{c}}, \mathbf{u}_{\mathrm{dec}}) + \mathbf{j}_{\mathrm{c\_c}}(\mathbf{x}_{\mathrm{c}}, \mathbf{x}_{\mathrm{g}}, \mathbf{d}_{\mathrm{g}}(\mathbf{u})\right) \tag{1.11}$$

or, more concisely, by further aggregating into general nonlinear vector-valued rate of change functions $\mathbf{f}_{\mathrm{g}}$ and $\mathbf{f}_{\mathrm{c}}$,

$$\dot{\mathbf{x}}_g = \mathbf{f}_g(\mathbf{x}_g, \mathbf{x}_c, \mathbf{u}, \mathbf{d}_g(\mathbf{u}), \mathbf{d}) \tag{1.12}$$

$$\dot{\mathbf{x}}_c = \mathbf{f}_c(\mathbf{x}_g, \mathbf{x}_c, \mathbf{d}_g(\mathbf{u}), \mathbf{u}_{dec}) \tag{1.13}$$

In this way, it becomes clear that apart from discrete handling on the crop, the influence of the controls on the crop is indirect via the greenhouse states $\mathbf{x}_g$ and the direct throughput $\mathbf{d}_g(\mathbf{u})$. We may therefore see the greenhouse as the instrument to control the crop.

The main reason to represent the greenhouse-crop system in the form of Equations 1.1 and 1.2 or Equations 1.8 and 1.9 is that it clearly brings out the central role of the greenhouse-crop interactive processes photosynthesis, respiration, and transpiration. In a slightly different form, this is represented again in Figure 1.4. The difference from Figure 1.3 is that $\mathbf{j}_{g\_c}$ now is no longer treated as a physical flow but rather as an information flow that enters both the greenhouse and the crop compartment. A distinctive feature of these elementary crop processes is that they are fast as compared with crop growth and development. In fact, they are usually assumed to be instantaneous. We will, when convenient, denote photosynthesis, respiration, and evapotranspiration as the "fast crop processes."

If greenhouse and crop are modeled as two compartments, with the fast processes in between, there are two ways to draw the subsystem boundaries in Figure 1.4, as shown in Figure 1.5.

The most natural way is to take photosynthesis and transpiration as part of the crop. In that case, the greenhouse states and the direct throughput component of the external inputs (i.e., photosynthetic active radiation) are the inputs for the crop model. The (net) CO_2 and the water vapor fluxes appear as an output, which are taken by the greenhouse model as (disturbance) input. The other way is to incorporate photosynthesis and transpiration as part of the greenhouse model. Then, the greenhouse model has the crop state as (disturbance) input, in addition to the control **u** and the external input **d**. The fast process fluxes are an output, which are taken by the crop model as inputs. At first sight, this may not seem to be a very logical choice, but in approaches that concentrate on the control of the greenhouse, without considering the crop explicitly, it is necessary to incorporate (simple) models for CO_2 uptake and evapotranspiration, and this is provided by this scheme. Moreover, fast processes are kept in the fast compartment, which will turn out to be an advantage in later time scale decompositions.

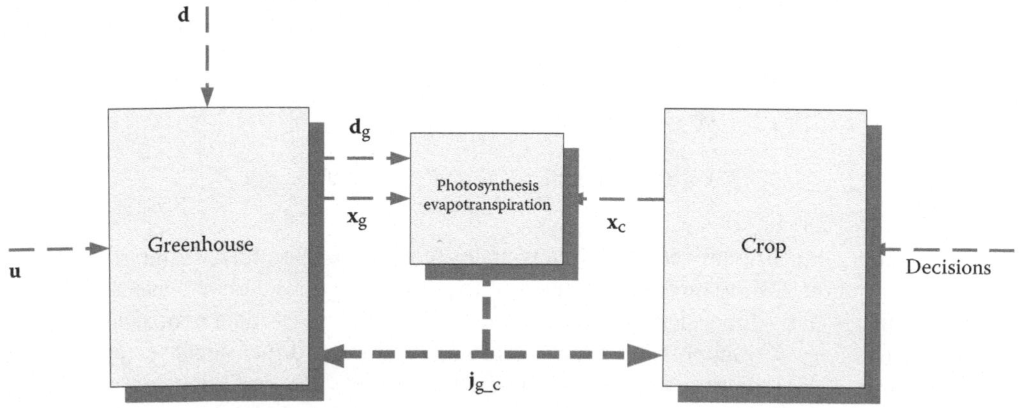

FIGURE 1.4 Central role of photosynthesis and evapotranspiration.

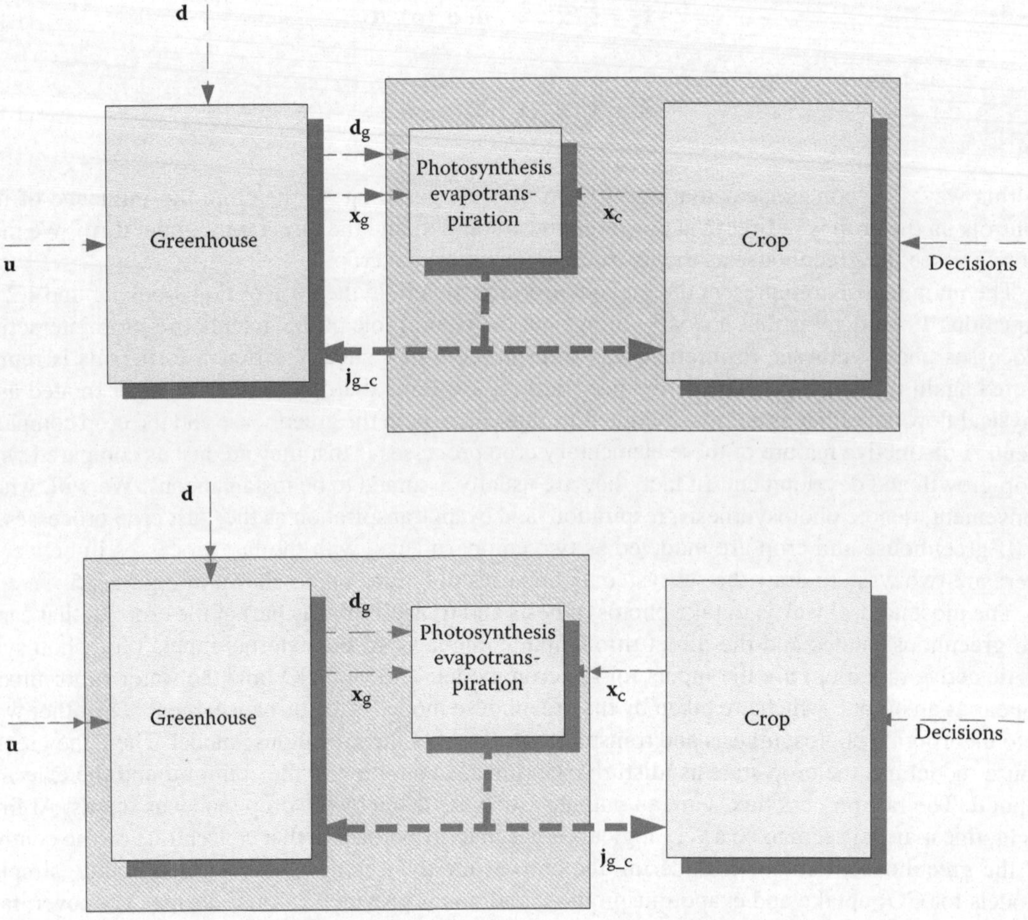

FIGURE 1.5 Two ways of assigning photosynthesis and transpiration. Top: to crop model. Bottom: to greenhouse model.

1.6 GENERAL STATE–SPACE REPRESENTATION

In systems theory terms, the dynamics of the combined greenhouse-crop system as briefly described earlier can be represented by the following general state–space description:

$$\dot{\mathbf{x}}(t) = \mathbf{f}(\mathbf{x}(t), \mathbf{u}(t), \mathbf{d}(t))$$
$$\mathbf{y}(t) = \mathbf{g}(\mathbf{x}(t), \mathbf{u}(t), \mathbf{d}(t)) \tag{1.14}$$
$$\mathbf{x} \in \Re^{n_x}, \quad \mathbf{u} \in \Re^{n_u}, \quad \mathbf{d} \in \Re^{n_d}, \quad \mathbf{y} \in \Re^{n_y}$$

where $\mathbf{x}(t)$ is an n_x-dimensional vector of system states (e.g., air temperature, air moisture content, air CO_2 concentration, assimilate carbon content of the crop, structural carbon content of the crop, fruit weight), $\mathbf{u}(t)$ is an n_u-dimensional vector of control inputs (e.g., heat input or mixing valve position, window opening, CO_2 supply rate, screen position), $\mathbf{d}(t)$ is an n_d-dimensional vector of external disturbances (e.g., solar radiation, outside air temperature, outside CO_2 concentration, wind speed), and $\mathbf{y}(t)$ is an n_y-dimensional vector of outputs (e.g., air temperature, relative humidity, crop dry and fresh weight). The physical model in Equations 1.1 and 1.2 usually provides us with a natural choice

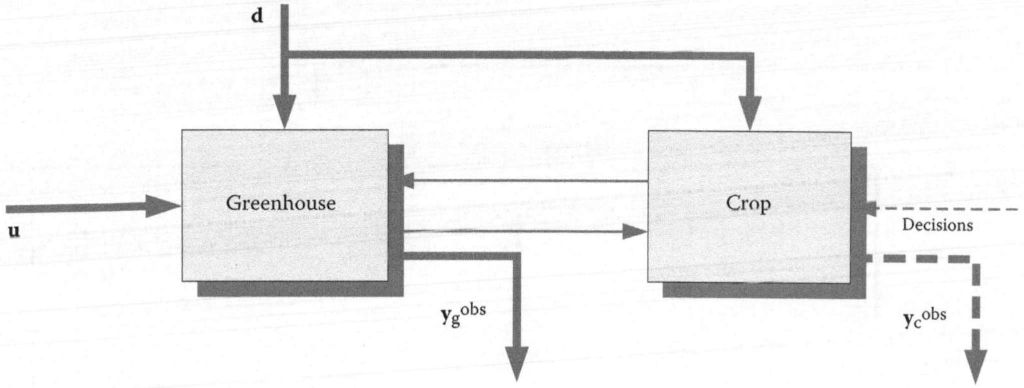

FIGURE 1.6 Greenhouse-crop I/O information flow.

of suitable state variables, but without giving details at this stage, it should be noted that there is not a single unique choice for the states and hence for the variables chosen in Equations 1.8 and 1.9. The exact specification is required in each particular case. The functions **f** and **g** are vector-valued functions of dimensions n_x and n_y, respectively, where **f** specifies the rate of change of the states and **g** how the output variables of interest depend on the states and the inputs.

The input–output information flow in a greenhouse-crop system can be represented schematically as shown in Figure 1.6. Also, the measured outputs are indicated formally. The vector $\mathbf{y}_g^{obs}$ is used to indicate the observed measurements of the greenhouse, obtained from instruments. Similarly, the observations on the crop are indicated by $\mathbf{y}_c^{obs}$. There may also be measurements on the exchange processes **j**, for instance, on photosynthesis. These can be taken as part of the vector $\mathbf{y}_g^{obs}$ or $\mathbf{y}_c^{obs}$, whatever is most convenient in the spirit of Figure 1.5. If the schedule of Figure 1.5 is used, measurements of CO_2 assimilation, photosynthesis, and crop transpiration are components of $\mathbf{y}_g^{obs}$, thus collecting all measurements that can be done in an automatic fashion. Most observations on the crop are made visually by the grower or by measuring the weight of pruned leaves and harvested product. These are not automated, which is the reason to show them as a dashed line in Figure 1.6.

In the design of controller solutions, the view on inputs and outputs is very determining for the chosen solution. This will be clear from the discussion on hierarchical control below. The issue is elaborated further in Chapter 5.

1.7 HIERARCHICAL COMPUTERIZED CONTROL

The control of the greenhouse-crop system by modern computerized controllers has a hierarchical structure as depicted in Figure 1.7.

In Figure 1.7, there are three major entities. At the top is the actual physical greenhouse-crop system. As explained before, it experiences the instantaneous influence of the weather, which in control terms is an uncontrollable external input signal. The actual values at time t are indicated by $\mathbf{d}(t)$. The position of the actuators, for example window openings and mixing valves, is the control input. The instantaneous value at any time is indicated by $\mathbf{u}(t)$. The observations obtained from the sensors are, as before, $\mathbf{y}_g^{obs}$ and those from the crop $\mathbf{y}_c^{obs}$. These output variables are manifest to the climate controller as (sampled data) inputs.

The state variables that appear when the greenhouse-crop system is modeled using physical principles are not manifest to the climate control computer. Only the observed outputs are available. Meaningful and operational interpretations of the relation between the states and the outputs can only be given if we start to model the system, but for the hierarchical scheme here, this is not necessary. This is very much in line with practice, where a large proportion of climate computers work

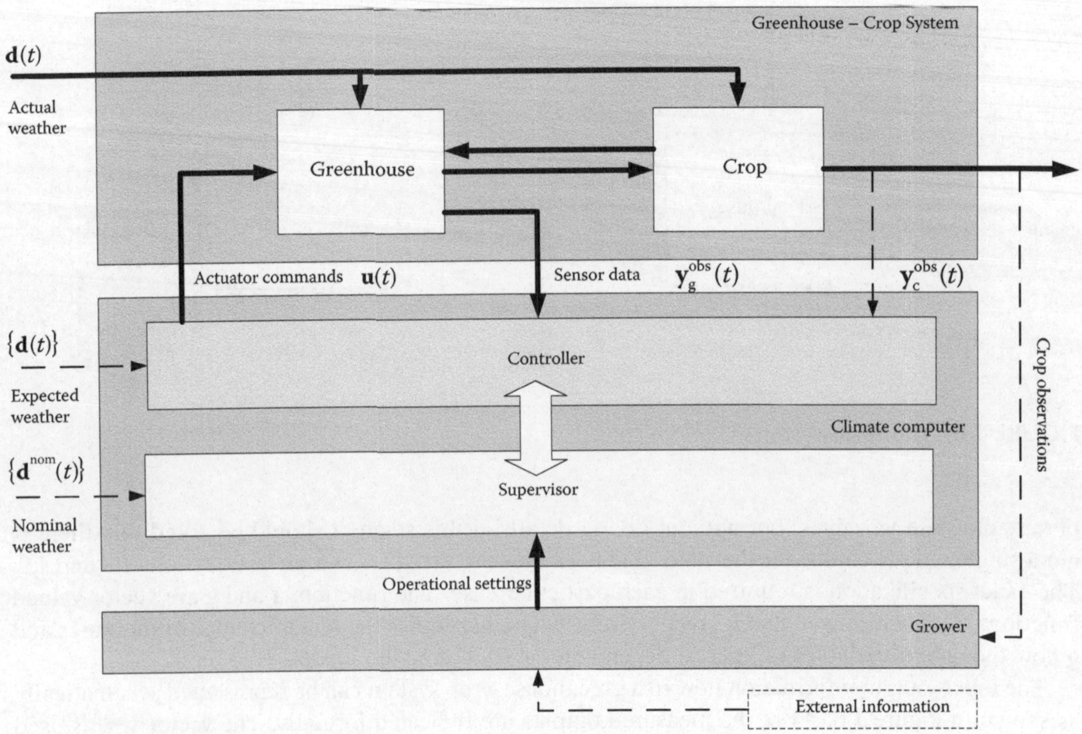

FIGURE 1.7 Hierarchical framework for greenhouse cultivation control.

without explicit models for greenhouse and crop. We will later see that if we want to be optimal, modeling of the system will become necessary.

The second major component of the overall system is the climate computer. We use this term here for any kind of automatic controller. In principle, within the climate computer, there are two levels: an operational level that performs the actual control and a strategic level that serves as a kind of supervisor. The major characteristic of the operational part is that it takes observations on the greenhouse physics, and possibly the crop, and returns control variables in the form of actuator commands. This system acts on the time scale of minutes. The fact that there are sometimes distributed local controllers, for instance, a controller that operates the windows to the desired position, is not important for the current discussion and is ignored at this stage. The operational control may or may not use actual and forecasted values of the weather. A sequence of future values is indicated in the scheme by curly brackets. In fact, the notation $\{\mathbf{d}(t)\}$ is equivalent to $\mathbf{d}(t)$, $t_0 \leq t \leq t_f$, where t_0 represents the current time and t_f the final time.

The operational controller receives "supervisory" information from the tactical level. The task of the supervisor is to translate the grower information on the tactical level in some way to information that can be used on the operational level. The kind of information exchange between these two varies from system to system and will be the main theme of the discussions in later chapters on control. On the tactical level, long-term weather expectations may or may not be used.

At the basis is the grower. The grower observes the crop and decides on corrective actions if he feels the need for it. These decisions are based on external information, such as market prices, blueprints, and his own experience. The grower interacts with the greenhouse climate computer via settings. In a classical greenhouse climate controller, these are upper and lower bounds of day and night temperatures, upper bounds on relative humidity, window opening enhancement at high radiation, and many more. This can easily amount to several hundreds of settings.

Hence, computerized control is an intrinsic part of present-day modern greenhouses. The functions of a current hierarchical climate computer can be summarized as follows:

(1) It takes care of realizing a suitable protected environment despite fluctuations of external weather (controller function, operational level on scale of minutes).
(2) It acts as a program memory and supervisory layer, which can be operated by the grower as a tool to steer his cultivation (supervisory function, tactical level on longer time scales).

1.8 CURRENT STATUS OF COMPUTERIZED CONTROL

The controller algorithms that can be found in current climate computers often have been designed in a heuristic way, starting from switching rules to decide about heating and ventilation, and supplemented with single loop proportional controllers. Temperature control, humidity control, and carbon dioxide control interact in a way that is not constant but is dependent on whether the system is in heating or cooling mode. Moreover, a set of decision rules is needed to resolve conflicts between the temperature and the humidity controller because the ventilation actuators serve to release surplus heat as well as surplus moisture. To leave room for the controllers, usually there is an operation band, which can be defined by the grower. On top of this, automatic adaptations are made to allow higher temperatures when the solar radiation is higher. The grower can adjust the settings and desired trajectories in accordance with his observations on the status of the crop based on his experience and skill. Also, he decides on risks of condensation of moisture on fruits or on overheating of plants by setting constraints to humidity or by operating a fog system. Finally, the main algorithm can be overruled by safety considerations, for example, in the case of rain or stormy weather.

Although highly successful, the computer systems in use today leave much to be improved. First, from the point of view of low-level controller performance, it is unlikely that desirable characteristics, such as overshoot, rise time, suppression of oscillations, and offset, can be handled in a systematic and insightful way in the heuristic rule-based assembly of separate loops found in today's controller programs. Second, the computer's function as a memory for programmable trajectories introduces a very large number of user adjustable settings to define them. Modifications in trajectory definitions have a definite effect on the energy and other resources consumption as well as on the growth and development of the crop, but the exact effect is unknown to the grower and is only inferred from experience. Third, despite current energy management overlays, there is little information about the economics of the operation and about the grower-accessible factors that determine the economics. If a grower is making changes in settings, the consequences for the process and its economy are essentially unknown.

In the scientific community, several efforts have been made to improve this situation. It is also the main motivation for writing this book. In principle, the best operation strategy is achieved by calculating control actions on the basis of optimization of an explicitly formulated and well-conceived goal function that combines expected benefits, costs, and risks. Hence, the problem discussed and solved in this book is as follows:

> Given the actual external input variables and expectations about them in the near future and given the currently observed output variables of the greenhouse, how can the control inputs be chosen such that over a specified cultivation period an explicitly formulated benefit function is maximized.

Obviously, instead of maximizing a benefit function, a cost function can be minimized. The goal function, be it benefit or cost, is free to be formulated by the ultimate user and can be anything that the grower wishes to achieve. This will be elaborated in great detail in the chapters to follow. Also, a mathematical formulation of the problem is postponed to later chapters. The interested reader who cannot wait may wish to jump to the introductory sections of Chapter 5. What is important here is that we adopt the idea of optimality as the leading principle for providing control solutions to the cultivation as a whole, not just for the control of greenhouse climate.

1.9 HOW IS THIS BOOK ORGANIZED?

In this chapter, we have briefly defined greenhouse cultivation and outlined the problem. As we focus on finding an optimal solution, we start in Chapter 2 with an appetizer example. An extremely simplified problem is used to illustrate the basic principles of open-loop optimal control. In Chapter 3, the optimal control methodology is worked out in somewhat larger detail. We choose to first discuss the open-loop problem arising from the assumption that the models are perfect and that all external variables are fully known in advance. This part is relevant as it elucidates some properties of optimal control that are relevant in a true feedback application. The latter is the topic of Chapter 4, where the loop is closed to counteract the effect of uncertainties, in model behavior as well as in future disturbance inputs. The main line of the methodology worked out in this book is to first solve, offline, a dynamic optimization problem on the scale of a full season, using smooth nominal external weather, and next to solve online a model-predictive optimal control problem to counteract uncertainties in the model and to exploit the possibilities offered by the actual weather. Unlike many similar approaches, a distinctive characteristic is that we use an economic criterion on both levels. The target is to maximize the profit to the grower. Another distinctive characteristic is the way the offline seasonal problem is connected to the online control problem. It will turn out that the costates of the slow crop variables are serving as the linking pin, which is quite different from the usual setpoint control. Chapter 5 summarizes the optimal control framework as outlined in Chapters 2–4 and sets out to see how historical developments fit into this framework on the basis of an extensive review of the relevant literature.

The series of chapters that follow are particular applications that underline the approach and discuss a set of issues that need to be solved before optimal control can be applied in practice. As models play a crucial role, the first part of each chapter is devoted to modeling the most important physical and biological phenomena in a form that is suitable for use in (optimal) control. The latter addition is important because most crop models described in the literature are not intended for control and are therefore often not in a suitable form. Our purpose is to present simple yet relevant models, and gradually a generic pattern can be recognized. Having defined the models, in each chapter there is next the definition of the respective goal functions. Finally, the problem is solved, often in open loop as well as, ultimately, in closed loop. Sample cases are presented on a single harvest crop (lettuce (Chapter 6)), a continuous harvest crop (tomato (Chapter 7)), and an elaborate modern greenhouse (the solar greenhouse (Chapter 8)). Each chapter discusses at some length a number of issues encountered when implementing optimal control, offers solutions, and describes the results in some detail. In all chapters, the main theme of solving first an offline seasonal problem and connecting it to the online control via the costates is recurring.

Finally, in Chapter 9, on retrospect, an overview is given of the developments that can be expected in the greenhouse industry and its consequences for control. The need for advanced controllers is expected to grow. A discussion is devoted to potential showstoppers for the actual application of optimal control in practice and what can be done about it. Scientific and technological challenges are summarized. This final chapter is intended as a stimulus, incentive, and source of inspiration to scientists and developers to bring the ultimate goal of optimality in greenhouse cultivation control closer worldwide.

REFERENCE

Hanan, J.J. 1998. *Greenhouses—Advanced Technology for Protected Horticulture*. Boca Raton: CRC Press.

2 Introduction to Optimal Control of Greenhouse Climate

2.1 INTRODUCTION AND MOTIVATION

The advantages of using optimal instead of conventional greenhouse climate control can be summarized as follows. An optimal control approach to greenhouse climate control fully exploits scientific quantitative knowledge concerning the greenhouse, the greenhouse equipment, and the crop. These are all captured in a mathematical dynamic model. Furthermore the goals of a grower, which usually come down to maximizing profit, are also stated quantitatively and explicitly in terms of a mathematical cost function that is maximized. This cost function is based on auction prices obtained for the crop as well as the costs associated with greenhouse climate management, such as heating costs. The latter costs are often underestimated by growers that focus on the welfare of the crop. Optimal control reveals that crop welfare may be retained against less operating costs such as heating. Sometimes a slight loss of crop quality may save a lot of operating costs leading also to higher profits. These outcomes are partly due to the fact that the optimal controller cleverly exploits weather predictions and measurements. The tuning of an optimal greenhouse control system is performed by changing something in the order of ten settings that all have a clear meaning and interpretation. Conventional greenhouse climate controllers usually have several hundreds of settings the meaning of which is usually not very transparent. Growers often use only a few of these settings. In general, however, no two growers use the same settings to control their greenhouses.

The control of greenhouse climate is characterized by the fact that several processes, such as crop growth and greenhouse climate change, occur on different time scales. The development of the crop occurs on a time scale of weeks or months, whereas most of the greenhouse climate changes on a daily basis. Both greenhouse climate and crop growth are influenced by light, which may change on a time scale of seconds or minutes, especially on cloudy days, which occur quite often in The Netherlands. The different time scales complicate a control system design. The control system becomes computationally very expensive as well as inaccurate. In overcoming these problems, short- and long-term objectives have to be separated and assessed against one another. Optimal control enables a quantitative approach to this problem that is again very transparent and based on quantitative scientific knowledge that relates to these different time scales.

What is the meaning of the word *optimal* in *optimal control*? It means that given the mathematical model of the system and given the cost function, an optimal controller computes the best control, i.e., the control that maximizes the cost function. In practice the optimal controller will not be truly optimal because the mathematical model will not be an exact description of the system but only an approximation. Also the cost function may not perfectly describe the actual goals. So in practice the optimality depends critically on the accuracy of the mathematical model and the cost function. They should therefore be selected with care.

2.2 A SIMPLE ILLUSTRATIVE EXAMPLE

The optimal control of any system, in our case a greenhouse, is based on two things. First, it is based on a mathematical dynamic model of the system. In our case the system is the greenhouse, including its equipment, the crop, and also the outside weather. Second, it is based on a mathematical cost function that is either maximized or minimized. In our case the cost function is profit, which must be maximized. The profit equals the money obtained from selling the crops minus the costs required for maintaining a favorable greenhouse climate.

The following example is deliberately kept very simple, and therefore does not meet the requirements of accuracy stated at the end of the last section. The illustrative example is only meant to illustrate the main ideas and problems associated with optimal greenhouse climate control.

Example 1

Consider the following mathematical dynamic model of the greenhouse, and the crop:

$$\dot{W} = c_1 ITW, \tag{2.1}$$

$$\dot{T} = c_2\left(T_o - T\right) + c_3 H, \tag{2.2}$$

where,
W (kg m^{-2}) denotes dry weight of the crop
I (W m^{-2}) denotes light intensity of the light entering the greenhouse
T (°C) denotes the greenhouse air temperature
T_o (°C) denotes the outside temperature
H (W m^{-2}) denotes the heat input from the greenhouse heating system
c_1, c_2, c_3 are constants

Equation 2.1 states that the increase of crop weight is positively proportional to both light and temperature. Equation 2.2 is a simple description of how the greenhouse temperature T changes due to the outside temperature T_o and the heat input H obtained from the greenhouse heating system. It actually is a very simple heat balance equation. The information flow diagram of the system is shown in Figure 2.1. The constant c_3 represents the heat efficiency of the heating system in the greenhouse.

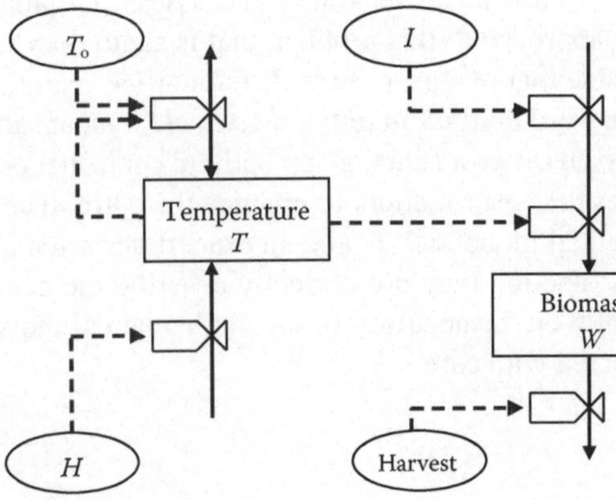

FIGURE 2.1 Information flow diagram.

Note that in this simple model there is no heat contribution from the light. A close inspection of the equations reveals that biomass increase is stimulated by elevated temperatures, provided there is light. On the other hand, increasing the temperature costs energy, so there will be a trade off. If there is no light, heating will make no sense.

The cost function J(€ m^{-2}) that is meant to represent profit reads as follows:

$$J = c_5 W(t_f) - \int_{t_0}^{t_f} c_4 H \, dt. \tag{2.3}$$

The first term $c_5 W(t_f)$ on the right in Equation 2.3 represents the money obtained from selling the harvested crops at the end of the growing period that starts at time t_0 (h) and ends at time t_f (h). As a result the constant c_5 represents the auction price for one unit of dry weight $W(t_f)$. The integral on the right represents the costs of heating the greenhouse. In this simple example heating costs with are the only costs associated with greenhouse climate control. As a result the constant c_4 represents the costs associated with one unit of heating H. In addition to Equations 2.1 through 2.3 to obtain an optimal control problem the initial conditions of the system, i.e., $W(t_0)$ and $T(t_0)$, have to be specified. These may be considered part of the systems model.

2.3 GENERAL FORMULATION OF OPTIMAL CONTROL PROBLEMS

To analyze and solve optimal control problems, they are represented in a general form called the *state–space form*. This form distinguishes between fundamentally different types of variables and enables the use of standard software to solve the optimal control problem. In this section the optimal control Example 1 (which covers Equations 2.1 through 2.3) including the initial conditions $W(t_0)$ and $T(t_0)$ will be represented in state–space form. To do this we need to first recognize the *state variables* of the system described by Equations 2.1 and 2.2. State variables are variables of which *time derivatives* appear in the system Equations 2.1 and 2.2, which means W and T are state variables. Time derivatives of state variables are also state variables up to (and thus not including) the highest order time derivative that appears in the equations. Since the highest order time derivative of both W and T in Equations 2.1 and 2.2 is the first-time derivative, W and T are the only state variables. State variables are always denoted by the symbol x. Therefore, we obtain

$$x_1 = W, x_2 = T. \tag{2.4}$$

The other variables in Equations 2.1 and 2.2, except for those that are constant, are called *input variables* or *inputs* of the system. Two types of inputs are distinguished: *control inputs* that can be manipulated versus *external inputs* that are determined by external conditions. Control inputs are represented by the symbol u. Therefore,

$$u_1 = H, \tag{2.5}$$

because the heating can be manipulated. External inputs are represented by the symbol d. Therefore,

$$d_1 = I, d_2 = T_o, \tag{2.6}$$

since both the light I and the outside temperature T_o are not constant and cannot be manipulated but are determined by external conditions. Now the remaining variables are constants. These constants are called the *parameters* of the system denoted by the symbol p. Therefore, we obtain

$$p_1 = c_1, \; p_2 = c_2, \; p_3 = c_3. \tag{2.7}$$

Using the new state–space notations (2.4 through 2.7), the systems model (2.1 and 2.2) can be represented by,

$$\dot{x}_1 = p_1 d_1 x_2, \tag{2.8}$$

$$\dot{x}_2 = p_2\left(d_2 - x_2\right) + p_3 u_1. \tag{2.9}$$

In the optimal control literature, it is common practice to employ vector notation. Introducing the vectors

$$\mathbf{x} = \begin{bmatrix} x_1 \\ x_2 \end{bmatrix}, \ \dot{\mathbf{x}} = \begin{bmatrix} \dot{x}_1 \\ \dot{x}_2 \end{bmatrix}, \ \mathbf{u} = u_1, \ \mathbf{d} = \begin{bmatrix} d_1 \\ d_2 \end{bmatrix}, \ \mathbf{p} = \begin{bmatrix} p_1 \\ p_2 \\ p_3 \end{bmatrix}, \tag{2.10}$$

Equations 2.8 and 2.9 can be rewritten as

$$\dot{\mathbf{x}} = \begin{bmatrix} p_1 d_1 x_2 \\ p_2(d_2 - x_2) + p_3 u_1 \end{bmatrix} \tag{2.11}$$

The right-hand side of Equation 2.11 is called a *vector function* because it is a *vector* that is a function of other variables that are collected in the vectors **x**, **u**, **d**, and **p**. Let us denote this vector function by **f**(**x**, **u**, **d**, **p**). Then Equation 2.11 reads

$$\dot{\mathbf{x}} = \mathbf{f}\left(\mathbf{x}, \mathbf{u}, \mathbf{d}, \mathbf{p}\right), \tag{2.12}$$

where

$$\mathbf{f}(\mathbf{x}, \mathbf{u}, \mathbf{d}, \mathbf{p}) = \begin{bmatrix} p_1 d_1 x_2 x_1 \\ p_2(d_2 - x_2) + p_3 u_1 \end{bmatrix}. \tag{2.13}$$

Since the vectors **x**, **u**, **p**, **d** in Equation 2.12 can have arbitrary dimensions, and since the vector function **f**(**x**, **u**, **d**, **p**) can be selected arbitrarily, Equation 2.12 is a *general state–space representation* of a system. By specifying the vector function **f**(**x**, **u**, **d**, **p**), we specify the system. The general state–space system representation in Equation 2.12 is encountered in most of the optimal control and systems literature. The initial conditions $W(t_0)$ and $T(t_0)$ of the system, 2.1 and 2.2, in state–space form are represented by,

$$\mathbf{x}(t_0) = \begin{bmatrix} x_1(t_0) \\ x_2(t_0) \end{bmatrix} = \begin{bmatrix} W(t_0) \\ T(t_0) \end{bmatrix}. \tag{2.14}$$

In summary, a general and complete system representation in state–space form of any system is given by Equation 2.12 with initial conditions $\mathbf{x}(t_0)$. By specifying the vector function **f**(**x**, **u**, **d**, **p**) and the vector $\mathbf{x}(t_0)$, you specify the system and its initial conditions, respectively. Given the inputs $\mathbf{u}(t)$, $\mathbf{d}(t)$, $t_0 \le t \le t_f$ to the system, Equation 2.12 together with its associated initial condition $\mathbf{x}(t_0)$ determine the *system behavior* $\mathbf{x}(t)$, $t_0 \le t \le t_f$, i.e., the solution to Equation 2.12. Although analytical solutions of Equation 2.12 cannot be obtained in general, numerical solutions can be obtained by means of *numerical integration*, for which many software tools are available. Numerical integration tools are also needed to solve general optimal control problems.

The dimensions of the vectors **x**, **u**, **d**, **p** are denoted by n_x, n_u, n_d, n_p respectively. From Equation 2.12 observe that the dimension of the vector function $\mathbf{f}(\mathbf{x}, \mathbf{u}, \mathbf{d}, \mathbf{p})$, which will often be referred to as just **f**, equals n_x. The dimension n_x of the state vector x is also called *the dimension of the system.*

The notation of the cost function follows in a straightforward manner from the state–space notation of the system. However, in optimal control it is usually assumed that the cost function is minimized instead of maximized. By reversing the sign of the cost function, maximization can be replaced with minimization. With this in mind the cost function (2.3) that has to be maximized now turns into the following cost function that is minimized:

$$J = -p_5 x_1\left(t_f\right) + \int_{t_0}^{t_f} p_4 u_1 \, dt, \tag{2.15}$$

where

$$p_4 = c_4, \quad p_5 = c_5. \tag{2.16}$$

To express that the cost function depends on the control input trajectory $\mathbf{u}(t)$, $t_0 \le t \le t_f$, the cost function J is often written as $J(\mathbf{u}(t))$. Mathematically, $J(\mathbf{u}(t))$ is a cost functional since it is a function of another function (mathematically a *function* is a function of a variable). Introducing

$$\Phi(\mathbf{x}(t_f)) = -p_5 x_1(t_f) \tag{2.17}$$

and

$$L(\mathbf{x}, \mathbf{u}, \mathbf{d}, \mathbf{p}) = p_4 u_1, \tag{2.18}$$

the cost function (2.15) reads

$$J(\mathbf{u}(t)) = \Phi\left(\mathbf{x}\left(t_f\right)\right) + \int_{t_0}^{t_f} L(\mathbf{x}, \mathbf{u}, \mathbf{d}, \mathbf{p})\, dt. \tag{2.19}$$

Equation 2.19 is a *general* representation of a cost function because $\Phi(\mathbf{x}(t_f))$ can be an *arbitrary scalar* function, which also applies to $L(\mathbf{x}, \mathbf{u}, \mathbf{d}, \mathbf{p})$. Because $\Phi(\mathbf{x}(t_f))$ depends solely on the terminal state $\mathbf{x}(t_f)$ of the system, it is called the *terminal costs*. In our example these are the negative costs (benefit) $-p_5 x_1(t_f)$ of selling the crop after harvesting it at the terminal time t_f. Because $L(\mathbf{x}, \mathbf{u}, \mathbf{d}, \mathbf{p})$ represents costs that occur while "running" from the initial time t_0 to the final time t_f, these are called the *running costs*. In our example these are the costs $p_4 u_1$ associated with greenhouse heating.

To summarize, a *general* optimal control problem reads as follows. Given the system

$$\dot{\mathbf{x}} = \mathbf{f}(\mathbf{x}, \mathbf{u}, \mathbf{d}, \mathbf{p}), \tag{2.20}$$

with initial conditions,

$$\mathbf{x}(t_0) = \mathbf{x}_0, \tag{2.21}$$

and given the external input trajectory,

$$\mathbf{d}(t),\ t_0 \le t \le t_f, \tag{2.22}$$

find the control input trajectory,

$$\mathbf{u}(t),\ t_0 \le t \le t_f, \tag{2.23}$$

that minimizes the functional cost,

$$J(\mathbf{u}(t)) = \Phi\left(\mathbf{x}\left(t_f\right)\right) + \int_{t_0}^{t_f} L(\mathbf{x}, \mathbf{u}, \mathbf{d}, \mathbf{p})\,dt. \tag{2.24}$$

To specify completely an optimal control problem, one has to provide the following:

1. Systems model $\mathbf{f}(\mathbf{x}, \mathbf{u}, \mathbf{d}, \mathbf{p})$ with the associated parameter values
2. Initial time t_0 and terminal time t_f
3. Initial conditions $\mathbf{x}_0$ of the system
4. External input trajectory $\mathbf{d}(t)$, $t_0 \le t \le t_f$
5. Running costs $L(\mathbf{x}, \mathbf{u}, \mathbf{d}, \mathbf{p})$
6. Terminal costs $\Phi(\mathbf{x}(t_f))$

Example 2

For the simple illustrative optimal control Example 1 described here and in the previous section, the specifications mentioned above are

(1)
$$\mathbf{f}(\mathbf{x},\mathbf{u},\mathbf{d},\mathbf{p}) = \begin{bmatrix} p_1 d_1 x_2 \\ p_2(d_2 - x_2) + p_3 u_1 \end{bmatrix},\ \mathbf{p} = \begin{bmatrix} 7.50\times10^{-8} \\ 1.00 \\ 0.10 \\ 4.55\times10^{-4} \\ 136.4 \end{bmatrix}, \tag{2.25}$$

(2)
$$t_0 = 0,\ t_f = 48, \tag{2.26}$$

(3)
$$\mathbf{x}_0 = \begin{bmatrix} x_1(t_0) \\ x_2(t_0) \end{bmatrix} = \begin{bmatrix} 0 \\ 10 \end{bmatrix}, \tag{2.27}$$

(4)
$$\mathbf{d}(t) = \begin{bmatrix} d_1(t) \\ d_2(t) \end{bmatrix} = \begin{bmatrix} \max(0, 800\bullet\sin(4\pi t/t_f - 0.65\pi)) \\ 15+10\bullet\sin(4\pi t/t_f - 0.65\pi) \end{bmatrix},\ t_0 \le t \le t_f, \tag{2.28}$$

(5) $$L(\mathbf{x}, \mathbf{u}, \mathbf{d}, \mathbf{p}) = p_4 u_1, \tag{2.29}$$

(6) $$\Phi(\mathbf{x}(t_f)) = -p_5 x_1. \tag{2.30}$$

2.4 BENEFITS AND DIFFICULTIES ASSOCIATED WITH OPTIMAL CONTROL

By means of solutions to the simple illustrative optimal control example presented in the previous section, benefits and difficulties associated with the optimal control of greenhouse climate will be demonstrated. Solutions are obtained from optimal control algorithms that will be briefly discussed in the next section. For the parameter values specified in the previous section, Figure 2.2 represents the solution to the optimal greenhouse climate control problem. Several variables in these figures have been scaled to obtain a clear visualization.

The optimal control input $H(t) = u_1(t)$, $0 \le t \le 48$ is a so called *bang-bang control* because the heat input is either maximal or minimal. From Figure 2.2 it then follows that the maximum value of the heat input is assumed to be 100 W m^{-2}, whereas the minimum is obviously 0 W m^{-2}. The associated minimal costs are computed to be –3.30 € m^{-2} so the profit amounts to 3.30 € m^{-2}. The fact that the *maximum profit* and the associated *optimal heat input* can be calculated even in advance is a very attractive property of optimal control in comparison with conventional greenhouse climate control. If the maximum or minimum value of the heat input $H(t) = u_1(t)$, $0 \le t \le 48$ changes, so does the optimal heat input, and the associated minimal costs. This reveals that the maximum and minimum values of the heat input u_1 must be considered part of the optimal control problem formulation. General bounds on control inputs are described by,

$$u_i^{\min} \le u_i \le u_i^{\max},\ i = 1, 2, \ldots, n_u, \tag{2.31}$$

where $u_i^{\max}$ and $u_i^{\min}$ represent known upper and lower bounds on the associated control input u_i, respectively.

As to the optimal heat input from Figure 2.2, observe that maximum heating is applied roughly when there is light, which is to be expected from inspection of Equation 2.1. Also from Figure 2.2, observe that at very low levels of light it is not beneficial to switch the heater on. It can be deduced from optimal control theory that for our illustrative optimal control problem the optimal heating will always be of the bang-bang type because both $\mathbf{f}(\mathbf{x}, \mathbf{u}, \mathbf{d}, \mathbf{p})$ and $L(\mathbf{x}, \mathbf{u}, \mathbf{d}, \mathbf{p})$ are linear functions of $\mathbf{u}$.

Given any greenhouse systems model and any associated cost function, optimal control is able to compute the optimal control and associated maximum profit. Therefore, we can change, for instance, the parameters of the problem such as the heating costs p_4 and the auction price p_5, and see how this affects our optimal control and maximum profit. This information is highly interesting and relevant to both growers and legislators who, for instance, want to reduce energy consumption. As an example, Table 2.1 lists the minimum costs $J(\mathbf{u}(t))$ against several parameter values of the heating costs p_4 and the auction price p_5, where all the other optimal control problem parameters and data are as specified before.

As expected, with increasing heating costs, the profit $-J(\mathbf{u}(t))$ decreases and with an increasing auction price, and the profit $-J(\mathbf{u}(t))$ increases. The interesting thing is that we can tell exactly how much and that for every case we can compute the optimal heat input.

To compute the optimal control $\mathbf{u}(t)$, $t_0 \le t \le t_f$, we have to specify the external inputs $\mathbf{d}(t)$ over the time horizon $t_0 \le t \le t_f$, of interest because the external inputs influence the system behavior. In the case of optimal greenhouse climate control, according to Equation 2.6 the external inputs $\mathbf{d}$ are weather conditions such as outside temperature and light (solar radiation). Depending on the time horizon of interest, these are difficult to predict accurately. Like the mathematical model and the cost function, for the control to be approximately optimal, the predictions of the external inputs

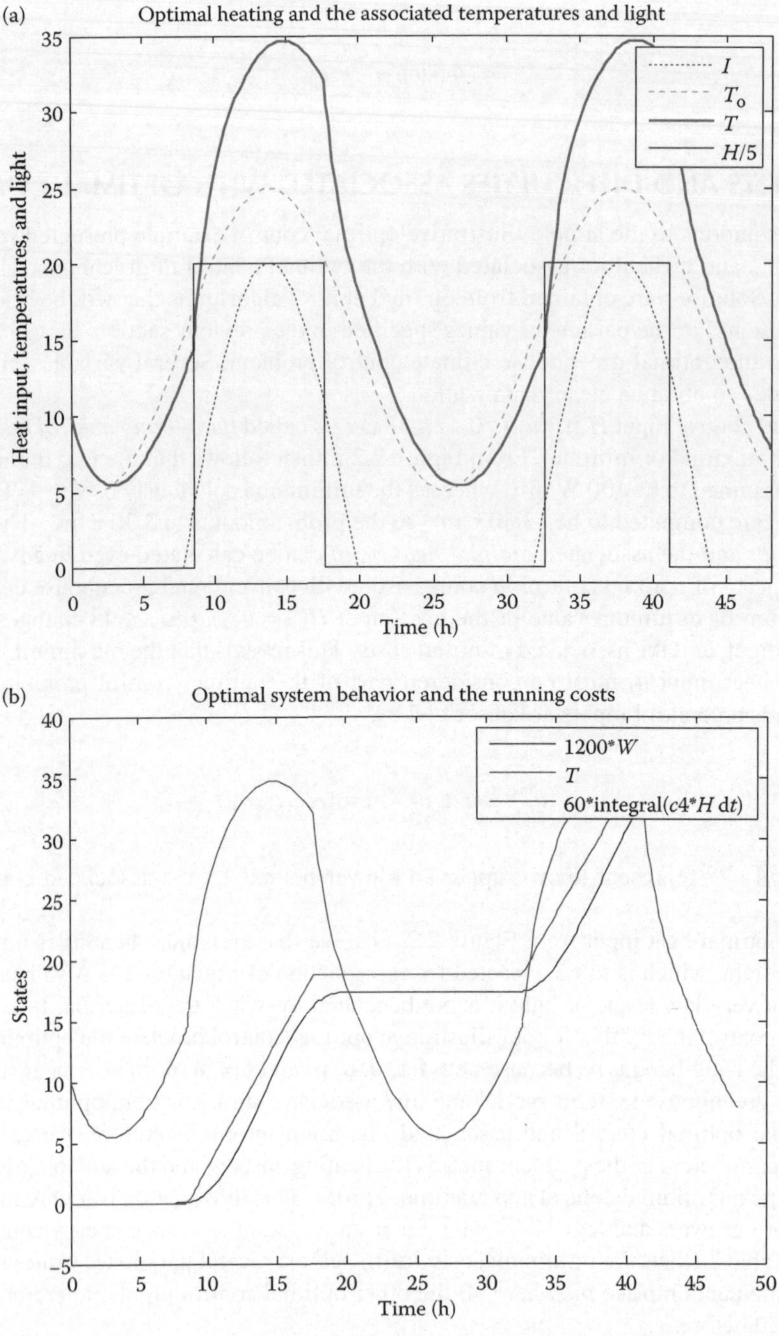

FIGURE 2.2 (a) Optimal heating, associated temperatures, and light. (b) Optimal system behavior and running costs.

$\mathbf{d}(t)$, $t_0 \le t \le t_f$ have to be accurate. The optimal control of greenhouse climate therefore uses weather predictions to specify $\mathbf{d}(t)$, $t_0 \le t \le t_f$. These predictions will however not always be accurate. This degrades the optimality of the control.

The processes involved in greenhouse climate control occur on very different time scales. Although in our simple example a horizon of two days was considered, in actual practice crops are grown over a period of months. Whereas crop growth occurs on a large time scale of months, the

TABLE 2.1
Minimum Costs as a Function of Various Parameters

$J(u(t))$	p_4	$J(u(t))$	p_5
–3.5126e+000	2.2727e–004	–9.3070e–001	4.5455e+001
–3.4669e+000	2.5000e–004	–1.4239e+000	6.8182e+001
–3.4251e+000	2.7273e–004	–2.0255e+000	9.0909e+001
–3.3823e+000	2.9545e–004	–2.6562e+000	1.1364e+002
–3.3400e+000	3.1818e–004	–3.3000e+000	1.3636e+002
–3.3002e+000	3.4091e–004	–3.9524e+000	1.5909e+002
–3.2625e+000	3.6364e–004	–4.6057e+000	1.8182e+002
–3.2251e+000	3.8636e–004	–5.2672e+000	2.0455e+002
–3.1890e+000	4.0909e–004	–5.9295e+000	2.2727e+002
–3.1531e+000	4.3182e–004		
–3.1194e+000	4.5455e–004		

greenhouse climate changes on a daily basis and light may even change on a time scale of minutes or even seconds!

The facts mentioned above cause the following problems. A preferably accurate weather prediction over months is required whereas the computation of optimal controls over this period should be performed taking time steps on the smallest time scale of minutes or seconds. As to predicting the weather over a period of months, we can usually do no better than taking averaged weather recorded over several identical periods in previous years. Over a shorter period of, say, one or two days, we can usually come up with much better weather predictions. Computing optimal controls over a period of months taking time steps at the small time scale level of seconds or minutes is computationally very expensive. Moreover, due to the large differences in time scales these computations may become inaccurate.

Research on optimal greenhouse climate control performed over the last two decades has suggested the following solutions to these problems. The optimal greenhouse climate control problem should be decomposed into separate optimal control problems on two or three different time scales. The solutions of these separate optimal control problems can be obtained relatively easily and quickly. Together these solutions produce an optimal control that takes into account all relevant time scales and uses accurate weather predictions on a time scale of one or two days, and averaged weather on the long run. This seems to be the best approach to preserve optimality as much as possible while obtaining optimal controllers that can be implemented in practice because they are not too expensive computationally.

We end this section by illustrating that ordinary approaches to deal with the problem of optimal control for systems having different time scales lead to a significant loss of optimality in the case of optimal greenhouse climate control. This loss is very much prevented by the approach described above. The general approach is to presume the systems differential equations to be in equilibrium (static) on the small time scale(s). Equation 2.1 describes the dry weight accumulation of the crop that occurs on a much larger time scale than the greenhouse temperature changes described by Equation 2.2. Presuming Equation 2.2 to be in equilibrium we obtain,

$$0 = c_2(T_o - T) + c_3 H, \tag{2.32}$$

which implies,

$$T = T_o + \frac{c_3}{c_2} H. \tag{2.33}$$

Substitution of Equation 2.33 into Equation 2.1 gives

$$\dot{W} = c_1 L \left(T_o + \frac{c_3}{c_2} H \right). \tag{2.34}$$

Equation 2.34 is used to replace the state–space Equations 2.1 and 2.2 and is meant to approximately describe the dry weight accumulation on the large time scale of months presuming that on the small time scale of days or minutes the system is approximately in equilibrium as described by Equation 2.32. In this way a new optimal control problem is obtained that can be solved using numerical integration taking time steps at the level of the large time scale of months to prevent excessive computation. Taking time steps on the large time scale level of months, however, presumes that the light L and the outside temperature T_o vary hardly on the level of the small time scale of days or minutes. This is clearly not so and that is the reason why in the case of greenhouse climate control this approach results in a serious loss of optimality. To demonstrate this we compute the optimal control for the system 2.34 taking the average value of L, and T_o over $0 \le t \le 48$ to apply at every time $0 \le t \le 48$. It turns out that the optimal heating in this case is everywhere equal to zero. The associated minimum costs are computed to be –1.88 € m^{-2}. Next we apply this optimal control to the "real system" described by Equations 2.1 and 2.2 with the real light $L = d_1$ and outside temperature $T_o = d_2$ patterns given by Equation 2.28. Then the costs (Equation 2.3) are computed to be –2.79 € m^{-2}, whereas the optimal costs computed earlier was –3.30 € m^{-2}.

Computing the optimal costs using Equation 2.34 and average values for light and outside temperature leads to a loss of profit of 43%. Applying the optimal control computed in this manner to the real system with the real external light and temperature conditions still leads to a loss of profit of 15.5%.

3 Open-Loop Optimal Control

3.1 INTRODUCTION

When exactly is a control called optimal? What exactly is needed to compute optimal controls? To answer these questions, a summary of optimal control theory is presented. Along with it, by a simple optimal greenhouse control example, important interpretations are provided. Next, optimal control algorithms are described and classified. These algorithms generate so-called open-loop optimal controls. The methodology to obtain these controls in open loop is also known as *dynamic optimization*. In practice, the control of greenhouses also requires feedback. Feedback control is the topic of the next chapter.

3.2 OPTIMAL CONTROL THEORY

There are several excellent and classic books on the theory of optimal control (e.g., Bryson and Ho, 1969; Bryson, 1999; Lewis, 1986; Stengel, 1994). Here, we summarize the major line of thought to serve as a basis for understanding the main theme developed in this book.

Recall the general optimal control problem described by Equations 3.1 through 3.5, for convenience, repeated here.

Given the system,

$$\dot{\mathbf{x}} = \mathbf{f}(\mathbf{x}, \mathbf{u}, \mathbf{d}, \mathbf{p}), \tag{3.1}$$

with initial conditions,

$$\mathbf{x}(t_0) = \mathbf{x}_0, \tag{3.2}$$

and given the external input trajectory,

$$\mathbf{d}(t),\ t_0 \le t \le t_f, \tag{3.3}$$

find the control input trajectory,

$$\mathbf{u}(t),\ t_0 \le t \le t_f, \tag{3.4}$$

that minimizes the cost functional,

$$J(\mathbf{u}(t)) = \Phi\big(\mathbf{x}(t_f)\big) + \int_{t_0}^{t_f} L(\mathbf{x}, \mathbf{u}, \mathbf{d}, \mathbf{p})\mathrm{d}t. \tag{3.5}$$

This problem is an optimization problem because the cost functional $J(\mathbf{u}(t))$ given by Equation 3.5 has to be minimized by choosing suitably the control input trajectory $\mathbf{u}(t)$, $t_0 \le t \le t_f$. Moreover, this optimization problem has dynamic constraints specified by Equation 3.1 that represents the dynamic systems model. Equation 3.1, that is, $\dot{\mathbf{x}} = \mathbf{f}(\mathbf{x}, \mathbf{u}, \mathbf{d}, \mathbf{p})$, describes n_x dynamic constraints $\dot{x}_i = f_i(\mathbf{x}, \mathbf{u}, \mathbf{d}, \mathbf{p})$, $i = 1, 2, \ldots, n_x$, that have to be satisfied at each time $t_0 \le t \le t_f$.

Optimization problems without constraints can be solved by repeatedly calculating several values of the cost functional and several associated first derivatives. According to Lagrange theory, optimization problems with constraints can be solved by transforming them into optimization problems without constraints that have the same solution. This is done by introducing one additional variable associated to each constraint. At every time $t_0 \le t \le t_f$, we have n_x additional variables because for the optimal control problem (3.1 through 3.5) we have n_x constraints represented by $\dot{\mathbf{x}} = \mathbf{f}(\mathbf{x}, \mathbf{u}, \mathbf{d}, \mathbf{p})$ that have to be satisfied at every time $t_0 \le t \le t_f$. These variables are denoted by $\lambda_i(t)$, $i = 1, 2, \ldots, n_x$ and are called *Lagrange multipliers*. Each $\lambda_i(t)$ is associated with the constraint $\dot{x}_i = f_i(\mathbf{x}, \mathbf{u}, \mathbf{d}, \mathbf{p})$, $i = 0, 1, \ldots, n_x$ at every time $t_0 \le t \le t_f$. The vector,

$$\boldsymbol{\lambda}(t) = \begin{bmatrix} \lambda_1(t) \\ \lambda_2(t) \\ \vdots \\ \lambda_{n_x}(t) \end{bmatrix}, \quad t_0 \le t \le t_f, \tag{3.6}$$

of Lagrange multipliers is also called a Lagrange multiplier. According to Lagrange theory, the optimal control problem (3.1 through 3.5) with dynamic constraints has the same solution as the following problem without constraints.

Given the external input trajectory $\mathbf{d}(t)$, $t_0 \le t \le t_f$, find the optimal control input trajectory $\mathbf{u}(t)$, $t_0 \le t \le t_f$, that minimizes the cost functional,

$$J'(\mathbf{u}(t)) = \Phi\left(\mathbf{x}\left(t_f\right)\right) + \int_{t_0}^{t_f} L(\mathbf{x}, \mathbf{u}, \mathbf{d}, \mathbf{p}) + \boldsymbol{\lambda}^{\mathrm{T}}(\dot{\mathbf{x}} - \mathbf{f}(\mathbf{x}, \mathbf{u}, \mathbf{d}, \mathbf{p}))\mathrm{d}t. \tag{3.7}$$

Observe that the augmented cost functional $J'(\mathbf{u}(t))$ in 3.7 equals the original one (3.5), with $\boldsymbol{\lambda}^{\mathrm{T}}(\dot{\mathbf{x}} - \mathbf{f}(\mathbf{x}, \mathbf{u}, \mathbf{d}, \mathbf{p}))$ added to the integrand. Observe that if the constraint (3.1), that is, $\dot{\mathbf{x}} = \mathbf{f}(\mathbf{x}, \mathbf{u}, \mathbf{d}, \mathbf{p})$, is satisfied at every time $t_0 \le t \le t_f$, the augmented cost function $J'(\mathbf{u}(t))$ equals the original one. By applying mathematical techniques, we obtain the following necessary optimality conditions for the solution of the unconstrained optimization problem and therefore also of our original optimal control problem (3.1 through 3.5),

$$\dot{\mathbf{x}} = \mathbf{f}(\mathbf{x}, \mathbf{u}, \mathbf{d}, \mathbf{p}), \; t_0 \le t \le t_f, \; \mathbf{x}\left(t_0\right) = \mathbf{x}_0, \tag{3.8}$$

$$-\dot{\boldsymbol{\lambda}} = \left(\frac{\partial \mathbf{f}}{\partial \mathbf{x}}\right)^{\mathrm{T}} \boldsymbol{\lambda} + \left(\frac{\partial L}{\partial \mathbf{x}}\right)^{\mathrm{T}}, \; t_0 \le t \le t_f, \; \boldsymbol{\lambda}(t_f) = \left(\left.\frac{\partial \Phi}{\partial \mathbf{x}}\right|_{t=t_f}\right)^{\mathrm{T}}, \tag{3.9}$$

$$0 = \left(\frac{\partial \mathbf{f}}{\partial \mathbf{u}}\right)^{\mathrm{T}} \boldsymbol{\lambda} + \left(\frac{\partial L}{\partial \mathbf{u}}\right)^{\mathrm{T}}, \; t_0 \le t \le t_f. \tag{3.10}$$

A large class of numerical algorithms to solve optimal control problems is based on these necessary optimality conditions. In Equations 3.9 and 3.10, derivatives of vector functions like $\frac{\partial \mathbf{f}}{\partial \mathbf{x}}$ appear. Recall that $\frac{\partial \mathbf{f}}{\partial \mathbf{x}}$ is a matrix of which element i,j is by definition $\frac{\partial f_i}{\partial x_j}$. Observe that Equations 3.8 through 3.10 only contain function values and values of first derivatives associated to the optimization problem.

Equation 3.8 represents the system dynamics that determine the system behavior $\mathbf{x}(t)$ of the n_x state variables at each time $t_0 \le t \le t_f$. Next, it represents the initial conditions of the system. Equation 3.9 represents the dynamics of what is called *the adjoined system*. It determines the behavior of what is called *the costate* $\boldsymbol{\lambda}(t)$, that is, the n_x Lagrange multipliers that are adjoined to the constrained optimization problem to turn it into an unconstrained one. Next, it specifies the terminal condition of this adjoined system. Finally, Equation 3.10 represents a coupling between Equations 3.8 and 3.9, that is, the system and the adjoined system at each time $t_0 \le t \le t_f$.

Equation 3.10 can often be used to express the control input $\mathbf{u}(t)$ as an explicit function of both the state $\mathbf{x}(t)$ and the costate $\boldsymbol{\lambda}(t)$, that is,

$$\mathbf{u}(t) = \mathbf{h}(\mathbf{x}(t), \boldsymbol{\lambda}(t)),\ t_0 \le t \le t_f. \tag{3.11}$$

Substitution of 3.11 into 3.8 and 3.9 eliminates $\mathbf{u}(t)$ and leaves a problem where we have to find the initial costate $\boldsymbol{\lambda}(t_0)$ and the terminal state $\mathbf{x}(t_f)$ such that 3.8 and 3.9 are satisfied. Finding these vectors at the two time boundaries, t_0 and t_f, is called *a two-point boundary value problem.*

A control input $\mathbf{u}(t)$, $t_0 \le t \le t_f$, that satisfies the necessary optimality conditions (3.8 through 3.10) is called *optimal*. Optimality is indicated by an asterisk (*). Therefore, $\mathbf{u}^*(t)$, $\mathbf{x}^*(t)$, and $\boldsymbol{\lambda}^*(t)$, $t_0 \le t \le t_f$ denote an optimal control input history (trajectory) and the associated optimal state and costate histories (trajectories). Furthermore, $J^* = J(\mathbf{u}^*(t))$ indicates the minimal costs.

The optimal costate has an interesting interpretation that will play a major role in the decomposition of optimal control problems to deal with different time scales. It can be shown that,

$$\boldsymbol{\lambda}^*(t) = \left(\frac{\partial J^*}{\partial \mathbf{x}}\right)^{\mathrm{T}},\quad t_0 \le t \le t_f. \tag{3.12}$$

Therefore, the optimal costate may be interpreted as the marginal value of the state $\mathbf{x}$ in terms of the optimal costs J^* at any time $t_0 \le t \le t_f$ along the optimal solution. In other words, it can be interpreted as the cost sensitivity to state perturbations from the optimal trajectory at any time t (Stengel 1994, p. 207).

To determine the necessary optimality conditions (3.8 through 3.10), it is beneficial to first determine $\frac{\partial \mathbf{f}}{\partial \mathbf{x}}, \frac{\partial L}{\partial \mathbf{x}}, \frac{\partial \mathbf{f}}{\partial \mathbf{u}}, \frac{\partial L}{\partial \mathbf{u}}, \left.\frac{\partial \Phi}{\partial \mathbf{x}}\right|_{t=t_f}$.

Example 1

For the optimal control problem specified in Example 2 of Chapter 2 by Equations 2.25 through 2.30, we obtain,

$$\frac{\partial \mathbf{f}}{\partial \mathbf{x}} = \begin{bmatrix} 0 & p_1 d_1 \\ 0 & -p_2 \end{bmatrix},\ \frac{\partial L}{\partial \mathbf{x}} = \begin{bmatrix} 0 & 0 \end{bmatrix},\ \frac{\partial \mathbf{f}}{\partial \mathbf{u}} = \begin{bmatrix} 0 \\ p_3 \end{bmatrix},\ \frac{\partial L}{\partial \mathbf{u}} = [p_4],\ \left.\frac{\partial \Phi}{\partial \mathbf{x}}\right|_{t=t_f} = \begin{bmatrix} p_5 & 0 \end{bmatrix}. \tag{3.13}$$

Therefore, the necessary optimality conditions (3.8 through 3.10) are,

$$\begin{bmatrix} \dot{x}_1 \\ \dot{x}_2 \end{bmatrix} = \begin{bmatrix} p_1 d_1 x_2 \\ p_2(d_2 - x_2) + p_3 u_1 \end{bmatrix}, \ 0 \le t \le 10, \ \begin{bmatrix} x_1(0) \\ x_2(0) \end{bmatrix} = \begin{bmatrix} 0 \\ 10 \end{bmatrix}, \tag{3.14}$$

$$\begin{bmatrix} \dot{\lambda}_1 \\ \dot{\lambda}_2 \end{bmatrix} = \begin{bmatrix} 0 & 0 \\ p_1 d_1 & -p_2 \end{bmatrix} \begin{bmatrix} \lambda_1 \\ \lambda_2 \end{bmatrix} + \begin{bmatrix} 0 \\ 0 \end{bmatrix} = \begin{bmatrix} 0 \\ p_1 d_1 \lambda_1 - p_2 \lambda_2 \end{bmatrix}, \ \begin{bmatrix} \lambda_1(48) \\ \lambda_2(48) \end{bmatrix} = \begin{bmatrix} p_5 \\ 0 \end{bmatrix}, \tag{3.15}$$

$$0 = \begin{bmatrix} 0 & p_3 \end{bmatrix} \begin{bmatrix} \lambda_1 \\ \lambda_2 \end{bmatrix} + p_4 = p_3 \lambda_2 + p_4, \ 0 \le t \le 48. \tag{3.16}$$

From Equation 3.16, observe that $\lambda_2(t) = \frac{-p_4}{p_3}$, $0 \le t \le 48$. From Equation 3.15, $\dot{\lambda}_1 = 0$, $0 \le t \le 48$, so $\lambda_1(t)$ is constant as well. However, this is incompatible with $\dot{\lambda}_2 = p_1 d_1 \lambda_1 - p_2 \lambda_2$, which must also be satisfied according to Equation 3.15. Therefore, the necessary optimality conditions (3.8 through 3.10) cannot be satisfied.

The outcome of Example 1 seems to contradict the fact that we obtained numerical solutions for the optimal control problem described in this section. However, the optimal control problem in this section had the additional property that the control input has an upper and lower bound. For problems having this additional property, the necessary optimality condition (3.10) does not apply and must be replaced by one that is more general. To introduce this generalized necessary optimality condition, it is convenient to first introduce the so-called Hamiltonian $H(\mathbf{x}, \mathbf{u}, \boldsymbol{\lambda}, \mathbf{d}, \mathbf{p})$ associated to an optimal control problem,

$$H(\mathbf{x}, \mathbf{u}, \boldsymbol{\lambda}, \mathbf{d}, \mathbf{p}) = L(\mathbf{x}, \mathbf{u}, \mathbf{d}, \mathbf{p}) + \boldsymbol{\lambda}^{\mathrm{T}} \mathbf{f}(\mathbf{x}, \mathbf{u}, \mathbf{d}, \mathbf{p}), \ t_0 \le t \le t_{\mathrm{f}}. \tag{3.17}$$

The necessary optimality condition (3.10) is replaced by the following one that also applies to cases where the control input is in any way restricted, such as by upper and lower bounds,

$$H(\mathbf{x}^*, \mathbf{u}^*, \boldsymbol{\lambda}^*, \mathbf{d}, \mathbf{p}) \le H(\mathbf{x}^*, \mathbf{u}, \boldsymbol{\lambda}^*, \mathbf{d}, \mathbf{p}), \ t_0 \le t \le t_{\mathrm{f}}. \tag{3.18}$$

In Equation 3.18, the difference between the left-hand and the right-hand side is that u on the right-hand side represents any admissible control input, that is, any control input that satisfies the restrictions imposed on it, such as upper and lower bounds. On the other hand, $\mathbf{u}^*$ on the left-hand side is the optimal control that must also be admissible. Stated in words, Equation 3.18 demands that the value of the Hamiltonian along an optimal solution $\mathbf{u}^*(t)$, $\mathbf{x}^*(t)$, and $\boldsymbol{\lambda}^*(t)$ should be minimal with respect to all admissible controls. Equation 3.18 is known as Pontryagin's minimum principle. This principle and the associated Hamiltonian have the following interesting interpretation. Using Equations 3.12 and 3.8, the Hamiltonian (3.17) can be written as,

$$H(\mathbf{x}, \mathbf{u}, \boldsymbol{\lambda}, \mathbf{d}, \mathbf{p}) = L(\mathbf{x}, \mathbf{u}, \mathbf{d}, \mathbf{p}) + \frac{\partial J^*}{\partial \mathbf{x}} \frac{\mathrm{d}\mathbf{x}}{\mathrm{d}t} = L(\mathbf{x}, \mathbf{u}, \mathbf{d}, \mathbf{p}) + \frac{\mathrm{d}J^*}{\mathrm{d}t}, \ t_0 \le t \le t_{\mathrm{f}}. \tag{3.19}$$

Multiplying Equation 3.19 with an infinitely small time increment $\mathrm{d}t$, we obtain,

$$H(\mathbf{x}, \mathbf{u}, \boldsymbol{\lambda}, \mathbf{d}, \mathbf{p})\mathrm{d}t = L(\mathbf{x}, \mathbf{u}, \mathbf{d}, \mathbf{p})\mathrm{d}t + \mathrm{d}J^*, \ t_0 \le t \le t_{\mathrm{f}}, \tag{3.20}$$

where,

$$dJ^* = J^*_{t,t_f} - J^*_{t+dt,t_f}. \quad (3.21)$$

Hence, over every infinitely small time interval $(t, t + dt)$, $t_0 \le t \le t_f$, Equation 3.20 describes the running costs $L(\mathbf{x}, \mathbf{u}, \mathbf{d}, \mathbf{p})dt$ associated with this time interval together with the contribution dJ^* over this time interval to the costs $J^*_{t_0,t_f}$. Clearly, these together must be minimal to guarantee $J^*_{t_0,t_f}$ to be minimal as stated by Pontryagin's minimum principle.

Example 2

Reconsider Example 1 in this section with additional upper and lower bounds on the heat input,

$$0 \le u_1(t) \le 100, \; t_0 \le t \le t_f, \quad (3.22)$$

then Equation 3.18 reads,

$$p_4 u_1^* + \begin{bmatrix} \lambda_1^* & \lambda_2^* \end{bmatrix} \begin{bmatrix} p_1 d_1 x_2^* \\ p_2\left(d_2 - x_2^*\right) + p_3 u_1^* \end{bmatrix} \le p_4 u_1 + \begin{bmatrix} \lambda_1^* & \lambda_2^* \end{bmatrix} \begin{bmatrix} p_1 d_1 x_2^* \\ p_2\left(d_2 - x_2^*\right) + p_3 u_1 \end{bmatrix}, \quad (3.23)$$

which after multiplying out, collecting the terms in u_1, and equating identical terms on both sides becomes

$$\left(p_4 + \lambda_2^* p_3\right) u_1^* \le \left(p_4 + \lambda_2^* p_3\right) u_1. \quad (3.24)$$

According to Equation 3.24, the optimal control is indeed bang-bang, that is, either maximal or minimal as determined by the following so-called switching rule,

$$\begin{aligned} u_1^*(t) &= 100 \quad \text{if } p_4 + \lambda_2^* p_3 < 0, \\ u_1^*(t) &= 0 \quad \text{if } p_4 + \lambda_2^* p_3 > 0. \end{aligned} \quad (3.25)$$

For obvious reasons in Equation 3.25, $p_4 + \lambda_2^* p_3$ is called the *switching function* associated with the control input u_1.

For optimal control problems where both $\mathbf{f}(\mathbf{x}, \mathbf{u}, \mathbf{d}, \mathbf{p})$ and $L(\mathbf{x}, \mathbf{u}, \mathbf{d}, \mathbf{p})$ are linear functions of $\mathbf{u}$, the switching function for each control input u_i, $i = 1, 2, \ldots, n_u$, equals $\dfrac{\partial H}{\partial u_i}$.

Having introduced the Hamiltonian (3.17), the necessary optimality conditions (3.9 and 3.10) can be written more compactly,

$$-\dot{\boldsymbol{\lambda}} = \left(\frac{\partial H}{\partial \mathbf{x}}\right)^{\mathrm{T}}, \; t_0 \le t \le t_f, \; \boldsymbol{\lambda}(t_f) = \left(\left.\frac{\partial \Phi}{\partial \mathbf{x}}\right|_{t=t_f}\right)^{\mathrm{T}}, \quad (3.26)$$

$$\left(\frac{\partial H}{\partial \mathbf{u}}\right)^{\mathrm{T}} = 0,\ t_0 \le t \le t_{\mathrm{f}}. \tag{3.27}$$

Observe that Equation 3.27 is a necessary condition for Equation 3.18 to hold in the special case that the control input $\mathbf{u}(t)$ is not in any way restricted, that is, in the case that any control is admissible.

In summary, the necessary optimality conditions for a general optimal control problem including restrictions on the control input are given by,

$$\dot{\mathbf{x}} = \mathbf{f}(\mathbf{x}, \mathbf{u}, \mathbf{d}, \mathbf{p}),\ t_0 \le t \le t_{\mathrm{f}},\ \mathbf{x}\left(t_0\right) = x_0, \tag{3.28}$$

$$-\dot{\boldsymbol{\lambda}} = \left(\frac{\partial H}{\partial \mathbf{x}}\right)^{\mathrm{T}},\ t_0 \le t \le t_{\mathrm{f}},\ \boldsymbol{\lambda}(t_{\mathrm{f}}) = \left(\left.\frac{\partial \Phi}{\partial \mathbf{x}}\right|_{t=t_{\mathrm{f}}}\right)^{\mathrm{T}}, \tag{3.29}$$

$$H\left(\mathbf{x}^*, \mathbf{u}^*, \boldsymbol{\lambda}^*, \mathbf{d}, \mathbf{p}\right) \le H\left(\mathbf{x}^*, \mathbf{u}, \boldsymbol{\lambda}^*, \mathbf{d}, \mathbf{p}\right),\ t_0 \le t \le t_{\mathrm{f}}. \tag{3.30}$$

In the special case that there are no restrictions on the control input $\mathbf{u}(t)$, $t_0 \le t \le t_{\mathrm{f}}$, Equation 3.30 may be replaced with,

$$\left(\frac{\partial H}{\partial \mathbf{u}}\right)^{\mathrm{T}} = 0,\ t_0 \le t \le t_{\mathrm{f}}. \tag{3.31}$$

Many numerical algorithms to solve optimal control problems are based on the necessary optimality conditions (3.28 through 3.31). These algorithms are considered in the next section.

3.3 OPTIMAL CONTROL ALGORITHMS

It is common terminology to distinguish between direct and indirect methods to solve optimal control problems numerically. Indirect methods are based on the necessary optimality conditions presented in the previous section, whereas direct methods take a nonlinear programming (NLP) approach to solve optimal control problems. Indirect methods exploit the specific structure of an optimal control problem that is reflected in the necessary optimality conditions. Exploiting the structure of a problem in general provides solution methods and algorithms that are computationally more efficient.

Direct methods transform the optimal control problem into a general NLP problem, thus not exploiting the specific structure of the optimal control problem. Although the structure is lost in the transformation, much of it can be retained by computing the derivatives required by any NLP algorithm in a suitable manner. This manner is again dictated by the necessary optimality conditions presented in the previous section. Most of the optimal control software that is commercially available nowadays takes this approach. Probably this is because NLP algorithms are particularly well developed and can handle almost any type of constraint. Indirect methods are less well developed and have difficulties in handling several types of constraints.

In this section, the indirect methods and the associated algorithms are considered first. Their treatment facilitates the explanation of how optimal control problems can be transformed into NLP problems. This transformation and the associated NLP algorithms together with the special computation of derivatives are considered next.

3.3.1 Indirect Methods

Of the four necessary optimality conditions (3.28 through 3.30), three are generally easy to satisfy a priori. Starting from a solution that satisfies three optimality conditions a priori, indirect solution methods iterate toward satisfying the fourth condition. Therefore, there are basically four types of indirect methods. Each one is characterized by the optimality condition that is not satisfied a priori. We will describe and discuss briefly the most common of these four types, namely, the one that does not satisfy Equation 3.30 a priori. The algorithms of this type are described by the following steps.

1. Start with a guess of the optimal control trajectory: $\mathbf{u}(t)$, $t_0 \le t \le t_f$.
2. Starting from $\mathbf{x}(t_0)$ given by Equation 3.28, numerically integrate the systems model $\dot{\mathbf{x}} = \mathbf{f}(\mathbf{x}, \mathbf{u}, \mathbf{d}, \mathbf{p})$ also given by 3.28, forward in time using the known external input $\mathbf{d}(t)$ and control input $\mathbf{u}(t)$, $t_0 \le t \le t_f$. Store $\mathbf{x}(t)$, $t_0 \le t \le t_f$.
3. From $\mathbf{x}(t_f)$ computed under step 2, compute $\boldsymbol{\lambda}(t_f) = \left.\frac{\partial \Phi}{\partial \mathbf{x}}\right|_{t=t_f}$. Starting from $\boldsymbol{\lambda}(t_f)$, numerically integrate the adjoined system $-\dot{\boldsymbol{\lambda}} = \left(\frac{\partial H}{\partial \mathbf{x}}\right)^{\mathrm{T}}$, given by Equation 3.29, backward in time using $\mathbf{u}(t)$, $\mathbf{d}(t)$, $\mathbf{x}(t)$, $t_0 \le t \le t_f$. Store $\boldsymbol{\lambda}(t)$, $t_0 \le t \le t_f$.
4. Using $\mathbf{u}(t)$, $\mathbf{d}(t)$, $\mathbf{x}(t)$, $\boldsymbol{\lambda}(t)$, $t_0 \le t \le t_f$, compute $\left(\frac{\partial H}{\partial \mathbf{u}}\right)^{\mathrm{T}} = \left(\frac{\partial \mathbf{f}}{\partial \mathbf{u}}\right)^{\mathrm{T}} \boldsymbol{\lambda} + \left(\frac{\partial L}{\partial \mathbf{u}}\right)^{\mathrm{T}}$, $t_0 \le t \le t_f$.
5. Find the scalar $\alpha > 0$ that defines a new control trajectory $\mathbf{u}'(t) = \mathbf{u}(t) - \alpha\left(\frac{\partial H}{\partial \mathbf{u}}\right)^{\mathrm{T}}$, $t_0 \le t \le t_f$, such that $J(\mathbf{u}'(t))$ is (approximately) minimal. At any time $t_0 \le t \le t_f$, where u_i', $i = 1, 2, \ldots, n_u$ exceeds one of its bounds, set it to this bound. Compute $\Delta J = J(\mathbf{u}(t)) - J(\mathbf{u}'(t)) \ge 0$. Replace $\mathbf{u}(t)$ with $\mathbf{u}'(t)$, $t_0 \le t \le t_f$.
6. If the cost function improvement ΔJ falls repeatedly below a small tolerance $\varepsilon > 0$, go to the next step; otherwise, go to step 2.
7. For $i = 1, 2, \ldots, n_u$, verify if $\frac{\partial H}{\partial u_i}$ is "small" at all times $t_0 \le t \le t_f$, where u_i is not at a bound. If so, $\mathbf{u}^*(t) := \mathbf{u}(t)$, $\mathbf{x}^*(t) := \mathbf{x}(t)$, and $\boldsymbol{\lambda}^*(t) := \boldsymbol{\lambda}(t)$, $t_0 \le t \le t_f$.

Step 1 is an important step because if the guess of the optimal control is poor, the algorithm needs more iterations (computation time) to find the solution or it may not find a solution at all. Also, depending on the control guess, the algorithm may find a local solution.

In performing step 2, we satisfy a priori Equation 3.28.

In performing steps 3 and 4, we satisfy a priori Equation 3.29. Therefore, only Equation 3.30 is not satisfied a priori.

Steps 5 and 6 improve the satisfaction of Pontryagin's minimum principle, that is, Equation 3.30.

Step 5 uses gradients, that is, first derivatives $\left(\frac{\partial H}{\partial \mathbf{u}}\right)^{\mathrm{T}}$ to improve the control. It searches for a control improvement in the direction $\left(-\frac{\partial H}{\partial \mathbf{u}}\right)^{\mathrm{T}}$, which complies with Pontryagin's minimum principle to minimize H with respect to $\mathbf{u}$. This type of algorithm is therefore called a gradient type algorithm. Searching in the gradient direction is not very efficient, especially if one is near the optimal solution. More efficient algorithms such as conjugate gradient algorithms during each iteration do not just use the current value $\left(\frac{\partial H}{\partial \mathbf{u}}\right)^{\mathrm{T}}$ but also values obtained from previous iterations to compute a search direction. Basically, these more advanced algorithms try to estimate higher derivatives to improve the search direction. This intends to improve the speed of convergence of the algorithm.

Simple versions of the algorithm do not search for α but use a fixed small value for α. The search for α is called *a line search*. Although the line search itself takes additional computation time, it generally improves the efficiency of the algorithm. By evaluating the cost function $J(\mathbf{u}'(t))$ over a wide range of α's, we can try to find a global minimum. Implementing such a global line search increases the chances of finding a global solution of the optimal control problem.

Step 7 checks whether the improvement of the costs is sufficient to justify continuation of the algorithm. It should be noted that the improvement of the costs can be very small during some iterations. Therefore, one only stops iterating when several consecutive iterations have failed to provide significant improvement of the costs.

Step 8 is a step that verifies whether Pontryagin's minimum principle is approximately satisfied. In that case, the optimal control and the associated state and costate trajectories are obtained.

The advantages of this most common type of indirect method compared with the other three can be roughly stated as follows. The algorithm requires an initial guess of the control input and no guesses for the costate, whereas the other types of indirect methods do. Usually by experience or simulation experiments, one can come up with a control that provides reasonable costs, in other words with a reasonable initial guess of the optimal control. It is generally more difficult to come up with reasonable guesses for the costate. Moreover, very often the other algorithms turn out to be highly sensitive to initial guesses of the costate. This causes convergence problems and makes them less robust. Several of the other direct methods use Equation 3.11 to turn the problem into a two-point boundary value problem to be solved. For optimal greenhouse climate control problems, the systems model and cost function can be quite large and complicated. Then, finding Equation 3.11 from the optimality condition (3.30 or 3.31) is difficult if not impossible.

To perform steps 2 and 3, numerical integration has to be performed. Numerical integration is performed by taking a finite number of small time steps from t_0 to t_f or vice versa and by computing the state $\mathbf{x}(t)$ and costate $\boldsymbol{\lambda}(t)$ at these intermediate times. During step 5, to compute the costs $J(\mathbf{u}(t)) = \Phi\left(\mathbf{x}\left(t_f\right)\right) + \int_{t_0}^{t_f} L(\mathbf{x}, \mathbf{u}, \mathbf{d}, \mathbf{p})\,dt$ for different control histories $\mathbf{u}(t) = \mathbf{u}'(t)$, the following augmented system is integrated numerically,

$$\begin{bmatrix} \dot{x}_1 \\ \dot{x}_2 \\ \vdots \\ \dot{x}_{n_x} \\ \dot{J}_\mathrm{r} \end{bmatrix} = \begin{bmatrix} f_1(\mathbf{x}, \mathbf{u}, \mathbf{d}, \mathbf{p}) \\ f_2(\mathbf{x}, \mathbf{u}, \mathbf{d}, \mathbf{p}) \\ \vdots \\ f_{n_x}(\mathbf{x}, \mathbf{u}, \mathbf{d}, \mathbf{p}) \\ L(\mathbf{x}, \mathbf{u}, \mathbf{d}, \mathbf{p}) \end{bmatrix}, \quad \begin{bmatrix} x_1(t_0) \\ x_2(t_0) \\ \vdots \\ x_{n_x}(t_0) \\ J_\mathrm{r}(t_0) \end{bmatrix} = \begin{bmatrix} x_0 \\ 0 \end{bmatrix}. \tag{3.32}$$

Note that Equation 3.32 is identical to the systems model with initial conditions (3.28) augmented with the differential equation,

$$\dot{J}_\mathrm{r} = L(\mathbf{x}, \mathbf{u}, \mathbf{d}, \mathbf{p}), \; J_\mathrm{r}(t_0) = 0. \tag{3.33}$$

From Equation 3.32, observe that $J_\mathrm{r}(t_f) = \int_{t_0}^{t_\mathrm{f}} L(\mathbf{x}, \mathbf{u}, \mathbf{d}, \mathbf{p})dt$ represents precisely the running costs that is required to compute $J(\mathbf{u}(t)) = \Phi\left(\mathbf{x}\left(t_\mathrm{f}\right)\right) + \int_{t_0}^{t_\mathrm{f}} L(\mathbf{x}, \mathbf{u}, \mathbf{d}, \mathbf{p})dt$. Also $J_r(t)$, $t_0 \le t \le t_\mathrm{f}$, may be viewed as an additional state variable that represents the running costs $\int_{t_0}^{t} L(\mathbf{x}, \mathbf{u}, \mathbf{d}, \mathbf{p})dt$ at any time $t_0 \le t \le t_f$. Introducing $J_r(t)$ as an additional state variable, a new optimal control problem formulation that is equivalent to the original one is obtained, where the system is now the augmented system (3.32) and where the cost function equals,

$$J(\mathbf{u}(t)) = \Phi(\mathbf{x}(t_f)) + J_r(t_f). \tag{3.34}$$

From Equation 3.34, observe that this augmented equivalent optimal control problem has no running costs because J_r has become a state variable now. Therefore, Equation 3.32 may also be viewed as a way to transform any optimal control problem with running costs into one without running costs. This is known as the Mayer formulation of the optimal control problem.

During every small time step involved in numerical integration, the control input and the external input are either assumed constant or interpolated in some manner. In both cases, this implies that the control and the external inputs that are continuous functions in principle are parameterized, that is, described by a finite number of parameters. The parameterization of control inputs is the key to transforming optimal control problems into NLP problems.

3.3.2 Direct Methods and Control Parameterization

The objective of any optimal control problem is to find an optimal control trajectory $\mathbf{u}(t)$, $t_0 \le t \le t_f$. To specify an arbitrary control trajectory $\mathbf{u}(t)$, $t_0 \le t \le t_f$, infinitely many parameters are needed in principle because there are infinitely many times t in between t_0 and t_f. Fourier analysis is an example of the fact that arbitrary time functions can be approximated by using only finitely many parameters, namely the Fourier coefficients. The approximation of arbitrary control trajectories $\mathbf{u}(t)$, $t_0 \le t \le t_f$, by finitely many parameters is called *control parameterization*. Numerical integration that is used by indirect methods to solve optimal control problems actually also uses control parameterization as explained at the end of the previous section. One of the most simple parameterizations of the control input is the following piecewise constant parameterization,

$$\mathbf{u}(t) = \mathbf{u}(t_k),\ t_k \le t \le t_{k+1},\ k = 0, 1, \ldots, N-1,\ t_N = t_f. \tag{3.35}$$

Equation 3.35 approximates $\mathbf{u}(t)$, $t_0 \le t \le t_f$, using a finite number of vectors $\mathbf{u}(t_i)$, $i = 0, 1, \ldots, N$, and so using a finite number of parameters. The approximations are staircase functions. Equation 3.35 is called a piecewise constant control parameterization. By varying the parameters, that is, the components of the vectors $\mathbf{u}(t_k)$, $k = 0, 1, \ldots, N-1$, many different control histories are obtained, although each is restricted to being a staircase function (see Figure 3.1).

For every choice of parameters $\mathbf{u}(t_k)$, $k = 0, 1, \ldots, N-1$, using 3.35 to determine the control input at any time $t_0 \le t \le t_f$, we can calculate the associated costs $J(\mathbf{u}(t))$, given by Equation 3.34, through numerical integration of Equation 3.32. Observe that this computation ensures that the dynamic constraints, that is, the systems model and its initial conditions, are automatically satisfied. This demonstrates that $J(\mathbf{u}(t))$ can be viewed as a (complicated, nonlinear) computable function of the variables $\mathbf{u}(t_k)$, $k = 0, 1, \ldots, N-1$. In other words, by using the control parameterization (3.35), the optimal control problem is transformed into a problem where the function J of the parameters $\mathbf{u}(t_k)$, $k = 0, 1, \ldots, N-1$, has to be minimized. In general, any type of control parameterization can be used to transform an optimal control problem into a (nonlinear) function minimization problem.

If the control inputs $\mathbf{u}(t)$, $t_0 \le t \le t_f$, are in any way restricted, these restrictions have to be translated into restrictions on the parameters that approximate the control input trajectory. If the control parameterization (3.35) is used and if the restrictions concern lower and upper bounds, then the translation is particularly easy,

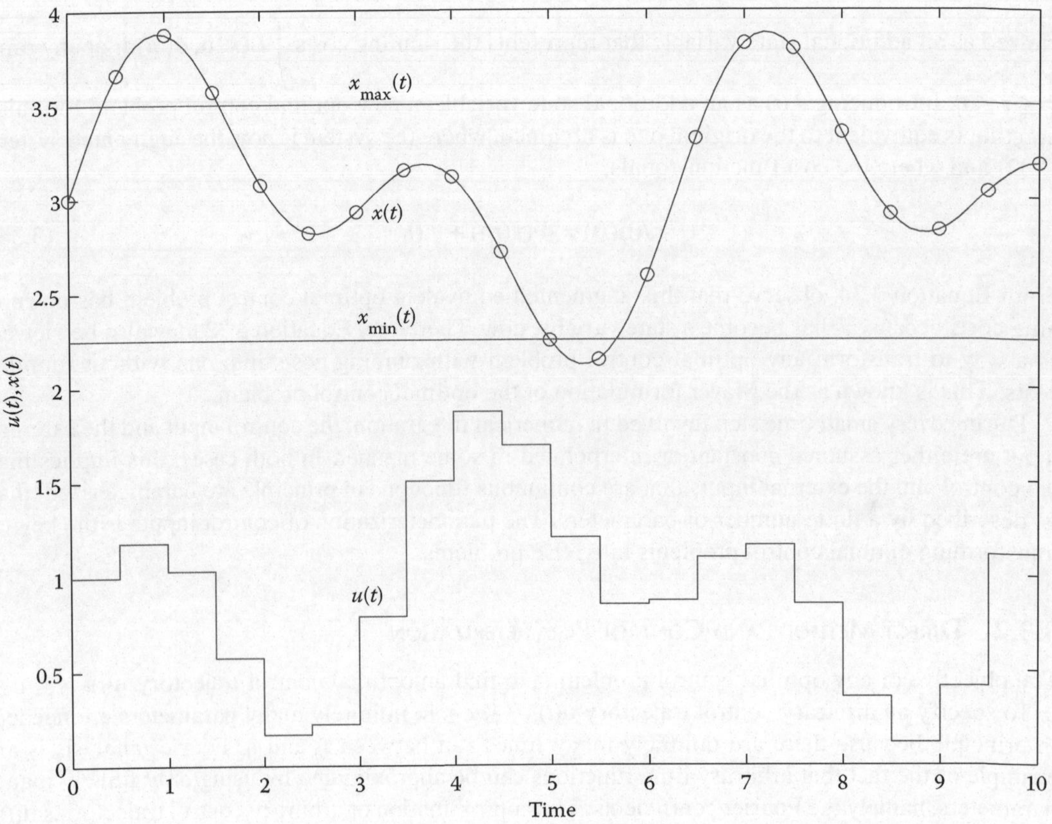

FIGURE 3.1 Control parameterization and evaluation of constraints.

$$u_i^{\min} \le u_i\left(t_k\right) \le u_i^{\max},\ i = 1, 2, \ldots, n_u,\ k = 0, 1, \ldots, N-1. \tag{3.36}$$

If the parameters $\mathbf{u}(t_k)$, $k = 0, 1, \ldots, N-1$, of the nonlinear functions J to be minimized are in any way restricted, such as by 3.36, then the function minimization problem is called an NLP problem. A general NLP problem reads as follows.

Minimize the scalar function,

$$J(\mathbf{s}), \tag{3.37}$$

with respect to the variables collected in the vector $\mathbf{s}$ while satisfying the following two types of general constraints,

$$E_i(\mathbf{s}) = 0, i = 1, 2, \ldots, n_E, \tag{3.38}$$

$$I_i(\mathbf{s}) \le 0, i = 1, 2, \ldots, n_I. \tag{3.39}$$

In Equations 3.38 and 3.39, $\mathbf{E}(\mathbf{s})$ and $\mathbf{I}(\mathbf{s})$ are arbitrary vector functions of dimension n_E and n_I. The index i refers to their components. Equation 3.38 therefore describes n_E equality constraint in **s**, whereas Equation 3.39 describes n_I inequality constraint in **s**. Any number and type of equality and inequality constraints in **s** can be written in the form of 3.38 and 3.39.

Using the control parameterization (3.36), any optimal control problem with bounds on the control inputs (3.36) can be represented as an NLP problem (3.37 and 3.39) as follows,

$$\mathbf{s} = \begin{bmatrix} u(t_0) \\ u(t_1) \\ \vdots \\ u(t_{N-1}) \end{bmatrix}, I(\mathbf{s}) = \begin{bmatrix} u(t_0) - u^{\max} \\ u(t_1) - u^{\max} \\ \vdots \\ u(t_{N-1}) - u^{\max} \\ u^{\min} - u(t_0) \\ u^{\min} - u(t_1) \\ \vdots \\ u^{\min} - u(t_{N-1}) \end{bmatrix}. \tag{3.40}$$

The cost function $J(\mathbf{s})$ is computed through numerical integration of Equation 3.32 using the control parameterization (3.35).

Observe that n_s, the dimension of the vector **s**, of to be optimized variables equals Nn_u and that n_I, the dimension of the vector function $\mathbf{I}(\mathbf{s})$, of constraints equals $2Nn_u$. The more accurate the control parameterization (3.35) becomes, the larger the N, hence the larger the number of variables to be optimized as well the number of constraints of the NLP problem.

Numerical algorithms to solve NLP problems are very well established. Most of them require repeated computations of

$$\frac{\partial J}{\partial \mathbf{s}}, \frac{\partial \mathbf{E}}{\partial \mathbf{s}}, \frac{\partial \mathbf{I}}{\partial \mathbf{s}}, \tag{3.41}$$

for several different values of **s**. To compute the derivatives (3.41), the algorithms supply two options. Either the user specifies analytical expressions or a computational procedure to compute the derivatives, or the algorithm uses finite differences to automatically approximate the derivatives. The former is generally more accurate (unless mistakes are made), whereas the latter does not require any additional effort from the user because only function evaluations of $J(\mathbf{s})$, $\mathbf{E}(\mathbf{s})$, and $\mathbf{I}(\mathbf{s})$ are needed that are performed automatically by the algorithm.

Although the specific structure of an optimal control problem is lost in the transformation into an NLP problem, much of it can be retained by supplying a computation of the derivatives (3.41) on the basis of the necessary optimality conditions (3.28 through 3.30) that do carry the specific structure of an optimal control problem. This derivative computation is based on the following equality that approximately holds for any optimal control problem with a parameterized control (3.35),

$$\frac{\partial J}{\partial \mathbf{u}(t_k)} \approx \int_{t_k}^{t_{k+1}} \frac{\partial H}{\partial \mathbf{u}} dt \approx \frac{(t_{k+1} - t_k)}{2} \left(\left.\frac{\partial H}{\partial \mathbf{u}}\right|_{t=t_k} + \left.\frac{\partial H}{\partial \mathbf{u}}\right|_{t=t_{k+1}} \right), \; k = 0, 1, \ldots, N-1. \tag{3.42}$$

Observe from Equation 3.42 that the derivatives $\dfrac{\partial J}{\partial \mathbf{u}(t_k)}$, $t_0 \le t \le t_f$, can be computed using steps 2 and 3 used by indirect methods because these compute $\dfrac{\partial H}{\partial \mathbf{u}}$, $t_0 \le t \le t_f$. This computation of $\dfrac{\partial H}{\partial \mathbf{u}}$,

$t_0 \le t \le t_f$, is based on the necessary optimality conditions (3.28 through 3.30). By using it, the efficiency of the NLP algorithm increases seriously as compared with the automatic computation of derivatives by finite differences. This is especially so if the number of time steps N or the number of control inputs N_u or both are large.

Notoriously difficult constraints that is regularly encountered in practice and cannot be handled easily by indirect methods are the upper and lower bounds on state variables,

$$x_i^{\min} \le x_i(t) \le x_i^{\max},\ t_0 \le t \le t_f,\ i = 1, 2, \ldots, n_x. \tag{3.43}$$

If these state constraints are approximated by (see Figure 3.1)

$$x_i^{\min} \le x_i\left(t_k\right) \le x_i^{\max},\ i = 1, 2, \ldots, n_x,\ k = 0, 1, 2, \ldots, N, \tag{3.44}$$

they can be implemented as

$$\mathbf{I}(\mathbf{s}) = \begin{bmatrix} \mathbf{x}(t_0) - \mathbf{x}^{\max} \\ \mathbf{x}(t_1) - \mathbf{x}^{\max} \\ \vdots \\ \mathbf{x}(t_N) - \mathbf{x}^{\max} \\ \mathbf{x}^{\min} - \mathbf{x}(t_0) \\ \mathbf{x}^{\min} - \mathbf{x}(t_1) \\ \vdots \\ \mathbf{x}^{\min} - \mathbf{x}(t_N) \end{bmatrix}, \tag{3.45}$$

where $\mathbf{x}(t_k)$, $k = 0, 1, \ldots, N$, is computed through numerical integration of 3.28 using the control parameterization (3.35).

Observe from Equations 3.44 and 3.45 and Figure 3.1 that the constraint (3.43) is only implemented at times t_k, $k = 0, 1, \ldots, N$, so it may be violated in between. This constitutes the approximation. Also observe that the larger we take N, the more accurate the approximation becomes at the expense of additional constraints in 3.45.

If the constraints represented by the vector functions **E**, **I** do not depend on the control input **u**, such as Equation 3.45, their components may be interpreted as cost functions (3.5) without running costs and with t_f replaced by t_k, that is,

$$J(\mathbf{u}(t)) = \Phi(\mathbf{x}(t_k)),\ t_0 \le t \le t_k. \tag{3.46}$$

Therefore, the derivatives $\dfrac{\partial \mathbf{E}}{\partial \mathbf{s}}, \dfrac{\partial \mathbf{I}}{\partial \mathbf{s}}$ associated with such equality and inequality constraints can be computed similar to 3.42.

Applying the above interpretation (3.46) to the vector function (3.45), the derivative of every component of 3.45 can be computed similar to 3.42. By executing steps 2 and 3 used by indirect methods, with t_f replaced by t_k, $\frac{\partial H}{\partial \mathbf{u}}$, $t_0 \le t \le t_k$ is obtained. In that case, the following terminal values $\boldsymbol{\lambda}(t_k)$ apply to step 3,

$$\boldsymbol{\lambda}(t_k) = \left(\left. \frac{\partial \Phi(\mathbf{x})}{\partial \mathbf{x}} \right|_{t=t_k} \right)^{\mathrm{T}} = \left(\left. \frac{\partial \left(x_i - x_i^{\max} \right)}{\partial \mathbf{x}} \right|_{t=t_k} \right)^{\mathrm{T}}, \ i = 1, 2, \ldots, n_x, \ k = 0, 1, \ldots, N,$$

$$\boldsymbol{\lambda}(t_k) = \left(\left. \frac{\partial \Phi(\mathbf{x})}{\partial \mathbf{x}} \right|_{t=t_k} \right)^{\mathrm{T}} = \left(\left. \frac{\partial \left(x_i^{\min} - x_i \right)}{\partial \mathbf{x}} \right|_{t=t_k} \right)^{\mathrm{T}}, \ i = 1, 2, \ldots, n_x, \ k = 0, 1, \ldots, N. \quad (3.47)$$

The first terminal value $\boldsymbol{\lambda}(t_k)$ in Equation 3.47 has components that are all zero except for $\lambda_i(t_k) = 1$ and applies to the first half of 3.45. The second has components that are all zero except for $\lambda_i(t_k) = -1$ and applies to the second half of 3.45.

Other than the simple piecewise constant control parameterization, 3.35 may be used to transform optimal control problems into NLP problems. Often, spline functions or other polynomial functions are used, which are more smooth and may therefore more accurately and efficiently approximate continuous optimal control inputs. The piecewise constant control parameterization (3.35) has the following advantage. Because Equation 3.42 only applies to piecewise constant control parameterizations, these enable the derivative computations treated in this section that seriously speed up the NLP algorithm because they exploit the special structure of an optimal control problem. Moreover, as a by-product of these computations, we also obtain the optimal costate $\boldsymbol{\lambda}^*(t)$, $t_0 \le t \le t_f$. This is important because the costate has the interesting interpretation (3.12) and is moreover required for the decomposition of optimal control problems into separate ones to accommodate for different time scales. Also, the optimal controller is generally implemented by a digital computer that is connected to the system by analog-to-digital and digital-to-analog converters. If the digital-to-analog converters act as zero-order hold circuits, which they generally do, the control is truly piecewise constant. Then, 3.35 becomes an exact description of the control input.

To end this section, we draw attention to an NLP approach that uses polynomial approximations for the state as well. In that case, the dynamic constraints (3.1) are translated into equality constraints $\mathbf{E}(\mathbf{s})$. These constraints concern the polynomial coefficients that determine the state and the control trajectories $\mathbf{x}^*(t)$ and $\mathbf{u}^*(t)$, $t_0 \le t \le t_f$. These constraints are evaluated at a finite number of times in between t_0 and t_f. In this case, the associated NLP problem no longer requires numerical integration to perform function evaluations. Therefore, the computational efficiency may increase, depending on the number and nature of the equality constraints that replace the dynamic constraints.

In summary, indirect methods to solve optimal control problems fully exploit the structure of an optimal control problem that is reflected in the necessary optimality conditions used to solve the problem. In general, this provides algorithms that are computationally most efficient. Algorithms of this type are characterized by steps 1–7 described in this section. Step 5, which determines the control improvement during each iteration of the algorithm, is a critical step with respect to computational efficiency. Such a step is also performed by direct methods that use NLP solvers. One might roughly conclude that NLP algorithms are better developed with respect to this step than indirect methods to solve optimal control problems. The drawback of losing the specific structure of an optimal control problem through the transformation into an NLP problem can be largely compensated for by using user-supplied derivative computations that do exploit this specific structure. Together with the fact that several types of optimal control constraints can be handled more easily by the NLP approach, this probably explains why most commercially available software tools used to solve optimal control problems are of this type.

Recently, we investigated the MATLAB® optimal control toolbox called PROPT, which is commercially available from TOMLAB (Ross and Fahroo, 2004; Rutquist and Edvall, 2009). It uses polynomial approximations of both the state and control and NLP. Through a number of examples that have been added to this toolbox, one of them concerning optimal greenhouse climate control, we discovered that it performs very efficiently (solutions are obtained in a few seconds) and is highly user friendly (students with only basic knowledge of optimal control easily use it).

Commercially available optimal control software in principle only requires that the user specify the optimal control problem data including the constraints. In general, however, the convergence of optimal control algorithms depends critically on the initialization of the algorithm and the conditioning of the problem. Ill conditioning may, for instance, occur because of widely different time scales. Therefore, it is advisable to start with a simplified problem and perform some analysis, for example, by simulations that may be verified against experimental results, to come up with a reasonable initialization. Next, after critically judging and examining the solutions, one may proceed in one or several steps toward the whole problem.

REFERENCES

Bryson, A.E. 1999. *Dynamic Optimization*. Addison-Wesley.

Bryson, A.E., and Y.C. Ho. 1969. *Applied Optimal Control*. New York: Halsted Press.

Lewis, F.L. 1986. *Optimal Control*.

Ross, I.M., and F. Fahroo. 2004. Pseudospectral knotting methods for solving optimal control problems. *Journal of Guidance, Control, and Dynamics* 27: 3.

Rutquist, P.E., and M.M. Edvall. 2009. Lightning fast solutions to your optimal control problems! Tomlab PROPT Users Guide, November 25, 2009.

Stengel, R.F. 1994. *Optimal Control and Estimation*. Mineola, NY: Dover.

4 Closed-Loop Optimal Control

4.1 INTRODUCTION

What happens if we apply the optimal controls $\mathbf{u}(t)$, $t_0 \leq t \leq t_f$, computed in the previous chapter, directly to the greenhouse? Will the greenhouse climate behavior $\mathbf{x}(t)$, $t_0 \leq t \leq t_f$ be optimal? In other words, will $\mathbf{x}(t) = \mathbf{x}^*(t)$, $t_0 \leq t \leq t_f$ hold? The most simple and straightforward answer would be *yes*. Unfortunately, this answer is only correct if all elements of the optimal control problem formulation are perfect descriptions of reality. Most optimal control problem formulations like the ones for greenhouses are not. The initial condition,

$$\mathbf{x}(t_0) = \mathbf{x}_0, \tag{4.1}$$

and the dynamic systems model,

$$\dot{\mathbf{x}}(t) = \mathbf{f}(\mathbf{x}(t), \mathbf{u}(t), \mathbf{d}(t), \mathbf{p}), \tag{4.2}$$

are at best very good approximations of reality. When controlling a greenhouse the external input trajectory $d(t)$, $t_0 \leq t \leq t_f$, needed to solve the optimal control problem, will not be perfect, either, because it involves *weather predictions*. The cost function

$$J(\mathbf{u}(t)) = \Phi\left(\mathbf{x}\left(t_f\right), t_f\right) + \int_{t_0}^{t_f} L(\mathbf{x}(t), \mathbf{u}(t), \mathbf{d}(t), \mathbf{p})\, dt, \tag{4.3}$$

is a description of the control objectives and as such can be perfect. Because the dynamic model, the initial condition and the external input trajectory $\mathbf{d}(t)$, $t_0 \leq t \leq t_f$ are at best very good approximations $\mathbf{x}(t) = \mathbf{x}^*(t)$, $t_0 \leq t \leq t_f$ will not hold. Stated differently,

$$\Delta\mathbf{x}(t) = \mathbf{x}(t) - \mathbf{x}^*(t) \neq 0, \tag{4.4}$$

for almost any $t_0 \leq t \leq t_f$. The associated important question now is: How large are the state perturbations (deviations) $\Delta\mathbf{x}(t)$? This clearly depends on the magnitude of the errors just mentioned and the way these propagate. Without going into details, practical applications require a feedback control mechanism to limit the state perturbations in Equation 4.4. Observe that in Equation 4.4 $\mathbf{x}(t)$ represents actual system behavior. To obtain information about actual system behavior measurements must be performed on the system. Variables that are measured are also called outputs of the system. In general they are functions of the current state, input, external input and the parameters of the system. This is described by,

$$\mathbf{y}(t) = \mathbf{g}(\mathbf{x}(t), \mathbf{u}(t), \mathbf{d}(t), \mathbf{p}). \tag{4.5}$$

In Equation 4.5 $\mathbf{y}$ is a n_y dimensional column vector, its components being the measured variables (outputs), and $\mathbf{g}$ is the associated vector function. For control purposes the most favorable situation is,

$$\mathbf{y}(t) = \mathbf{x}(t), \tag{4.6}$$

meaning that all state variables (the complete state) are measured. When Equation 4.6 applies, control engineers call this "having full state information." In most practical circumstances, however, instead of the special case Equation 4.6, the general description Equation 4.5 applies. In that case state estimators are applied to obtain estimates $\hat{\mathbf{x}}(t)$ of $\mathbf{x}(t)$. These state estimators are designed using Equations 4.1, 4.2, and 4.5. State estimation will be briefly treated in Section 4.2.

Having $\mathbf{x}(t)$ the actual system behavior or estimates $\hat{\mathbf{x}}(t)$ of these, one can compute the state perturbations $\Delta\mathbf{x}(t) = \mathbf{x}(t) - \mathbf{x}^*(t)$. In contrast to the optimal control computations performed so far this computation must be done online because only then the actual system behavior becomes manifest and can be measured. Knowing the actual state of the system one has to adjust the optimal control $\mathbf{u}^*(t)$ to limit the state perturbations. This adjustment can only be performed online. The adjustment is called *feedback control.* A control system with feedback is called *a feedback control system.*

In the next sections, we address the issue of feedback control system design. In Section 4.3, we address a general, very common, and attractive feedback control system design. But we argue that this design is generally unsuitable for greenhouse climate control. In Section 4.3.1, the feedback control system design for greenhouse climate control is presented. It is based on receding horizon control. Receding horizon control enables a suitable feedback. Moreover, it overcomes the nasty problem of dealing with widely different time scales. This is a characteristic of most greenhouse optimal control problems that tends to destroy the accuracy of numerical solutions.

4.2 STATE ESTIMATION

Mathematical descriptions of practical problems are hardly ever exact. In dealing with the actual control of systems, it is therefore necessary to monitor the actual behavior of the system on-line in some manner. As explained in the previous section, this enables one to compute or correct the controls. The most favorable situation is the special case where we have complete state information as described by Equation 4.6. In the general case described by Equation 4.5 a state estimator is required to generate estimates $\hat{\mathbf{x}}(t)$ of the actual state $\mathbf{x}(t)$. These estimates are later used for control computations. State estimators are based on Equations 4.1, 4.2, and 4.5 that mathematically describe the system and the measurements. A widely used state estimators is called the *Kalman filter,* named after its discoverer, who played a leading role in systems theory between 1955 and 1970. The Kalman filter is described by,

$$\dot{\hat{\mathbf{x}}}(t) = \mathbf{f}\left(\hat{\mathbf{x}}(t), \mathbf{u}(t), \mathbf{d}(t), \mathbf{p}\right) + \boldsymbol{K}(t)\left(\mathbf{y}(t) - \hat{\mathbf{y}}(t)\right), \ \hat{\mathbf{x}}(t_0) = \hat{\mathbf{x}}_0, \tag{4.7}$$

$$\hat{\mathbf{y}}(t) = \mathbf{g}\left(\hat{\mathbf{x}}(t), \mathbf{u}(t), \mathbf{d}(t), \mathbf{p}\right). \tag{4.8}$$

Equation 4.7 is a differential equation that determines the state estimates $\hat{\mathbf{x}}(t)$ with initial value $\hat{\mathbf{x}}_0$, an estimate of the initial state. This differential equation at each time $t_0 \le t \le t_f$ has the measured values $\mathbf{y}(t)$ as an input. In Equation 4.8 $\hat{\mathbf{y}}(t)$ represents estimates of these measurements based on the current estimate $\hat{\mathbf{x}}(t)$ of the state and the current inputs $\mathbf{u}(t)$, $\mathbf{d}(t)$. Observe that if the actual measurements $\mathbf{y}(t)$ equal estimates $\hat{\mathbf{y}}(t)$ the term $\boldsymbol{K}(t)(\mathbf{y}(t) - \hat{\mathbf{y}}(t))$ in Equation 4.7 is equal to zero. Then the propagation of the state estimate $\hat{\mathbf{x}}(t)$ is entirely determined by $\dot{\hat{\mathbf{x}}}(t) = \mathbf{f}(\hat{\mathbf{x}}(t), \mathbf{u}(t), \mathbf{d}(t), \mathbf{p})$, in other words the systems model (4.2). If not, then the term $\boldsymbol{K}(t)(\mathbf{y}(t) - \hat{\mathbf{y}}(t))$ corrects the model. This makes sense because there is a *difference* between the expected (estimated) measurements $\hat{\mathbf{y}}(t)$ and the actual ones $\mathbf{y}(t)$. Besides $\mathbf{y}(t) - \hat{\mathbf{y}}(t)$, which is called the *innovation*, the magnitude of the correction is determined by the entries of the $n_x \times n_y$ matrix $\boldsymbol{K}(t)$ in Equation 4.7. This matrix is called the *Kalman gain* of the state estimator. Observe that apart from the Kalman gain $\boldsymbol{K}(t)$, the state estimator (Kalman filter) is entirely determined by the equations describing the systems model and the measurements, namely Equations 4.1, 4.2, and 4.5. The Kalman gain $K(t)$ entirely determines the state estimator design. Many techniques are available to determine it (see, for instance, Lewis, 1986). We will not go into detail here but just mention that most of these techniques require as an input measures of errors in the model the initial state and the measurements.

Example 1

Reconsider Example 1 from Chapter 2 that concerns our heavily simplified greenhouse. For this example,

$$\dot{\mathbf{x}}(t) = \mathbf{f}(\mathbf{x}(t), \mathbf{u}(t), \mathbf{d}(t), \mathbf{p}) = \begin{bmatrix} p_1 d_1(t) x_2(t) \\ p_2(d_2(t) - x_2(t)) + p_3 u_1(t) \end{bmatrix}. \tag{4.9}$$

Recall that x_2 represents the greenhouse temperature. Suppose this temperature is measured. This is described by,

$$y(t) = \mathbf{g}(\mathbf{x}(t), \mathbf{u}(t), \mathbf{d}(t), \mathbf{p}) = x_2(t). \tag{4.10}$$

For this example, the Kalman filter in Equations 4.7 and 4.8 becomes,

$$\begin{bmatrix} \dot{\hat{x}}_1(t) \\ \dot{\hat{x}}_2(t) \end{bmatrix} = \begin{bmatrix} p_1 d_1(t) \hat{x}_2(t) \\ p_2 \left(d_2(t) - \hat{x}_2(t)\right) + p_3 u_1(t) \end{bmatrix} + \begin{bmatrix} K_{11}(t) \\ K_{21}(t) \end{bmatrix} \left(y(t) - \hat{x}_2(t)\right), \tag{4.11}$$

with initial condition $\begin{bmatrix} \hat{x}_1(t_0) \\ \hat{x}_2(t_0) \end{bmatrix}$.

4.3 LINEAR QUADRATIC FEEDBACK

Feedback control system design concerns the online adjustment of the (optimal) control to the actual system behavior. In this section, we consider the computation of control corrections:

$$\Delta\mathbf{u}(t) = \mathbf{u}(t) - \mathbf{u}^*(t), \tag{4.12}$$

intended to limit the state perturbations $\Delta\mathbf{x}(t) = \mathbf{x}(t) - \mathbf{x}^*(t)$. Clearly the computation of the control correction $\Delta\mathbf{u}(t)$ at any time $t_0 \le t \le t_f$ requires $\Delta\mathbf{x}(t)$ or estimates $\Delta\hat{\mathbf{x}}(t)$ of it. A simple very common feedback control law is the following one:

$$\Delta\mathbf{u}(t) = -L(t)\Delta\mathbf{x}(t). \tag{4.13}$$

Equation 4.13 represents a linear feedback control law in which $L(t)$ is a known time-variable matrix of dimensions $n_u \times n_x$. The feedback control law in Equation 4.13 is actually an optimal feedback control law because it solves a so-called linear quadratic (LQ) optimal control problem that is associated with the original optimal control problem from which $\mathbf{u}^*(t)$ was computed. Details concerning this association and the LQ optimal control problem can be found in Athans (1971) and Van Willigenburg and De Koning (2006). The feedback control law in Equation 4.13 is highly attractive because it requires just a single multiplication of the stored feedback matrix $L(t)$ with the online measured (estimated) state perturbation $\Delta\mathbf{x}(t)$. Therefore, it is most suitable for online implementation.

In practice the linear feedback control law in Equation 4.13 performs well only if the errors are modest. The errors associated with optimal greenhouse climate control are not modest in general. This is mainly due to significant errors in the weather prediction. This is especially serious because the weather, such as the light conditions, heavily determine the optimal control. Therefore, greenhouse climate control requires another type of feedback control system design. This design is treated in the next section.

4.3.1 Feedback by Receding Horizon Control

4.3.1.1 The Problem of Widely Different Time Scales

The more the time scales of a system differ, the more inaccurate and inefficient numerical integration becomes. This causes optimal control computations to become inaccurate and inefficient. This problem can be largely overcome by employing suitable time scale decompositions of the optimal control problem. In the context of greenhouse cultivation, the method was analyzed in some detail in van Henten (1994), applied in a real experiment by Tap (2000), and described in concise form in Van Henten and Bontsema (2009). Time scale decomposition results in separate optimal control problems that can be solved accurately and efficiently. Together these produce an approximately optimal control. Moreover, the time scale decomposition presented in this section introduces feedback.

Reconsider the optimal control problem given by Equations 4.1 through 4.3. Suppose the system in Equation 4.2 operates on two time scales that lie wide apart. This implies that part of the system's first-order differential equations $\dot{x}_i = f_i(\mathbf{x}, \mathbf{u}, \mathbf{d}, \mathbf{p})$, $i = 1, 2, \ldots, n_x$ produce derivatives $\dot{x}_i$ that are small relative to x_i, whereas others produce derivatives $\dot{x}_i$ that are large relative to x_i. Let us collect the former state variables that vary slowly into the vector $\mathbf{x}^{\mathrm{s}}$ and let us collect the latter state variables that vary fast into the vector $\mathbf{x}^{\mathrm{f}}$. Separating the state variables in this manner, the following system description applies that forms the basis of a time scale decomposition,

$$\dot{\mathbf{x}}^{\mathrm{s}}(t) = \mathbf{f}^{\mathrm{s}}\left(\mathbf{x}^{\mathrm{s}}(t), \mathbf{x}^{\mathrm{f}}(t), \mathbf{u}(t), \mathbf{d}(t), \mathbf{p}\right), \tag{4.14}$$

$$\dot{\mathbf{x}}^{\mathrm{f}}(t) = \mathbf{f}^{\mathrm{f}}\left(\mathbf{x}^{\mathrm{s}}(t), \mathbf{x}^{\mathrm{f}}(t), \mathbf{u}(t), \mathbf{d}(t), \mathbf{p}\right). \tag{4.15}$$

In Equation 4.14, $\mathbf{f}^{\mathrm{s}}$ collects the components of $\mathbf{f}$ associated with the states $\mathbf{x}^{\mathrm{s}}$ that vary slowly, whereas $\mathbf{f}^{\mathrm{f}}$ collects the components associated with the states $\mathbf{x}^{\mathrm{f}}$ that vary fast. Singular perturbation theory reveals that if there would be no external inputs $\mathbf{d}$ the optimal control varies only slowly, i.e., at the level of the large time scale except near the initial and terminal time t_0, t_{f}. This implies that except near the time boundaries t_0, t_{f} we may consider the "fast dynamics" in Equation 4.15 that operate on the small time scale to be in equilibrium:

$$\mathbf{0} = \mathbf{f}^{\mathrm{f}}(\mathbf{x}^{\mathrm{s}}(t), \mathbf{x}^{\mathrm{f}}(t), \mathbf{u}(t), \mathbf{d}(t), \mathbf{p}). \tag{4.16}$$

With the approximation in Equation (4.16) the dynamic system in Equations 4.14 and 4.15 turns into what is called a *differential algebraic system* because Equation 4.16 represents algebraic equations. Neglecting the fast transients near the time boundaries t_0, t_{f}, an optimal control problem for the differential algebraic system in Equation 4.14, Equation 4.16 is obtained that approximates the original one and no longer suffers from two time scales. To see this, observe that one can solve $\mathbf{x}^{\mathrm{f}}(t)$ from Equation 4.16, knowing $\mathbf{x}^{\mathrm{s}}$, $\mathbf{u}$, $\mathbf{d}$ at all times $t_0 \le t \le t_{\mathrm{f}}$. This effectively eliminates $\mathbf{x}^{\mathrm{f}}(t)$ from Equation 4.14, which then becomes a differential equation containing just the slow state $\mathbf{x}^{\mathrm{s}}(t)$. From the slow state $\mathbf{x}^{\mathrm{s}}(t)$, the fast state $\mathbf{x}^{\mathrm{f}}(t)$ can be recovered through Equation 4.16.

Example 2

Reconsider Example 1 from Chapter 2 that concerns our heavily simplified greenhouse. For this example,

$$\begin{bmatrix} \dot{x}_1(t) \\ \dot{x}_2(t) \end{bmatrix} = \dot{\mathbf{x}}(t) = \mathbf{f}(\mathbf{x}(t), \mathbf{u}(t), \mathbf{d}(t), \mathbf{p}) = \begin{bmatrix} p_1 d_1(t) x_2(t) \\ p_2(d_2(t) - x_2(t)) + p_3 u_1(t) \end{bmatrix}. \tag{4.17}$$

Recall that $x_1(t)$ is the crop dry weight and $x_2(t)$ the greenhouse temperature. Obviously the changes of the greenhouse temperature $x_2(t)$ occur much faster than those of the crop dry weight $x_1(t)$. Therefore, we may select $x^s = x_1$, $x^f = x_2$. Then from Equations 4.7, 4.14, and 4.15 we obtain,

$$\dot{x}_1(t) = \dot{x}^s(t) = f^s\left(x^s(t), x^f(t), \mathbf{u}(t), \mathbf{d}(t), \mathbf{p}\right) = p_1 d_1(t) x_2(t), \tag{4.18}$$

$$\dot{x}_2(t) = \dot{x}^f(t) = f^f\left(x^s(t), x^f(t), \mathbf{u}(t), \mathbf{d}(t), \mathbf{p}\right) = p_2\left(d_2(t) - x_2(t)\right) + p_3 u_1(t). \tag{4.19}$$

Assuming the fast state to be in equilibrium, we obtain from Equation 4.16,

$$0 = p_2(d_2(t) - x_2(t)) + p_3 u_1(t). \tag{4.20}$$

From Equation 4.20, we obtain,

$$x_2(t) = d_2(t) + \frac{p_3}{p_2} u_1(t). \tag{4.21}$$

Observe that Equation 4.21 expresses the fast state $x^f(t) = x_2(t)$ explicitly in terms of the slow state $x^s(t) = x_1(t)$, the control $\mathbf{u}(t)$, the external inputs $\mathbf{d}(t)$, and the parameters $\mathbf{p}$. So there is no need to solve Equation 4.20 numerically. The example is special in the sense that the fast state $x^f(t) = x_2(t)$ does not at all depend on the slow state $x^s(t) = x_1(t)$ but only on the control and external inputs $\mathbf{u}(t)$, $\mathbf{d}(t)$. Finally, using Equations 4.18 and 4.21,

$$\dot{x}_1(t) = \dot{x}^s(t) = f^s\left(x^s(t), x^f(t), \mathbf{u}(t), \mathbf{d}(t), \mathbf{p}\right) = p_1 d_1(t)\left(d_2(t) + \frac{p_3}{p_2} u_1(t)\right). \tag{4.22}$$

Equation 4.22 represents a dynamic systems model that only contains the slow state $x^s(t) = x_1(t)$. Knowing the slow state $x^s(t) = x_1(t)$, the fast state is recovered from Equation 4.21.

The approach described above provides an accurate approximation except near the time boundaries t_0, t_f. Optimal control methods have been proposed that take into account the fast transient behavior near these time boundaries that improve the accuracy though often only marginally. With respect to fast transients the situation becomes entirely different when the system is affected by external inputs that vary fast, i.e., at the level of the small time scale. These external inputs cause the approximation in Equation 4.16, that assumes the fast dynamics to be in equilibrium, to be violated *permanently* and not just near the time boundaries t_0, t_f. Optimal greenhouse climate control problems are of this type due to weather variations that act as external inputs that vary fast, i.e., at the level of the small time scale. In these cases the approach presented in the next subsection has been proposed in the literature. This approach realizes feedback that deals appropriately with the different types of uncertainty associated with the model and the external inputs. Moreover, it resolves the problem of widely different time scales.

4.3.1.2 Feedback Design for Optimal Greenhouse Climate Control

The feedback design for optimal greenhouse climate control consists of two major computations. The first one concerns an offline computation, whereas the second concerns the online feedback using information obtained from the first. The next two subsections describe the two major computations.

4.3.1.2.1 *First Major Computation*

Solve offline the following optimal control problem that only considers the slow dynamics, presuming the fast dynamics are in equilibrium despite the fast variations of the external inputs.
Given the system,

$$\dot{\mathbf{x}}^{\mathrm{s}}(t) = \mathbf{f}^{\mathrm{s}}\left(\mathbf{x}^{\mathrm{s}}(t), \mathbf{x}^{\mathrm{f}}(t), \mathbf{u}(t), \mathbf{d}(t), \mathbf{p}\right), \ \mathbf{x}^{\mathrm{s}}(t_0) = \mathbf{x}_0^{\mathrm{s}}, \tag{4.23}$$

$$\mathbf{0} = \mathbf{f}^{\mathrm{f}}(\mathbf{x}^{\mathrm{s}}(t), \mathbf{x}^{\mathrm{f}}(t), \mathbf{u}(t), \mathbf{d}(t), \mathbf{p}), \tag{4.24}$$

and the external input trajectory,

$$\mathbf{d}(t) = \bar{\mathbf{d}}(t), \ t_0 \leq t \leq t_{\mathrm{f}}, \tag{4.25}$$

find the control trajectory,

$$\mathbf{u}(t), \ t_0 \leq t \leq t_{\mathrm{f}}, \tag{4.26}$$

that minimizes the original cost functional in Equation 4.3, i.e.,

$$J(\mathbf{u}(t)) = \Phi\left(\mathbf{x}\left(t_{\mathrm{f}}\right)\right) + \int_{t_0}^{t_{\mathrm{f}}} L\left(\mathbf{x}^{\mathrm{s}}(t), \mathbf{x}^{\mathrm{f}}(t), \mathbf{u}(t), \mathbf{d}(t), \mathbf{p}\right) \mathrm{d}t. \tag{4.27}$$

In Equation 4.25 $\bar{\mathbf{d}}(t)$, $t_0 \leq t \leq t_{\mathrm{f}}$ represents an external input trajectory containing only slow variations, i.e., variations at the level of the slow time scale. In the case of greenhouse climate control, it may be associated with averaged or filtered weather conditions recorded over one or several similar time periods. Denote the optimal control and associated state and costate trajectories of this optimal control problem by respectively,

$$\mathbf{u}^{\mathrm{s}*}(t), \mathbf{x}^{\mathrm{s}*}(t), \boldsymbol{\lambda}^{\mathrm{s}*}(t), \ t_0 \leq t \leq t_{\mathrm{f}}. \tag{4.28}$$

The optimal control problem (4.23 through 4.27) is called the *slow subproblem*. This is associated with the original optimal control problem (4.1 through 4.3). This ends the first major computation.

Example 3

Reconsider Example 2 that concerns our heavily simplified greenhouse. For this example,

$$\begin{bmatrix} \dot{x}_1(t) \\ \dot{x}_2(t) \end{bmatrix} = \dot{\mathbf{x}}(t) = \mathbf{f}(\mathbf{x}(t), \mathbf{u}(t), \mathbf{d}(t), \mathbf{p}) = \begin{bmatrix} p_1 d_1(t) x_2(t) \\ p_2(d_2(t) - x_2(t)) + p_3 u_1(t) \end{bmatrix}. \tag{4.29}$$

Recall that the slow state $x^s(t) = x_1(t)$ is the crop dry weight and that the fast state $x^f(t) = x_2(t)$ is the greenhouse temperature. Example 1 in Chapter 2 considered the optimal control of the greenhouse based on the following cost function:

$$J(\mathbf{u}(t)) = -p_5 x_1(t_f) + \int_{t_0}^{t_f} p_4 u_1(t)\,dt. \tag{4.30}$$

The first major computation described in this subsection for this example reads as follows. Equation 4.24 becomes,

$$0 = p_2(d_2(t) - x_2(t)) + p_3 u_1(t), \tag{4.31}$$

and is identical to Equation 4.21 of Example 2. From Equation 4.31 we obtain,

$$x_2(t) = d_2(t) + \frac{p_3}{p_2} u_1(t). \tag{4.32}$$

This is identical to Equation 4.21 of Example 2. The slow system dynamics in Equation 4.23 become,

$$\dot{x}_1(t) = f^s\left(x_1(t),\, x_2(t),\, \mathbf{u}(t),\, \mathbf{d}(t), \mathbf{p}\right),\; x_1(t_0) = x_0^1, \tag{4.33}$$

and using Equation 4.32,

$$\dot{x}_1(t) = f^s\left(x_1(t),\, x_2(t),\, \mathbf{u}(t),\, \mathbf{d}(t), \mathbf{p}\right) = p_1 d_1(t)\left(d_2(t) + \frac{p_3}{p_2} u_1(t)\right),\; x_1(t_0) = x_0^1. \tag{4.34}$$

Because the cost function in Equation 4.30 does not involve the fast state $x^f = x_2$, Equation 4.32 is not needed to eliminate it. Therefore, the cost function in Equation 4.30 remains unchanged:

$$J(\mathbf{u}(t)) = -p_5 x_1(t_f) + \int_{t_0}^{t_f} p_4 u_1(t)\,dt. \tag{4.35}$$

4.3.1.2.2 *Second Major Computation*

Solve online the following optimal control problem repeatedly.
Given the system,

$$\dot{\mathbf{x}}^f = \mathbf{f}^f\left(\mathbf{x}^{s*}, \mathbf{x}^f, \mathbf{u}, \mathbf{d}, \mathbf{p}\right),\; \mathbf{x}^f(t_s) = \hat{\mathbf{x}}^f(t_s), \tag{4.36}$$

and the external input trajectory,

$$\mathbf{d}(t) = \hat{\mathbf{d}}(t),\; t_s \le t \le t_s + h, \tag{4.37}$$

find the control trajectory,

$$\mathbf{u}(t),\; t_s \le t \le t_s + h, \tag{4.38}$$

that minimizes the cost functional,

$$J^{\mathrm{f}}(\mathbf{u}(t)) = \int_{t_{\mathrm{s}}}^{t_{\mathrm{s}}+h} \left(L\left(\mathbf{x}^{\mathrm{s}*}, \mathbf{x}^{\mathrm{f}}, \mathbf{u}, \mathbf{d}, \mathbf{p}\right) + \boldsymbol{\lambda}^{\mathrm{s}*^{\mathrm{T}}}(t)\mathbf{f}^{\mathrm{s}}\left(\mathbf{x}^{\mathrm{s}*}, \mathbf{x}^{\mathrm{f}}, \mathbf{u}, \mathbf{d}, \mathbf{p}\right)\right) \mathrm{d}t. \tag{4.39}$$

The optimal control problem in Equations 4.36 through 4.39 is called the *fast subproblem*. Observe that the cost functional in Equation 4.39 is determined entirely by the original optimal control problem as well as the solution to the slow subproblem. The initial time t_{s} of the fast subproblem, that is solved repeatedly online, increases with every repetition by an amount of T starting at t_0 and ending with $t_{\mathrm{f}} - T$. Here T represents the sampling period of the feedback control system. In Equations 4.36 through 4.39, h represents the time horizon of the fast subproblem that satisfies $h > T$. Denote the optimal control and associated state and costate trajectories by,

$$\mathbf{u}^{\mathrm{f}*}(t),\ \mathbf{x}^{\mathrm{f}*}(t),\ \boldsymbol{\lambda}^{\mathrm{f}*}(t) t_{\mathrm{s}} \le t \le t_{\mathrm{s}} + h. \tag{4.40}$$

Only the first part $\mathbf{u}^{\mathrm{f}*}(t)$, $t_{\mathrm{s}} \le t \le t_{\mathrm{s}} + T$ of the optimal control trajectory is supplied to the system. In Equation 4.36, $\hat{\mathbf{x}}^{\mathrm{f}}(t_{\mathrm{s}})$ represents an estimate of the fast states at time t_{s}. These may be obtained from a state estimator that employs measurements of the greenhouse climate. In Equation 4.37 $\hat{\mathbf{d}}(t)$, $t_{\mathrm{s}} \le t \le t_{\mathrm{s}} + h$ represents estimates of the external input. In the case of greenhouse climate control, these will be short-term weather predictions. The online computation of the estimate $\hat{\mathbf{x}}^{\mathrm{f}}(t_{\mathrm{s}})$ and the associated optimal control in Equation 4.40 must be performed in between times $t_{\mathrm{s}} - T$ and t_{s}. Therefore, the sampling period T of the control system is lower bounded by computational resources. Equations 4.36 and 4.39 depend on $\mathbf{x}^{\mathrm{s}*}$, the optimal state trajectory obtained from the slow subproblem. Due to modeling errors and inaccurate estimates of the fast states and external inputs, despite the feedback, the slow states $\mathbf{x}^{\mathrm{s}}$ may deviate significantly from $\mathbf{x}^{\mathrm{s}*}$. If, by taking crop measurements, more accurate online estimates $\hat{\mathbf{x}}^{\mathrm{s}}$ can be obtained for $\mathbf{x}^{\mathrm{s}}$, these might be employed instead of $\mathbf{x}^{\mathrm{s}*}$. Doing so, $\boldsymbol{\lambda}^{\mathrm{s}*}(t)$, $t_0 \le t \le t_{\mathrm{f}}$ is the only information used from the slow subproblem. This information, however, is highly relevant because from the interpretation associated to Equation 4.39 it can be seen that $\boldsymbol{\lambda}^{\mathrm{s}*}(t)$ represents the momentary marginal value of the slow states associated with the crop. Not considering the value of the crop, leave alone its variability in time, is a major problem when treating greenhouse (optimal) control problems only at the level of the small time scale.

Example 4

Reconsider once more the optimal control problem associated with the heavily simplified greenhouse characterized by,

$$\begin{bmatrix} \dot{x}_1(t) \\ \dot{x}_2(t) \end{bmatrix} = \dot{\mathbf{x}}(t) = \mathbf{f}(\mathbf{x}(t), \mathbf{u}(t), \mathbf{d}(t), \mathbf{p}) = \begin{bmatrix} p_1 d_1(t) x_2(t) \\ p_2(d_2(t) - x_2(t)) + p_3 u_1(t) \end{bmatrix}, \tag{4.41}$$

$$J(\mathbf{u}(t)) = -p_5 x_1(t_{\mathrm{f}}) + \int_{t_0}^{t_{\mathrm{f}}} p_4 u_1(t) \mathrm{d}t. \tag{4.42}$$

Recall that the slow state is the crop dry weight $x^{\mathrm{s}}(t) = x_1(t)$ and the fast state is the greenhouse temperature $x^{\mathrm{f}}(t) = x_2(t)$. From the solution of the associated slow subproblem described in the previous subsection, we obtain $\lambda^{s^*}(t) = \lambda_1^*(t)$, $t_0 \le t \le t_f$. The fast subproblem described in this subsection that has to be solved repeatedly on-line is characterized by

$$\dot{x}^f = f^f\left(x^{s^*}(t), x^f(t), \mathbf{u}(t), \mathbf{d}(t), \mathbf{p}\right) = p_2(d_2(t) - x_2(t)) + p_3 u_1(t), \tag{4.43}$$

$$\begin{aligned} J^f(u(t)) &= \int_{t_s}^{t_s+h} \left(L\left(x^{s^*}(t), x^f(t), \mathbf{u}(t), \mathbf{d}(t), \mathbf{p}\right) + \lambda^{s^*}(t) f^s\left(x^{s^*}(t), x^f(t), \mathbf{u}(t), \mathbf{d}(t), \mathbf{p}\right)\right) dt, \\ &= \int_{t_s}^{t_s+h} \left(p_4 u_1(t) + \lambda_1^*(t) p_1 d_1(t) x_2(t)\right) dt. \end{aligned} \tag{4.44}$$

A special feature of this problem is that neither $f^f(x^{s^*}(t), x^f(t), \mathbf{u}(t), \mathbf{d}(t), \mathbf{p})$ nor $L(x^{s^*}(t), x^f(t), \mathbf{u}(t), \mathbf{d}(t), \mathbf{p})$ depends on the optimal long-term slow state $x^{s^*}(t) = x_1^*(t)$. If they would depend on the optimal long-term slow state, then one should preferably replace $x^{s^*}(t) = x_1^*(t)$ with $\hat{x}^s(t) = \hat{x}_1(t)$ obtained from a state estimator.

4.3.2 Conclusions

The feedback control system design for optimal greenhouse climate control presented in this chapter has the following attractive and necessary features:

1. Both the slow and fast subproblems obtained from the time-scale decomposition no longer suffer from different time scales and therefore can be solved accurately and efficiently.
2. The separation into these two subproblems also enables the use of a long-term averaged weather prediction $\bar{\mathbf{d}}(t)$, $t_0 \le t \le t_f$ and a short-term more-accurate weather prediction $\hat{\mathbf{d}}(t)$, $t_s \le t \le t_s + h$.
3. Moreover, by repeatedly solving the fast subproblem, feedback is achieved, which is necessary to deal properly with errors associated with the model and the weather predictions.

Because the approach is motivated by singular perturbation methods applied to optimal control problems, it appears to be approximately optimal. The precise conditions under which they are valid as well as the proof of this remains an open area of research. An important source of suboptimality relates to the assumption that the fast dynamics are in equilibrium when solving the slow subproblem. Because the slow subproblem is solved offline one might consider not making this assumption by solving the original full problem with fast-varying, long-term weather predictions. This seriously increases the computation time and storage required for the weather data. Also it reintroduces the danger that these computations become inaccurate. What might be gained is a more accurate costate trajectory that determines how the crop and climate must be valued when solving the fast subproblem online. On the other hand fast-varying long-term weather predictions will never be very accurate. So what has been gained in principle may be lost in practice. Therefore, from a practical point of view the feedback and time-scale decompositions presented in this section appear to preserve optimality as much as possible.

Interesting design issues related to the optimal control approach described in this section involve the selection of both h and T. These must be related to dynamic properties of the system as well as to the accuracy of predictions of the state and the external inputs. Also the online computational resources may affect them.

REFERENCES

Athans, M. 1971. The role and use of the stochastic linear-quadratic-Gaussian problem in control system design. *IEEE Transactions on Automatic Control* AC-16 (6):529–552.

Lewis, F.L. 1986. *Optimal Estimation*. New York: John Wiley & Sons.

Tap, R.F. 2000. *Economics-based Optimal Control of Greenhouse Tomato Crop Production, Systems and Control*, Wageningen Agricultural University, Wageningen, The Netherlands.

Van Henten, E.J. 1994. *Greenhouse Climate Management: An Optimal Control Approach*, Agricultural University Wageningen, Wageningen.

Van Henten, E.J., and J. Bontsema. 2009. Time scale decomposition of an optimal control problem in greenhouse climate management. *Control Engineering Practice* 17 (1):88–96.

Van Willigenburg, L.G., and W.L.De Koning. 2006. On the synthesis of time-varying LQG weights and noises along optimal control and state trajectories. *Optimal Control Applications and Methods* 27 (3):137–160.

5 Greenhouse Cultivation Control Paradigms

5.1 INTRODUCTION

Ever since the existence of greenhouses, the control of the indoor environment has attracted the attention of engineers and scientists. This has led to a large number of approaches and proposed solutions. Differences arise partly because of differences in greenhouse layout, target crops, and weather conditions on the spot and certainly also partly because of a particular view to climate control, skills, and technical opportunities. Hence, it is not easy to oversee the field and judge the suitability of proposed solutions for any particular situation. Therefore, there is a need to put the various controller paradigms into a common framework, if possible. To provide such a framework is the purpose of this chapter.

Having argued in the previous chapters that the optimal control methodology provides, in principle, the most comprehensive way of solving the problem, it will turn out that solutions as proposed in the literature can all be viewed as parts or elements of the total solution. Hence, the optimal control paradigm provides a nice framework to put the various contributions from the literature into a common perspective. In this way, the strong and weaker points of the various proposed solutions can be analyzed against the background of the ultimate desire to be as close to the theoretical optimum as possible.

We begin this chapter by first briefly revisiting and summarizing the method as outlined in the previous chapters, supplemented with a number of implementation considerations. Next, on the basis of this overview, we will classify the various solutions in the literature in the remaining sections in the chapter.

5.2 OPTIMAL CONTROL REVISITED

5.2.1 Generic Problem Statement

Optimal control is achieved by solving the following problem. Given the system model with the initial state,

$$\dot{\mathbf{x}}(t) = \mathbf{f}\left(\mathbf{x}(t), \mathbf{u}(t), \mathbf{d}(t)\right), \ \mathbf{x}(t_0) = \mathbf{x}_0, \tag{5.1}$$

with output accessible to observation,

$$\mathbf{y}(t) = \mathbf{g}(\mathbf{x}(t), \mathbf{u}(t), \mathbf{d}(t)), \tag{5.2}$$

plus additional output representing auxiliary variables of interest,

$$\mathbf{z}(t) = \mathbf{h}(\mathbf{x}(t), \mathbf{u}(t), \mathbf{d}(t)), \tag{5.3}$$

find the control trajectory,

$$\mathbf{u}(t),\ t_0 \le t \le t_f, \tag{5.4}$$

that minimizes the goal function,

$$J\big(\mathbf{u}(t), t_0, t_f\big) = \Phi\big(\mathbf{x}(t_f), t_f\big) + \int_{\tau=t_0}^{\tau=t_f} L\big(\mathbf{z}(\tau)\big)\mathrm{d}\tau, \tag{5.5}$$

subject to the additional inequality constraints,

$$\mathbf{c}(\mathbf{x}(t), \mathbf{u}(t), \mathbf{d}(t)) \le 0,\ t_0 \le t \le t_f. \tag{5.6}$$

The input $\mathbf{d}(t)$ is introduced to account for external input variables that are measurable but not accessible to control. In greenhouse control, this pertains to the weather. Its inclusion is important because part of the optimal control task is to exploit the weather rather than to suppress its influence. Note that there is a subtle but important difference between viewing $\mathbf{d}(t)$ as an observable external input and viewing $\mathbf{d}(t)$ as an unobservable (stochastic) disturbance input.

In addition to the treatment in Chapter 3, here we have introduced variables of interest $\mathbf{z}(t)$. These are variables that, just like the customary outputs $\mathbf{y}(t)$, can be computed from the states, control inputs and external inputs, but without having a counterpart in the actually observed variables $\mathbf{y}^{\mathrm{obs}}(t)$. Examples of observational model outputs $\mathbf{y}(t)$ are temperature or relative humidity, for which sensors are available. Examples of variables of interest $\mathbf{z}(t)$ are the heating rate or the ventilation rate, for which usually no direct measurements are available. From a mathematical point of view, their introduction is not strictly necessary because by virtue of Equations 5.3 and 5.2, the formulation of the goal function in Equation 5.5 is completely equivalent with the formulation given earlier in Equation 3.5. Nevertheless, they are introduced here because of their role in practical implementations.

Another addition is the explicit formulation of constraint conditions in Equation 5.6. They are formulated as inequality constraints and can refer to input constraints—defined by the operating range of the actuators, state constraints such as maximum allowable temperatures, or output constraints such as maximum allowable relative humidity. In Equation 5.6, $\mathbf{c}(.)$ is a vector-valued function with as many elements as there are constraints. Note that the consideration of state constraints would not be necessary if the models were accurate over the entire space domain of interest. If, for instance, a high temperature would be detrimental to the crop, and this phenomenon was correctly captured in the models, then the optimal control algorithm would automatically avoid that high temperatures occur.

5.2.2 Open-Loop Solution of the Whole Problem

The open-loop solution can be calculated offline, that is, without taking actual data into account, except for an initial condition,

$$\mathbf{x}(t_0) = \mathbf{x}_0, \tag{5.7}$$

and an assumed trajectory, called the *nominal trajectory*, of the external inputs,

$$\mathbf{d}(t) = \mathbf{d}^{\mathrm{nom}}(t),\quad t_0 \le t \le t_f. \tag{5.8}$$

A particularly powerful solution method involves the formation of the Hamiltonian,

$$H(\mathbf{x}, \mathbf{u}, \boldsymbol{\lambda}, \mathbf{d}) = L(\mathbf{x}, \mathbf{u}, \mathbf{d}) + \boldsymbol{\lambda}^T \mathbf{f}(\mathbf{x}, \mathbf{u}, \mathbf{d}), \quad t_0 \le t \le t_f, \tag{5.9}$$

where $\boldsymbol{\lambda}$ is the adjoint variable or costate, and requiring, among others, that

$$\frac{\partial H}{\partial \mathbf{u}} = 0, \quad t_0 \le t \le t_f. \tag{5.10}$$

This method transforms the original problem of finding the optimal control trajectory by maximizing J over the full horizon to maximizing, at every time, the Hamiltonian with respect to the actual control. The requirement that the Hamiltonian is stationary with respect to infinitesimal control variations is part of a set of necessary conditions, as outlined in Chapter 3. It leads to locally optimal solutions but not necessarily to solutions that minimize J globally (Stengel, 1994).

Performing the optimization by one of the methods outlined in Chapter 4 leads to the following results:

- Control trajectory,

$$\mathbf{u}^*(t), \quad t_0 \le t \le t_f. \tag{5.11}$$

- State trajectory,

$$\mathbf{x}^*(t), \quad t_0 \le t \le t_f. \tag{5.12}$$

- Costate trajectory,

$$\boldsymbol{\lambda}^*(t), \quad t_0 \le t \le t_f. \tag{5.13}$$

- Goal function evolution,

$$J^*(t) \equiv J(\mathbf{u}^*(t)), \quad t_0 \le t \le t_f, \tag{5.14}$$

with ultimate value $J^*(t_f)$.

It should be noted that even if no use is made of a solution method that involves costates, the costate can still be computed afterward from the optimal solution by using Equation 3.9 or 3.12.

5.2.3 The Choice of the Weather

Ignoring for the moment the considerable problem of actually computing the offline full solution, it is clear that assumptions about the weather have a major effect on the result of the offline open-loop optimization. At this point, it is relevant to distinguish and define a number of different cases regarding the generation of the nominal weather input trajectory.

1. The most direct way is to take one actual realization as observed in the past. The associated solution is called here a *single realization* solution. It is useful for analysis purposes but does not give information about the variability among realizations. We denote the goal function at final time for the jth input realization by J_j^*, or simply J^* if there is just one realization.
2. Instead of a single realization, solutions can be generated for a number of realizations, which are then aggregated. The associated solution is called a solution with *multiple realizations*. The aggregate is obtained by averaging in the solution space. The mean of the

goal function values at final time then represents an expectation of the average achievable goal. It can be expressed as

$$\bar{J}^* = \frac{1}{N}\sum_{j=1}^{N} J_j^*. \tag{5.15}$$

In addition, an impression can be obtained of the variance. Input, state, and costate trajectories are an ensemble of individual trajectories.

3. Another option is to generate the nominal trajectory by averaging and smoothing the weather input over a number of realizations. Averaging or smoothing takes place in external input space to yield the smooth nominal input $\bar{\mathbf{d}}^{\mathrm{nom}}$. This method alters the frequency spectrum of the data as it reduces higher frequencies, but it increases the chances of finding a solution. The associated goal function is denoted by $J^*(\bar{\mathbf{d}}^{\mathrm{nom}})$. Methods and analyses that can help in generating input data from historical data, including reduced information such as daily minimum and maximum values, are described for instance by Alscher, Krug, and Liebig (2001); Jones, Jones, and Hwang (1990); Marsh and Albright (1991a); and Seginer and Jenkins (1987).

A solution that is obtained with one of the preliminary assumed input patterns is called an *a priori solution* because it can be computed in advance. When needed, the subscript "prior" is used. Afterward, when the real weather is known, the problem can be repeated with the actual disturbance input as observed. This is called an *a posteriori solution.* The a posteriori solution represents the result that could have been obtained had the true weather been known in advance. When needed, this is denoted by the subscript "post." We denote the pattern obtained with the real weather in as much detail as possible, that is, the pattern belonging to J^*_{post} according to method 1, the *dream pattern.* This name expresses that it is something to dream of but that it can never be achieved in practice.

5.2.4 Closed-Loop Solution of the Whole Problem

5.2.4.1 Online Solution by Repeated Optimization

A closed-loop online solution can theoretically be obtained if at time instant t the problem as defined by Equations 5.1 through 5.6 is solved again, but now with an estimated initial state and with a modified expectation of the weather and other observable disturbances. The state is estimated from the observation data as given in Equations 4.7 and 4.8. In online applications, sampled data are always used; hence, the state estimator has to be cast in sampled data form:

- Time update (prediction step before the observation is available):

$$\hat{\mathbf{x}}(t_k^-) = \hat{\mathbf{x}}(t_{k-1}^+) + \int_{\tau=t_{k-1}}^{t_k} \mathbf{f}\big(\mathbf{x}(\tau), \mathbf{d}(\tau), \mathbf{u}(\tau)\big)\,\mathrm{d}\tau, \tag{5.16}$$

  where $t_k = t_0 + kT$, and T is the sampling interval of the observations. Notation $\hat{\mathbf{x}}(t_k^-)$ denotes the predicted value at time t_k resulting from the state equations before the measurements are available. Remark that for the evaluation of the integral, it is necessary to make assumptions on the trajectory of $\mathbf{d}(t)$ over the interval (t_{k-1}, t_k) and likewise on the control $\mathbf{u}(t)$.
- Measurement update (correction step, once the observation $\mathbf{y}^{\mathrm{obs}}(t_k)$ is available):

$$\hat{\mathbf{x}}(t_k^+) = \hat{\mathbf{x}}(t_k^-) + \boldsymbol{K}(t_k)\left(\mathbf{y}^{\text{obs}}(t_k) - \mathbf{y}(t_k^-)\right), \tag{5.17}$$

where

$$\hat{\mathbf{y}}(t_k^-) = \mathbf{g}\left(\hat{\mathbf{x}}(t_k^-),\ \mathbf{u}(t_k),\ \mathbf{d}^{\text{obs}}(t_k)\right). \tag{5.18}$$

In Equation 5.17, a correction is made to the predicted state on the basis of the difference between the actual observation and the expected, predicted output at time t_k. Note that in the calculation of the latter, the actual control input and the actually observed external input are used (assuming, for convenience, the same sampling interval for the external inputs as for the system outputs). The Kalman gain matrix follows from the Kalman filter design, as mentioned in Chapter 4 (e.g., see Lewis, 1986).

In the equations above, the observations are denoted by $\mathbf{y}^{\text{obs}}$. There is a subtle difference between the observation data $\mathbf{y}(t)$ as generated by the model from the true $\mathbf{x}$, $\mathbf{u}$, and $\mathbf{d}$ via output Equation 5.2 and the actual observation data:

$$\mathbf{y}^{\text{obs}}(t_k) = \mathbf{y}(t_k) + \boldsymbol{\varepsilon}(t_k). \tag{5.19}$$

The error $\boldsymbol{\varepsilon}(t_k)$ contains the effects of errors in the readout function $\mathbf{g}$—usually small—plus measurement noise because of sampling errors and instrument noise. The measurement or sampling error $\boldsymbol{\varepsilon}(t_k)$ must not be confused with the innovation $\mathbf{y}^{\text{obs}}(t_k) - \mathbf{y}(t_k^-)$ in Equation 5.17. The latter uses the model-predicted output because the purpose of the filter is to correct for unknown modeling and input errors on the basis of observations. It is not the intention to correct for measurement and sampling errors of the output variables themselves. It follows that the state estimator will be better if the output observation error can be kept small.

The optimization is now performed by resetting the initial time to the current time, that is, $t_0 = t_k$, taking as initial condition,

$$\mathbf{x}(t) = \hat{\mathbf{x}}(t_k^+), \tag{5.20}$$

and minimizing the goal function,

$$J\left(\mathbf{u}(t),\ t_k,\ t_f\right) = \Phi\left(\mathbf{x}(t_f),\ t_f\right) + \int_{\tau=t_k}^{\tau=t_f} L\left(\mathbf{z}(\tau)\right)\mathrm{d}\tau, \tag{5.21}$$

subject to the constraints as before while using as external input the current value plus an actualized prediction. In an online situation, it is realistic to use the following input pattern,

$$\mathbf{d}(t) = \begin{cases} \mathbf{d}^{\text{obs}}(t), & t = t_k \\ \hat{\mathbf{d}}(t), & t_k < t \le t_k + h_d \\ \mathbf{d}^{\text{nom}}(t), & t > t_k + h_d \end{cases}, \tag{5.22}$$

that is, the observed input is used at actual time, an actualized forecast is used over a forecast interval h_d, and beyond that the nominal pattern is used.

The result of this optimization is a control pattern from current time t_k,

$$\mathbf{u}^*(t), \quad t_k \leq t \leq t_f, \tag{5.23}$$

and similarly optimal states and costates. In the implementation, it is usually assumed that the control is kept constant over a control interval. The applied control is the first calculated value of the optimized trajectory, that is, $\mathbf{u}^*(t_k)$.

The goal function value at final time $J^*(t_k, t_f)$ represents in this case the expected costs for the time to go and will gradually decrease if final time gets nearer. The a posteriori overall performance over the full cultivation time must be computed afterward by running the model with the actually applied controls and the actually observed external input values. To be able to do this, these data must be kept in store.

Feedback in this solution is obtained from the resetting of the initial state on the basis of data. If, instead of Equation 5.20, $\mathbf{x}(t) = \hat{\mathbf{x}}(t_k^-)$ is used, which is already computable at the previous time step, then the time interval (t_{k-1}, t_k) minus the time needed to evaluate Equation 5.16 is available to complete the optimization. Depending on the complexity of the problem, this may or may not be enough. To date, no real online applications of this approach have been reported for the greenhouse cultivation problem in the literature, probably because of these computational demands.

5.2.4.2 Online Solution by Using Stationarity of the Hamiltonian

The solution of an open-loop problem with nominal weather as outlined in Sections 5.2.2 and 5.2.3 results in time trajectories of the costates. If it appears that these time trajectories are insensitive to the weather, Equation 5.10 together with the state estimate according to Equations 5.16 through 5.18 and an actually measured weather $\mathbf{d}^{obs}(t_k)$ provides a closed feedback law, from which the required control can be computed without the need for searching an optimization space. If this were possible, it would be very attractive. The method is also known as tracking the necessary condition of optimality (Srinivasan et al. (2003) proposed it for optimal control of batch reactors).

The main difficulty in application to greenhouses is the assumption that the costates will be nearly invariant to the weather pattern and the actual state. Although the low frequent crop biomass states probably have costates that are also low frequent, this is most likely not the case for the greenhouse states. In a study on nitrate in lettuce, De Graaf (2006) explored the possibilities of the necessary condition of optimality method. An analysis was made of the typical open-loop patterns. It appears that in that study, during considerably long periods during a day, the controls are at the constraints, meaning that Equation 5.10 cannot be used as it does not apply when the controls are at the constraints. All that can be said at this stage is that more research is needed to really know the limitations and merits of online application (see also Section 9.4.7.3 in Chapter 9).

5.2.5 Time-Scale Decomposition

Because of widely different time scales in greenhouses, the task of computing the control as described in the previous sections can be horrendous. As outlined in Chapter 4, a time-scale decomposition is possible (see also Van Henten and Bontsema 2009). In open loop, it reduces the computational burden and increases the accuracy. Moreover, in closed loop, it offers ways to limit the online computations. It is composed of two steps: offline solution of a slow subproblem as discussed below in Section 5.2.5.1 and subsequently solving a fast subproblem online with feedback. In greenhouse control, the slow subproblem is related to the crop—or, if applicable, to seasonal energy storage—whereas the fast subproblem is related to the greenhouse dynamics, including crop evapotranspiration and photosynthesis. The issue here is, what information is conveyed from the offline subproblem to the online problem and in what way? This is the topic of Sections 5.2.5.2 through 5.2.5.5.

5.2.5.1 Offline Solution of the Slow Subproblem

Repeating Equations 4.23 and 4.24 in slightly different form, the following slow subproblem is solved offline, assuming the greenhouse to be pseudostatic:

$$\dot{\mathbf{x}}^{\mathrm{s}}(t) = \mathbf{f}^{\mathrm{s}}\left(\mathbf{x}^{\mathrm{s}}(t),\, \mathbf{x}^{\mathrm{f}}(t),\, \mathbf{u}(t),\, \mathbf{d}^{\mathrm{nom}}(t)\right), \tag{5.24}$$

$$\mathbf{0} = \mathbf{f}^{\mathrm{f}}(\mathbf{x}^{\mathrm{s}}(t),\, \mathbf{x}^{\mathrm{f}}(t),\, \mathbf{u}(t),\, \mathbf{d}^{\mathrm{nom}}(t)). \tag{5.25}$$

The original dynamics of the greenhouse Equation 4.15 acts as a filter that filters out very high frequencies, whereas Equation 5.25 has no filtering properties. Therefore, in evaluating Equation 5.25, smooth input signals must be used, as otherwise one would get unrealistic high frequencies in the fast system states. Often, smoothed nominal inputs $\bar{\mathbf{d}}^{\mathrm{nom}}$ derived from historical data are used, as discussed in Section 5.2.3.

The result of the calculation are optimal controls, slow states, and costates, that is,

$$\bar{\mathbf{u}}^*(t),\ \bar{\mathbf{x}}^{\mathrm{s}*}(t),\ \bar{\boldsymbol{\lambda}}^{\mathrm{s}*}(t), \quad t_0 \le t \le t_{\mathrm{f}}, \tag{5.26}$$

where the overbar denotes that the solution is approximate compared with the full solution. For simplicity, it will be left out in the sequel. In addition to Equation 5.26, we have an output equation that calculates observation variables according to

$$\mathbf{y}^*(t) = \mathbf{g}\left(\mathbf{x}^{\mathrm{s}*}(t),\, \mathbf{x}^{\mathrm{f}*}(t),\, \mathbf{u}^*(t),\, \mathbf{d}^{\mathrm{nom}}(t)\right), \tag{5.27}$$

where $\mathbf{x}^{\mathrm{f}*}(t)$ is obtained by solving

$$0 = \mathbf{f}^{\mathrm{f}}(\mathbf{x}^{\mathrm{s}*}(t),\, \mathbf{x}^{\mathrm{f}*}(t),\, \mathbf{u}^*(t),\, \mathbf{d}^{\mathrm{nom}}(t)). \tag{5.28}$$

Similarly, it would be possible to compute the optimal trajectory of variables of interest:

$$\mathbf{z}^*(t) = \mathbf{h}\left(\mathbf{x}^{\mathrm{s}*}(t),\, \mathbf{x}^{\mathrm{f}*},\, \mathbf{u}^*(t),\, \mathbf{d}^{\mathrm{nom}}(t)\right). \tag{5.29}$$

And finally, also, an a priori expectation of the goal function is obtained, $J^*_{\mathrm{nom}}(\mathbf{d}^{\mathrm{nom}}(t), t_0, t_{\mathrm{f}})$.

5.2.5.2 Online Implementation

From here on, there are two major pathways to achieve a practical online controller. These are as follows:

1. Use the output trajectories as setpoint to low-level controllers. This is the usual hierarchical scheme as encountered in industry, and it is the dominant approach in the greenhouse control literature (Section 5.2.5.3).
2. Use the slow costate as shadow prices and repeatedly solve online an optimal control problem on the basis of the same economic goal function as used on the level of the slow subproblem but over a shorter horizon. This leads to the receding horizon optimal controller (Section 5.2.5.4).

Both approaches can be seen as hierarchical solutions, in the sense that there is a decoupling between offline calculations and online control. Strictly speaking, the term *hierarchical* should perhaps be reserved for the first pathway because the link between activities on the two levels is rather loose, and various solutions on each level can be chosen without affecting the main structure. The second pathway may be better denoted by the term *decomposed*, which underlines that the two parts of the problem are closely interlinked. In essence, this is caused by the desire to achieve the economically best solution; changes in economic goal can be accommodated at both levels in an integrated fashion. We now analyze both solutions in somewhat greater detail.

5.2.5.3 Hierarchical Control, Setpoint Tracking

This approach is inspired by practice in industrial automation, where the calculated optimal output trajectories are used as setpoints for low-level controllers (Richalet et al., 1978). The information flow is shown in Figure 5.1.

The task of the low-level control system is to adjust the controls such that the offline-calculated optimal trajectory $\mathbf{y}_g^{sp^*}(t)$, or, in fact, any setpoint $\mathbf{y}_g^{sp}(t)$, is followed while suppressing disturbances. This approach has also been proposed for greenhouses (e.g., Arnold, 1988; Challa and Van Straten, 1991; Markert, 1990; Rodríguez et al., 2008; Reinisch et al., 1989; Schmidt et al., 1987; Sigrimis, Arvanitis, and Pasgianos, 2000; Tantau, 1991, 1993; Udink ten Cate, Bot, and Dixhoorn, 1978). In the calculation of the optimal trajectory, the greenhouse is nearly always treated as pseudostatic. This is based on the assumption that the greenhouse is fast with respect to the seasonal behavior of the crop. As the setpoints are computed on the basis of the assumed nominal $\mathbf{d}^{nom}(t)$, the low-level controller will strive to suppress any effects of deviations caused by the actual $\mathbf{d}(t)$. As such, it does not exploit the opportunities that might be present in the actual weather. Note that handling the online problem by controllers replaces the fast subproblem as formulated earlier by a feedback controller

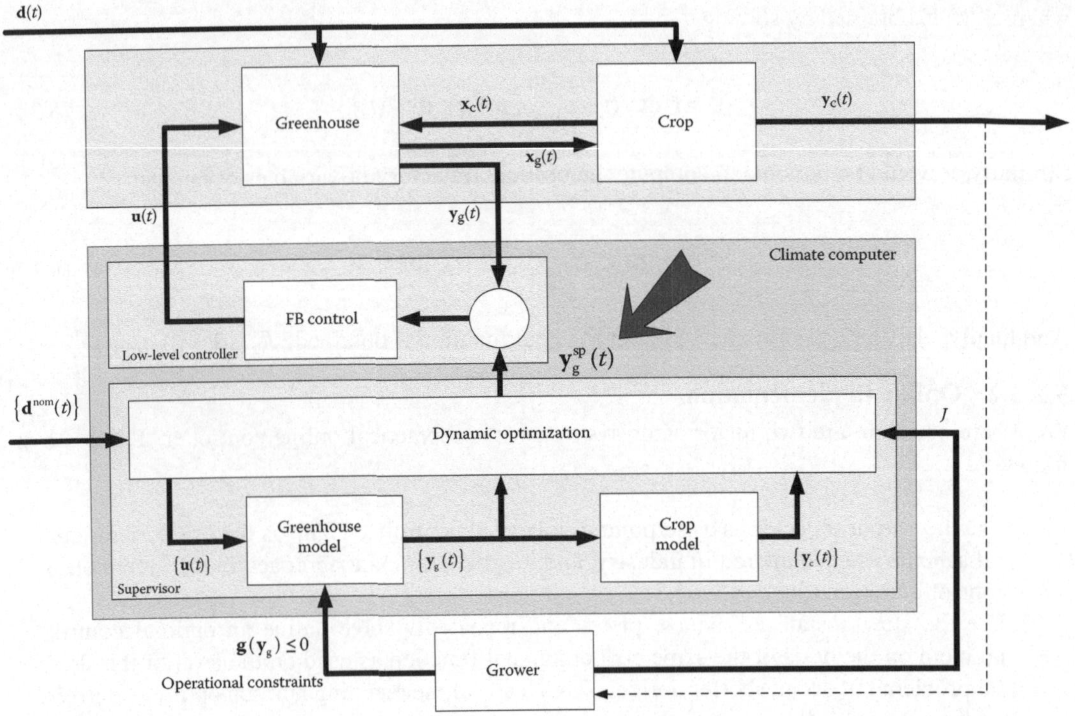

FIGURE 5.1 Hierarchical control. Information transfer via setpoints.

design problem. By doing so, there may not be an explicit goal function at the lowest level. Rather, the controller is designed on the basis of control performance considerations, such as setpoint tracking and disturbance rejection. If designed as a multivariable optimal controller, there is a goal function, but it is a quadratic criterion with no obvious relation to economics. In the hierarchical setup, on the low level, there is no obvious link any more to the economic goal function used in the slow subproblem. Also, the low-level controller does not exploit the external signals.

It should be noted that the low-level control may itself consist of a layered structure. An intermediate controller may generate setpoints for fast local control loops, such as, for instance, a flow controller.

5.2.5.4 Receding Horizon Optimal Control with Slow Costates as Inputs

A solution that does allow the exploitation of variations in the external weather is the receding horizon optimal control presented in Chapter 4. Although the calculated optimal states and controls provide useful insight on the nature of the problem, the evolution of the slow costates $\boldsymbol{\lambda}^{s*}(t)$, $t_0 \le t \le t_f$, is viewed as the main result of the optimization in this integrated solution (Gal, Seginer, and Angel, 1984; Ioslovich, Gutman, and Linker, 2009; Seginer, 1989, 2008; Van Henten, 1994; Van Henten and Bontsema, 2009; Van Straten, 1999). This is shown in Figure 5.2.

The optimal slow costates $\boldsymbol{\lambda}^{s*}(t)$ are used in the augmented goal function that runs over the shorter horizon, as clearly described in Section 4.1.2.2:

$$J^{(f)}(\mathbf{u}(t)) = \int_{t_k}^{t_k+h} \left(L\left(\mathbf{x}^{s*}(\tau), \mathbf{x}^{f}(\tau), \mathbf{u}(\tau), \hat{\mathbf{d}}(\tau)\right) + \boldsymbol{\lambda}^{s*}(\tau)^{T}\mathbf{f}^{s}\left(\mathbf{x}^{s*}(\tau), \mathbf{x}^{f}(\tau), \mathbf{u}(\tau), \hat{\mathbf{d}}(\tau)\right)\right)d\tau. \quad (5.30)$$

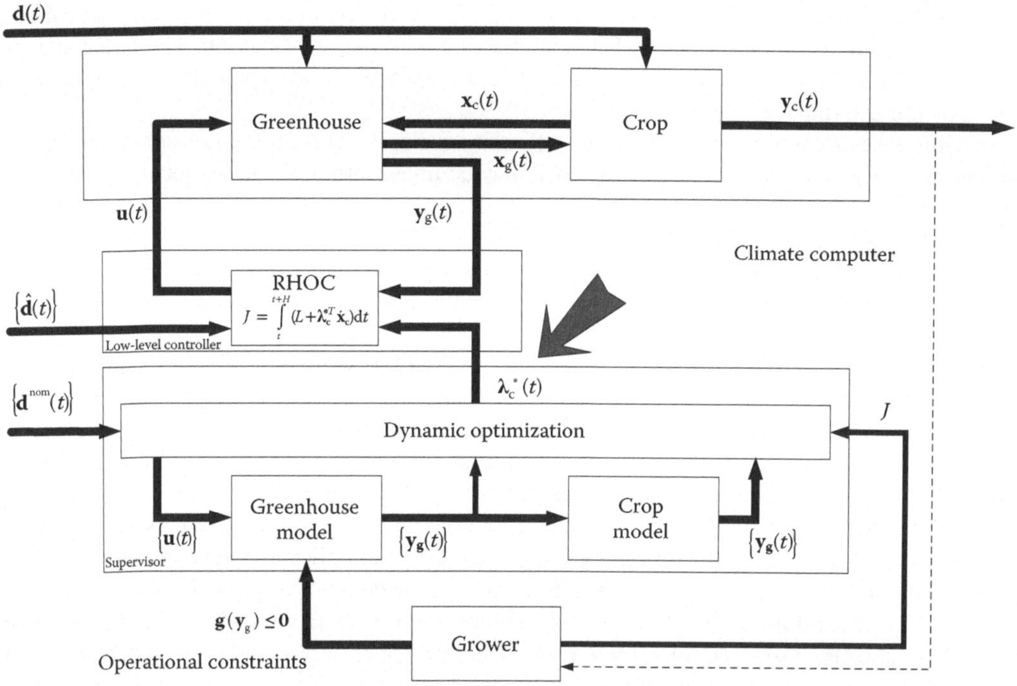

FIGURE 5.2 Decomposed control. Information transfer via slow costates.

The meaning of Equation 5.30 becomes more clear by substituting Equation 4.23, leading to

$$J^{(f)}(\mathbf{u}(t)) = \int_{t_k}^{t_k+h} \left(L\left(\mathbf{x}^{s^*}, \mathbf{x}^{f}, \mathbf{u}, \hat{\mathbf{d}}\right) + \boldsymbol{\lambda}^{s^*T} \frac{d\mathbf{x}^{s}}{dt} \right) dt, \tag{5.31}$$

where, for readability, obvious shorthand notation is used. We can see from this equation that a rate of change of the slow variables, read the crop, is valued by the slow costates $\boldsymbol{\lambda}^{s*}$. The costates thus act as a kind of instantaneous price for an increase in crop biomass components, and its value may vary over the season, as computed from the solution of the slow subproblem. It is also clear from Equation 5.31 that the receding horizon uses the short-term forecast of the weather $\hat{\mathbf{d}}(t)$.

The optimal control trajectory from the current time t_k until the time at the end of the horizon $t_k + h$ is usually approximated by a sequence of piecewise constant controls. The first control is applied, and at the next sampling instant the computation is repeated. To reduce computation time, the control can be parameterized by piecewise constant controls over intervals that are larger than the sampling interval of the observational data. Instead of a continuous control function or a large number of control values, in this way, it is only necessary to compute a limited number of control values. As the computation is repeated at the sampling instances, the actual control pattern still has the corresponding finer time resolution.

The online-calculated sequence of a priori short-term goal functions $J^{(f)*}(t_k)$, $t_0 \leq t_k \leq t_f - h$, is, in general, not equal to the actually realized sequence of short-term goal functions because these depend on the real weather. Moreover, as $J^{(f)}$ at any time pertains to a shorter horizon that moves in time, it provides no direct information on the value of the goal function that is achieved by the receding horizon controller over the full season. The realized goal $J_{\text{post}}^{\text{RHOC}}$ must be computed afterward by a simulation with the control input as actually applied and with weather as actually occurred. To be able to do this, these data must have been stored.

The receding horizon controller as presented here can be seen as a kind of model-predictive controller (MPC). It should be noted, however, that because of its economics-inspired goal function, it is quite different from the standard MPC that uses a quadratic goal function. The latter may provide good tracking and disturbance rejection but not necessarily good exploitation of opportunities offered by the external disturbances.

As in the case oriented on setpoint tracking (Section 5.2.5.3), the receding horizon controller may compute control actuator positions directly, or it may compute setpoints for fast local controllers.

5.2.5.5 Explaining the Difference: The Sailing Analogy

The difference between the hierarchical scheme with setpoint control and the decomposed scheme with receding horizon control and slow costates can be illustrated by the following analogy. Consider the sailboat in Figure 5.3.

The goal is to get as fast as possible from point A to point C while rounding the buoy B on starboard. The first step in both approaches is to make an open-loop calculation of the optimal path to sail under the prevailing wind (solid line). In the hierarchical approach using setpoints, this will be the path to track online. But will it lead to winning the match? If the wind is as forecasted and only small disturbances occur, the answer may be yes, although it becomes somewhat tricky near the constraint buoy B. If, on the other hand, the wind is gusty and makes large sweeps, then it is quite easy to see that chasing the trajectory is a particularly bad policy and may even be unfeasible. Instead, using the receding horizon control, the actual online control balances the achievable fastest speed over the near future against the need to get closer to point B and finally to point C. Therefore, the result will be better (dashed line). In view of the disturbances, it does not make much sense to be worried about the actions to be taken later, so a shorter horizon is justified, provided the long-term goal is accommodated by valuing a path that brings the target nearer in the long run. In short, in the

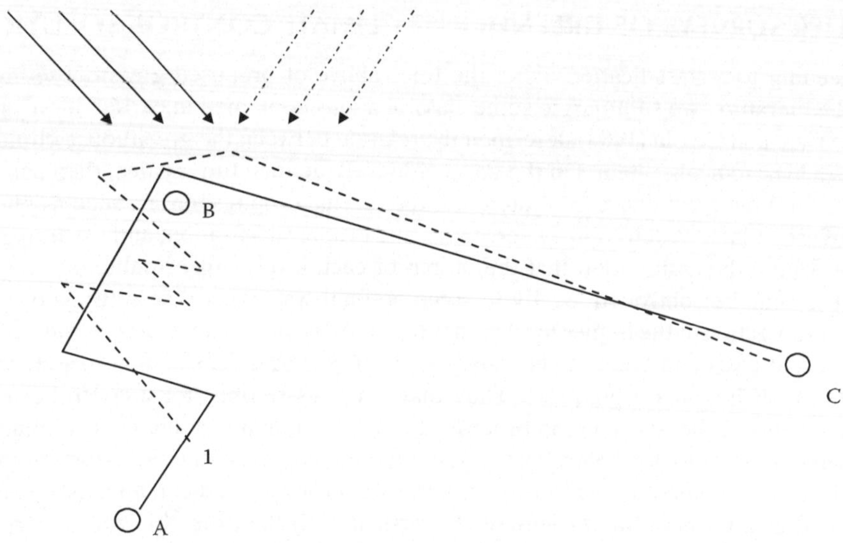

FIGURE 5.3 Optimal trajectories when sailing. Solid line: optimal trajectory with wind indicated by solid arrows. Dashed line: new optimal trajectory when at point 1 wind shifts as indicated by the dashed lines.

presence of large disturbances, it would be wise to exploit the opportunities and to give in when the circumstances are unfavorable, just waiting for better times to come.

Because in greenhouse cultivation the disturbances, especially solar radiation, are large, in short-term fluctuations as well as in daily averages, we may expect that just keeping a setpoint, for example, for temperature, will invoke an economic loss. To avoid this, a controller that preserves the economic goal function on each level is very likely achieving the best possible result. One can see that instead of suppressing the influence of the disturbance, it is exploited. In addition, because of the predictive nature of the receding horizon controller, anticipatory actions are taken in response to the weather forecast.

One may wonder about the analogy of the fast lowest level controllers in the sailing case. An example of this would be a servo that controls the rudder to take the position commanded by the wheel. If well designed and well tuned, such controllers are fast and accurate and might be considered to be ideal. In some cases, it may, however, be worthwhile to study losses because of the controller dynamics. For instance, in greenhouse heating with a pipe system, one could think of taking the pipe temperature as the control input for the glasshouse and leave it to a lower level controller to adjust the pipe temperature by manipulating the mixing valve according to the demand. However, since there is considerable inertia due to the heat capacity of the pipe system, the assumption of instantaneous response in this case can be questioned, and without precautions, there is no guarantee that the commanded pipe temperature can indeed be realized.

An important side remark is in order. In the presence of hard constraints, it may be risky to sail the optimal path because if there is a wind gust from a different angle near the constraint, the ship may be swept away, potentially resulting in a collision with the buoy or yielding a failed passage around the buoy and thus requiring an extra round. To avoid this, a safety margin is needed, at the expense of some sacrifice of optimality, to back off from the constraint. In greenhouses, it is well known that growers prefer to stay away from high moisture to assure that no condensation takes place on the crop, which is seen as a higher risk for fungal diseases, although this policy usually costs some energy. Remarkably enough, the aspect of risk avoidance has not been discussed extensively in the literature on optimal greenhouse cultivation so far.

5.3 EARLIER SURVEYS OF GREENHOUSE CLIMATE CONTROL SOLUTIONS

Before proceeding to a classification along the lines above of proposed greenhouse-crop control systems in the literature, we summarize some classifications contained in earlier surveys.

Udink ten Cate and Challa (1984) described the relation between the greenhouse climate and the final yield as a hierarchical system. On the basis of loosely defined time scales, they consider three different levels: (1) the greenhouse air system; (2) the diurnal crop responses, such as photosynthesis, storage of assimilates, respiration, water status, and evapotranspiration; and (3) crop growth and development. Under the assumption that the output of each level is measurable or computable by simulation, each can be controlled locally in some optimal way, where the settings or bounds are supplied by optimization at the higher level. Older literature is cited, which uses a trade-off between averaged expected costs and yield to generate a set of trajectories. These may include trajectories of process rates, such as photosynthesis. They may serve as settings for a controller on level 2, using measurements of the diurnal crop process rates as input, and instantaneous climate states as outputs to serve as settings for controllers at level 1. This approach is coined the *speaking plant* approach. The exact methodology of information transfer is not worked out, and, as the authors say, the scheme will lead to suboptimal solutions. Nevertheless, at the time, this was a large improvement over the rule-based heuristic controllers in use so far.

In an article emerging from his earlier work on dynamic optimization of crop growth, Seginer (1993) divided the literature into four classes. The classes distinguished are (1) expert systems or rule-based systems, (2) exploiting crop integration capacity, (3) instantaneous optimization, and (4) seasonal optimization. In classes 1 and 3, no information about the future is required in contrast to classes 2 and 4, where an assumption about the future daily or seasonal weather is needed. An explicit crop model is used in classes 3 and 4 only. Table 5.1 summarizes this.

In all cases, it is in effect assumed that there exist sufficiently fast low-level controllers that can realize the computed climate air conditions.

In passing, it is interesting to note that Seginer (1993) showed on the basis of simple examples and using a pseudostatic greenhouse how classes 2–4 can all be cast in the framework of the Hamiltonian methodology. In all these examples, the costate associated with the crop—be it the temperature integral, the development stage, or the biomass—plays a central role.

Albright et al. (2001) confronted the current state of greenhouse-crop cultivation with the needs of growing plants in space. They too started from a hierarchical decomposition, following Van Henten (1994), who adapted the general industrial scheme of Richalet et al. (1978) to the greenhouse industry. Level 0 refers to the actuators (with time scales measured in seconds), level 1 refers to the greenhouse climate (with time scales measured in minutes), and level 2 deals with crop growth and production (ranging from hours to weeks). The third level refers to decisions regarding production space, type of crops grown, and timing, which is relevant on the scale of a growing season or year. This level concerns business decisions by the grower and is not considered part of the control system. Partial solutions for each level as found in the literature are briefly reviewed, such as time-based operation, switching control systems, local P and PI loops, and model-based control. The

TABLE 5.1
Categories of Crop Cultivation Optimization

Future Information	Explicit Crop Model	
	No	**Yes**
No	Expert systems; rules	Instantaneous optimization
Yes	Crop integration	Seasonal optimization

Source: Based on Seginer, I., *Acta Horticult.*, 328, 79, 1993.

second large group concerns supervisory control, which integrates the current time, the midrange, and the longer time horizon. In the midrange, solutions that make use of the integrative capacity of the crop are discussed, and integrating knowledge-based control examples that try to make a link between long-term and current control are also reviewed.

The status of greenhouse climate control in China has been reviewed by Ding et al. (2009). Most research and development efforts have been directed to the hardware, such as programmable logic controllers (PLCs), microcontrollers, and (wireless) sensor networks, and to the performance of low-level control loops. The need for a development toward economic optimization on the basis of knowledge of the greenhouse and crop behavior is well recognized, but this goal is considered not easy to achieve.

5.4 CLASSIFICATION OF PROPOSED GREENHOUSE CLIMATE CONTROL SOLUTIONS

On the basis of the overview of the optimal control methodology above, we are now ready to classify the various contributions proposed in the literature. We used the hierarchy or decomposition in two levels as the major guideline. The following major categories are distinguished.

1. References that focus on the fast time scales, that is, online control of the greenhouse climate. This approach is related but not necessarily equal to the fast subproblem in optimal control. A major distinction in this class is between
   a. Realizing specified climate conditions (settings or setpoints)
   b. Accommodating the integrator capacity of the crop
   c. Controlling fast crop processes
2. References that focus on the slow time scale, that is, generating control strategies motivated by the behavior of the crop. This approach is related but not necessarily equal to the slow subproblem in optimal control. A major distinction in this class is between
   a. Solutions that use simple strategies or expert judgment
   b. Solutions based on dynamic optimization
3. References that discuss both levels in an integrated fashion. This approach is related but not equal to the full-fledged optimal control problem. The subclasses here are
   a. System integration based on reasoning, for example, expert systems
   b. Hierarchical control based on setpoints
   c. Truly integrated optimal or near-optimal control

In categories 3b and 3c, only references are discussed where implementation issues of the integration are discussed. Otherwise, as the slow and fast problems can be separated, they are mentioned under categories 1a and 2b.

In each category, there are many variants that depend on (1) the degree to which the solution accommodates or exploits common attributes of greenhouse-crop cultivation systems, (2) the differences in situation, for example, type of crop grown or economic constraints, and (3) the differences in methodology (model type, optimization method). A brief overview of these variants is given in the succeeding paragraphs as a reference, but we did not use these as the basis of categorizing the literature because they are less generic than the classification adopted in the previous paragraphs.

Differences by Attributes

- The greenhouse system is multivariable. Some investigations deal with partial control solutions for separate variables, for example, temperature, CO_2, or humidity, whereas others take interactions into account.

- Important external inputs can be monitored by automatic sensors. This allows feed-forward compensation. Some articles only discuss pure feedback, whereas others involve feed-forward compensation as well.
- External inputs can be forecasted over short horizons. Solutions may or may not anticipate future weather.
- Most physical variables in the greenhouse are monitored. Most references do not consider state estimation because the physical states can be often reconstructed directly from the measurements by inverting the algebraic output equation. Other approaches have expanded state variables for which no direct measurements are available, or they take measurement and modeling error into account to reconstruct the state. Although it is easy to obtain feedback from the greenhouse variables, the situation is less favorable regarding the crop. There are developments toward direct measurement of plant process rates, such as evapotranspiration or photosynthesis, but it is less easy to obtain useful information from the canopy.
- The solar radiation is an external disturbance input and a resource at the same time. A standard control view is to suppress the effect of variations in solar radiation on internal climate as opposed to solutions that try to exploit the variability.
- There are time-varying constraints on the effect of actuators. Actuator constraints are common, for example, maximum window opening. However, when the ventilation rate is seen as the control, its maximum is not known as it depends on both the window aperture and the wind speed. This issue has to do with the choice of the system boundary. For instance, the tube temperature rather than the heating valve can be seen as the input. In that case, the attainable maximum value is state dependent because it depends on the heat exchange rate with the greenhouse air.
- There are physically determined time-varying and interacting state constraints. The achievable temperature and moisture content in the greenhouse cannot be chosen freely because they are constrained by actuation limits and by saturation bounds dictated by saturation laws (Albright et al., 2001; Pasgianos et al., 2003; Takakura, 1974). This is relevant in control solutions that use greenhouse climate setpoints.

Situational Differences

- Crop type. There is a large variety in crop types in greenhouse cultivation that can have a profound effect on the type of solutions found. Important differences can be noted between single harvest crops, for example, seedlings, lettuce, flowers, and pot plants, and multiple harvest crops, for example, tomato, sweet pepper, and cucumber. These differences lead to differences in goal functions because there is a value to be summed or integrated during the cultivation in the multiple harvest crop, whereas there is a value only to the final state in the single harvest crop. The value of a crop may be from the whole plant (seedlings, lettuce, and pot plants) or from a part of the crop only (flowers, fruits, or even roots). The latter usually requires more elaborate models.
- Economic conditions. Proposed control solutions differ with respect to the control objectives. In solutions that focus on the greenhouse, the control objectives can be to minimize violation of user set bounds, to minimize controlled output error, to minimize sum of control effort and controlled output error, to minimize control effort within user set state conditions, and to minimize bound violations of integrated variables over time.

  In solutions that focus on the crop, objectives can be to maximize yield or photosynthetic rate or to optimize by trading off between yield benefits and instantaneous expected costs. The full-fledged optimal solutions try to maximize economic profit by trading off between expected long-term crop yield and integrated costs. Also, there may be different ways of dealing with constraints: hard constraints or soft constraints, where violation is translated in equivalent costs. Attention for other attributes of the ultimate control methodology is also causing differences, such as robustness, risk, ease of implementation, and information exchange with the grower.

- Greenhouse layout and design. It is clear that the available equipment and the greenhouse layout also yield differences in the actual outcome. In particular, the presence of short and long-term heat storage devices, units for cogeneration of heat and electricity, lighting equipment, sprinklers, or fogging units, air coolers, and heat pumps have a profound effect on the type of solutions and the ultimate economic result.

Differences by Methodology Used

- Further time-scale decompositions. Within the major time scales, a further distinction is often made. For instance, the heating pipe system has its own local controller, thus suppressing the influence of boiler temperature fluctuations, whereas the setpoint is generated by the greenhouse climate, thus leading to cascade control. Other examples are local flow controllers for heat exchangers.
- Model types. Another distinction that can be made is between the design stage and the actual implementation stage of a control system. In the design stage, almost always some kind of model is used. Design models can be of various types—for example, transfer function models, discrete time recursive models, mechanistic differential equations, and so forth. In the implementation, the actual controller may or may not contain the model. Ideal for implementation is a control law in closed form; that is, the actual control action is computed from an algebraic relation between known variables. The model used for the design is not in the controller, but its influence is expressed in the controller parameters. In other solutions, such as receding horizon optimal control, the model is used online as well.
- Optimization method. There are references where a solution of the control problem is offered without optimization. Others use optimization but differ in the solution method. In particular, it is relevant in the current discussion to distinguish between the use of Pontryagin's maximum principle/Hamiltonian approach, which naturally yields costates, and optimization methods that do not use this methodology. Even in the latter case, costates can always be computed afterward, although we did not find references where this is in fact done.
- Suboptimal approximate solutions. In fact, all solutions are suboptimal with respect to the dream pattern, but some authors have been looking on purpose for approximate solutions to avoid excessive online computations or for other reasons, for example, maintenance of models. These suboptimal approximate solutions form another class that can be distinguished.

5.4.1 Focus on Feedback Control of Fast Greenhouse and Fast Crop Subsystems

5.4.1.1 General Overview

The idea of feedback control is to maintain or satisfy given conditions that are supposed to be favorable to the crop. Pure single loop feedback control is rare in greenhouse climate control, as there is no clear-cut connection between the variables to be controlled and the available actuators. The windows, for instance, are used for both temperature control and humidity control. Conversely, modifying the temperature by heating also affects the relative humidity. Moreover, operational control constraints are often hit. Nevertheless, properly designed multivariable feedback/feed-forward controllers are relevant in a hierarchical setup, where the slow subproblem is used to compute desired state trajectories. The realization of these desired trajectories can then be done with standard control objectives, such as minimizing integral squared error.

Acceptable operational ranges or setpoints vary over time, most notably between day and night. Often, in a real-climate computer, these curves and bounds are defined by a number of definition points called settings. To the controller, it is immaterial how the settings are derived. In daily practice, the settings are often chosen heuristically by the grower on the basis of extension service advice, his own experience, and/or his visual observation of the crop status. Research by experimental stations has been directed to the improvement of climate trajectories and has led to an impressive set of "blueprints" for many crops. Settings or setpoint trajectories may also originate from seasonal optimizations, that is, the solution of the slow subproblem.

In the realization of the desired trajectories by the controller, there is no trade-off between benefits from selling the crop and operational costs. In this approach, the problem is reduced to a controller design problem. Hence, the full machinery of controller design theory can be used to design a controller.

Because the crop is not part of the control system, the economic result depends to a large extent on how the settings are chosen. Unless additional tools are provided, such as simulation aids or balance calculations, there is no clue about the extent of effects of settings changes.

5.4.1.2 Realizing a Given Greenhouse Climate

5.4.1.2.1 PI Control

Early designs of PI control on the basis of transfer function models have been described, for example, by Udink ten Cate (1987). In principle, transfer function descriptions can be obtained from experimentation. The theory assumes linearity. In commercial nurseries, P or PI control is commonly used in the control of greenhouse heating pipe temperature, which is part of cascade control of greenhouse air temperature. Nonlinear response behavior can be tackled by adjusting the controller gain to the actual operation point, known as gain scheduling. A recent example in relation to event-based sampling and wireless sensor networks is described by Pawlowski et al. (2009).

5.4.1.2.2 Forward Compensation, Pseudoderivative Feedback

Using models, it is also possible to provide feed-forward compensation (e.g., see Udink ten Cate, 1987). This requires the measurement of external inputs to be able to cancel their effects in advance. Static compensation of heat load by solar input can be achieved by turning the heater down. In practice, the inertia of the actuator system has to be taken into account. To cope with this and also with time delays in the loop, pseudoderivative feedback with load compensation has been proposed by Setiawan, Albright, and Phelan (2000).

5.4.1.2.3 Decoupling and Feedback Linearization

The design of feed-forward/feedback controller relies on linear models. Boulard and Baille (1993) linearized the heat and vapor balances to obtain linearized equations that allow the taking into account of the coupling between these systems because of ventilation and fogging. Decoupling between temperature and CO_2 loops was achieved by Linker, Gutman, and Seginer (1999), where temperature is controlled in a loop with the ventilation, and the CO_2 loop is conditional on the achieved ventilation rate.

The greenhouse dynamics is bilinear with respect to the ventilation rate as control. By writing the (scalar) system in the mixed linear/nonlinear control affine form,

$$\dot{x} = ax + f(x) + b(x)u, \tag{5.32}$$

and by defining a virtual model-based control u' via

$$u = \frac{u' - f(x)}{b(x)}, \quad b(x) \neq 0, \tag{5.33}$$

a system is obtained that is linear in the virtual control, and any linear controller design methodology can be used. This method was proposed by Berenguel et al. (2006) for ventilation control. A multivariable case including psychometric constraints was presented by Pasgianos et al. (2003). They also give an extensive account on coupling issues and constraints related to the psychometric properties. The same treatment can also be found in the study of Albright et al. (2001).

5.4.1.2.4 State or Output Feedback Multivariable Control

A state feedback controller takes the closed form, in discrete time, of

$$\boldsymbol{\delta}\mathbf{u}(t_k) = -\boldsymbol{K}\boldsymbol{\delta}\mathbf{x}(t_k), \tag{5.34}$$

where $\boldsymbol{\delta}\mathbf{x}(t_k) = \mathbf{x}(t_k) - \mathbf{x}^{\mathrm{ref}}(t_k)$ and $\boldsymbol{\delta}\mathbf{u}(t_k) = \mathbf{u}(t_k) - \mathbf{u}^{\mathrm{ref}}(t_k)$ are deviation variables from a trajectory that satisfies the state equation $\dot{\mathbf{x}}^{\mathrm{ref}} = \mathbf{f}(\mathbf{x}^{\mathrm{ref}}, \mathbf{u}^{\mathrm{ref}}, \mathbf{d}^{\mathrm{ref}})$, and $\boldsymbol{K}$ is a gain matrix. The gain matrix follows from the minimization of a quadratic goal function in conjunction with a linear state space system model (Kwakernaak and Sivan, 1972). Often, in practice, a stationary reference state is taken, equal to the nominal setpoint, leading to an implicit function from which an associated required stationary reference input can be selected under an assumed mean value of the external input. The correct selection of reference values is a much ignored issue in the control literature at large. However, by implementing 5.34 in incremental form and by ignoring changes in reference values over a control interval, the reference trajectories are not explicitly appearing anymore, that is,

$$\mathbf{u}(t_{k+1}) = \mathbf{u}(t_k) - \boldsymbol{K}(\mathbf{x}(t_{k+1}) - \mathbf{x}(t_k)). \tag{5.35}$$

The gain matrix $\boldsymbol{K}$ can be designed by pole placement of the closed loop, which gives the closed-loop response a particular dynamics, or by minimizing a quadratic error (linear quadratic designs). Such designs require linear or linearized models of the system. An application to greenhouse control was proposed by Van Henten (1989).

Equation 5.34 is based on the availability of the states. These have to be reconstructed from data. In most greenhouse applications, the reconstruction is rather straightforward as states or state-related outputs are measured. However, this assumes that the measurements are reliable representatives of the true states. Because spatial effects can be quite important, sensor information from a specific location may not be sufficiently representative for spatially averaged variables in a lumped model. Therefore, state reconstructors have also been proposed for greenhouse control, such as the reference quoted by Piñón et al. (2005) and Speetjens, Stigter, and Van Straten (2009).

A special case is constituted by the so-called true digital control design philosophy discussed in the next section.

5.4.1.2.5 Proportional-Integral-Plus Control

A special class of PI controllers derives from the true digital control philosophy originating from Young et al. (1987). These so-called proportional-integral-plus controller designs start from the idea that in a digital controller, such as encountered in the current industry, it is attractive to set up the input–output models directly in sampled data space; that is, the model is formulated as a discrete time autoregressive moving average model of the form

$$\begin{aligned} y(k) = {} & a_1 y(k-1) + a_2 y(k-2) + \ldots + a_n y(k-n) \\ & + b_1 u(k-1) + b_2 u(k-2) + \ldots + b_m u(k-m), \end{aligned} \tag{5.36}$$

where u and y are defined as deviation variables. Next, a state vector is defined as

$$\mathbf{x}(k) = \begin{bmatrix} y(k) & y(k-1) & \cdots & y(k-n+1) & u(k-1) & \cdots & u(k-m+1) & z(k) \end{bmatrix}, \tag{5.37}$$

where $z(k)$ is an integral of error term

$$z(k) = z(k-1) + (y_d(k) - y(k)), \tag{5.38}$$

with $y_d(k)$ as the reference value. This term is provided to cope with offsets because of load variations, as is also customary in ordinary PI controllers. A load variation arises if the external input deviates from the external input assumed in deriving the nominal operation point. Note that the external input is not appearing in Equation 5.36 but is contained in the derivation of the operating point with respect to which the deviation variables are defined. Because a nonzero external input deviation requires a nonzero control effort to keep the controlled state at the nominal zero point, integral action is needed.

Next, using Equations 5.37 and 5.38, Equation 5.36 can be written in nonminimal state space form as

$$\mathbf{x}(k) = \boldsymbol{F}\mathbf{x}(k-1) + \mathbf{g}u(k-1) + \mathbf{h}y_d(k), \tag{5.39}$$

where the matrix $\boldsymbol{F}$ and the vectors $\mathbf{g}$ and $\mathbf{h}$ follow easily by equating Equations 5.36 through 5.39. They are composed of the coefficients of Equation 5.36.

The controller then follows as a standard state feedback controller

$$\mathbf{u}(k) = -\mathbf{K}\mathbf{x}(k). \tag{5.40}$$

The gain vector $\mathbf{K}$ can be chosen by pole placement or by applying a linear quadratic optimization.

This approach has been applied to a scale model of the nutrient film technique (Young, Tych, and Chotai, 1991), to the free air carbon dioxide enrichment systems (Lees et al., 1998), and to the carbon dioxide enrichment in open top chambers (Taylor et al., 2000). Young et al. (1994) described an application for greenhouse temperature control and a multivariable expansion to control relative humidity and CO_2 as well. Preliminary results were reported, which show that relative humidity can be controlled within ±3 percent and temperature to within ±0.5°C.

5.4.1.2.6 *Robust Control*

Robust control aims at designing controllers that maintain certain properties, most notably stability and to some extent performance, under deviations from the design model. The application of the robust control design philosophy (e.g., see Morari and Zafiriou, 1989) has also been proposed for greenhouses (Bennis et al., 2008). The designer must set the expected uncertainty bounds in the system's responses in the frequency domain, or alternatively, the parameter uncertainty may be formulated explicitly (Linker, Gutman, and Seginer, 1999). The controller is intended to work in a stable fashion over a wide range of actual operation points, but such designs tend to be conservative, especially in situations like a greenhouse, where variability in external inputs can be exploited.

5.4.1.2.7 *Adaptive Control*

Another answer to uncertainty in greenhouse modeling is provided by various adaptive control philosophies (Aström and Wittenmark, 1994). In this case, adjustments are made either directly to the controller parameters or indirectly by adjusting parameters of the system model that is used to achieve the controller. An example of an adaptive temperature controller was presented by Arvanitis, Paraskevopoulos, and Vernardos (2000), where a pole-placement technique is used in a linear quadratic (LQ) setting on the basis of multirate recursive parameter estimation. Also, Rodríguez et al. (2008) referred to adaptive controller designs. Online parameter estimation on the basis of an extended Kalman filter for an adaptive optimal controller in a novel water recovering greenhouse in semiarid regions was reported by Speetjens (2008).

5.4.1.2.8 *Model-Predictive Control*

MPCs have gained quite some popularity in the chemical industry. The basics go back to Richalet et al. (1978). The idea is to use models to predict the future time course of the variable to be controlled and then compute the time trajectory of the control that minimizes a weighted sum between the quadratic difference of the output from a reference trajectory and a quadratic measure of the differential control effort. In discrete time, the goal function to be minimized at each actual discrete time instant t_k is

$$J\left(u(t_k),\ldots,u(t_{k+M-1})\right)=\sum_{j=k+1}^{k+N}\left(\left(\hat{y}(t_j)-y^{\mathrm{ref}}(t_j)\right)^2+\beta\Delta u(t_j)^2\right), \tag{5.41}$$

where $M < N$ is the control horizon, N is the prediction horizon, $\hat{y}(t_j)$ is the predicted output at future times t_j, $y^{\mathrm{ref}}(t_j)$ is a desired reference trajectory, and $\Delta u(t_j) = u(t_j) - u(t_{j-1})$ is the control move. The factor β is a design parameter that weighs the relative importance of tracking error versus control effort.

The MPC problem can be solved explicitly—thus leading to a closed feedback control law—if the model is cast in a control autoregressive integrative moving average form (e.g., see Camacho and Bordons, 1995). MPC can be seen as a special case of optimal control, where the goal function has the quadratic form of Equation 5.41. It provides solutions that are optimal from the point of view of control performance, but this is not necessarily optimal in an economic sense. Conversely, the receding horizon control introduced in Section 5.2.5.4 can be seen as a special case of an MPC, if one refrains from the rather restrictive custom to reserve the term MPC for the quadratic goal case.

One interesting property of MPC is that it can easily deal with constraints on inputs and outputs (Maciejowski, 2002). Often, in that case, no closed solutions can be found anymore, thus introducing the need for online optimization. The implementation of MPC is nearly always in a receding horizon fashion, where the prediction horizon is taken sufficiently far away to make sure that transient effects due to the control are leveled out.

Because MPC is a universal and flexible multivariable controller, its use has also been proposed for greenhouses. An early application, without using the term MPC, was performed by Diezemann et al. (1986). They used a static optimization at the top level, with time-series crop models and generated smooth weather, to obtain the setpoint trajectory. MPC is used to calculate the controls, which themselves may be setpoints for low-level PI controllers. More recent applications have been reported by Berenguel et al. (2006), Blasco et al. (2007), El Ghoumari, Tantau, and Serrano (2005), Piñón et al. (2005), and Ramírez-Arias et al. (2005).

5.4.1.3 Control of Greenhouse Climate within Operational Bounds

5.4.1.3.1 *Heuristic Solutions*

Basically, most commercial control solutions are heuristic. The design of the actual control loops varies greatly. Historically, control loops were set up as separate loops for temperature and humidity, but because the actuators for these variables partly coincide, for example, window opening for both ventilation and cooling, heuristic solutions are needed to prioritize the operation of the ventilators for either temperature or humidity control. An example of this type of control action can be found in the book of Bakker et al. (1995). Other heuristic additions were made, for example, incrementing the window opening under high radiation. The effect of this is that the temperature in the greenhouse is allowed to be higher and that there is more ventilation. From what we know about the behavior of the crop, this makes sense: with more solar radiation, there is more evapotranspiration and also more photosynthesis. The latter calls for a faster rate of conversion of assimilates into structural biomass, which is achieved by a higher temperature. The degree of action is entirely determined by the grower and depends on his experience. In passing, we note that an optimal control scheme with a correct crop biomass model would automatically render this behavior, without the need for grower interaction.

5.4.1.3.2 *Multiobjective Controllers*

Xu, Hu, and Zhu (2009) pointed out that control within operational bounds can in fact be seen as a multiobjective optimization problem. They argue that the operation of the heating, fogging, and ventilation controls is associated to costs and that each needs to be minimized while maintaining favorable conditions for the crop. The latter is achieved by constraints on the states. The cost functions for each are taken as quadratic functions. Minimization with an evolutionary algorithm within the predefined constraint region leads to a Pareto optimal front. They then propose to pick from this front the control that has the lowest energy cost. The method, coined *multiobjective compatible control*, has been evaluated in simulation for piecewise constant control intervals of 15 minutes over a prediction horizon up to 75 minutes. Control intervals of twenty minutes with a prediction horizon up to 80 minutes in a similar approach were used by Hu, Xu, and Hu (2009). Here, the authors minimize a quadratic difference from a desired temperature and moisture content, separately, and from the resulting Pareto front, they take the sum of the quadratic input costs to select. The idea to see the greenhouse control problem as a multiobjective control has also been developed by Zhang (2008) as a case to illustrate a general algorithm based on analogy with the immune system, coined *multiobjective immune algorithm*. Simple discrete time models are used to describe the temperature, CO_2, and crop biomass inside the greenhouse in response to the controls and external conditions. Moisture effects are not considered. The external conditions are modeled by sinusoidal patterns with an additive random component. The multiple objectives are to maximize monetary value of the crop and to minimize heating and CO_2 costs within a constrained domain over a specific horizon. At each control step, the algorithm computes a Pareto optimal front, but as only one actual control can be applied, in the end it is necessary to use a decision rule to obtain one single implementable control vector. In the presented case, the sum of the instantaneous monetary value of crop minus the heating and CO_2 costs was taken. Unfortunately, no comparison was made with standard optimal control, and the benefits of multiobjective control over standard optimal control with a single goal function were not discussed.

5.4.1.4 Greenhouse Climate Control with Cost Minimization

Although economics is often given as motive to design the advanced controllers described in the previous sections, economics is not explicitly included, and there is no guarantee that the economically best result is indeed achieved. Instead of designing the controller on the basis of pure control criteria, control solutions have been proposed, where the controller does have an economic goal function. The overall structure of this approach is illustrated in Figure 5.4.

5.4.1.4.1 *Minimizing Costs within Operational Bounds*

Basically, in this approach, a new optimal control problem is formulated, with crop demands as constraints and cost of energy and other resources as part of the goal function. There are no states associated with the crop. Some authors use the Hamiltonian approach to solve the resulting optimization problem. The usual way of thinking is to try to save energy (Bailey and Seginer, 1989; Marsh and Albright, 1991b) or to minimize CO_2 costs (Challa and Schapendonk, 1986; Ioslovich et al., 1995; Seginer et al., 1986), that is, to minimize costs while satisfying conditions related to crop yield. The long-term crop aspects are covered by blueprints and the like. Gutman et al. (1993), building on Seginer (1988), used the Hamiltonian approach to strive for minimization of heating costs by exploiting deviations allowed from the standard blueprints expressed in temperature sums on the basis of perfect weather conditions. The result is a nonlinear MPC. Assuming that a value is attached to the photosynthetic rate, Trigui, Barrington, and Gauthier (2001) proposed to use the Hamiltonian approach to optimize a short-term goal function with costates for the fast variables.

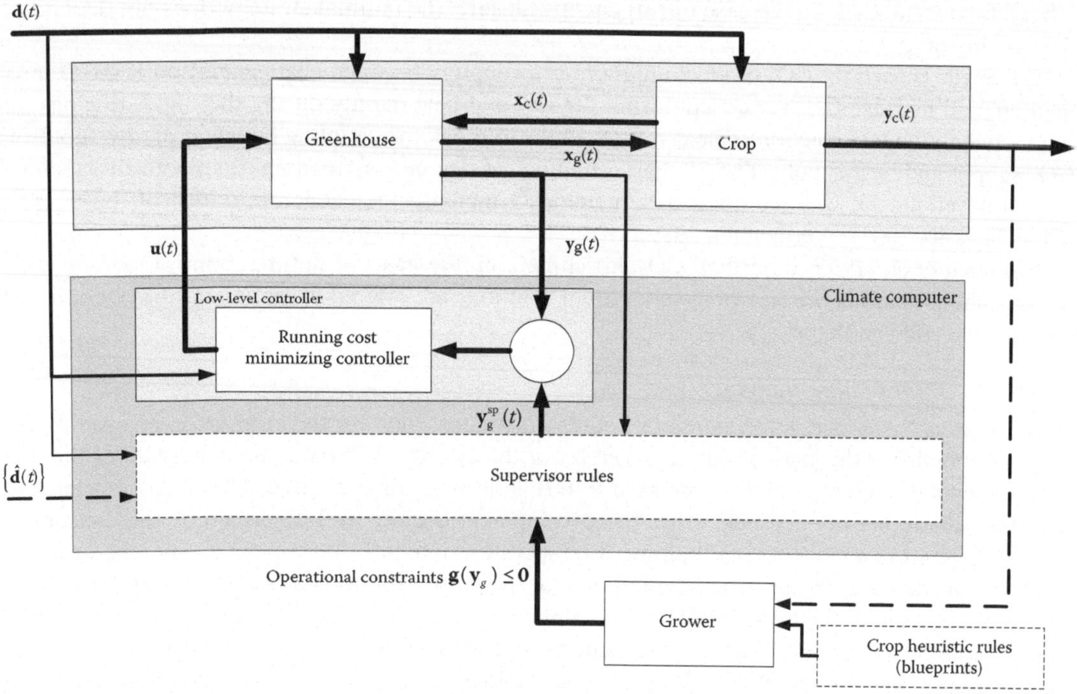

FIGURE 5.4 Hierarchical control with focus on running cost minimization.

Partial or approximate solutions are summarized in the paragraphs that follow.

5.4.1.4.2 CO_2 Enrichment

As photosynthesis rate depends on CO_2 and usual ambient concentrations are well below photosynthetic saturation levels, an idea to raise crop productivity that has been around for a long time is to increase the CO_2 concentration in the greenhouse (e.g., see Wittwer and Robb, 1964). An early heuristic automatic controller was described by Harper, Mitchell, and Pallas (1979), who based the strategy on ambient light level and internal greenhouse temperature. A method to supplement the CO_2 concentration to ambient levels on the basis of predictions by crop models and greenhouse balance equations was described by Kläring et al. (2007). These methods do not use any economic optimization. In the 1980s, publications appear that balance costs of CO_2 production against an assumed value of photosynthetic rate. Initially, instantaneous optimization is used, for example, Challa and Schapendonk (1986), who used numerical procedures to find the optimum, or Critten (1991), who based the optimization on analytical expressions. An implementation experiment was reported by Van Meurs and Van Henten (1994). Ferentinos, Albright, and Ramani (2000), who were interested in producing a fixed lettuce head in a fixed amount of time, tried to trade off the cost of daily supplementary light against CO_2 dosage cost, which is essentially based on the idea that less light can be to some degree compensated by more CO_2. The desired daily CO_2 concentration is computed on the basis of an empirical model that relates it to the target daily light integral. An online procedure on the basis of an hourly iterative calculation that maximizes the margin between crop values—derived from the actual photosynthetic rate—and CO_2 cost for pure CO_2 dosage under the prevailing weather conditions was presented by Chalabi et al. (2002a).

The use of more advanced optimization theory in CO_2 enrichment was reviewed by Chalabi (1992), who exploited the integration capacity of the crop (see below). An analytical expression for the CO_2 setpoint is derived, which depends directly on the actual ventilation rate and further on the past history of the CO_2 concentration and the light. The analysis unites earlier attempts in the literature, which either hinged on the assumption of full canopy closure or on the requirement to reach a fixed final state. Only in the case of full canopy closure, the optimal strategy does not depend on the past history.

If flue gas is used for CO_2 production, there is a coupling between heat generation and CO_2 production, and because CO_2 is needed during the day and heat mainly during the night, this has led in practice to the introduction of heat storage tanks to break the coupling between the two controls CO_2 supply and heat supply. The increased number of degrees of freedom, thanks to this technological development, was reported to tremendously increase the economic return of the nursery (Chalabi et al., 2002b; Vermeulen, 1989).

Ioslovich et al. (1995) described CO_2 enrichment on the basis of optimal control methodology for greenhouses in hot areas, where suboptimal solutions are offered to realize online operation of CO_2 dosage and ventilators.

5.4.1.4.3 Temperature Integral

In the approaches above, information or beliefs on what is good for the crop are expressed in settings or bounds for the state or output variables of the system. Another approach assumes that the requirements of the crop can be expressed in terms of integrals over time. This idea goes back to the observation that in open field crops the development is quite well described by the concept of degree-days. In other words, maintaining a temperature integral over a period, say, one or a few days, is considered to create correct conditions for the crop. This boils down to striving for a mean temperature, without tight control of the actual temperature. Lower values can be compensated by higher values at a later time, which then can be used to achieve cost savings. Experimental evidence of temperature integration in case of a greenhouse tomato crop was provided, among others, by De Koning (1990) and Hurd and Graves (1984).

Bailey and Chalabi (1994) described shifting some of the heating during the day to periods with low wind—and hence less heat loss to the environment—while maintaining a temperature integral. On the basis of this, Chalabi, Bailey, and Wilkinson (1996) published an algorithm to achieve a running twenty-four-hour temperature integral within constraints set by the grower using weather forecasts received daily. In the same class belongs optimal use of daily heat storage buffers, which allow the decoupling of heat demand from heat supply and flue gas CO_2 use (e.g., see Chalabi et al., 2002b). Other studies on minimizing resources with temperature integration were reported by Körner and Challa (2003a, 2003b), Lacroix (1999), and Sigrimis, Anastasiou, and Rerras (2000). Gutman et al. (1993) described a solution to control the temperature integral over a limited 96-hour horizon within predefined bounds while minimizing heating costs in a situation without humidity constraints. The problem was solved with linear programming, but the article also contains an in-depth analysis of the solution using costates and Pontryagin's maximum principle. Seginer, Gary, and Tchamitchian (1994) presented a detailed analysis for assessing the effects of day and night temperatures on crop yield and stated that "deviations from the optimal temperature may often be tolerated, provided that the mean temperature is maintained."

The idea that lower temperatures can be compensated by higher ones later on has led to commercially available controllers. In one Dutch controller available on the market, weather forecasts are used in conjunction with a greenhouse climate balance and a calculation of the evapotranspiration to compute the most favorable heating temperature trajectory, that is, the trajectory that minimizes the heating costs, having the temperature sum over the day as constraint. Another commercial controller does not use forecasts but compensates surpluses or shortages on the next day.

Plant physiology suggests that it is actually important to maintain the source–sink balance, meaning that the desired temperature integral should be increased with higher light sum over a day. This happens naturally in an outdoor environment, but it is interesting to note that because of the poor control power of practical greenhouses, this behavior occurs more or less inadvertently in greenhouses as well. This observation questions the utility of tight temperature control. A time-averaged analysis of crop models on the basis of structural and nonstructural biomass (assimilates) shows that balanced growth boils down to assuming that over periods of days there is no net accumulation of nonstructural biomass (Van Straten, 1999). Any time evolution of CO_2, temperature, and light that keeps the time-integrated growth the same would yield the same biomass. Control of the source–sink balance is a form of rate control to be discussed later.

It should be mentioned that simple crop models describing biomass dynamics by integrals cannot describe development aspects. In ornamental crops, there may be effects on stem elongation and flowering if the temperature pattern is changed. Such changes can be triggered by switching to a regime with higher night than day temperatures, so called negative "difference" (DIF). A strategy that combines more or less heuristic modifications of the playground for cost minimization by introducing dynamic bounds taking this into account is worked out for chrysanthemum by Körner and Challa (2003a, 2003b, 2003c).

5.4.1.4.4 *Light Integral*

A special case arises if the purpose is to produce a crop with predefined properties over a fixed cultivation time at all times during the year despite variation in solar input. Albright, Both, and Chiu (2000) discussed this problem, where the idea is to produce a contracted number of lettuce heads of a specific weight per unit time. This is achieved by maintaining, at a constant temperature, a daily light integral using supplementary lighting, which, in principle, does not require optimization. The original strategy is compared with a new strategy on the basis of minimizing the distance to a reference trajectory by Seginer, Albright, and Ioslovich (2006), which can be viewed as a setpoint tracking controller. As mentioned, an optimization problem results when trading off light integral against CO_2, as studied by Ferentinos, Albright, and Ramani (2000).

5.4.1.4.5 *Thermal Screens*

Bailey (1988) and Bailey and Chalabi (1994) briefly described the operation of thermal screens on the basis of a trade-off between energy savings versus instantaneous loss of production that is assumed proportional to the light intensity reduction. The optimum is obtained by simulation. Seginer and Albright (1980) found the break-even point by equating the rate of cost savings of heat loss prevention due to closing the curtains to the rate of cost increase due to prolonged production time. The latter includes space occupation as well as potential effects of price changes over the expanded time. A simple crop production model is used, which is linear in temperature integral and light integral, and the costs are computed from a static energy balance within humidity constraints.

5.4.1.5 Controlling Fast Crop Processes: The "Speaking Plant"

Instead of setpoints or short-term integral values, one could also consider the fast physiological crop processes as the variable to be controlled. Then the task of the online controller will be to realize predefined rates of, for example, photosynthesis and transpiration. For this approach to work, it is necessary to measure these variables, or, alternatively, to reconstruct them from available signals using models (soft sensor). Ehret et al. (2001) presented an overview of current crop rate measuring possibilities. The idea of using observed crop processes as the variables to be controlled is also known under the name "speaking plant" concept (Udink ten Cate, Bot, and Dixhoorn, 1978). An early application was described by Hashimoto (1980), who proposed to use the electrical capacitance of the stem and the leaf temperature as indicators of the short-term plant growth. An interpretation of the concept from different perspectives was given by Hashimoto (1989). The concept has received quite some attention in Japan in the frame of the so-called plant factories, where there is more control over the environment because of the use of artificial lighting. This is also relevant for manned space missions. There is also a tendency to use online monitoring of crop responses as early warning systems to detect unintended crop stress conditions (e.g., see Dekock et al., 2006).

5.4.1.5.1 *Transpiration Control*

Stanghellini and Van Meurs (1992) experimented with a method to keep transpiration of the crop at a desired level by manipulating the greenhouse temperature and ventilation rate, arguing that sustaining a "minimal" rate of transpiration is required to prevent the effects of calcium deficiencies such as blossom-end rot in tomato. They motivate their work by suggestions made in the plant production literature that control of crop processes rather than of environmental variables might be a more direct way to prevent unsatisfactory crop development and to satisfy quality requirements. Transpiration is calculated from measured variables in the greenhouse using an extensive transpiration model.

Schmidt (1996) presented heuristic control rules in combination with an adaptive model to control leaf temperature and to keep plant evapotranspiration within specific bounds. For rapid online measurement of canopy transpiration, the models were adapted by the measurement of photosynthetic rate and a gas exchange method not further specified. Canopy temperature measurements by an infrared sensor were used as a direct indicator of crop behavior in the study of Langton et al. (2002). They proposed control strategies on the basis of the difference between canopy temperature and air temperature.

Körner and Challa (2004) described a process-related control method inspired by relative humidity effects on crop state, such as calcium deficiency, plant water stress, and crop-related processes such as crop growth, crop development, and airborne fungal diseases (e.g., powdery mildew, chrysanthemum white rust, and gray mould) on chrysanthemums. Tantau and Lange (2003) also elaborated the idea that control can be used to reduce pesticide use.

5.4.1.5.2 Photosynthesis

An example of a combination of a photosynthetic model with heuristics in determining the online operation of the controls is the IntelliGrow system, as described by Aaslyng et al. (2003). The core of this system is a table that relates, at each irradiation level, the photosynthesis rate to temperature and CO_2. At each radiation level, the desired photosynthetic rate is heuristically set equal to 80 percent of the maximum attainable rate, and next the CO_2 and temperature setpoints are selected that can be realized with the lowest energy input on the basis of a simple static balance. No specific details are given on the role of humidity, perhaps not being too important in the reported application, that is, potted roses.

5.4.1.5.3 Assimilate Balance

The principle of maintaining the assimilate source–sink balance was used in a simulation study by Elings et al. (2006). An elaborate crop model, applicable to tomato and sweet pepper, is used to assess the actual balance between assimilate production, mainly driven by light and CO_2, and to assimilate consumption by growth, which is mainly temperature driven. The assimilate balance is maintained on a daily basis, using the short-term integration capacity of the crop. Results suggest that it is indeed advantageous to correlate daily temperature setpoint to light sum. Moreover, the freedom to chose a temperature trajectory within the specified integral offers some room for energy savings by shifting some of the heating and ventilation to more favorable periods during the day.

Morimoto and Hashimoto (2000) described an attempt to control the instantaneous assimilate balance in the production of hydroponic tomato seedlings. The ratio of leaf length and stem diameter, which can both be measured online, is taken as predictor of balanced growth.

5.4.1.5.4 Rate of Development

On the basis of a model of the effect of light and temperature on the rate of leaf development in cucumber, Schapendonk et al. (1984) developed a static optimization scheme to find the cultivation temperature under the prevailing light conditions where the marginal benefits from leaf production balance the marginal value of energy input.

5.4.2 Focus on Strategies Driven by Slow Crop Processes

The major motivation for approaches in this category is the view that greenhouse cultivation is primarily concerned with producing a valuable crop. The climate control is an instrument to achieve this. Information on the crop is taken into account either by using expert rules (see Section 5.4.3.1) or by using crop models. The majority of publications in this category try to minimize the difference between expected costs and benefits obtained from the crop production. As the focus is on the crop, the greenhouse is incorporated only to obtain an estimate of the expected cost of operation. The common assumption here is that on the time scale of the crop, the greenhouse dynamics can be ignored. This problem is equivalent to the slow subproblem of optimal control. The results of the solution of the slow subproblem are often denoted by the term control strategy because it pertains to the long term. The issue of how to actually use the strategic information online is left to the control engineer. This will be discussed in Sections 5.4.3.2 through 5.4.3.7.

5.4.2.1 Assessing Economics by Simulation or Local Optimization

If a crop model, a greenhouse model, a control method, and a model for prices and external weather are available, it is possible to compute the economic result of various heuristic setpoint strategies by simulation. This is the approach taken by Jones, Jones, and Hwang (1990) for tomato. The extensive tomato crop model TOMGRO was used. The greenhouse, although calibrated in a dynamic fashion, was finally used only to compute the required control fluxes and associated costs to maintain a preset setpoint. It was demonstrated how settings for heating temperature (i.e., the temperature level below which the heating is switched on) affect the economic result, but no detailed time trajectories have been investigated.

One step further is a development described by Marsh and Albright (1991a, 1991b) for production of hydroponic lettuce. This entails the calculation of a daytime temperature setpoint for a specific day by maximizing the difference between predicted value of the crop at harvest time using a fixed temperature (25°C) and nominal daily solar radiation for the remaining part of the season and the energy costs for that day. The crop model used was SUCROS, and the calculation of the energy costs was based on a simple pseudostatic heat balance for the greenhouse according to the following rephrased algorithm during daytime hours, executed for every hour:

$$u_{\mathrm{q}}^{\mathrm{dem}} = U\left(T_{\mathrm{g}}^{\mathrm{sp}} - T_0\right) - b\tau_{\mathrm{r}} I - q_{\mathrm{al\_g}}, \tag{5.42}$$

$$\begin{aligned} &\text{if } u_{\mathrm{q}}^{\mathrm{dem}} > 0: \quad u_{\mathrm{q}} = u_{\mathrm{q}}^{\mathrm{dem}}, \quad \varphi_{\mathrm{g\_o}}^{\mathrm{vent}} = 0 \\ &\text{if } u_{\mathrm{q}}^{\mathrm{dem}} \le 0: \quad u_{\mathrm{q}} = 0, \quad \varphi_{\mathrm{g\_o}}^{\mathrm{vent,\,dem}} = \frac{1}{\rho c_{\mathrm{p,a}}}\left(\frac{b\tau_{\mathrm{r}} I + q_{\mathrm{al\_g}}}{T_{\mathrm{g}}^{\mathrm{sp}} - T_0} - U\right), \end{aligned} \tag{5.43}$$

$$\begin{aligned} &\text{if } \varphi_{\mathrm{g\_o}}^{\mathrm{vent,\,dem}} \le \varphi_{\mathrm{g\_o}}^{\mathrm{vent,\,max}}: \quad \varphi_{\mathrm{g\_o}}^{\mathrm{vent}} = \varphi_{\mathrm{g\_o}}^{\mathrm{vent,\,dem}}, \quad T_{\mathrm{g}} = T_{\mathrm{g}}^{\mathrm{sp}} \\ &\text{if } \varphi_{\mathrm{g\_o}}^{\mathrm{vent,\,dem}} > \varphi_{\mathrm{g\_o}}^{\mathrm{vent,\,max}}: \quad \varphi_{\mathrm{g\_o}}^{\mathrm{vent}} = \varphi_{\mathrm{g\_o}}^{\mathrm{vent,\,max}}, \quad T_{\mathrm{g}} = T_{\mathrm{o}} + \frac{b\tau_{\mathrm{r}} I + q_{\mathrm{al\_g}}}{U + \rho c_{\mathrm{p,\,a}}\varphi_{\mathrm{g\_o}}^{\mathrm{vent,\,max}}}, \end{aligned} \tag{5.44}$$

where $T_{\mathrm{g}}^{\mathrm{sp}}$ and T_{o} are the greenhouse air setpoint temperature and the outdoor temperature, respectively (K), u_q is the heat input from the heater (W m^{-2}[gh]), $U = U_{\mathrm{g\_o}} \dfrac{A_{\mathrm{g\_o}}}{A_s}$ is the overall heat transfer coefficient (W m^{-2}[gh] K^{-1}) between greenhouse compartment and outdoor environment with respect to the greenhouse projected area A_s (m^2[gh]), b is a positive factor <1 correcting the heat input from the sun for evapotranspiration (–), τ_{r} is the roof transmissivity (–), I is the solar radiation intensity expressed in energy units (W m^{-2}[gh]), $q_{\mathrm{al\_g}}$ is the heat flux associated with lighting equipment (W m^{-2}[gh]), $\varphi_{\mathrm{g\_o}}^{\mathrm{vent}}$ is the volumetric flux of air due to ventilation (m^3[air] m^{-2}[gh] s^{-1}), and

$p = \frac{\rho_a}{A_s}$. The superscript "dem" stands for "demand". The heat demand belonging to the actual ambient conditions and the daytime temperature setpoint to be optimized are computed in Equation 5.42. If negative, it means there is a need for cooling, which is achieved by opening the ventilation windows. If opening the windows can satisfy the temperature setpoint, it is implemented. Otherwise, the maximum ventilation is applied, and the temperature goes up beyond the setpoint according to Equation 5.44.

This resulting problem is a simple optimization problem with just a single decision variable (daytime temperature), which is easily found. Note that it is not required to compute future operational costs because they are not influenced by the decision for the day. This method has been coined as the sequential control search by Seginer and Sher (1993), where the method was applied to the tomato model TOMGRO and where also the sensitivity of the method for assumptions about the assumed greenhouse temperatures in the future and for future weather was investigated. Although, of course, the future costs do depend on the assumed temperature setpoint, the day-by-day optimal control decisions appear to be virtually independent of the choice of future setpoints or weather, and consequently, the realized objective function is hardly affected by knowledge of the long-term weather. A numerical analysis for lettuce in the study of Seginer and McClendon (1992) suggests that this simplified sequential control search approach gives about the same state trajectories as the whole slow problem.

Optimal selection of a temperature and CO_2 concentration out of a set of predefined day/night temperature and CO_2 setpoints to manage crop cultivation was described by Alscher, Krug, and Liebig (2001). An empirical regression type crop model is used, and the economic criterion encompasses benefit from selling the crop minus the direct costs and the so-called opportunity costs associated with a longer occupation of the greenhouse when the crop lags behind.

It has to be noted that in all these studies, the effect of humidity control on ventilation and costs has not been taken into account. This biases the solutions and the economic result.

5.4.2.2 Optimal Strategies Using Dynamic Optimization

Seminal articles in this category have been written by Seginer (1980) in cooperation with several coauthors. In general, for the purpose of the analysis, very simple crop models are striven for. One of the earliest contributions was described by Seginer (1980), which we represent here in slightly paraphrased form. To start with, it is assumed that a stage in the development of a crop can approximately be described by

$$\frac{\mathrm{d}W}{\mathrm{d}t} = G(\mathbf{z})g(W), \tag{5.45}$$

where W is the biomass. $G(z)$ is equal to the relative growth rate as observed in early stages of crop growth, where $g(W) = W$, and to absolute growth rate at mature stages, where $g(W) = 1$. The vector $\mathbf{z} = [\mathbf{x}_f(\mathbf{u}, \mathbf{d})\ \mathbf{d}]^T$ represents the crop environmental conditions inside the greenhouse, for example, temperature, light intensity, and so forth. The desired crop biomass increase is from W_A to W_B. The time needed to realize this increment can be deduced from the equality

$$\int_{W_A}^{W_B} \frac{1}{g(W)} \mathrm{d}W = \int_{t_A}^{t_B} G(\mathbf{z})\,\mathrm{d}t. \tag{5.46}$$

If the desired biomass increment is fixed and the model is known, the left-hand side is a known constant γ. Hence, any change in greenhouse states expressed in $\mathbf{z}$ leads to a different harvest time via the right hand side integral. This is supposed to generate costs because of space usage and also can give losses or benefits because of changes in market prices. The latter can be especially significant for timely crops, such as carnations for Mother's Day. We will denote these terminal costs by Φ for later use. The net costs (benefits) must be balanced against the cost of operating the equipment, which here is assumed to be directly proportional to the corrective fluxes needed to keep the greenhouse at the desired trajectory of $\mathbf{z}(t)$. This amounts to assuming that the greenhouse is pseudostatic. Under further assumptions, the problem can be solved by linear programming, but this has not been worked out. A simple example is given for the optimal operation of a thermal screen.

The basic methodology along this line of thoughts has been worked out further by Gal, Seginer, and Angel (1984). The greenhouse is modeled by a set of pseudostatic equations:

$$\mathbf{h}(\mathbf{x}^{\mathrm{f}}, \mathbf{u}, \mathbf{d}) = \mathbf{0}. \tag{5.47}$$

The associated costs are denoted by $L(\mathbf{u})$, and hence the total costs to be minimized by the choice of $\mathbf{u}(t)$ are given by

$$J(\mathbf{u}) = \Phi(t_{\mathrm{B}}, t_{\mathrm{A}}) + \int_{t_{\mathrm{A}}}^{t_{\mathrm{B}}} L(\mathbf{u})\,\mathrm{d}t. \tag{5.48}$$

The condition that one wishes to reach, in this case a specific crop increment as given in Equation 5.46, is handled as a constraint:

$$\int_{t_{\mathrm{A}}}^{t_{\mathrm{B}}} G(\mathbf{u}, \mathbf{d})\,\mathrm{d}t = \gamma, \tag{5.49}$$

where γ is a constant and where 5.47 is used to eliminate $\mathbf{x}^{\mathrm{f}}$. This problem is equivalent to finding the control sequence $\mathbf{u}^*(t)$ that minimizes, at any time $t_{\mathrm{A}} \le t \le t_{\mathrm{B}}$, the Hamiltonian defined by

$$H = L(\mathbf{u}(t)) - \lambda G(\mathbf{u}(t), \mathbf{d}(t)), \tag{5.50}$$

where the termination time t_{B} is determined by

$$\int_{t_{\mathrm{A}}}^{t_{\mathrm{B}}} G(\mathbf{u}^*, \mathbf{d})\,\mathrm{d}t = \gamma. \tag{5.51}$$

Although this is a problem with fixed crop increment requirements and variable time, it is perhaps the first article that introduces the idea of a Lagrange multiplier in greenhouse climate control, which has its equivalent in the dynamic costate of optimal control with fixed final time. It is shown that the Lagrange multiplier λ represents the marginal cost worth paying for an additional unit of growth rate. A proposal has been made to realize online control on the basis of a strategy that uses the optimal Lagrange multiplier. Also, the sensitivity of the solution to permutations in the assumed weather is discussed. The proposed control methodology has been applied in a simulation study by Seginer et al. (1986) for optimal CO_2 enrichment.

In a later article, Seginer (1989) extended the case of fixed final state and variable final time described above and applied it to the optimal cultivation of tomato seedlings, where the separation of the crop model (5.45) into the product of mutually independent terms is quite a good approximation. It was shown that the optimal solution above, that is, the solution that generates control actions to minimize costs at the expense of longer cultivation duration under unfavorable weather, actually applies to a situation where production space is not limiting the grower. The level of the product market price then does not influence the result—which could be deduced from the fact that Φ does not appear in the equivalent optimal problem (Equation 5.49 or 5.50)—although variation in the expected final price results in changes in the policy. On the other hand, it was shown that when the grower is constrained by area, it makes sense to apply a higher intensity of cultivation, that is, heating and CO_2 dosage, when the market price is higher.

The breakdown of the crop biomass model as a product of a biomass-dependent term and a term that solely depends on the greenhouse environmental conditions, as in Equation 5.45, was also the basis of a further development for a two-stage crop described by Seginer and Ioslovich (1998), with continuous harvest in the second generative stage. In the notation used here, the development can be summarized as follows. Let Equation 5.45 now describe the time course of the total biomass, that is, the biomass still on the plant plus the biomass already harvested. The accumulated yield at final time can formally be described by

$$Y(t_f) = \int_{t_0}^{t_f} f_F \frac{dW}{dt} dt, \tag{5.52}$$

where f_F is some function of biomass or time, expressing which fraction of the biomass produced is going into fruits. Assume that this has a price p_F, then the goal function to be maximized is

$$J = \int_{t_0}^{t_f} \left[p_F f_F G(\mathbf{z}) g(W) - L(\mathbf{u}) \right] dt. \tag{5.53}$$

Hence, the associated Hamiltonian is

$$H = \lambda g(W) G(\mathbf{z}) + p_F f_F g(W) G(\mathbf{z}) - L(\mathbf{u}), \tag{5.54}$$

or by combining terms,

$$H = NG(\mathbf{z}) - L(\mathbf{u}). \tag{5.55}$$

The term N is called a transformed costate and is defined by comparing Equations 5.54 and 5.55 as

$$N = (\lambda + p_F f_F) g(W). \tag{5.56}$$

In the vegetative period, when f_F is zero, the transformed costate is the product of the original costate $\lambda(t)$ and the biomass-dependent term $g(W(t))$, which is associated to leaf area index. It was shown that, during this period, N is constant. A constant N implies that the costate itself varies (decreases) over time when the canopy closes. This confirms the results found numerically by Van Henten (1994) for lettuce. A constant N also implies that when the vegetative biomass does not increase anymore (i.e., $g(W) \approx 1$), the costate would become more or less constant, depending on the fraction of biomass that is allocated to the fruits. The assertion is that the transformed costate, being essentially constant, may be a better candidate to transfer long-term information to the daily control than the costate itself.

Ioslovich, Gutman, and Linker (2009) expanded the idea to three stages. Tomato growth was divided into vegetative, mixed vegetative–generative, and generative stage. As discussed by Seginer and Ioslovich (1998), part of the analysis deals with the transition between stages. Ultimately, the transition depends on the number of accumulated effective degree-days (i.e., the integral of the temperature when the temperature reaches values above a threshold). The actual parameters are derived from comparison with the calibrated TOMGRO model. The length of the vegetative stage depends on the controls but not on the accumulated biomass at the moment that the required effective degree-days are reached. The end of the mixed period is defined by the maximum amount of vegetative mass plus green fruits that can be sustained for the assimilate partitioning function assumed, which is different for each stage and is also derived from TOMGRO. Here again, the green biomass at the end of the stage does not depend on the controls. Finally, in the reproductive stage, all assimilates are allocated to the fruits. This stage ends at the fixed final time. Next, an analytical solution is obtained for the seasonal optimal control problem with the aim to maximize the economic yield for the grower, using a pseudostatic greenhouse and a constant fixed periodic weather. The solution is derived by evaluating the Krotov–Bellman sufficient conditions. This method requires the proposition of a Krotov–Bellman function for each stage and requires continuity at the stage transitions. The authors show that, again, N is constant throughout the growing season. They propose to use the optimal N as a basis for online control, but no details are given. The idea was also put forward by Seginer (2008), who coins the term *cultivation intensity* to denote the transformed costate. He also shows two examples on how this might be used online to accommodate variable weather or inaccurate models.

In the approaches above, there is only one single costate in contrast to the two costates associated with the leaves and fruits in the study of Van Straten, Van Willigenburg, and Tap (2002), where under realistic weather conditions, the values were reported to vary over the season.

The separation of the basic crop model as independent products as in 5.45 was also used by Chalabi (1992) for dynamic optimization of CO_2 dosage already described earlier.

5.4.2.2.1 Other Studies on the Nature of Costates and Optimal Control Solutions

Seginer et al. (1991) presented costate trajectories for a two-state lettuce crop, where the states are biomass and leaf area ratio using smoothed nominal synthetic periodic weather, in a comparative study on the effects of approximate optimization schemes. In this study, the effect of various crop models and two nonoptimizing strategies on the ultimate economic result in lettuce cultivation was evaluated. The problem formulation is, as before, directed to fixed final crop weight and variable final time. The costate for leaf area ratio decreases to zero in about ten days, suggesting that in the beginning, it makes sense to achieve canopy closure as soon as possible.

Van Henten (1994) and Van Henten and Bontsema (1991) also computed optimal temperature and CO_2 trajectories for lettuce cultivation, both with single-state (biomass) or two-state (structural and nonstructural biomass) crop models. They used long-year smooth averaged daily weather patterns as well as actually observed weather as hind-cast weather forecast (cf. Chapter 6). The computations showed that the optimal temperature and CO_2 profiles with real weather are strongly fluctuating as compared with a calculation based on long-year averages, although the trend is similar. This shows that setpoints generated by open-loop calculations with real weather do not qualify as good setpoints for online control. However, it does not mean that solutions where setpoints are obtained from smoothed weather will also result in solutions that are much further away from the actual posterior optimum. All costate trajectories show a decreasing trend toward zero in about twenty days, equally suggesting that initial investment in dry matter production is worthwhile. Although not said explicitly, this amounts to on average higher greenhouse temperatures at the beginning of the growing season, and lower ones later on, consistent with the findings of Seginer et al. (1991). However, as shown by De Graaf (2006) in a study on optimal control of nitrate in lettuce, the actual optimal online greenhouse temperatures are fluctuating quite heavily throughout the season, as driven by the actual weather, which is also more consistent with the findings of Gal, Seginer, and Angel (1984), who stated that the control, that is, the temperature setpoint, should depend on the actual weather only. Although this may be the case, the results of Seginer et al.

(1991) suggest that—for the particular models and cost evaluations chosen—the ultimate economic result in the case of lettuce cultivation is not much different when applying dynamic temperature setpoints versus a single, optimized, fixed setpoint. The effect of the cultivation system choice on the optimal control policy was studied by Ioslovich and Seginer (2002) for the same case of nitrate concentration in lettuce but then assuming constant weather. When it is possible to continuously adjust the spacing, the optimal control policy turns out to be to maintain the canopy density constant, with a nonunique alternating temperature and a constant ventilation temperature. With constant spacing, the optimal temperature and the nutrient supply theoretically show a bang-bang behavior.

The use of a pseudostatic greenhouse assumption in the offline optimization implies that the true costs of resource use are an approximation. The rapid adjustment of crop-related fluxes to the actual weather, in conjunction with the greenhouse dynamics, leads to changes in the greenhouse climate that force the system away from equilibrium. It is likely that trying to follow optimal temperature and CO_2 settings computed on the basis of assumed weather patterns will lead to the need of large control inputs. Hence, the true costs will almost certainly be quite different from those assumed in the seasonal optimization. A similar argument applies to the application of static balances. Because in reality the greenhouse can only respond to changes in control inputs with some sluggishness, the actual patterns will be different from those assumed in the optimization. Ioslovich et al. (1995) showed that if a constant optimal control solution exists to the dynamic problem, it is also optimal to the static problem obtained by setting the time derivatives equal to zero. However, when the derivatives are nonzero, the existence of a static solution may be questioned. They also show that the optimality of quasi–steady-state approximations degrades as the frequency of the external disturbance inputs increases. Some ideas of the sensitivity of the goal function to actual disturbances may also be obtained from the study of Trigui, Barrington, and Gauthier (2001). Van Henten (2003) investigated the sensitivity of the optimal control solution to parameter uncertainties and model errors. Both the model description of crop growth and production and the outside climate conditions have a strong impact on the performance. Humidity control plays a dominant role in economic optimal greenhouse climate management, emphasizing the need for an accurate description of humidity effects on crop growth and production. The study revealed that the dynamic response times in the greenhouse climate do not seem to be limiting factors for economic optimal greenhouse climate control. A study by Tap et al. (1993) suggests that it is, however, important to take the fast dynamics of the weather into account.

A numerical evaluation of various dynamic losses for a simple greenhouse/crop model was described by Van Straten and Van Willigenburg (2008), who compared, among others, receding horizon control with PI control using setpoints. In the simple model used, losses on the order of 5 to 10 percent were reported.

Most studies have been restricted to temperature and CO_2 alone. However, rapid weather fluctuations as well as large deviations from assumed patterns will cause quite a different humidity regime in the greenhouse. Because humidity needs to be maintained within certain limits, this leads to control actions not foreseen in the optimization, and again the actual costs can be quite different from those assumed. De Halleux and Gauthier (1998) and Tap, Van Straten, and Van Willigenburg (1998) have shown that in temperate climate zones, the humidity constraints have a considerable effect on the value of the goal function.

5.4.3 Integration, Application, and Implementation

This section describes publications reporting on the integration of both levels of control. The separation considered in the previous sections is somewhat arbitrary, but the emphasis here is on implementable ideas. First, knowledge-based solutions without optimization are briefly mentioned, whereas the remainder of this section deals with solutions involving economically optimal control.

5.4.3.1 Expert Systems

This class of greenhouse management solutions is based on the idea that formalizing experience and expert knowledge could lead to improved—hence more economical—greenhouse operation. This approach does not focus on dynamic optimization and will therefore not be discussed extensively here, but some notable developments need to be mentioned. An early survey is provided by Martin-Clouaire, Schotman, and Tchamitchian (1996).

Artificial intelligence decision support tools, in synergism with low-level process controllers or schedulers, were investigated by Sigrimis, Arvanitis, and Pasgianos (2000). They built an environment that incorporates a native fuzzy knowledge–based system plus a number of procedural control functions that allows flexible interfacing with the greenhouse operation.

Tchamitchian et al. (2006) discussed the expert system SERRISTE, which is a decision-making system to generate daily climate settings for greenhouse-grown tomatoes. The system is set up as a constraint satisfaction problem using a mathematical formalism to express expert practices and knowledge as constraint rules. An important notion used is the concept of crop vigor, which is a conglomerate of assessments made by visual inspection of the grower on vegetative and generative balance, crop head, butt formation, and the like. Although this is not a scientific concept per se, it has to be recognized that it takes into account aspects of crop development that are not encapsulated by current crop models. A practical test against common practice revealed similar harvest rates but 5 to 20 percent lower energy consumption.

Constraint satisfaction was also the basic approach of Schotman (2000), who was focusing on preventing blossom-end rot in tomatoes.

Lafont and Balmat (2002) described a controller on the basis of fuzzy rules. Fuzzy rules must be deduced from practical experience, which implies that the existing control must already be fairly successful. Also ideas have been put forward to mimic a grower's behavior by machine learning (Kurata and Eguchi, 1990) or by neural nets (Seginer, 1997). Kolokotsa et al. (2010) described a system of two fuzzy controllers designed on the basis of rules aiming at minimizing CO_2 supply and energy and water use, but the setpoints need to be determined by the user and no crop optimization is involved.

It should be noted that the expected benefit of expert systems could be evaluated by simulation, but there is within this framework no way to know how far these solutions are away from the optimum. Moreover, to do such simulations in a trustworthy way, a comprehensive model is needed, the lack of which was the very motive to use an expert system.

5.4.3.2 Implementation of Optimal Control—Overview

The possible outcomes of the optimization procedure for the slow problem are as follows:

- The control sequences. They are less suitable for online control as they are not robust against variations in external inputs.
- The trajectories of the fast states. Many authors consider these as setpoints for low-level online controllers.
- The trajectories of the slow states. These could be used as targets to track, if observations on the crop can be obtained.
- The costates of the slow variables. They serve as biomass shadow prices on the level of the online control.

5.4.3.3 Direct Application of Computed Controls

Direct application of the controls is frequently used in robotics. The necessary feedback is provided by a linear quadratic controller that makes small adjustments to the control to compensate for observed state deviations (Athans, 1971). The scheme was proposed for greenhouses as well (Challa and Van Straten, 1991) but abandoned later because this approach is not robust against variations in external inputs when they are large and when the goal is to exploit them. In an application to tomato seedlings, Pucheta et al. (2006), using a two-state crop model derived from TOMGRO and a static greenhouse model, directly applied the calculated ventilation and heating control and solved the robustness issue by frequent recalculation of the slow problem by feeding back observed crop information.

5.4.3.4 Hierarchical Control with Settings

Here we discuss some references where authors deal with both the long-term solution and the online control on the basis of setpoints. Admittedly, the distinction with Section 5.4.2.2 is somewhat arbitrary, but here we list primarily contributions that deal with implementation aspects, taking into account all relevant variables.

A method that uses a hierarchical decomposition with offline optimization, a nine-day discretization, and a crop model on the basis of an input–output time-series model derived from an elaborate cucumber model has been described and tested by Schmidt et al. (1987). They solved the optimization problem with the help of Lagrange multipliers and repeated the optimization periodically to correct for linearization errors and changes in weather. The variables used at the lower level are nighttime temperature, daytime temperature, and CO_2 concentration. Yield improvements of 15 percent as compared with standard heuristic control are reported. The idea of a so-called control model of the crop in the form of an input–output discrete time model calibrated on the basis of a more elaborate mechanistic model has been further exploited by Arnold (1988). He also uses repeated optimization of the slow problem to cope with variations in weather. The control model for the crop is adjusted for the actually observed crop yield. He also proposes to split the seasonal period into more tractable shorter partial periods, using temperature integrals as constraints. The temperature integrals are obtained from a seasonal optimization that is performed only once. Reinisch et al. (1989) described an application in simulation to cucumber and tomatoes. The ideas have been further elaborated by Markert (1990), who divided the season into twenty-four unequal periods. Period-wise constant optimized temperature sums for cucumber are obtained as the result of the slow subproblem and then translated with the help of the actual outdoor temperature and solar radiation into settings for a conventional greenhouse control system.

Rodríguez et al. (2008) described a hierarchical setup using sequential programming to solve a seasonal optimization maximizing biomass dry weight minus operational costs via the calculation of temperature setpoints. On the lower level, they used an MPC designed on the basis of a hybrid representation of the system that represents the switching between heating and ventilation dynamics. The temperature setpoints are modified to stay within bounds dictated by the humidity balance.

5.4.3.5 Implementations of Optimal Control Using Meta-Information

Once a seasonal optimization has been done, it might be possible to see repeating patterns that can be used to build a kind of meta-model suitable for online control.

Seginer and McClendon (1992) ran dynamic optimizations for several years of historical weather data and then trained a neural network to arrive at temperature setpoints depending on the actual state and the actual weather. Similar ideas on greenhouse ventilation using experimental data are found in the study of Seginer, Boulard, and Bailey (1994). An overview of the usefulness of nonlinear mappings, be it from dynamic optimal control computations or from mimicking an experienced grower, was presented by Seginer (1997).

In a study on controlling nitrate in lettuce, De Graaf (2006) translated the offline optimal control solutions in a set of rules to be followed online. On the basis of observed crop evolutions, advices are transmitted on a daily basis to the greenhouse.

Van Henten et al. (2006) described how a model for sweet pepper fruit formation (Buwalda et al., 2006) is used to generate settings for a standard advanced climate control computer to eliminate the waves in fruit production that are observed in the market because of biological synchronization of the production among growers. Such waves adversely affect market prices. Attempts are described to control sweet pepper in two counterphase compartments or to level out the onset of fruits via optimal control of the climate.

5.4.3.6 Tracking the Slow Variables—Crop Development

A model of floral development rate of the pot plant *Cineraria* was proposed by Nilsson and Nybrant (1992) to ensure that a plant bears flowers at a predefined moment. The method was made adaptive by Nilsson and Nybrant (1996). As such, this approach is a model-based replacement of the "blue-print" approach. Empirical evidence on how daily mean temperature affects the development and flowering rate allows the construction of an ideal development line, using nominal expected day length and radiation sums. Online, if the real development—to be observed by the grower—is ahead or behind the desirable development, the daily mean temperature is adjusted.

5.4.3.7 Integrated Optimal Control

Here we discuss solutions that use optimization on both levels.

Pohlheim and Heissner (1996, 1997) presented a computation of the heating, ventilation, and CO_2 dosage rate by optimizing an economic goal function online. The value of the crop is evaluated by simply integrating the photosynthetic rate—derived from an empirical regression equation expression—using a fixed price, thus avoiding the need to solve the slow subproblem. The control horizon is two hours. Control parameterization is used, having piecewise linear controls over a fifteen-minute interval, which yields an optimization problem with twenty-seven variables to be found. This problem is solved by a genetic algorithm. A computation time of fifteen to thirty minutes is reported, thus making a receding horizon control impossible. Instead, a new control computation is done for the next horizon of two hours, using the last calculated states as initial condition. Simulations are presented using smooth ten-year–averaged weather.

The receding horizon control of Equation 5.31 was implemented in a real application for a tomato production during the reproductive stage, as described by Tap (2000), Tap, Van Straten, and Van Willigenburg (1997), and Tap, Van Willigenburg, and Van Straten (1996). An experimental comparison of an optimally controlled greenhouse with a traditionally controlled greenhouse was made, showing that yield and quality are comparable. Computation of the economic benefit by simulation shows that 10 to 15 percent savings are easily obtained. More details on this study are provided in Chapter 7. Van Ooteghem (2007) applied the receding horizon optimal control methodology to a design of a novel solar greenhouse with long-term heat storage in an aquifer in conjunction with a heat pump. More details on this application are given in Chapter 8.

5.5 DISCUSSION AND CONCLUSION

The optimal control methodology offers a convenient framework for classification of the numerous solutions to the greenhouse cultivation control as described in the literature. The methodology hinges on two notions: the time-scale decomposition and the desire to exploit the opportunities offered by the natural variation of the sun. The method, however, does rely on the availability of realistic models for crop and greenhouse. Moreover, most feedback available is from observations on the greenhouse climate, whereas automatic feedback from the crop is not easy because of lack of instrumentation to measure biomass and development online. Because of this lack, constraints on climate conditions are still needed to prevent the system to move toward areas with unknown effects. It is likely that this results in larger conservatism than necessary. On the other hand, it has to be recognized that certain aspects related to threats such as infections or diseases, when occurring, may lead to such large losses that some sacrifice of optimality to avoid risk is imperative as long as accurate models to describe these threats are lacking. So far, this aspect has been hardly studied. In addition to these limitations, there are only very few examples of real applications; hence, it is a task of considerable interest to study and alleviate the difficulties that might lead to this reluctance of the industry to implement optimal cultivation (Van Straten, Challa, and Buwalda, 2000). This topic will be discussed further in Chapter 9.

REFERENCES

Aaslyng, J.M., J.B. Lund, N. Ehler, and E. Rosenqvist. 2003. IntelliGrow: A greenhouse component-based climate control system. *Environmental Modelling and Software* 18 (7): 657–666.

Albright, L.D., A.J. Both, and A.J. Chiu. 2000. Controlling greenhouse light to a consistent daily integral. *Transactions of the American Society of Agricultural Engineers* 43 (2): 421–431.

Albright, L.D., R.S. Gates, K.G. Arvanitis, and A.E. Drysdale. 2001. Environmental control for plants on earth and in space. *IEEE Control Systems Magazine* 21 (5): 28–47.

Alscher, G., H. Krug, and H.-P. Liebig. 2001. Optimisation of CO_2 and temperature control in greenhouse crops by means of growth models at different abstraction levels: I. Control strategies, growth models and input data. *Gartenbauwissenschaft* 66 (2): 105–114.

Arnold, E. 1988. Zur optimalen Steuerung zeitdiskreter dynamischer Prozesse mittels nichtlinearer Optimierung mit Anwendungen auf die Klimasteuerung van Gewächshäusern [On the optimal control of time-discrete processes with application to climate control in greenhouses]. PhD dissertation, Technische Hochschule Illmenau, Illmenau (GDR).

Arvanitis, K.G., P.N. Paraskevopoulos, and A.A. Vernardos. 2000. Multirate adaptive temperature control of greenhouses. *Computers and Electronics in Agriculture* 26 (3): 303–320.

Aström, K.J., and B. Wittenmark. 1994. *Adaptive Control*. Englewood Cliffs, NJ: Addison-Wesley.

Athans, M. 1971. The role and use of the stochastic linear-quadratic-Gaussian problem in control system design. *IEEE Transactions on Automatic Control* AC-16 (6): 529–552.

Bailey, B.J. 1988. Control strategies to enhance the performance of greenhouse thermal screens. *Journal of Agricultural Engineering Research* 40 (3): 187–198.

Bailey, B.J., and Z.S. Chalabi. 1994. Improving the cost effectiveness of greenhouse climate control. *Computers and Electronics in Agriculture* 10 (3): 203–214.

Bailey, B.J., and I. Seginer. 1989. Optimum control of greenhouse heating. *Acta Horticulturae* 245: 512–518.

Bakker, J.C., G.P.A. Bot, H. Challa, and N.J. Van de Braak. 1995. *Greenhouse Climate Control—An Integrated Approach*. Wageningen: Wageningen Pers.

Bennis, N., J. Duplaix, G. Enéa, M. Haloua, and H. Youlal. 2008. Greenhouse climate modelling and robust control. *Computers and Electronics in Agriculture* 61 (2): 96–107.

Berenguel, M., F. Rodríguez, J.L. Guzmán, D. Lacasa, and J. Pérez-Parra. 2006. Greenhouse diurnal temperature control with natural ventilation based on empirical models. *Acta Horticulturae* 719: 57–64.

Blasco, X., M. Martinez, J.M. Herrero, C. Ramos, and J. Sanchis. 2007. Model-based predictive control of greenhouse climate for reducing energy and water consumption. *Computers and Electronics in Agriculture* 55 (1): 49–70.

Boulard, T., and A. Baille. 1993. A simple greenhouse climate control model incorporating effects of ventilation and evaporative cooling. *Agricultural and Forest Meteorology* 65 (3–4): 145–157.

Buwalda, F., E.J. Van Henten, A. De Gelder, J. Bontsema, and J. Hemming. 2006. Toward an optimal control strategy for sweet pepper cultivation: I. A dynamic crop model. *Acta Horticulturae* 718: 367–374.

Camacho, E.F., and C. Bordons. 1995. *Model Predictive Control in the Process Industry*. Berlin, Germany: Springer Verlag.

Chalabi, Z.S. 1992. A generalized optimization strategy for dynamic CO_2 enrichment in a greenhouse. *European Journal of Operational Research* 59 (2): 308–312.

Chalabi, Z.S., B.J. Bailey, and D.J. Wilkinson. 1996. A real-time optimal control algorithm for greenhouse heating. *Computers and Electronics in Agriculture* 15 (1): 1–13.

Chalabi, Z.S., A. Biro, B.J. Bailey, D.P. Aikman, and K.E. Cockshull. 2002a. Optimal control strategies for carbon dioxide enrichment in greenhouse tomato crops: Part I. Using pure carbon dioxide. *Biosystems Engineering* 81 (4): 421–431.

Chalabi, Z.S., A. Biro, B.J. Bailey, D.P. Aikman, and K.E. Cockshull. 2002b. Optimal control strategies for carbon dioxide enrichment in greenhouse tomato crops: Part II. Using the exhaust gases of natural gas fired boilers. *Biosystems Engineering* 81 (3): 323–332.

Challa, H., and A.H.C.M. Schapendonk. 1986. Dynamic optimisation of CO_2 concentration in relation to climate control in greenhouses. In *Carbon Dioxide Enrichment of Greenhouse Crops*, ed. K.B.A.H.Z. Enoch. Boca Raton, FL: CRC Press.

Challa, H., and G. Van Straten. 1991. Reflections about optimal climate control in greenhouse cultivation. In *Mathematical and Control Applications in Agriculture and Horticulture*, ed. H. Hashimoto and W. Day. Oxford: Pergamon Press.

Critten, D.L. 1991. Optimization of CO_2 concentration in greenhouses: A modelling analysis for the lettuce crop. *Journal of Agricultural Engineering Research* 48 (4): 261–271.

De Graaf, S.C. 2006. *Low Nitrate Lettuce Cultivation in Greenhouses—Optimal Control in the Presence of Measurable Disturbances*. Wageningen: Wageningen University.

De Halleux, D., and L. Gauthier. 1998. Energy consumption due to dehumidification of greenhouses under northern latitudes. *Journal of Agricultural Engineering Research* 69 (1): 35–42.

De Koning, A.N.M. 1990. Long-term temperature integration of tomato. Growth and development under alternating temperature regimes. *Scientia Horticulturae* 45 (1–2): 117–127.

Dekock, J., J.M. Aerts, D. Berckmans, K. Janssen, P. Bleyaert, K. Vermeulen, K. Steppe, R. Lemeur, J. Westra, and T. Rieswijk. 2006. Crop monitoring by means of an on-line early warning system. *Acta Horticulturae* 718: 183–188.

Diezemann, M., R. Mahlendorf, H. Schmeil, and U. Engmann. 1986. Anwendung neuer Steueralgorithmen zur energieoptimalen Klimaführung in Gewächshäusern [Application of New Control Algorithms for Energy Optimal Greenhouse Climate Control]. *Tagungsbericht, Akademie der Landwirtschaftswissenschaften der DDR, Berlin* 238: 183–190.

Ding, W., X. Wang, Y. Li, and J. Wang. 2009. Review on environmental control and simulation models for greenhouses. *Transactions of the Chinese Society of Agricultural Machinery* 40 (5): 162–168.

Ehret, D.L., A. Lau, S. Bittman, W. Lin, and T. Shelford. 2001. Automated monitoring of greenhouse crops. *Agronomie* 21 (4): 403–414.

El Ghoumari, M.Y., H.J. Tantau, and J. Serrano. 2005. Non-linear constrained MPC: Real-time implementation of greenhouse air temperature control. *Computers and Electronics in Agriculture* 49 (3): 345–356.

Elings, A., H.F. De Zwart, J. Janse, L.F.M. Marcelis, and F. Buwalda. 2006. Multiple-day temperature settings on the basis of the assimilate balance: a simulation study. *Acta Horticulturae* 718: 227–232.

Ferentinos, K.P., L.D. Albright, and D.V. Ramani. 2000. Optimal light integral and carbon dioxide concentration combinations for lettuce in ventilated greenhouses. *Journal of Agricultural Engineering Research* 77 (3): 309–315.

Gal, S., I. Seginer, and A. Angel. 1984. Optimal control of greenhouse climate: methodology. *European Journal of Operational Research* 17 (1): 45–56.

Gutman, P.O., P.O. Lindberg, I. Ioslovich, and I. Seginer. 1993. A non-linear optimal greenhouse control problem solved by linear programming. *Journal of Agricultural Engineering Research* 55: 335–351.

Harper, L.A., B.W. Mitchell, and J.E. Pallas Jr. 1979. CO_2 controller for greenhouses. *Transactions of the American Society of Agricultural Engineers* 22 (3): 649–652.

Hashimoto, Y. 1980. Computer control of short term plant growth by monitoring leaf temperature. *Acta Horticulturae* 106: 139–146.

Hashimoto, Y. 1989. Recent strategies of optimal growth regulation by the speaking plant concept. *Acta Horticulturae* 260: 115–122.

Hu, H., L. Xu, and Q. Hu. 2009. Model-based compromise control of greenhouse climate using Pareto optimization. 2009 World Summit on Genetic and Evolutionary Computation, 2009 GEC Summit. In *Proceedings of the 1st ACM/SIGEVO Summit on Genetic and Evolutionary Computation, GEC'09, Shanghai*. New York: Association for Computing Machinery (ACM), 217–222.

Hurd, R.G., and C.J. Graves. 1984. The influence of different temperature patterns having the same integral on the earliness and yield of tomatoes. *Acta Horticulturae* 148: 547–554.

Ioslovich, I., P.O. Gutman, and R. Linker. 2009. Hamilton–Jacobi–Bellman formalism for optimal climate control of greenhouse crop. *Automatica* 45 (5): 1227–1231.

Ioslovich, I., and I. Seginer. 2002. SE—Structures and environment: Acceptable nitrate concentration of greenhouse lettuce: Two optimal control policies. *Biosystems Engineering* 83 (2): 199–115.

Ioslovich, I., I. Seginer, P.O. Gutman, and M. Borshchevsky. 1995. Sub-optimal CO enrichment of greenhouses. *Journal of Agricultural Engineering Research* 60 (2): 117–136.

Jones, P., J.W. Jones, and Y. Hwang. 1990. Simulation for determining greenhouse temperature setpoints. *Transactions of the American Society of Agricultural Engineers* 33 (5): 1722–1728.

Kläring, H.P., C. Hauschild, A. Heißner, and B. Bar-Yosef. 2007. Model-based control of CO_2 concentration in greenhouses at ambient levels increases cucumber yield. *Agricultural and Forest Meteorology* 143 (3–4): 208–216.

Kolokotsa, D., G. Saridakis, K. Dalamagkidis, S. Dolianitis, and I. Kaliakatsos. 2010. Development of an intelligent indoor environment and energy management system for greenhouses. *Energy Conversion and Management* 51 (1): 155–168.

Körner, O., and H. Challa. 2003a. Corrigendum to "Design for an improved temperature integration concept in greenhouse cultivation": [*Computers and Electronics in Agriculture* 39 (2003) 39–59]. *Computers and Electronics in Agriculture* 39 (3): 257.

Körner, O., and H. Challa. 2003b. Design for an improved temperature integration concept in greenhouse cultivation. *Computers and Electronics in Agriculture* 39 (1): 39–59.

Körner, O., and H. Challa. 2003c. Temperature integration and DIF in cut chrysanthemum. *Journal of Horticultural Science & Biotechnology* 78 (3): 335–342.

Körner, O., and H. Challa. 2004. Temperature integration and process-based humidity control in chrysanthemum. *Computers and Electronics in Agriculture* 43 (1): 1–21.

Kurata, K., and N. Eguchi. 1990. Machine learning of fuzzy rules for crop management in protected cultivation. *Transactions of the American Society of Agricultural Engineers* 33 (4): 1360–1368.

Kwakernaak, H., and R. Sivan. 1972. *Linear Optimal Control Systems*. New York: Wiley-Interscience.

Lacroix, R.R. 1999. Simulation-based control of enclosed ecosystems—A case study: Determination of greenhouse heating setpoints. *Canadian Agricultural Engineering* 41 (3): 175–183.

Lafont, F., and J.F. Balmat. 2002. Optimized fuzzy control of a greenhouse. *Fuzzy Sets and Systems* 128 (1): 47–59.

Langton, F.A., J.S. Horridge, M.D. Holdsworth, and P.J.C. Hamer. 2002. Control and optimization of the greenhouse environment using infra-red sensors. *Acta Horticulturae* 633: 145–152.

Lees, M.J., C.J. Taylor, P.C. Young, and A. Chotai. 1998. Modelling and PIP control design for open-top chambers. *Control Engineering Practice* 6 (10): 1209–1216.

Lewis, F.L. 1986. *Optimal Estimation*. New York: John Wiley & Sons.

Linker, R., P.O. Gutman, and I. Seginer. 1999. Robust controllers for simultaneous control of temperature and CO_2 concentration in greenhouses. *Control Engineering Practice* 7 (7): 851–862.

Maciejowski, J.M. 2002. *Predictive Control with Constraints*. Harlow, UK: Prentice Hall.

Markert, A. 1990. Aggregation Planzenphysiologischer Gewächstumsmodelle und Berechnung van Steuerstrategien für das Gewächshausinnenklima mittels Verfahren der nichtlinearen Optimierung [Crop-physiological crop growth models aggregated in the computation of control strategies for the greenhouse climate by non-linear optimisation procedures]. PhD dissertation, Technischen Hochschule Ilmenau, Ilmenau (GDR).

Marsh, L.S., and L.D. Albright. 1991a. Economically optimum day temperatures for greenhouse hydroponic lettuce production: Part I. A computer model. *Transactions of the American Society of Agricultural Engineers* 34 (2): 550–556.

Marsh, L.S., and L.D. Albright. 1991b. Economically optimum day temperatures for greenhouse hydroponic lettuce production: Part II. Results and simulations. *Transactions of the American Society of Agricultural Engineers* 34 (2): 557–562.

Martin-Clouaire, R., P. J. Schotman, and M. Tchamitchian. 1996. A survey of computer-based approaches for greenhouse climate management. *Acta Horticulturae* 406: 409–423.

Morari, M., and E. Zafiriou. 1989. *Robust Process Control*. Englewood Cliffs, NJ: Prentice-Hall, Inc.

Morimoto, T., and Y. Hashimoto. 2000. AI approaches to identification and control of total plant production systems. *Control Engineering Practice* 8 (5): 555–567.

Nilsson, L.E.G., and T.G. Nybrant. 1992. Control of flower development of potplants in greenhouses. In *Agricultural Engineering International Conference, AgEng '92, Uppsala*. Uppsala, Sweden: Jordbrukstekniska Institut, 1–8.

Nilsson, L.E.G., and T.G. Nybrant. 1996. Adaptive control of floral development. *Acta Horticulturae* 406: 73–80.

Pasgianos, G.D., K.G. Arvanitis, P. Polycarpou, and N. Sigrimis. 2003. A nonlinear feedback technique for greenhouse environmental control. *Computers and Electronics in Agriculture* 40 (1–3): 153–177.

Pawlowski, A., J.L. Guzmán, F. Rodríguez, M. Berenguel, J. Sánchez, and S. Dormido. 2009. Simulation of greenhouse climate monitoring and control with wireless sensor network and event-based control. *Sensors* 9 (1): 232–252.

Piñón, S., E.F. Camacho, B. Kuchen, and M. Peña. 2005. Constrained predictive control of a greenhouse. *Computers and Electronics in Agriculture* 49 (3): 317–329.

Pohlheim, H., and A. Heissner. 1996. Anwendung genetischer Algorithmen zur optimalen Steuerung des Gewächshausklimas. *VDI-Berichte*. Düsseldorf, Germany: Deutches Ingenieur-Verlag, 1282: 799–809.

Pohlheim, H., and A. Heissner. 1997. Optimal control of greenhouse climate using a short time climate model and evolutionary algorithms. In *Proceedings of the 3rd International IFAC Workshop on Mathematical and Control Applications in Agriculture and Horticulture, Hannover, Germany,* 113–118.

Pucheta, J.A., C. Schugurensky, R. Fullana, H. Patiño, and B. Kuchen. 2006. Optimal greenhouse control of tomato-seedling crops. *Computers and Electronics in Agriculture* 50: 70–82.

Ramírez-Arias, A., F. Rodríguez, J.L. Guzmán, M.R. Arahal, M. Berenguel, and J.C. López. 2005. Improving efficiency of greenhouse heating systems using model predictive control. In *Proceedings of the 16th IFAC World Congress, Prague*. Elsevier (CD-ROM).

Reinisch, K., E. Arnold, A. Markert, and H. Puts. 1989. Development of strategies for temperature and CO_2 control in the greenhouse production of cucumbers and tomatoes based on modelbuilding and optimisation. *Acta Horticulturae* 260: 67–75.

Richalet, J., A. Rault, J.L. Testud, and J. Papon. 1978. Model predictive heuristic control: applications to industrial processes. *Automatica* 14 (5): 413–428.

Rodríguez, F., J.L. Guzmán, M. Berenguel, and M. R. Arahal. 2008. Adaptive hierarchical control of greenhouse crop production. *International Journal of Adaptive Control and Signal Processing* 22 (2): 180–197.

Schapendonk, A.H.C.M., H. Challa, P.W. Broekharst, and A.J. Udink Ten Cate. 1984. Dynamic climate control; an optimization study for earliness of cucumber production. *Scientia Horticulturae* 23 (2): 137–150.

Schmidt, M., K. Reinisch, H. Puta, A. Markert, P. Augustin, and A. Heissner. 1987. Determining climate strategies for greenhouse cucumber production by means of optimization. In *Proceedings of the 10th IFAC World Congress on Automatic Control, Munich (FRG)*. Laxenburg, Austria: International Federation of Automatic Control, 350–555.

Schmidt, U. 1996. Greenhouse climate control with a combined model of greenhouse and plant by using online measurement of leaf temperature and transpiration. *Acta Horticulturae* 406: 89–98.

Schotman, P.J. 2000. Improving support for greenhouse climate management: An exploration of a knowledge-based system. PhD dissertation, Wageningen University, The Netherlands.

Seginer, I. 1980. Optimizing greenhouse operation for best aerial environment. *Acta Horticulturae* 106: 169–178.

Seginer, I. 1988. Optimal control techniques for energy conservation in controlled environments. *CNRE Bulletin (FAO)* 23: 160–169.

Seginer, I. 1989. Optimal greenhouse production under economic constraints. *Agricultural Systems* 29 (1): 67–80.

Seginer, I. 1993. Crop models in greenhouse climate control. *Acta Horticulturae* 328: 79–97.

Seginer, I. 1997. Some artificial neural network applications to greenhouse environmental control. *Computers and Electronics in Agriculture* 18 (2–3): 167–186.

Seginer, I. 2008. Co-state variables as strategic set-points for environmental control of greenhouses. *Acta Horticulturae* 797: 69–74.

Seginer, I., and L.D. Albright. 1980. Rational operation of greenhouse thermal-curtains. *Transactions of the American Society of Agricultural Engineers* 23 (5): 1240–1245.

Seginer, I., L.D. Albright, and I. Ioslovich. 2006. Improved strategy for a constant daily light integral in greenhouses. *Biosystems Engineering* 93 (1): 69.

Seginer, I., A. Angel, S. Gal, and D. Kantz. 1986. Optimal CO enrichment strategy for greenhouses: A simulation study. *Journal of Agricultural Engineering Research* 34: 285–304.

Seginer, I., T. Boulard, and B.J. Bailey. 1994. Neural network models of the greenhouse climate. *Journal of Agricultural Engineering Research* 59 (3): 203–216.

Seginer, I., C. Gary, and M. Tchamitchian. 1994. Optimal temperature regimes for a greenhouse crop with a carbohydrate pool: A modelling study. *Scientia Horticulturae* 60 (1–2): 55–80.

Seginer, I., and I. Ioslovich. 1998. Seasonal optimization of the greenhouse environment for a simple two-stage crop growth model. *Journal of agricultural engineering research* 70 (2): 145–155.

Seginer, I., and B.M. Jenkins. 1987. Temperature exposure of greenhouses from monthly means of daily maximum and minimum temperatures. *Journal of Agricultural Engineering Research* 37 (3–4): 191–208.

Seginer, I., and R.W. McClendon. 1992. Methods for optimal control of the greenhouse environment. *Transactions of the American Society of Agricultural Engineers* 35 (4): 1299–1307.

Seginer, I., and A. Sher. 1993. Optimal greenhouse temperature trajectories for a multi-state-variable tomato model. In *The Computerized Greenhouse*, ed. Y. Hashimoto, G.P.A. Bot, W. Day, H.-J. Tantau, and H. Nonami, 153–172. San Diego: Academic Press.

Seginer, I., G. Shina, L.D. Albright, and L.S. Marsh. 1991. Optimal temperature setpoints for greenhouse lettuce. *Journal of Agricultural Engineering Research* 49: 209–226.

Setiawan, A., L.D. Albright, and R.M. Phelan. 2000. Application of pseudo-derivative-feedback algorithm in greenhouse air temperature control. *Computers and Electronics in Agriculture* 26 (3): 283–302.

Sigrimis, N., A. Anastasiou, and N. Rerras. 2000. Energy saving in greenhouses using temperature integration: A simulation survey. *Computers and Electronics in Agriculture* 26 (3): 321–341.

Sigrimis, N.A., K.G. Arvanitis, and G.D. Pasgianos. 2000. Synergism of high and low level systems for the efficient management of greenhouses. *Computers and Electronics in Agriculture* 29 (1–2): 21–39.

Speetjens, S.L. 2008. *Towards Model Based Adaptive Control for the Watergy Greenhouse*. Wageningen: Wageningen University.

Speetjens, S.L., J.D. Stigter, and G. Van Straten. 2009. Towards an adaptive model for greenhouse control. *Computers and Electronics in Agriculture* 67 (1–2): 1–8.

Srinivasan, B., D. Bonvin, E. Visser, and S. Palanki. 2003. Dynamic optimization of batch processes II. Role of measurements in handling uncertainty. *Computers & Chemical Engineering* 27 (1): 27–44.

Stanghellini, C., and W. Th. M. Van Meurs. 1992. Environmental control of greenhouse crop transpiration. *Journal of Agricultural Engineering Research* 51 (C): 297–311.

Stengel, R.F. 1994. *Optimal Control and Estimation*. New York: Dover Publications.

Takakura, T. 1974. Direct digital control of plant growth: I. Design and operation of the system. *Transactions of the American Society of Agricultural Engineers* 17 (6): 1150–1154.

Tantau, H.J. 1991. Optimal control for plant production in greenhouses. In *Mathematical and Control Applications in Agriculture and Horticulture*, ed. H. Hashimoto and W. Day. Oxford: Pergamon Press.

Tantau, H.J. 1993. Optimal control for plant production in greenhouses. In *The Computerized Greenhouse*, ed. Y. Hashimoto, G.P.A. Bot, W. Day, H.-J. Tantau, and H. Nonami. San Diego: Academic Press, 139–152.

Tantau, H.J., and D. Lange. 2003. Greenhouse climate control: An approach for integrated pest management. *Computers and Electronics in Agriculture* 40 (1–3): 141–152.

Tap, R.F. 2000. *Economics-Based Optimal Control of Greenhouse Tomato Crop Production, Systems and Control*. Wageningen, The Netherlands: Wageningen Agricultural University.

Tap, R.F., G. Van Straten, and L.G. Van Willigenburg. 1997. Comparison of classical and optimal control of greenhouse tomato crop production. In *Mathematical and Control Applications in Agriculture and Horticulture, Proceedings of the 3rd IFAC Workshops, Hanover, Germany*, ed. A. Munack and H.-J. Tantau. Pergamon-Elsevier, 1–6.

Tap, R.F., L.G. Van Willigenburg, and G. Van Straten. 1996. Experimental results of receding horizon optimal control of greenhouse climate. *Acta Horticulturae* 406: 229–238.

Tap, R.F., L.G. Van Willigenburg, G. Van Straten, and E.J. Van Henten. 1993. Optimal control of greenhouse climate: Computation of the influence of fast and slow dynamics. In *Proceedings of the 12th IFAC Triennal World Congress, Sydney, Australia*. Australia: Institution of Engineers, 10: 321–324.

Taylor, C.J., P.C. Young, A. Chotai, A.R. McLeod, and A.R. Glasock. 2000. Modelling and proportional-integral-plus control design for free air carbon dioxide enrichment systems. *Journal of Agricultural Engineering Research* 75 (4): 365–374.

Tchamitchian, M., R. Martin-Clouaire, J. Lagier, B. Jeannequin, and S. Mercier. 2006. SERRISTE: A daily set point determination software for glasshouse tomato production. *Computers and Electronics in Agriculture* 50: 25–47.

Trigui, M., S. Barrington, and L. Gauthier. 2001. A strategy for greenhouse climate control: Part II. Model Validation. *Journal of Agricultural Engineering Research* 79 (1): 99–105.

Udink ten Cate, A.J. 1987. Analysis and synthesis of greenhouse climate controllers. In *Computer Applications in Agricultural Environments*, ed. K.G.J.A. Clark and R.A. Saffell. London: Butterworths.

Udink ten Cate, A.J., G.P.J. Bot, and J.J. Dixhoorn. 1978. Computer control of greenhouse climates. *Acta Horticulturae* 87: 265–272.

Udink ten Cate, A.J., and H. Challa. 1984. On optimal control of the crop growth system. *Acta Horticulturae* 148: 267–277.

Van Henten, E.J. 1989. Model based design of optimal multivariable climate control systems. *Acta Horticulturae* 248: 301–306.

Van Henten, E.J. 1994. *Greenhouse Climate Management: An Optimal Control Approach*. Wageningen: Agricultural University Wageningen.

Van Henten, E.J. 2003. Sensitivity analysis of an optimal control problem in greenhouse climate management. *Biosystems Engineering* 85 (3): 355–364.

Van Henten, E.J., and J. Bontsema. 1991. Optimal control of greenhouse climate. In *Proceedings of the 1st International IFAC ISHS Workshop on Mathematical and Control Applications in Agriculture and Horticulture, Matsuyama, Japan*. Published for the International Federation of Automatic Control by Pergamon Press, Oxford, England, 27–32.

Van Henten, E.J., and J. Bontsema. 2009. Time-scale decomposition of an optimal control problem in greenhouse climate management. *Control Engineering Practice* 17 (1): 88–96.

Van Henten, E.J., F. Buwalda, H.F. De Zwart, A. De Gelder, J. Hemming, and J. Bontsema. 2006. Toward an optimal control strategy for sweet pepper cultivation: II. Optimization of the yield pattern and energy efficiency. *Acta Horticulturae* 718: 391–398.

Van Meurs, W. Th. M., and E.J. Van Henten. 1994. An experiment on the optimization of CO_2 in greenhouse climate control. *Acta Horticulturae* 366: 201–208.

Van Ooteghem, R.J.C. 2007. *Optimal Control Design for a Solar Greenhouse, Systems and Control*. Wageningen: Wageningen University.

Van Straten, G. 1999. Acceptance of optimal operation and control methods for greenhouse cultivation. *Annual Reviews in Control* 23: 83–90.

Van Straten, G., H. Challa, and F. Buwalda. 2000. Towards user accepted optimal control of greenhouse climate. *Computers and Electronics in Agriculture* 26 (3): 221–238.

Van Straten, G., and L.G. Van Willigenburg. 2008. On evaluating optimality losses of greenhouse climate controllers. In *Proceedings of the 17th World Congress, IFAC, Seoul, Korea*. IFAC-PapersOnLine, Paper No. 2944.

Van Straten, G., L.G. Van Willigenburg, and R.F. Tap. 2002. The significance of crop co-states for receding horizon optimal control of greenhouse climate. *Control Engineering Practice* 10 (6): 625–632.

Vermeulen, P.C.M. 1989. A method to calculate the economic possibilities of heat storage and CO_2 distribution in glasshouses. *Acta Horticulturae* 248: 141–150.

Wittwer, S.H., and W.M. Robb. 1964. Carbon dioxide enrichment of greenhouse atmospheres for food crop production. *Economic Botany* 18 (1): 34–56.

Xu, L., H. Hu, and B. Zhu. 2009. Energy-saving control of greenhouse climate based on MOCC strategy. 2009 World Summit on Genetic and Evolutionary Computation, 2009 GEC Summit. In *Proceedings of the 1st ACM/SIGEVO Summit on Genetic and Evolutionary Computation, GEC'09, Shanghai*, New York: Association for Computing Machinery (ACM), 645–650.

Young, P., M.A. Behzadi, C.L. Wang, and A. Chotai. 1987. Direct digital and adaptive control by input–output state variable feedback pole assignment. *International Journal of Control* 46 (6): 1867–1881.

Young, P.C., M.J. Lees, A. Chotai, W. Tych, and Z.S. Chalabi. 1994. Modelling and PIP control of a glasshouse micro-climate. *Control Engineering Practice* 2 (4): 591–604.

Young, P.C., W. Tych, and A. Chotai. 1991. Identification, estimation and control of glasshouse systems. In *Mathematical and Control Applications in Agriculture and Horticulture,* ed. H. Hashimoto and W. Day. Oxford: Pergamon Press.

Zhang, Z. 2008. Multiobjective optimization immune algorithm in dynamic environments and its application to greenhouse control. *Applied Soft Computing* 8 (2): 959–971.

6 A Seminal Case: Lettuce*

6.1 INTRODUCTION

One of the first developments in the Wageningen school on optimal greenhouse control goes back to the thesis of Van Henten (1994). The main aspects of the approach and important results are described in this chapter.

At the time of this development in the early 1990s, it was observed that greenhouse climate control in practice was essentially based on the realization of climate strategies originating from horticultural research and the grower's own experience (as it is still today). The motivation of the research was that, in principle, greenhouse climate control can be made more efficient by explicitly balancing the benefits associated with the sellable crop against the operating costs of manipulating the climate conditioning equipment. While this idea was not new, and had been promoted by others, most of this earlier work was largely focusing on either the crop, ignoring the greenhouse dynamics, or on optimizing the greenhouse dynamics, while considering the conditions for the crop as given. In this chapter, the two aspects are combined in an integrated fashion. In addition, the early analyses were largely based on simple analytical models or approximated solution methods.

Although optimal control methodology was known from the 1950s, the advent of fast computers and new software, in combination with more advanced climate models and crop models, renewed the interest in investigating the possibilities and intricacies of optimal greenhouse cultivation control.

So as not to complicate the analysis, a vegetative crop, lettuce, was selected as a case. Although lettuce is not considered economically as one of the important crops in protected cultivation, it has been used to illustrate the principle of the optimal control approach. It is a single harvest crop which, from the point of view of modeling and optimal control, is much easier to deal with than multiple harvests crops like tomatoes and cucumber. However, the optimal control methodology is equally applicable to the latter type of crops as well.

This chapter will focus on optimal climate control for a lettuce crop production system, graphically depicted in Figure 6.1. A simple one-state crop model will be used to describe dry matter production of the lettuce crop. The greenhouse model is also relatively simple, only considering the air temperature, humidity, and CO_2 concentration as state variables. Heating, natural ventilation, and CO_2 supply are taken into account as control inputs. External conditions include outdoor solar radiation, temperature, humidity, CO_2 concentration, and wind speed. The models were calibrated and validated against real data.

Another focus of discussion here is the calculation of open-loop control strategies, albeit with realistic weather realizations, and with prices for the produce that fluctuate with the seasons, as observed in practice. This is a stepping stone toward real online optimal control.

The models of crop production and greenhouse climate will be described first. Then, the optimal control problem will be defined. Based on these premises, four case studies will be devoted to the following:

1. An economic interpretation and analysis of the optimal control problem
2. A comparison of optimal control versus control strategies implemented by the grower to assess the potential improvements of using the optimal control approach

* Notation: Chapter-specific symbols are defined in the text and are listed together with parameter values in Table 6.1.

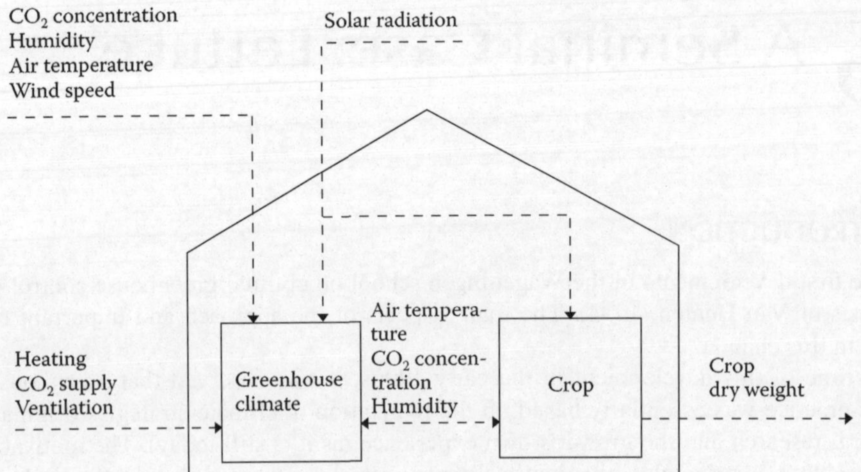

FIGURE 6.1 A schematic diagram of a lettuce crop production system.

3. A sensitivity analysis in which the relative sensitivity of the performance to variations in the parameter values is determined
4. A time-scale decomposition of the optimal control problem

6.2 MODELS

The lettuce crop production process is described by a four state variable dynamic model. The model describes the evolution in time of the dry matter content of the lettuce crop W in kg[dm] m^{-2}[gh], the carbon dioxide concentration in the greenhouse $C_{CO_2,a}$ in kg[CO_2] m^{-3}[gh], the air temperature in the greenhouse T_a^C in °C and the humidity content of the greenhouse air $C_{H_2O,a}$ in kg[H_2O] m^{-3}, with the equations:

$$\frac{dW}{dt} = c_{\alpha\beta}\varphi_{CO_2,a\_c} - c_{W,c\_a}W2^{(0.1T_a^C-2.5)} \qquad \text{(kg[dm] m}^{-2}\text{[gh] m}^{-1}\text{)} \quad (6.1)$$

where $c_{\alpha,\beta}$ is a yield factor, φ_{CO_2,a_c} is the gross canopy photosynthesis rate in kg[CO_2] m^{-2}[gh] s^{-1}, c_{W,c_a} in s^{-1} is the respiration rate expressed in terms of the amount of respired dry matter,

$$\frac{dC_{CO_2\_a}}{dt} = \frac{1}{c_{CO_2}^{cap}}\left[-\varphi_{CO_2,a\_c} + c_{CO_2,c\_a}W2^{\left(0.1T_a^C-2.5\right)} + u_{CO_2} - \varphi_{CO_2,g\_o}\right] \qquad \text{(kg[CO}_2\text{] m}^{-2}\text{[gh] s}^{-1}\text{)} \quad (6.2)$$

where $c_{CO_2}^{cap}$ is the greenhouse volume in m^3 [air] m^{-2}[gh], c_{CO_2,c_a} in s^{-1} is the respiration coefficient expressed in terms of the amount of carbon dioxide produced, u_{CO_2} is the supply rate of carbon dioxide in kg[CO_2] m^{-2}[gh] s^{-1}, and ϕ_{CO_2,g_o} is the mass exchange of carbon dioxide through the vents in kg[CO_2] m^{-2}[gh] s^{-1},

TABLE 6.1
Symbols, Parameters, and Their Nominal Values in Order of Appearance

W		kg[dm] m^{-2}[gh]	Dry matter content of the lettuce crop
$C_{CO_2,a}$		kg[CO_2] m^{-3}[gh]	Carbon dioxide concentration in the greenhouse
T_a^C		°C	Air temperature in the greenhouse
$C_{H_2O,a}$		kg[H_2O] m^{-3}[gh]	Humidity content of the greenhouse air
$c_{\alpha\beta}$	0.544		Yield factor
φ_{CO_2,a_c}		kg[CO_2] m^{-2}[gh] s^{-1}	Gross canopy photosynthesis rate
c_{W,c_a}	2.65×10^{-7}	s^{-1}	Respiration rate
$c_{CO_2}^{cap}$	4.1	m^3[air] m^{-2}[gh]	Greenhouse volume
c_{CO_2,c_a}	4.87×10^{-7}	s^{-1}	Respiration coefficient
u_{CO_2}		kg[CO_2] m^{-2}[gh] s^{-1}	Supply rate of carbon dioxide
ϕ_{CO_2,g_o}		kg[CO_2] m^{-2}[gh] s^{-1}	Mass exchange of carbon dioxide through the vents
c_a^{cap}	30000	J m^{-2}[gh] °C^{-1}	Effective heat capacity of the greenhouse air
u_q		W m^{-2}[gh]	Energy supply by the heating system
$q_{g\_o}^{trans,vent}$		W m^{-2}[gh]	Energy exchange with the outdoor air by ventilation and transmission through the cover
$q_{o\_g}^{rad}$		W m^{-2}[gh]	Heat load by solar radiation
$c_{H_2O}^{cap}$		m^3[air] m^{-2}[gh]	Greenhouse volume
φ_{H_2O,c_a}		kg[H_2O] m^{-2}[gh] s^{-1}	Canopy transpiration
φ_{H_2O,g_o}		kg[H_2O] m^{-2}[gh] s^{-1}	Mass exchange of water vapor through the vents
$c_{LAI,W}$	53	m^2[L] kg^{-1}[dw]	Effective canopy surface
$c_{I_0}^{phot}$	3.55×10^{-9}	kg[CO_2] J^{-1}	Light use efficiency
I_0		W m^{-2}[gh]	Solar radiation outside the greenhouse
$c_{CO_2,1}^{phot}$	5.11×10^{-6}	m s^{-1} °C^{-2}	Temperature influence on gross canopy photosynthesis
$c_{CO_2,2}^{phot}$	2.30×10^{-4}	m s^{-1} °C^{-1}	Temperature influence on gross canopy photosynthesis
$c_{CO_2,3}^{phot}$	6.29×10^{-4}	m s^{-1}	Temperature influence on gross canopy photosynthesis
c_{Γ}^{phot}	5.2×10^{-5}	kg[CO_2] m^{-3}[air]	Carbon dioxide compensation point
u_v		m^3[air] m^{-2}[gh] s^{-1}	Ventilation rate through the vents in
c_{leak}	0.75×10^{-4}	m s^{-1}	Ventilation leakage through the cover
$C_{CO_2,o}$		kg[CO_2] m^{-3}	Carbon dioxide concentration outside the greenhouse
c_{cap}^{vent}	1290	J m^{-3}[gh] °C^{-1}	Heat capacity per volume unit of greenhouse air
$c_{g\_o}^{trans,vent}$	6.1	W m^{-2}[gh] °C^{-1}	Overall heat transfer through the cover
T_o^c		°C	Outside air temperature
$q_{o\_g}^{rad}$		W m^{-2}[gh]	Energy input to the greenhouse system by solar radiation
$c_{o\_g}^{rad}$	0.2	-	Heat load coefficient due to solar radiation
$c_{c\_a}^{evap}$	3.6×10^{-3}	m s^{-1}	Mass transfer coefficient for water vapor between leaf and air
$c_{H_2O,1}^{sat}$	9348	J m^{-3}	Parameterizes saturation water vapor pressure
$c_{H_2O,2}^{sat}$	17.4	-	Parameterizes saturation water vapor pressure
$c_{H_2O,3}^{sat}$	239	°C	Parameterizes saturation water vapor pressure

(continued)

TABLE 6.1 (Continued)
Symbols, Parameters, and Their Nominal Values in Order of Appearance

c_R	8314	J K^{-1} $kmol^{-1}$	Gas constant
c_T	273.15	K	Parameterizes the conversion of temperature from °C to K
$C_{H_2O,o}$		kg[H_2O] m^{-3}[air]	Humidity concentration outside the greenhouse
$c_{W,1}^{income}$	1.8	Dfl m^{-2}[gh]	Parameterizes gross income of lettuce cultivation
$c_{W,2}^{income}$	16	Dfl kg^{-1}[dw] m^{-2}[gh]	Parameterizes gross income of lettuce cultivation
c_q^{cost}	6.35×10^{-9}	Dfl J^{-1}	Energy price
$c_{CO_2}^{cost}$	42×10^{-2}	Dfl kg^{-1}[CO_2]	Price of carbon dioxide
$u_{CO_2}^{min}$	0	kg[CO_2] m^{-2} s^{-1}	Lower bound on CO_2 supply rate
$u_{CO_2}^{max}$	1.2×10^{-6}	kg[CO_2] m^{-2} s^{-1}	Upper bound on CO_2 supply rate
u_q^{min}	0	W m^{-2}	Lower bound on energy input by the heating system
u_q^{max}	150	W m^{-2}	Upper bound on energy input by the heating system
u_v^{min}	0	m s^{-1}	Lower bound on the ventilation rate
u_v^{max}	7.5×10^{-3} V_{wind}	m s^{-1}	Upper bound on the ventilation rate
$T_a^{c,max}$	6.5	°C	Lower bound on the air temperature
$T_a^{c,min}$	40	°C	Upper bound on the air temperature
$C_{CO_2\_a}^{min}$	0	kg[CO_2] m^{-3}	Lower bound on the CO_2 concentration in the greenhouse
$C_{CO_2,a}^{max}$	2.75×10^{-3}	kg[CO_2] m^{-3}	Upper bound on the CO_2 concentration in the greenhouse
$R_{H_2O,a}^{min}$	0	%	Lower bound on relative humidity
$R_{H_2O,a}^{max}$	90	%	Upper bound on relative humidity
$C_{H_2O,a}^{sat}$		kg[H_2O] m^{-3}[air]	Saturation water vapor pressure
$c_{H_2O,4}^{sat}$	10998	J m^{-3}	Parameterizes the saturation water vapor pressure
$P_x(t)$		Dfl m^{-2}[gh] s^{-1}	Penalty functions
$c^{penalty}$		Dfl m^{-2}[gh] s^{-1}	Weighting factor
k			Counter
P_T			Penalty on T_a^C
P_{CO_2}			Penalty on $C_{CO_2,a}$
P_{H_2O}			Penalty on $C_{H_2O,a}$
λ_W		Dfl kg^{-1}[dw]	Costate of W
$\lambda_{CO_2}^f$		Dfl m kg^{-1}[CO_2]	Costate of $C_{CO_2,a}$
λ_T^f		Dfl m^{-2}[gh] °C^{-1}	Costate of T_a^C
$\lambda_{H_2O}^f$		Dfl m kg^{-1}[H_2O]	Costate of $C_{H_2O,a}$
$c_{H_2O}^{penalty}$			Weighting factor in penalty on $C_{H_2O,a}$
$c_{CO_2\_o}$	1		Perturbation parameter on $C_{CO_2,o}$
c_{I_0}	1		Perturbation parameter on I_0
$c_{H_2O\_o}$	1		Perturbation parameter on $C_{H_2O,o}$
c_{T_0}	1		Perturbation parameter on T_0

$$\frac{dT_a^c}{dt} = \frac{1}{c_a^{cap}}\left[u_q - q_{g\_o}^{trans,vent} + q_{o\_g}^{rad}\right] \qquad (^\circ C\ s^{-1}) \quad (6.3)$$

where c_a^{cap} is the effective heat capacity of the greenhouse air in J m^{-2}[gh] °C^{-1}, u_q is the energy supply by the heating system in W m^{-2}[gh], $q_{g\_o}^{trans,\,vent}$ is the energy exchange with the outdoor air by ventilation and transmission through the cover in W m^{-2}[gh], and $q_{o\_g}^{rad}$ is the heat load by solar radiation in W m^{-2}[gh],

$$\frac{dC_{H_2O,a}}{dt} = \frac{1}{c_{H_2O}^{cap}}\left[\varphi_{H_2O,c\_a} - \varphi_{H_2O,g\_o}\right] \qquad (kg[H_2O]\ m^{-3}\ [gh]\ s^{-1}) \quad (6.4)$$

where $c_{H_2O}^{cap}$ is the greenhouse volume in m^3[air] m^{-2}[gh], φ_{H_2O,c_a} is the canopy transpiration in kg[H_2O] m^{-2}[gh] s^{-1}, and φ_{H_2O,g_o} is the mass exchange of water vapor through the vents in kg[H_2O] m^{-2}[gh] s^{-1}.

The gross photosynthesis rate φ_{CO_2,a_c} in kg[CO_2] m^{-2}[gh] s^{-1} is described by:

$$\varphi_{CO_2,a\_c} = \left(1 - e^{-c_{LAI,W}W}\right)\frac{c_{I_0}^{phot} I_0\left(-c_{CO_2,1}^{phot}\left(T_a^c\right)^2 + c_{CO_2,2}^{phot}T_a^c - c_{CO_2,3}^{phot}\right)\left(C_{CO_2,a} - c_\Gamma^{phot}\right)}{c_{I_0}^{phot} I_0 + \left(-c_{CO_2,1}^{phot}\left(T_a^c\right)^2 + c_{CO_2,2}^{phot}T_a^c - c_{CO_2,3}^{phot}\right)\left(C_{CO_2,a} - c_\Gamma^{phot}\right)} \qquad (kg[CO_2]\ m^{-2}[gh]\ s^{-1}) \quad (6.5)$$

where $c_{LAI,W}$ is the effective canopy surface in m^2[L] kg^{-1}[dw], $c_{I_0}^{phot}$ is the light use efficiency in kg[CO_2] J^{-1}, I_0 is the solar radiation outside the greenhouse in W m^{-2}[gh], $c_{CO_2,1}^{phot}$ in m s^{-1} °C^{-2}, $c_{CO_2,2}^{phot}$ in m s^{-1} °C^{-1}, and $c_{CO_2,3}^{phot}$ in m s^{-1} parameterize the temperature influence on gross canopy photosynthesis, and c_Γ^{phot} is the carbon dioxide compensation point in kg[CO_2] m^{-3}[air]. The mass transfer of carbon dioxide due to ventilation and leakage φ_{CO_2,g_a} in kg[CO_2] m^{-2}[gh] s^{-1} is defined by

$$\varphi_{CO_2,g\_o} = \left(u_v + c_{leak}\right)\left(C_{CO_2,a} - C_{CO_2,o}\right) \qquad (kg[CO_2]\ m^{-2}[gh]\ s^{-1}) \quad (6.6)$$

where u_v is the ventilation rate through the vents in m^3[air] m^{-2}[gh] s^{-1}, c_{leak} is the leakage through the cover in m^3[air] m^{-2}[gh] s^{-1}, and $C_{CO_2,o}$ is the carbon dioxide concentration outside the greenhouse in kg[CO_2] m^{-3}. The energy transfer between the indoor environment and the outdoor environment due to ventilation and transmission $q_{g\_o}^{trans,\,vent}$ in W m^{-2}[gh] is covered by the equation

$$q_{g\_o}^{trans,vent} = \left(c_{cap}^{vent}u_v + c_{g\_o}^{trans}\right)\left(T_a^C - T_o^C\right) \qquad (W\ m^{-2}[gh]) \quad (6.7)$$

in which $c_{cap,a}^{vent}$ is the heat capacity per volume unit of greenhouse air in J m^{-3}[gh] °C^{-1}, $c_{g\_o}^{trans}$ in W m^{-2}[gh] °C^{-1} parameterizes the overall heat transfer through the cover, T_o^C in °C stands for the outside air temperature. The energy input to the greenhouse system by solar radiation $q_{o\_g}^{rad}$ in W m^{-2}[gh], is described by

$$q_{o\_g}^{rad} = c_{o\_g}^{rad}I_0 \qquad (W\ m^{-2}[gh]) \quad (6.8)$$

where $c_{\mathrm{o\_g}}^{\mathrm{rad}}$ is the heat load coefficient due to solar radiation. Canopy transpiration $\varphi_{\mathrm{H_2O,c\_a}}$ in kg[H$_2$O] m^{-2}[gh] s^{-1} is governed by the equation

$$\varphi_{\mathrm{H_2O,c\_a}} = \left(1 - e^{-c_{\mathrm{LAI,W}} W}\right) c_{\mathrm{c\_a}}^{\mathrm{evap}} \left(\frac{c_{\mathrm{H_2O,1}}^{sat}}{c_R\left(T_{\mathrm{a}}^{\mathrm{c}} + c_{\mathrm{T}}\right)} e^{\frac{c_{\mathrm{H_2O,2}}^{\mathrm{sat}} T_{\mathrm{a}}^{\mathrm{c}}}{T_{\mathrm{a}}^{\mathrm{c}} + c_{\mathrm{H_2O,3}}^{\mathrm{sat}}}} - C_{\mathrm{H_2O,a}} \right) \quad (\mathrm{kg[H_2O]\ m^{-2}[gh]\ s^{-1}}) \tag{6.9}$$

in which the term $\frac{c_{\mathrm{H_2O,1}}^{\mathrm{sat}}}{c_R\left(T_{\mathrm{a}}^{\mathrm{c}} + c_{\mathrm{T}}\right)} e^{\frac{c_{\mathrm{H_2O,2}}^{\mathrm{sat}} T_{\mathrm{a}}^{\mathrm{C}}}{T_{\mathrm{a}}^{\mathrm{c}} + c_{\mathrm{H_2O,3}}^{\mathrm{sat}}}}$ in kg[H$_2$O] m^{-3} represents the saturated water vapor content at canopy temperature, i.e., air temperature $T_{\mathrm{a}}^{\mathrm{C}}$, $c_{\mathrm{c\_a}}^{\mathrm{evap}}$ is the mass transfer coefficient in m s^{-1}, $c_{\mathrm{H_2O,1}}^{\mathrm{sat}}$ in J m^{-3}, $c_{\mathrm{H_2O,2}}^{\mathrm{sat}}$ and $c_{\mathrm{H_2O,3}}^{\mathrm{sat}}$ in °C parameterize the saturation water vapor pressure, c_{R} is the gas constant in J K^{-1} kmol^{-1}, and c_{T} parameterizes the conversion of temperature from °C to K. The mass transfer of water vapor by ventilation $\phi_{\mathrm{H_2O,g\_o}}$ in kg[H$_2$O] m^{-2}[gh] s^{-1} is described by

$$\varphi_{\mathrm{H_2O,g\_o}} = \left(u_{\mathrm{v}} + c_{\mathrm{leak}}\right)\left(C_{\mathrm{H_2O,a}} - C_{\mathrm{H_2O,o}}\right) \quad (\mathrm{kg[H_2O]\ m^{-2}[gh]\ s^{-1}}) \tag{6.10}$$

in which $C_{\mathrm{H_2O,o}}$ in kg[H$_2$O] m^{-3}[air] is the humidity concentration outside the greenhouse.

Model parameters are listed in Table 6.1. The model, though being of rather simple structure was found to describe measured data rather well. For a more detailed description and verification of this model, the reader is referred to Van Henten (1994).

6.3 THE OPTIMAL CONTROL PROBLEM

In this example, for the single harvest crop lettuce, the net economic revenue of the controlled crop production process is described by

$$J = \Phi\left(\mathbf{x}\left(t_{\mathrm{f}}\right)\right) - \int_{t_0}^{t_{\mathrm{f}}} \left(c_{\mathrm{q}}^{\mathrm{cost}} u_{\mathrm{q}}(t) + c_{\mathrm{CO_2}}^{\mathrm{cost}} u_{\mathrm{CO_2}}(t)\right) \mathrm{d}t \quad (\mathrm{Dfl\ m^{-2}[gh]}) \tag{6.11}$$

where $\Phi(\mathbf{x}(t_{\mathrm{f}}))$ in Dfl m^{-2}[gh] is the gross income obtained at harvest time t_{f} when selling the harvested product at the auction, and $c_{\mathrm{q}}^{\mathrm{cost}} u_{\mathrm{q}}(t) + c_{\mathrm{CO_2}}^{\mathrm{cost}} u_{\mathrm{CO_2}}(t)$ is the running costs of the climate conditioning equipment in Dfl m^{-2}[gh] s^{-1}.

In The Netherlands, lettuce is sold at auctions in grades based on the fresh weight and on the quality of the produce. Despite the fact that quality aspects have a significant effect on the value of the produce, quantitative relations between the greenhouse climate and crop quality that are needed to derive optimal greenhouse climate control strategies are not well developed. Therefore, in this study quality aspects were neglected and attention was focused on a quantitative relation between the harvest weight and auction price of a lettuce crop. Analysis of the auction price of lettuce in the period 1985–1990 revealed a linear relationship $\Phi(\mathbf{x}(t_{\mathrm{f}})) = c_{\mathrm{W,1}}^{\mathrm{income}} + c_{\mathrm{W,2}}^{\mathrm{income}} W(t_{\mathrm{f}})$, parameterized by $c_{\mathrm{W,1}}^{\mathrm{income}}$ in Dfl m^{-2}[gh] and $c_{\mathrm{W,2}}^{\mathrm{income}}$ in Dfl kg^{-1}[dw] m^{-2}[gh], between the auction price and the harvest weight of lettuce W in kg[dw] m^{-2}[gh]. The positive correlation found means that a higher harvest weight obtained, for example, by taking suitable climate control measures is rewarded by a higher gross economic return.

The running costs of the climate conditioning equipment are assumed to be linearly related to the amount of energy u_q in W m^{-2}[gh] and the amount of carbon dioxide u_{CO_2} in kg[CO_2] m^{-2}[gh] s^{-1} put into the system. These running costs are parameterized by the energy price c_q^{cost} in Dfl J^{-1} and the price of carbon dioxide $c_{CO_2}^{cost}$ in Dfl kg^{-1}[CO_2], respectively. Their values are given in Table 6.1. It is assumed that no costs were associated with natural ventilation used for cooling and dehumidification. The contributions of the electrical equipment used for greenhouse climate conditioning, such as pumps and valves, to the operating costs are ignored. Furthermore, it is assumed that other production factors, such as the nutrient and water supply, screening and those not directly related to greenhouse climate control, such as labor input, pest and disease control, do not affect the control strategies. Consequently, they are not included in the performance criteria.

Equation 6.11 states that the net economic return of the crop production process to be optimized is defined as the difference between the gross economic return of the crop production process and the operating costs of the climate conditioning equipment integrated over the whole growing period starting at the planting date t_0 and ending at harvest time t_f.

Physical limitations on the control inputs u_{CO_2}, u_q, and u_v are represented by the linear inequality constraints $u_{CO_2}^{min} \le u_{CO_2} \le u_{CO_2}^{max}$, $u_q^{min} \le u_q \le u_q^{max}$, $u_v^{min} \le u_v \le u_v^{max}$, respectively. Bounds are also imposed on the temperature in the greenhouse T_a^C, the carbon dioxide concentration $C_{CO_2,a}$, and the humidity level $C_{H_2O,a}$ to prevent the control system from driving the process into unfavorable conditions for crop growth and development. These bounds are represented by the linear inequality constraints $T_a^{C,min} \le T_a^C \le T_a^{C,max}$, $C_{CO_2,a}^{min} \le C_{CO_2,a} \le C_{CO_2,a}^{max}$, and $C_{H_2O,a}^{min} \le C_{H_2O,a} \le C_{H_2O,a}^{max}$. In fact, these bounds represent the limitations of the rather simple crop growth model used in this research, because the adverse effect of unfavorable climate conditions on crop growth and development should have been covered by the dynamic crop growth model. In the example considered, bounds are imposed on the relative humidity instead of the absolute humidity. This requires the transformation $C_{H_2O,a}^{min} = \frac{R_{H_2O,a}^{max}}{100} C_{H_2O,a}^{sat}(T_a^C)$ and $C_{H_2O,a}^{max} = \frac{R_{H_2O,a}^{max}}{100} C_{H_2O,a}^{sat}(T_a^C)$ with $R_{H_2O,a}^{min}$ and $R_{H_2O,a}^{max}$ being the lower and upper bound on the relative humidity, respectively, and the saturation water vapor pressure $C_{H_2O,a}^{sat}$ in kg[H_2O] m^{-3}[air] is a function of the temperature T_a^C:

$$C_{H_2O,a}^{sat} = \frac{c_{H_2O,4}^{sat}}{c_R(T_a^C + c_T)} \exp\left(\frac{c_{H_2O,2}^{sat} T_a^C}{T_a^C + c_{H_2O,3}^{sat}}\right) \quad (\text{kg[H}_2\text{O] m}^{-3}\text{[air]}) \quad (6.12)$$

where $c_{H_2O,4}^{sat}$ in J m^{-3} parameterizes the saturation water vapor pressure together with the parameters $c_{H_2O,2}^{sat}$ and $c_{H_2O,3}^{sat}$.

To deal with the inequality constraints on the state variables (Pierre, 1969), the performance criterion of Equation 6.11 is extended with penalty functions $P_x(t)$ in Dfl m^{-2}[gh] s^{-1}, having the general form

$$P_x(t) = c^{penalty}\left[\frac{2x(t) - x^{min} - x^{max}}{x^{max} - x^{min}}\right]^{2k} \quad (\text{Dfl m}^{-2}\text{ [gh]}) \quad (6.13)$$

where $c^{penalty}$ Dfl m^{-2}[gh] s^{-1} is a weighting factor, x is a state variable, x^{min}, and x^{max} are the lower and the upper bounds put on the state variable, respectively, and the exponent k forces the penalty function to attain values near zero between the bounds and very steep slopes close to the bounds when $k = 1, 2, \ldots, \infty$. In this way, the controlled system is prevented from traversing the bounds. To guarantee consistency in the units used, the penalty is expressed in Dfl m^{-2}[gh] s^{-1}. In a way this is a slightly artificial construction, though one may argue that by modifying the weighting parameter

c^{penalty} the grower is able to express his attitude toward taking risks when the health of the crop is considered.

After adding penalties for violations of the constraints on the temperature, carbon dioxide concentration, and humidity, P_T, P_{CO_2}, and P_{H_2O} respectively, the resulting performance measure has the following form:

$$J=\left(c_{W,1}^{\text{income}}+c_{W,2}^{\text{income}}W\left(t_f\right)\right)-\int_{t_0}^{t_f}\left(c_q^{\text{cost}}u_q(t)+c_{CO_2}^{\text{cost}}u_{CO_2}(t)+P_{CO_2}(t)+P_T(t)+P_{H_2O}(t)\right)\mathrm{d}t \quad (\text{Dfl m}^{-2}\,[\text{gh}]) \tag{6.14}$$

With the preliminaries presented above, the optimal control problem is defined as to find optimal control strategies for the control variables u_{CO_2}, u_q, and u_v over the time-interval $t \in [t_b, t_f]$, maximizing the performance criterion of Equation 6.14, subject to the differential equation constraints of Equations 6.1 through 6.4 and the linear inequality constraints on the controlled variables.

For the example considered, the Hamiltonian H has the following form:

$$H=-c_{CO_2}^{\text{cost}}u_{CO_2}-c_q^{\text{cost}}u_q+\lambda_W\left\{c_{\alpha\beta}\varphi_{CO_2,a\_c}-c_{W,c\_a}W2^{\left(0.1T_a^c-2.5\right)}\right\}+$$

$$\lambda_{CO_2}^f\left\{\frac{1}{c_{CO_2}^{\text{cap}}}\left[-\varphi_{CO_2,a\_c}+c_{CO_2,c\_a}W2^{\left(0.1T_a^c-2.5\right)}+u_{CO_2}-\varphi_{CO_2,g\_o}\right]\right\}+\lambda_T\left\{\frac{1}{c_a^{\text{cap}}}\left[u_q-q_{g\_o}^{\text{trans,vent}}+q_{o\_g}^{\text{rad}}\right]\right\}$$

$$+\lambda_{H_2O}\left\{\frac{1}{c_{H_2O}^{\text{cap}}}\left[\varphi_{H_2O,c\_a}-\varphi_{H_2O,g\_o}\right]\right\}-P_{CO_2}-P_T-P_{H_2O} \quad (\text{Dfl m}^{-2}\,[\text{gh}]) \tag{6.15}$$

in which λ_W in Dfl kg^{-1}[dw], $\lambda_{CO_2}^f$ in Dfl m kg^{-1}[CO_2], λ_T^f in Dfl m^{-2}[gh] °C^{-1}, and $\lambda_{H_2O}^f$ in Dfl m kg^{-1}[H_2O] are so-called adjoint variables or costates related to the state variables W, $C_{CO_2,a}$, T_a^C and $C_{H_2O,a}$. The dynamics of the costates are described by the general equation:

$$-\dot{\boldsymbol{\lambda}}=\frac{\partial H}{\partial \mathbf{x}} \quad (\text{Dfl m}^{-2}\,[\text{gh}]\,\text{s}^{-1}) \tag{6.16}$$

where $\boldsymbol{\lambda}$ is the costate and $\mathbf{x}$ is the state variable. The costates express the marginal value of a change in the associated state variables. If a costate is positive, an increment of the associated state variable will have a positive effect on the final net economic return, and vice versa. The Hamiltonian can be seen as a momentary profit rate in which current costs are balanced against future revenues. In this way the Hamiltonian is a great source of information for interpretation of the optimal control problem and the results of the sensitivity analysis in Sections 6.4.1 and 6.4.3.

The Maximum Principle of Pontryagin asserts that to maximize the performance criterion in Equation 6.14 it is sufficient to maximize the Hamiltonian at all time instants in the optimization interval, i.e.,

$$H(\mathbf{x}^{s*},\mathbf{x}^{f*},\mathbf{u}^*,\boldsymbol{\lambda}^{s*},\boldsymbol{\lambda}^{f*},t)\geq H(\mathbf{x}^{s*},\mathbf{x}^{f*},\mathbf{u},\boldsymbol{\lambda}^{s*},\boldsymbol{\lambda}^{f*},t) \tag{6.17}$$

in which $\mathbf{x}^{s*}$, $\mathbf{x}^{f*}$, $\mathbf{u}^*$, $\boldsymbol{\lambda}^{s*}$, and $\boldsymbol{\lambda}^{f*}$ are the optimal values of the states, control inputs, and costates. Here $\mathbf{x}^s$ and $\boldsymbol{\lambda}^s$ refer to the slow state W and its associated costate λ_W; $\mathbf{x}^f$, and $\boldsymbol{\lambda}^s$ refer to the fast state variables $C_{CO_2,a}$, $C_{H_2O,a}$, and T_a^C and their respective costates.

In the iterative solution, the state and costate equations were simulated in double-precision with an integration time step of half a minute using a fourth order Runge–Kutta algorithm described by Press et al. (1986). A modified steepest ascent algorithm exploiting the gradient information $\partial H/\partial \mathbf{u}$ was used for the iterative solution of the optimal control problem (Kirk, 1970; Van Henten, 1994).

6.4 OPTIMAL CONTROL CASE STUDIES

6.4.1 Analysis of the Optimal Control Problem

Building on the work of Van Henten (1994), in this section the structure and particular characteristics of the optimal control problem are investigated.

The performance criterion J, Equation 6.14, expresses a net economic return of crop production, evaluated over the whole production period. In dynamic optimization, the performance criterion J is replaced by a local performance criterion, the Hamiltonian, Equation 6.15. The Hamiltonian accounts for the momentary contribution of a control action to the running costs L and long-term revenues from a change in the state due to a control action, $\boldsymbol{\lambda}^T\mathbf{f}$. Here $\mathbf{f}$ is the generalized system equation. The revenues of an investment in a change of the state are determined by the costate or marginal value of a state variable $\boldsymbol{\lambda}$.

The marginal value of a state variable at a certain moment during the growing period is equivalent with the sensitivity of the economic return at harvest time, i.e., J, to a small variation of that particular state at that time in the growing period. In view of that interpretation, the absolute value and sign of a costate are of special interest. Given the definition of the performance criterion as in Equation 6.14, a positive costate value indicates benefits from a positive change in the associated state, a negative costate value indicates that losses will result from an increment of the state at that particular moment in time. A large (small) positive costate value indicates large (small) benefits from a positive change in the associated state. If at a given moment in time a costate equals zero, the performance of the controlled process is not affected by a change of the associated state variable at that time.

So, from the value and sign of the costate it can be deduced whether a change in the associated state will result in an improvement or a loss of economic performance of the controlled process. The economic feasibility of a controlled change of the state, however, is determined by the stationarity condition

$$\frac{\partial H}{\partial \mathbf{u}} = -\frac{\partial L}{\partial \mathbf{u}} + \left(\frac{\partial \mathbf{f}}{\partial \mathbf{u}}\right)^T \boldsymbol{\lambda} = \mathbf{0} \qquad (6.18)$$

This equation asserts that an optimal greenhouse climate control action is characterized by the fact that its marginal contribution to the gross economic return expressed by $(\partial f/\partial \mathbf{u})^T\boldsymbol{\lambda}$ just outweighs its marginal contribution to the momentary operating costs $(\partial L/\partial \mathbf{u})$. This notion is depicted in Figure 6.2 for $\lambda = 1$ and $\lambda = 2$. Because $\partial f/\partial u > 0$, a higher marginal value yields a higher optimal control and vice versa.

For unconstrained optimal control problems, the stationarity condition coincides with the Maximum Principle of Pontryagin, which asserts that a maximizing control should satisfy

$$H(\mathbf{x}^*, \mathbf{u}^*, \boldsymbol{\lambda}^*, t) \geq H(\mathbf{x}^*, \mathbf{u}, \boldsymbol{\lambda}^*, t) \qquad (6.19)$$

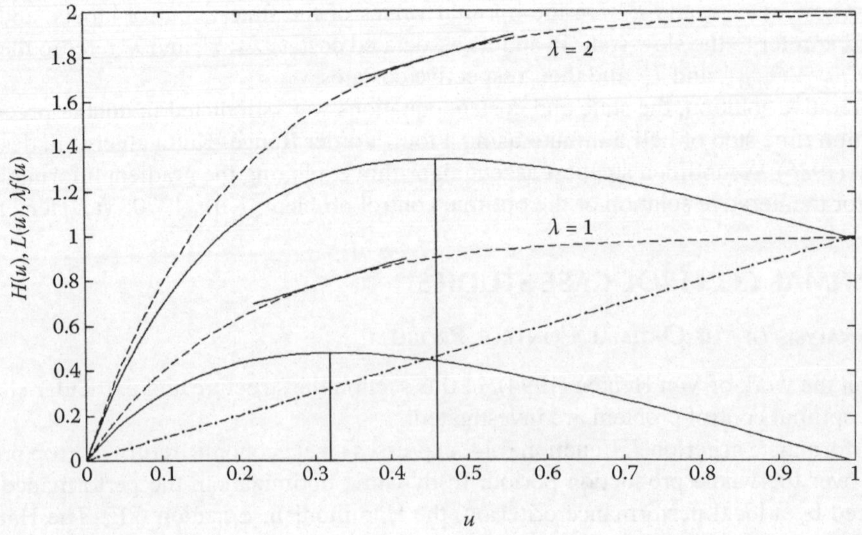

FIGURE 6.2 Optimal control: the balance between additional gross income $\lambda \partial f/\partial u$ and additional costs $\partial L/\partial u$ yields the optimal net return, (–) denotes the Hamiltonian H, (–·) denotes the operating costs L, and (---) denotes the gross income λf for $\lambda = 1$ and $\lambda = 2$. The tangents at the gross income curve indicate where $\partial f/\partial u = \partial L/\partial u$, i.e., where H is maximized, the maximum of H is indicated by vertical lines.

However, if the control is constrained by, e.g., inequality constraints of the form $\mathbf{0} = \mathbf{u}^{\min} \le \mathbf{u} \le \mathbf{u}^{\max}$, then a control that satisfies the maximum principle does not automatically satisfy the stationarity condition. This can be seen as follows. Assume that at a certain moment during the growing period the marginal value of the state is zero, i.e., $\lambda = 0$, then we can see in Figure 6.2 that the maximizing control will be $u = u^{\min} = 0$ because any increase in the control will yield smaller values of the Hamiltonian. While Equation 6.19 has been satisfied, Equation 6.18 is not satisfied by the maximizing control, because $-\partial L/\partial u < 0$, which contradicts the requirement defined in Equation 6.18. An identical argument will show the limitation of the stationarity condition if an upper bound $u^{\max}$ is encountered. Still, for controls lying between the constraint boundaries, both Equations 6.18 and 6.19 constitute the same necessary condition for optimality of the control and may be used simultaneously in an analysis for understanding of the optimal control problems.

The partial derivatives of the Hamiltonian (Equation 6.15) with respect to the control inputs being the basis of the stationarity condition reveal the following properties. The economic optimal operation of the carbon dioxide supply is solely determined by the marginal value of the carbon dioxide concentration in the greenhouse, i.e., $\lambda^{\mathrm{f}}_{\mathrm{CO_2}}$. However, because the ventilation affects all three greenhouse climate states—carbon dioxide concentration, air temperature, and humidity—the operation of the ventilation windows is affected by their respective costates $\lambda^{\mathrm{f}}_{\mathrm{CO_2}}$, $\lambda^{\mathrm{f}}_{\mathrm{T}}$, and $\lambda^{\mathrm{f}}_{\mathrm{H_2O}}$. The dry matter production is not affected by the control inputs as can be seen in Equation 6.1, and therefore the marginal value of crop dry weight λ_W does not influence the operation of the climate conditioning equipment directly. However, λ_W is indirectly involved in the operation of the climate conditioning equipment by its impact on the marginal values of the carbon dioxide concentration, air temperature, and humidity through the equation:

$$-\dot{\boldsymbol{\lambda}}^{\mathrm{f}} = \frac{\partial H}{\partial \mathbf{x}^{\mathrm{f}}} \tag{6.20}$$

Concentrating on carbon dioxide supply and heating first, Equations 6.2 and 6.3 reveal the trivial observation that carbon dioxide supply increases the carbon dioxide concentration and heating increases the temperature in the greenhouse. Both effects are effectively summarized by saying that $\partial \mathbf{f}/\partial \mathbf{u} > \mathit{0}$ in which $\mathbf{f}$ is the general system equation or model. If the costates related with the carbon dioxide concentration and air temperature are less than zero, i.e., $\lambda^{\mathrm{f}}_{\mathrm{CO_2}}, \lambda^{\mathrm{f}}_{\mathrm{T}} < 0$, then the marginal contribution of the control to the economic return at the end of the growing period will be less than zero, i.e., $(\partial \mathbf{f}/\partial \mathbf{u})\boldsymbol{\lambda}^{\mathrm{f}} < \mathbf{0}$, because $\partial \mathbf{f}/\partial \mathbf{u} > \mathit{0}$. Additionally, application of carbon dioxide and heating energy to the greenhouse yields higher running costs of the climate conditioning equipment, i.e., $\partial L/\partial \mathbf{u} > \mathbf{0}$. Consequently, using the arguments in line with the Maximum Principle of Pontryagin, this implies that the control input resulting in the largest value of the Hamiltonian will be $u_{\mathrm{CO_2}}, u_{\mathrm{q}} = 0$. Summarizing, carbon dioxide and heating energy will not be supplied to the greenhouse if the marginal contribution of the carbon dioxide concentration and air temperature in the greenhouse to the final economic return is less than zero.

However, if $\lambda^{\mathrm{f}}_{\mathrm{CO_2}} > 0$ and $\lambda^{\mathrm{f}}_{\mathrm{T}} > 0$, i.e., if the carbon dioxide and air temperature positively affect the final economic return, application of carbon dioxide and heating energy to the greenhouse may be economically feasible. Then, referring to Figure 6.2 and the interpretation of the stationarity condition given at the beginning of this section, the marginal contribution of the control to the economic return should outweigh its marginal contribution to the costs. From the Hamiltonian in Equation 6.15 and the necessary conditions in Equations 6.18 and 6.19 it can be deduced that for an optimizing control $u_{\mathrm{CO_2}}$ and u_{q} larger than zero, the costates should have values $\lambda^{\mathrm{f}}_{\mathrm{CO_2}} \geq c^{\mathrm{cost}}_{\mathrm{CO_2}} c^{\mathrm{cap}}_{\mathrm{CO_2}}$ and $\lambda^{\mathrm{f}}_{T} \geq c^{\mathrm{cost}}_{q} c^{\mathrm{cap}}_{a}$. This requirement illustrates that if the running costs of the climate conditioning equipment, i.e., $c^{\mathrm{cost}}_{\mathrm{CO_2}}$ and $c^{\mathrm{cost}}_{\mathrm{q}}$, increase, the marginal contribution of the carbon dioxide concentration and air temperature to the final economic return, need to be higher before actual operation of the climate conditioning equipment becomes economically feasible. In other words, the control strategies have an energy conservation attitude in situations with high energy costs. Alternatively, a reduction of the running costs will yield a more generous operation of the climate conditioning equipment. Also, $\lambda^{\mathrm{f}}_{\mathrm{CO_2}} \geq c^{\mathrm{cost}}_{\mathrm{CO_2}} c^{\mathrm{cap}}_{\mathrm{CO_2}}$ and $\lambda^{f}_{T} \geq c^{\mathrm{cost}}_{\mathrm{q}} c^{\mathrm{cap}}_{\mathrm{a}}$ indicate that for economic optimal greenhouse climate operation the mass and heat capacity of the greenhouse ($c^{\mathrm{cap}}_{\mathrm{CO_2}}$ and $c^{\mathrm{cap}}_{\mathrm{a}}$), i.e., the greenhouse volume, should be kept as small as possible, because the equations show that with a higher energy and mass capacity the marginal contribution of the carbon dioxide concentration and the air temperature to the final economic return need to be higher before heating and carbon dioxide is supplied to the greenhouse. With large mass and heat capacities, it will require more energy and carbon dioxide to attain a unit increase of the carbon dioxide concentration in the greenhouse.

For the ventilation, the analysis yields a slightly different picture because no operating costs are related with the ventilation and the impact of ventilation on the indoor climate is different from the influence of the heating and carbon dioxide supply. Generally, the air temperature, carbon dioxide, and humidity levels in the greenhouse are higher than outside the greenhouse and a higher ventilation rate results in reduction of the air temperature, carbon dioxide concentration, and humidity in the greenhouse, i.e., $\partial \mathbf{f}/\partial \mathbf{u} < \mathit{0}$. Sometimes, however, it may happen that due to a very high crop photosynthesis rate, the carbon dioxide concentration in the greenhouse air is lower than the carbon dioxide concentration in the outside air (Schapendonk and Gaastra, 1984). Then, a higher ventilation rate will yield a higher carbon dioxide concentration in the greenhouse, i.e., $\partial \mathbf{f}/\partial \mathbf{u} > \mathit{0}$.

If the carbon dioxide concentration and air temperature have a positive contribution to the final economic return, i.e., $\lambda^{\mathrm{f}}_{\mathrm{CO_2}}, \lambda^{\mathrm{f}}_{\mathrm{T}} > 0$, for an optimal ventilation rate the marginal contribution of the control to the final economic return, i.e., $(\partial \mathbf{f}/\partial \mathbf{u})\boldsymbol{\lambda}^{\mathrm{f}}$, should be larger or equal to zero. If the carbon dioxide concentration and air temperature in the greenhouse exceed the outside conditions, any increment in the ventilation rate will yield a loss of carbon dioxide and energy from the inside air ($\partial \mathbf{f}/\partial \mathbf{u} < \mathit{0}$). Thus the ventilation rate which yields the largest value of the Hamiltonian is $u_{\mathrm{v}} = 0$. If, however, $\lambda^{\mathrm{f}}_{\mathrm{CO_2}} > 0$ and the carbon dioxide concentration in the outside air exceeds the concentration in the greenhouse, then ventilation may be used to open a "cheap" carbon dioxide source to improve the inside climate. The optimal ventilation rate will equalize the inside carbon dioxide concentration with the outside concentration such that $\partial \mathbf{f}/\partial \mathbf{u} = \boldsymbol{0}$, thus, satisfying the stationarity condition.

A positive value of the humidity costate $\lambda^{f}_{H_2O}$ is not considered here, because it will be shown in Section 6.4.3 that the marginal value of humidity does not exceed zero in the control problem considered in this research.

If the marginal value of the carbon dioxide concentration, air temperature, and humidity is less than zero, i.e., $\lambda^{f}_{CO_2}, \lambda^{f}_{T}, \lambda^{f}_{H_2O} < 0$, and the inside climatic conditions exceed the outside conditions, the previous line of reasoning will yield exactly the opposite result. If for instance $\lambda^{f}_{T} < 0$, a reduction of the air temperature will yield a positive contribution to the performance. If the inside temperature is higher than the outside temperature, an increment in the ventilation flux will contribute to a reduction of the air temperature, $\partial \mathbf{f}/\partial \mathbf{u} < \boldsymbol{0}$. Because the marginal contribution of the control to the final economic return should be larger or equal to zero, the ventilation rate will be increased until $\lambda^{f}_{T} < 0$, or the inside temperature equals the outside temperature, thus resulting in $\partial \mathbf{f}/\partial \mathbf{u} = \boldsymbol{0}$. For the carbon dioxide concentration and humidity, similar arguments can be used.

The previous analysis has shed some light on the operation of the climate conditioning equipment depending on the marginal values of carbon dioxide, air temperature, and humidity. What remains is to get insight into how the costates $\lambda^{f}_{CO_2}$, λ^{f}_{T}, and $\lambda^{f}_{H_2O}$ are affected for instance by the state of the crop and the outside climatic conditions.

In the control problem considered, Equation 6.14 shows that the gross economic return is solely determined by the state of the crop at harvest time, i.e., $W(t_f)$. It is expected that the costates associated with the air temperature, carbon dioxide concentration, and humidity will be determined by their effect on dry matter production and consequently the final gross economic return.

In this research it is assumed that humidity does not directly affect crop growth. And if, for the moment, the humidity constraint is neglected, for instance by setting $c^{\text{penalty}}_{H_2O} = 0$, humidity is eliminated from the Hamiltonian and does not affect the economic performance of the controlled process anymore. Consequently, $\lambda^{f}_{H_2O} = 0$. The effect of constraints on the costates will be investigated below.

Figure 6.3 shows the response of the rate of dry matter production, i.e., Equation 6.1, to the carbon dioxide concentration in the greenhouse at different light levels. During the day, any increment

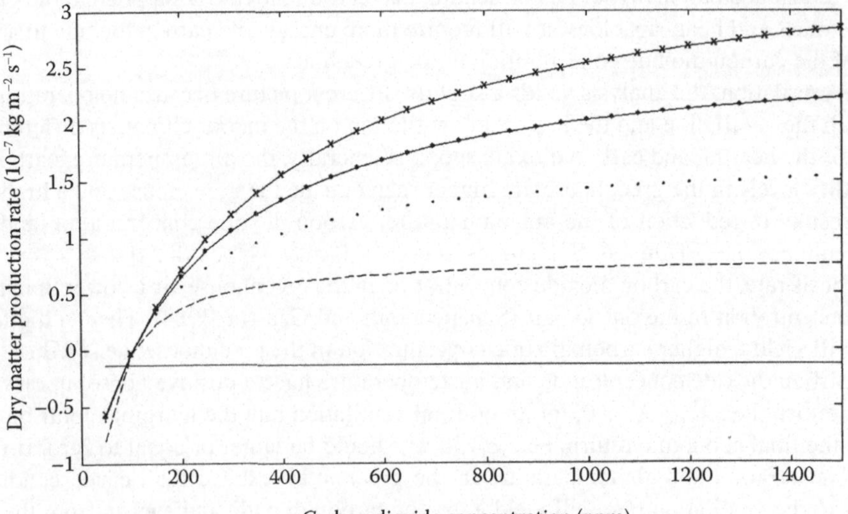

FIGURE 6.3 The response of the dry matter production rate to the carbon dioxide concentration, after canopy closure, under different light regimes (–: 0 W m^{-2}, ---: 50 W m^{-2}, ···: 100 W m^{-2}, –·–: 150 W m^{-2}, –x–: 200 W m^{-2}).

in carbon dioxide concentration contributes to an increase in dry matter production which suggests that during the day $\lambda^{f}_{CO_2} > 0$. If the radiation level increases, an increment in the carbon dioxide level in the greenhouse yields an even higher dry matter production rate, which indicates that $\lambda^{f}_{CO_2}$ is positively related to the radiation level. At night the carbon dioxide concentration does not affect dry matter production because the photosynthesis rate is zero and the marginal value of carbon dioxide is then zero.

In Figure 6.4, the response of the crop growth to air temperature is shown. This rate has an optimum response to temperature during the day. Below the optimum temperature, an increment in the air temperature contributes to a higher dry matter production and consequently to a higher gross economic return. So, during the day, the marginal value of the air temperature is expected to be larger than zero, i.e., $\lambda^{f}_{T} < 0$. However, for an air temperature exceeding the optimum temperature, the marginal value of the air temperature will be less than zero because any increment in the air temperature will yield a reduction in the dry matter production rate. Observe that at low temperatures the slope of the response curve is positively correlated with the radiation level, which suggests a positive correlation of λ^{f}_{T} with radiation. At night a temperature rise will result in higher respiratory losses of dry matter, thus affecting the gross economic return in a negative way. Therefore, the marginal value of air temperature is expected to be less than zero at night. Consequently, additional heating does not seem to be profitable at night. This seems to contradict standard horticultural practice where during lettuce production the greenhouse is heated at night to achieve a minimum temperature between 7 and 10°C. This indicates a potential flaw in the crop growth model used in this research.

In the example considered, the marginal value of the humidity, i.e., $\lambda^{f}_{H_2O}$, is completely determined by the constraint imposed on the humidity level in the greenhouse. If constraints are imposed on the humidity level in the greenhouse, i.e., $c^{penalty}_{H_2O} > 0$, the humidity affects the performance through a penalty function. If the humidity stays within the constraint boundaries, the penalty on the performance will be almost zero and consequently the marginal value of humidity will be approximately zero. However, if the humidity level approaches, for instance, the constraint boundary $C^{max}_{H_2O,a}$, then any increase in the humidity level will yield a reduction of the performance criterion J, which will result in a distinct negative marginal value of humidity, i.e., $\lambda^{f}_{H_2O} < 0$ at that time.

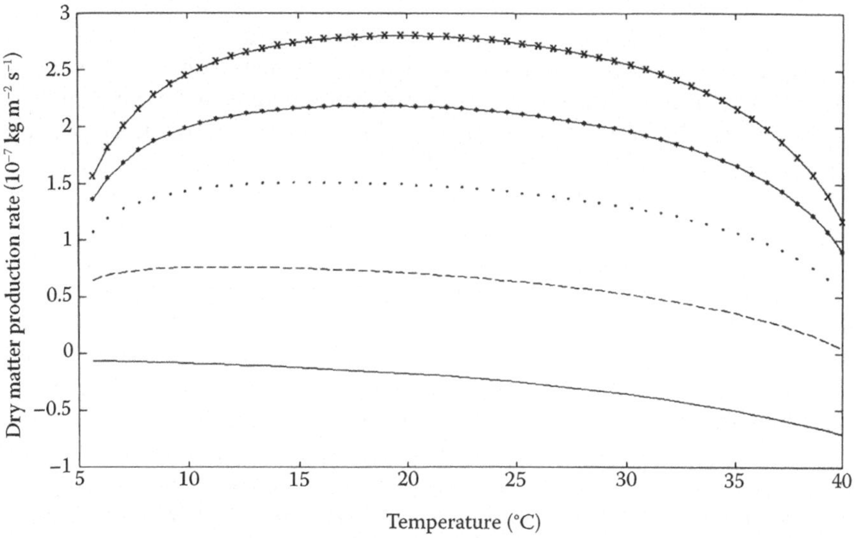

FIGURE 6.4 The response of the dry matter production rate to air temperature, after canopy closure, under different light regimes (–: 0 W m^{-2}, ---: 50 W m^{-2}, ···: 100 W m^{-2}, –·–: 150 W m^{-2}, –x–: 200 W m^{-2}).

For the carbon dioxide concentration and air temperature, the constraints have the same impact on the respective costates. Constraint violations will have a strong negative influence on the marginal value.

It has been shown that the costates λ_T^f and $\lambda_{CO_2}^f$ uniquely determine the operation of heating and CO_2 supply, respectively. In the operation of the ventilation windows, however, contradicting objectives are encountered. It may occur during the day that $\lambda_{CO_2}^f > 0$ demands a reduction in ventilation rate to reduce losses of carbon dioxide to the outside air, whereas at the same time $\lambda_{H_2O}^f < 0$ demands a higher ventilation rate to reduce the humidity level in the greenhouse and consequently to reduce the penalty on the performance of the controlled process. Obviously, a trade-off between both objectives will result in an optimum ventilation rate under the given circumstances.

Until now, the effect of single variable analysis of the operation of the climate conditioning equipment was performed. A closer inspection of the costate equations for $\lambda_{CO_2}^f$, λ_T^f, and $\lambda_{H_2O}^f$ will reveal that the marginal values of the various state variables are determined by a complex interaction of effects of state variables, costates, and external inputs. A clear example is the equation describing the dynamics of λ_T^f. Analysis of this equation shows, for instance, that the humidity constraint has a positive effect on the marginal value of the air temperature. A higher temperature will result in a reduction of the humidity constraint penalty. This positive effect suggests that heating may be used as a means to reduce the relative humidity in the greenhouse. When humidity control is considered, this positive costate of air temperature results in contradicting objectives with respect to ventilation. Then the necessary conditions in Equations 6.18 and 6.19 applied to the Hamiltonian in Equation 6.15 show that the positive λ_T^f induces a reduction of ventilation whereas, at the same time the negative $\lambda_{H_2O}^f$ will demand a higher ventilation rate. Possibly, heating and ventilation will be used simultaneously to achieve the single objective of controlling humidity in the greenhouse.

6.4.2 Comparison of Optimal Control with Climate Control Supervised by a Grower

Following Van Henten (1994) and Van Henten et al. (1997), a comparison is made between optimal control strategies and control strategies implemented by growers.

6.4.2.1 Materials and Methods

In an experimental greenhouse at DLO-Institute of Agricultural Engineering (IMAG-DLO) a lettuce crop was grown from January 21, 1992, until March 17, 1992. The four-span Venlo-type experimental greenhouse was oriented East–West and had a floor area of approximately 300 m^2. The roof consisted of single glass panes with 20 half pane ventilation windows on lee and windward sides. A hot water heating system consisting of four pipes per span was mounted parallel to the gutters at a height of approximately 2.0 m. In the greenhouse, a distribution network of one hose per span was used to supply carbon dioxide from a storage tank.

Lettuce plants were sown and grown at a nursery in peat blocks and then planted at a density of 18 plants per square meter of soil in a recirculating nutrient film technique system (NFT) consisting of 13 gutters per two spans. The commonly grown lettuce cultivar "Norden" was used.

Using an updated version of the IMAG-DLO computer control system implemented on a Digital PDP-11/73 (Van Meurs, 1980), the greenhouse climate was controlled according to the rules followed in normal horticultural practice. During the first few days of the cultivation period, the day and night temperature setpoints were 14°C. Then, the night temperature was lowered to 10°C, whereas the day air temperature setpoint was at least 14°C and increased dependent on the solar radiation level. During the day, carbon dioxide was supplied to a maximum concentration of 750 ppm depending on the amount of solar radiation and the opening of the ventilators. With a separate computer, the nutrient solution was controlled to have an electrical conductivity (EC) of around 2.3 mS and a pH of around six. At regular intervals during the growing period, the grower was advised by a commercial extension service. In this way the crop was grown using standard horticultural practice.

Using a data logging system connected to the greenhouse climate computer, measurements of the indoor climate, outdoor climate, and actuators of the climate conditioning system were recorded. The measurements of the indoor climate included single spot measurements of air temperature and humidity for which dry and wet bulb thermometers (ventilated and radiation shielded) were used. A spatial average value of the carbon dioxide concentration in the greenhouse was measured with a Siemens infrared absorption spectrometer.

Recordings of the actuators of the climate control system included the mean value of the inlet and outlet temperatures of the heating system, the valve position of the carbon dioxide supply system and the window aperture of both lee and windward side ventilation windows.

Outside the greenhouse, solar radiation, air temperature, relative humidity, and wind speed were measured with a Kipp solarimeter, dry and wet bulb thermometers (ventilated and radiation shielded), and a cup anemometer, respectively.

The measurements of the actuators, the indoor climate, and crop growth obtained during the greenhouse experiment in early 1992, represent the behavior of the controlled process using conventional greenhouse climate strategies defined by the grower. These data have been used as a reference in the comparison with optimal control strategies obtained by simulation.

This case study focused on the slow dynamics in the crop production process. Therefore, the data of the actuators, indoor climate, and outdoor climate were averaged over periods of half an hour. The performance of the grower's approach to greenhouse climate management was evaluated by simulating the system Equations 6.1 to 6.4, neglecting the dynamics of the greenhouse climate, and by calculating the value of the performance measure (Equation 6.14) and using the energy price and the actual price obtained at the auction.

The performance of the greenhouse climate control strategy of the grower was compared with two optimization runs in which complete knowledge about the auction price, the energy price, and the outdoor climate for the whole growing period was considered. In the first simulation "Optimal 1," the optimal control problem was solved with time-invariant constraints on air temperature, humidity, and carbon dioxide concentration. In greenhouse practice, the operation of the ventilators is determined to a certain extent by requirements on the humidity level in the greenhouse. Because using a time-invariant constraint on the relative humidity does not reflect practical management of humidity in greenhouses, in a second run, hereafter referred to as "Optimal 2," optimal control trajectories were calculated for heating and carbon dioxide supply only, omitting the humidity constraint and using the measured ventilation regime used by the grower, to control the humidity level in the greenhouse.

6.4.2.2 Results

In Figure 6.5, performance data of the three control approaches are presented on a relative basis. The data include the simulated harvest weight, the energy consumption, the carbon dioxide consumption, and the net economic return which, in the context of the present research, is defined as the difference between the value of the crop at harvest time and the climate conditioning costs integrated over the whole growing period.

In terms of net economic return, a considerable difference in performance is observed between the grower's climate control strategies and the optimal control ones. The results of the Optimal 1 simulation experiment, in which the relative humidity was limited by an upper bound of 90%, indicate a higher dry matter production and less energy and carbon dioxide consumption compared to the grower's results. In simulation run Optimal 2, in which the ventilation strategy of the grower was used to control the relative humidity in the greenhouse, the energy and carbon dioxide consumption are smaller than the counterpart results of the grower. Although fresh weight production in this simulation was approximately the same as that of the grower, the reduced carbon dioxide and energy consumption yielded a higher net economic return. These simulations show that with optimal control, energy and carbon dioxide are used more efficiently.

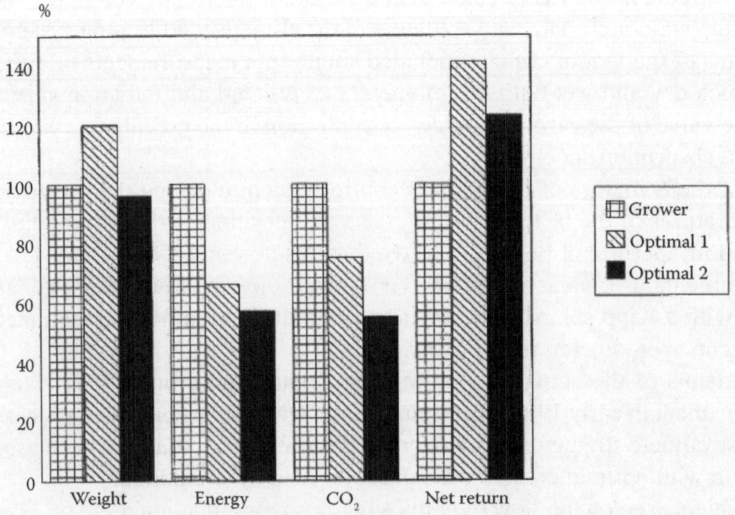

FIGURE 6.5 The overall performance of the controlled crop production process using the grower's climate control strategies (Grower), optimal strategies using a humidity constraint (Optimal 1), and optimal strategies using the ventilation regime of the grower (Optimal 2).

Table 6.2 gives the total carbon dioxide and energy consumptions and ventilation exchanges over the full production cycle for the three different control strategies. In simulation Optimal 1, a lower ventilation rate during the day was calculated than was used by the grower. However, at night, optimal ventilation was much higher. Although in simulation Optimal 1 less carbon dioxide was consumed than the grower had used, the reduced ventilation rate during the day resulted in a higher carbon dioxide concentration in the greenhouse, thus yielding the observed higher fresh weight production. Another distinct difference between the optimal and grower's climate control strategies is the reduced energy consumption during the day and, to a lesser extent, during the night (Table 6.2). Using the ventilation regime of the grower (Optimal 2) during both day and night, less energy and about half the amount of carbon dioxide were used.

Further insight into the differences between the grower's management strategies and the optimal control strategies in greenhouse climate management is obtained by comparing the measured and calculated control and state trajectories.

TABLE 6.2
Total Carbon Dioxide Consumption, Energy Consumption, and Ventilation During the Whole Growing Period in Early 1992 with Greenhouse Climate Control According to the Grower (Grower), Optimal Control with Humidity Constraint (Optimal 1), and Optimal Control without Humidity Constraint, Using the Measured Ventilation Trajectories Implemented by the Grower (Optimal 2)

	Carbon Dioxide Consumption (kg[CO_2] m^{-2})		Energy Consumption (MJ m^{-2})		Ventilation (m^3 m^{-2})	
	Day	Night	Day	Night	Day	Night
Grower	1.23	-	105	127	5439	2955
Optimal 1	0.94	-	43	110	3519	6298
Optimal 2	0.68	-	45	88	5439	2955

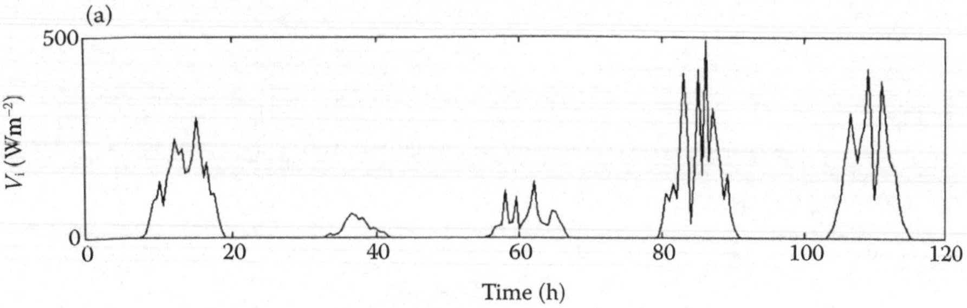

FIGURE 6.6 Solar radiation (I_o) over a 5-day period.

Averaged measured data of solar radiation are presented in Figure 6.6 for a period of 5 representative days. The time course of carbon dioxide supply, ventilation air exchange, and heat consumption of the grower's experiment and those corresponding to the Optimal 1 simulation are shown in Figure 6.7. The simulated greenhouse climate variables driven by the external climatic conditions of Figure 6.6 and control inputs of Figure 6.7 are presented in Figure 6.8.

Figures 6.6 through 6.8 can help clarify some of the differences found in Figure 6.5 and Table 6.2. For example, Figure 6.7a shows that, using optimal control strategies, the carbon dioxide supply responds to the solar radiation in a different way than that using grower's control strategies. Due

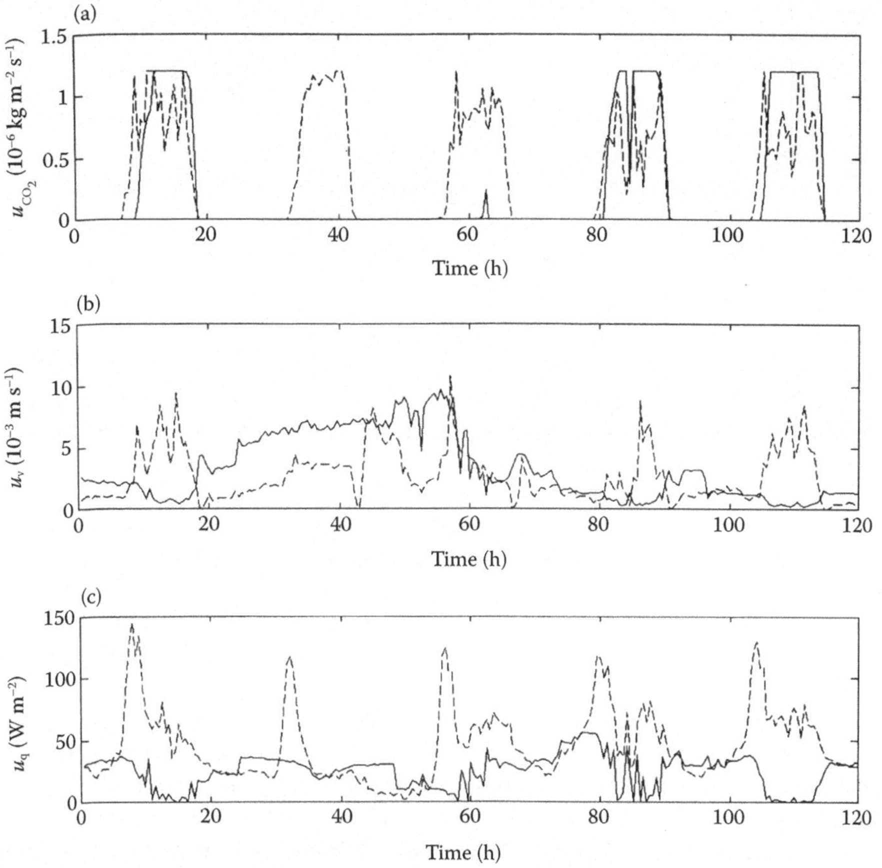

FIGURE 6.7 Trajectories of carbon dioxide supply rate u_{CO2} (a), ventilation rate u_v (b), and heating energy u_q (c), corresponding to the grower (---) and to optimal control strategy Optimal 1 (—).

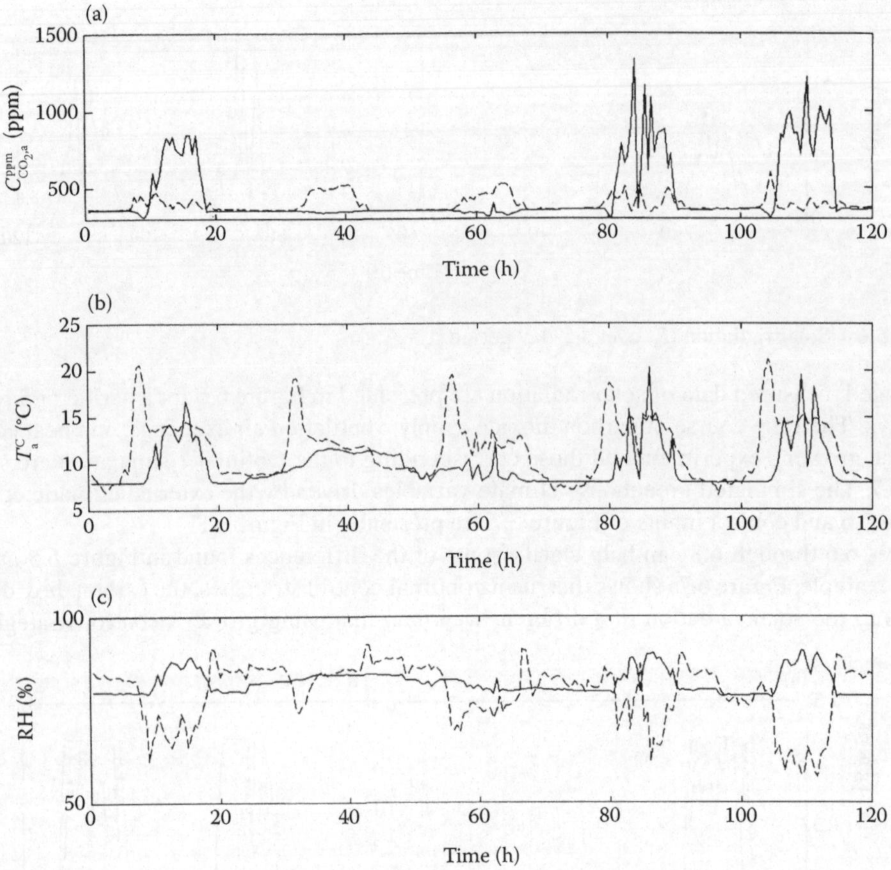

FIGURE 6.8 Trajectories of (a) carbon dioxide concentration $C^{ppm}_{CO_2,a}$, (b) air temperature T^C_a, and (c) relative humidity RH, corresponding to the grower's climate control strategy (---) and to the optimal control strategy Optimal 1 (—).

to the unfavorable radiation conditions as well as the high ventilation rates during days 2 and 3, in the optimal control approach the supply of carbon dioxide to the greenhouse air was not considered profitable under these circumstances. In the greenhouse climate control strategy implemented by the grower, the carbon dioxide setpoint was adapted to the solar radiation as well as to the ventilation rate. The carbon dioxide supply in days 2 and 3 was more than that using optimal control strategies. On the contrary, the control strategy implemented by the grower did not enrich adequately with carbon dioxide under the favorable radiation conditions of days 4 and 5. During this period, the grower seemed to prefer a high ventilation rate to reduce the relative humidity in the greenhouse (see Figure 6.7b and c) and consequently reduced the carbon dioxide supply rate to prevent excessive losses of carbon dioxide to the outside air.

In Figure 6.8a it is shown that when using optimal control strategies under favorable circumstances, the carbon dioxide concentration in the greenhouse exceeded 1000 ppm whereas the grower used an upper limit of 750 ppm to limit the carbon dioxide consumption.

The high ventilation regime implemented by the grower during the day was mainly intended to reduce the humidity level in the greenhouse and consequently to prevent fungal diseases and physiological damage, such as marginal spot. A relative humidity as low as 60% can be seen in Figure 6.8c. In the optimal control approach, however, the ventilation rate was reduced during the day to achieve a more efficient use of the carbon dioxide supplied. Consequently, a higher relative humidity (90%) was encountered than in practice. At night the differences in the humidity levels were found to be rather small.

With regard to the air temperature in the greenhouse, the grower used a minimum value of 14°C during the day. On average, this setpoint was almost equivalent to the calculated optimal air temperature during the five days shown. Due to the high ventilation regime used by the grower during the day, more heating energy was needed to realize the air temperature setpoint which explains to a certain extent the high daytime energy consumption observed in Table 6.2. The optimal air temperature trajectories in Figure 6.8b also suggest that adjustment of the indoor temperature to the outside climatic conditions such as solar radiation may improve the efficiency of greenhouse climate management.

The difference in energy consumption at night between the grower's strategies and the optimal control strategies is partly explained by the fact that, especially during the first two weeks of the growing period, the grower used an air temperature setpoint of 14°C. Further analysis revealed that with the particular crop growth model used, heating is not considered profitable at night (Van Henten, 1994). The resulting high air temperature has a negative effect on the dry matter production due to increased maintenance respiration at high temperatures. Therefore at night the air temperature was determined by the lower bound constraint so that values as low as 7°C were simulated. Also in the optimal control approach, the heat pulse at sunrise, implemented by the grower to activate the crop was not considered economically feasible because the possible benefits in terms of crop quantity or quality of this approach were not described by the model used. Clearly, the heat pulse implemented by the grower contributed to a higher energy consumption.

In a qualitative sense, the optimal control strategies used in simulation Optimal 2 (not shown) yielded the same results as those calculated in simulation Optimal 1. The improved efficiency of greenhouse climate control was achieved by a more efficient use of carbon dioxide and a reduction of the energy consumption due to the absence of the heat pulse at sunrise and the lower air temperature at night. Although less heat energy was used during the day, even though the ventilation regime of the grower was adopted, only a slightly higher humidity level in the greenhouse was reached than with the optimal control strategies.

6.4.2.3 Discussion

The differences in efficiency between the optimal control strategies and the grower's control strategies for greenhouse climate management are significant and one may argue whether such improvements can be achieved in practice. The following observations are made.

Albeit the crop growth model used for the calculation of the optimal control strategies quite accurately simulates crop fresh weight production, it does not account for other aspects related to crop quality, such as head formation, and the occurrence of physiological damage and fungal diseases under humid conditions. These deficiencies may affect the favorable results of the optimization, for example, in the following way. In greenhouse management practice, the high humidity levels calculated in simulation "Optimal 1" (in which the time-invariant constraint on the humidity was imposed) may be unfavorable for the quality of a lettuce crop. However, in simulation Optimal 2 (in which the ventilation regime adopted by the grower was used to control the humidity level) it was shown that carbon dioxide as well as heating energy were still used more efficiently. Furthermore, the optimal carbon dioxide concentrations and air temperatures, calculated in Optimal 1, were reasonable and are not expected to have an adverse effect on lettuce growth. Therefore, the major trends of the results reported in this chapter are still expected to hold.

Still, the optimal control approach strongly relies on an appropriate model of the process to be controlled. Therefore, to expand the ideas presented in this chapter, further research in the field of modeling the greenhouse crop production process is required. But, also, feedback control based on moving horizon optimal control techniques can effectively deal with uncertainty and model inaccuracy in nonlinear optimal control problems (e.g., Yang and Polak, 1993).

The optimal control simulations represented an ideal situation because the control trajectories were calculated after the greenhouse experiment had ended using complete knowledge about the outside climatic conditions as well as the auction price. In practice, these external factors have to be predicted and inaccuracies in these predictions may reduce the benefits of optimal control suggested in this chapter. Still, they are not expected to alter the major trends of these results.

Although further research is needed, these uncertainty issues have already received attention in the literature. The potential of predicting auction prices was investigated by Van Henten (1994). An analysis of the auction price of lettuce revealed that the estimates of the parameters used to describe the linear relation between harvest weight and product price of lettuce showed a very distinct seasonal pattern with a high auto-correlation function which offers the opportunity to predict the auction price using for example an auto-regressive model. Favorable results of short term weather forecasting using time series analysis were reported by Brown et al. (1984) and Huang and Chalabi (1994). A potential benefit is expected from the conjunction of these short-term weather forecasts with long-term meteorological weather forecasts. Also, as stated earlier, moving horizon optimal control techniques can effectively deal with uncertainty in nonlinear optimal control problems (e.g., Yang and Polak, 1993). This was confirmed by Tap et al. (1996) and Tap (2000) who reported on the effectiveness of this approach in dealing with the uncertainty in the outdoor climate conditions.

It appears that constraints, especially the humidity ones, play an important role in optimal greenhouse climate control. Hence, a more accurate assessment of the effect of humidity and other microclimatic variables on the quality and quantity of crop production, either in terms of model equations or in terms of (time-variant) constraints, is required.

In the present analysis the greenhouse climate dynamics were neglected based on the premise that only the slow trends in the outside climatic conditions were considered. In reality, rapid fluctuations in the outside climatic conditions do occur and their impact on optimal greenhouse climate management has not been considered in this study. Investigation of the effect of this assumption, which is commonly made in greenhouse climate optimization (e.g., Critten, 1991; Seginer, 1991; Van Henten and Bontsema, 1991; Bailey and Chalabi, 1994), has not yet been conclusive (Tap et al., 1993; Van Henten, 1994; Ioslovich et al., 1995). Therefore the main results obtained in this study are believed to hold. However, when online optimal control of the greenhouse climate is considered, the rapid fluctuations in the process and thus, for reasons of stability, the greenhouse climate dynamics have to be accounted for. A computational framework based on a two time-scale decomposition of the greenhouse climate control described in Section 6.4.4 and Van Henten and Bontsema (1992, 1996, 2009) and Van Henten (1994), deals effectively with the greenhouse climate dynamics as well. The applicability of this approach has been confirmed by Tap et al. (1996) and Tap (2000).

In the past, the considerable computer power required for the numerical solution of the optimal control problem has been an obstacle for the practical online implementation of optimal greenhouse control strategies. Recently, online control experiments convincingly showed that with the current performance of relatively inexpensive digital computers, this obstacle is circumvented (Van Meurs and Van Henten, 1994; Tap et al., 1996).

6.4.2.4 Concluding Remarks

In this chapter a comparative analysis, supported by experimental work, was carried out to determine the potential improvement in economics and energy efficiency in using optimal control strategies in greenhouse climate management over a whole growing season of a lettuce crop. The results obtained support the conclusion that a considerable improvement in the efficiency of greenhouse climate management is possible. This improvement may well exceed 15%. Clearly, the final test of the merits of optimal control has to be obtained in full scale validation experiments in the greenhouse.

6.4.3 Sensitivity Analysis of the Optimal Control Problem

Following the work of Van Henten (1994, 2003), the sensitivity of the performance of the optimal control strategies with respect to parameter variations is investigated.

6.4.3.1 Materials and Methods

Using variational arguments, first-order approximations of the performance sensitivity were derived by Courtin and Rootenberg (1971) and Evers (1979, 1980). The performance sensitivity with respect to the values of the state variables at the beginning of the growing period $\mathbf{x}(t_b)$, equals:

$$\frac{\partial J}{\partial x_i(t_0)} = \lambda_i^*(t_0), \quad i = 1, \ldots, n \tag{6.21}$$

where $\lambda_i^*(t_0)$ is the optimal value of the costate at the start time t_b and n is the number of state variables. The performance sensitivity with respect to the model parameters $\mathbf{c}$ is

$$\frac{\partial J}{\partial c_j} = \int_{t_0}^{t_f} \frac{\partial H}{\partial c_j}\left(\mathbf{x}^*, \mathbf{u}^*, \mathbf{c}^{\text{nom}}, t\right)\mathrm{d}t, \quad j = 1, \ldots, m \tag{6.22}$$

in which $\mathbf{x}^*$ and $\mathbf{u}^*$ are the optimal state and control trajectories, respectively, $\mathbf{c}^{\text{nom}}$ denotes the nominal value of the model parameter and m, represents the total number of model parameters.

The calculation of the first-order sensitivity contained two steps. First of all, the open-loop optimal control problem was solved. Then secondly, the effect of small perturbations of the model parameters on the performance measure was evaluated using the above-mentioned first-order measure of the performance sensitivity with respect to parameter perturbations. For the model parameters, this first-order measure was obtained by integrating the partial derivatives of the Hamiltonian with respect to the model parameters over the whole optimization interval. In the actual computation, these partial derivatives can be calculated analytically or numerically with a central difference approximation (e.g., Gill et al., 1981). In this research, analytical derivatives were obtained and implemented in the simulation software based on FORTRAN. For the initial conditions of the state variables, the performance sensitivity was determined by the value of the associated costates at the starting time t_0.

To compare the impact of perturbations in the different model parameters and the initial conditions of the state variables on the system performance, it was considered to be more convenient to express the sensitivity as the fractional change in the performance criterion as a result of the fractional change in the parameter value, i.e., a relative sensitivity criterion. For every state variable and model parameter, the relative sensitivity measure was defined as

$$\frac{\partial J}{\partial x_i(t_0)} \frac{x_i^{\text{nom}}(t_0)}{J^*}, i = 1, \ldots, n, \text{ and } \frac{\partial J}{\partial c_j} \frac{c_j^{\text{nom}}}{J^*}, \quad j = 1, \ldots, m \tag{6.23}$$

By doing so, the interpretation of the results became straightforward. A relative sensitivity measure larger (less) than zero indicated that a small positive perturbation in the parameter resulted in an increase (decrease) of the value of the performance criterion. To be more precise, a value of the relative sensitivity measure of 1, indicated that a parameter change of 1% should result in a 1% change of the value of the performance criterion. For a relative performance sensitivity measure having a value larger or smaller than unity, the interpretation changed accordingly. As the first-order sensitivity analysis was based on a first-order Taylor series approximation of the change in performance due to a change in a parameter, the validity of the previous interpretation was limited to small parameter variations only. Still, the relative performance sensitivity measure should provide valuable insight into the contribution of certain model parameters to the control strategies calculated.

Besides the performance sensitivity with respect to variations in the initial conditions and model parameters, the performance sensitivity with respect to small perturbations in the external inputs was evaluated. This was accomplished by multiplying each external input with a time-invariant parameter, i.e., $c_{CO_2\_o}$, c_{I_0}, $c_{H_2O\_o}$ and c_{T_o}, having a nominal value of one. Clearly, the effect of this perturbation parameter was that throughout the whole optimization interval the external inputs are perturbed by the same amount. This may not sound fully realistic but it should give an impression of the relative importance of the external inputs on the performance of the optimal controlled process. In the sensitivity analysis, these perturbation parameters were treated as normal model parameters.

6.4.3.2 Results and Discussion

The results of the sensitivity analysis are presented in Table 6.3. These results show that the constraint on the relative humidity seems to play a primary role in the performance of the optimal control problem. The importance of the humidity state constraint is indicated by the large sensitivity measure of c_R, c_T, $c^{sat}_{H_2O,2}$, $c^{sat}_{H_2O,3}$ and $c^{sat}_{H_2O,4}$, parameterizing the saturation water vapor pressure used in the definition of the constraint on the relative humidity. Although these parameters are also involved in the description of the canopy transpiration, their effect on the performance of the control strategies through the canopy transpiration seems to be of secondary importance. This can be inferred from the sign of the sensitivity measure. Taking c_R as an example, this can be seen as follows. An increment in the value of c_R results in a reduction of the saturation water vapor pressure. When the humidity constraint is encountered during daytime, the costate $\lambda^f_{H_2O,a}$ takes large negative values as can be seen in Figure 6.9 and indicated in Section 6.4.1. This indicates the required reduction of the humidity in the greenhouse. Focusing on the humidity balance in the Hamiltonian equation, it can be seen that given $\lambda^f_{H_2O,a} < 0$ and assuming the saturation water vapor pressure to be larger than the absolute humidity level in the greenhouse, any reduction in the water vapor pressure would result in a larger value of the Hamiltonian, thus suggesting a positive effect of an increment in c_R on the performance measure. The sign of the sensitivity measure, however, is negative. Therefore another effect of c_R is dominating. Close to the humidity constraint, the partial derivative of the penalty function takes very large positive values. Then, any reduction of the saturation water vapor pressure will result in an increasing penalty, thus yielding the negative effect on the performance measure observed in the sensitivity analysis. Apparently, the penalty function related to the humidity constraint dominates the Hamiltonian, thus emphasizing the importance of an accurate definition of the humidity constraint in optimal greenhouse climate control.

As the constraint on the relative humidity is of such great importance in the control strategies, an accurate description of the humidity balance in the greenhouse, including processes like canopy transpiration, seems required. This is confirmed by the relatively large performance sensitivity of parameter $c^{evap}_{c\_a}$ expressing the mass transfer coefficient for evaporative water vapor transport from the leaves to the ambient air. Under equal circumstances, a small positive increment in this parameter will result in a higher canopy transpiration and, consequently, it will result in an earlier conflict with the humidity constraint. The accompanying increment in the value of the penalty results in no sensitivity measure in Table 6.3.

In this example of lettuce cultivation, the gross economic return is determined by the dry matter production. Table 6.3 shows that most of the crop-related parameters more or less affect the performance of the control strategies. The sensitivity analysis of a lettuce growth model reported by Van Henten and Van Straten (1994) revealed the importance of parameters such as $c_{\alpha\beta}$, $c_{LAI,W}$, $c^{phot}_{I_0}$ and $c^{phot}_{CO_2,2}$. Consequently, in the present study a significant performance sensitivity for perturbations in these parameters was expected as well. Apart from the parameter $c_{LAI,W}$, Table 6.3 shows the expected relatively large performance sensitivity for these parameters, thus emphasizing the fact that for optimal greenhouse climate control their accurate parameterization is required. The

TABLE 6.3
Relative Performance Sensitivity of the Model Parameter in Decreasing Order of Their Absolute Values

Parameter	Relative Sensitivity	Parameter	Relative Sensitivity
$c_{H_2O,2}^{sat}$	5.4178	$c_{LAI,W}$	–0.1542
$c_{H_2O,4}^{sat}$	4.5173	$c_{CO_2,3}^{phot}$	–0.1399
$c_{H_2O,3}^{sat}$	–3.9328	c_{leak}	–0.1116
$c_{CO_2\_o}$	1.6637	$c_{cap,q,v}$	–0.0958
$c_{\alpha\beta}$	1.7807	c_{T_o}	0.0963
c_{I_0}	1.2627	$c_{H_2O,cap}$	0.0919
$c_{I_0}^{rad}$	1.1783	$W(t_b)$	0.0600
$c_{H_2O\_o}$	–1.0804	$c_{CO_2,cap}$	–0.0500
c_R	–1.0800	c_{CO_2,c_a}	0.0148
c_T	–1.0349	c_T^{phot}	–0.0123
$c_{CO_2,2}^{phot}$	0.8742	$c_{a,cap}$	–0.0095
$c_{CO_2,1}^{phot}$	–0.3668	$c_{H_2O,1}^{sat}$	–0.0064
c_q	–0.3617	$T_a^c(t_b)$	0.0007
$c_{g\_o}^{trans,vent}$	–0.3418	$c_{o\_g}^{rad}$	0.0004
$c_{c\_a}^{evap}$	–0.3313	$C_{H_2O\_a}(t_b)$	–0.0003
c_{W,c_a}	–0.2772	$C_{CO_2\_a}(t_b)$	0.0001
c_{CO_2}	–0.1672		

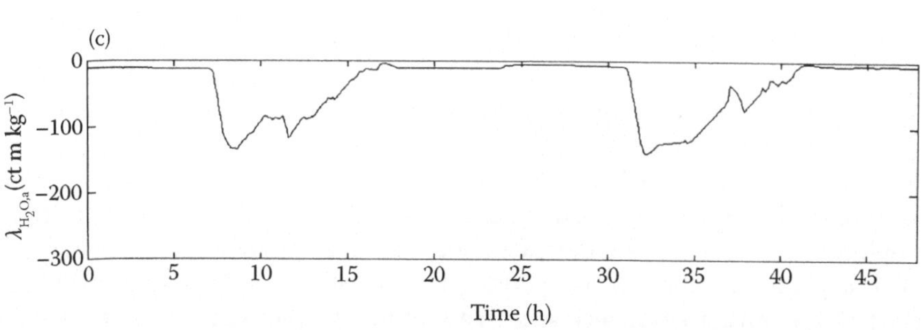

FIGURE 6.9 The costate trajectory of the humidity during 2 days of the 50 days optimization interval.

performance sensitivity for parameter $c_{LAI,W}$, however, is much less distinct than was expected. The reason for this is the fact that $c_{LAI,W}$ is also involved in the humidity balance of the greenhouse in which it describes the effective surface of the canopy. Before canopy closure, any increment of the canopy transpiration will result in more frequent conflicts with the humidity constraints. This has a negative effect on the performance measure, thus partly outweighing the positive effects of early canopy closure on dry matter production of the canopy.

Because the important role of the humidity has been discussed in some detail above, further analysis is focused on the parameters in the energy and carbon dioxide balances. Parameters that most clearly affect the dynamic behavior of the carbon dioxide concentration and air temperature in the greenhouse are the mass and heat capacities $c_{CO_2}^{cap}$ and c_a^{cap}, respectively. Table 6.3 reveals that the effect of a perturbation in their values on the performance is relatively small. The heat and mass capacities of the greenhouse air determine to a large extent the dynamic rate with which the greenhouse climate can be modified: a larger capacity will result in a longer response time. The small performance sensitivity suggests that the greenhouse climate system is fast enough to deal with fast fluctuations in the external inputs in an economic optimal fashion. This can be seen as follows. If large benefits can be obtained by rapid modifications of the greenhouse climate anticipating rapid fluctuations in the external inputs such as the solar radiation, any decrease in response time will contribute to a significant improvement of the economic performance. Then, a pronounced performance sensitivity with respect to these parameters is expected. Compared with, for instance, crop growth related parameters, however, $c_{CO_2}^{cap}$ and c_a^{cap} have a small impact on the performance. Apparently, the response time of the greenhouse climate is not a limiting factor in the economic optimal control of the crop production process. Or alternatively, the relatively small performance sensitivity to changes in the heat and mass capacity of the greenhouse air suggest that in economic optimal greenhouse climate control very fast modifications of the greenhouse climate do not contribute much to an improvement of the economic performance.

These observations are in line with the results of the sensitivity analysis of the two state variable crop growth model done by Van Henten and Van Straten (1994). Then it was found that crop growth is much more sensitive to changes in the long-term average of, for instance, the carbon dioxide concentration than to rapid fluctuations. Because in this example the performance of optimal greenhouse climate control is largely determined by the dry matter production, this would suggest a large performance sensitivity for parameters affecting the average indoor climate. In the present sensitivity analysis, this is confirmed by the significant sensitivity of the performance measure to a change in the carbon dioxide concentration in the outside air induced by $c_{CO_2\_o}$. Clearly, such a change does affect the long-term average carbon dioxide concentration in the greenhouse air, but not so much its dynamic rate of change.

In the sensitivity analysis of Van Henten and Van Straten (1994), it was concluded that during the day lettuce growth is not strongly influenced by the air temperature in the greenhouse. Due to this relatively low temperature sensitivity of crop growth and the comparably high heating costs, the greenhouse air is rarely heated during the day. Still, parameter $c_{g\_o}^{trans,vent}$, describing the energy losses to the outside air by transmission through the greenhouse cover and natural ventilation through the windows, shows a high performance sensitivity. During the major part of the growing period, heating energy is supplied to the greenhouse at night to satisfy the minimum temperature constraint. This minimum temperature constraint to a large extent determines the total energy consumption. Any reduction in the energy loss to the outside air will result in a reduction of the energy consumption required for heating the greenhouse. This results in a negative performance sensitivity. As the greenhouse climate is not exposed to rapid changes in the outside conditions during nighttime, economic optimal control does not require extremely fast modifications of the greenhouse air temperature. This explains the relatively low performance sensitivity of the heat capacity c_a^{cap}.

The very large positive performance sensitivity for perturbations in the solar radiation and outside carbon dioxide concentration is explained by the large sensitivity of crop growth for these climatic conditions. The large negative sensitivity for an increase in the outside humidity is due to the constraint on the relative humidity, which will then be more difficult to satisfy. The large performance sensitivities emphasize the need for accurate assessment, i.e., prediction and measurement, of these outside climatic conditions.

The total operating costs of the climate conditioning equipment (±1.00 Dfl m^{-2}[gh]) is relatively small compared with the gross economic return of the crop production (±5.25 Dfl m^{-2}[gh]). Therefore, a relatively small performance sensitivity for the operating costs expressed by the parameters $c_{CO_2}^{cost}$ and c_q^{cost} is found. Because in the simulations the overall heating costs exceeded the costs for carbon dioxide supply, the performance sensitivity for $c_{CO_2}^{cost}$ is smaller than for c_q^{cost}.

For all state variables the performance sensitivity to a perturbation in their initial conditions is relatively small. For the greenhouse climate variables this is explained by the fact that due to the fast system dynamics a small perturbation in the greenhouse climate has a very short lifetime that does not affect the performance of the system significantly. The low sensitivity of the initial dry weight of the crop, however, is unexpected. The start weight of the crop would be expected to have a significant positive effect on the final economic return. In this example, this still will be the case. In the optimization, however, rapid crop growth also has an adverse effect on the performance because the size of the crop has a positive influence on canopy transpiration and thus will result in humidity constraint violations. In this example, this negative effect partly outweighed the positive impact of a large start weight.

Finally, it is interesting to note that some of the insights obtained in this research are in line with experiences from horticultural practice. As in the example in this research, in Dutch horticultural practice humidity considerations play a dominant role in ventilation control. In practice, abundant ventilation is used as a risk aversion strategy against physiological disorder and diseases, but the underlying mechanisms are not yet very well understood. This research clearly emphasizes that a more prudent use of ventilation for humidity control can have a significant positive effect on the economic performance of greenhouse crop production. In the past few years this notion has been gaining interest in horticultural practice as well. Another example is the thermal heat loss through the greenhouse cover, in this research represented by the parameter $c_{g\_o}^{trans,vent}$. The high performance sensitivity confirms that growers rightly choose to reduce energy losses through the greenhouse cover by means of thermal isolation.

6.4.3.3 Concluding Remarks

With respect to the methodology used in this research, the following conclusions are drawn.

First of all, it was found that the first-order sensitivity analysis is a simple and straightforward way to obtain deeper insight into the operation of the optimal control problem and the relative importance of the model parameters, the initial conditions of the state variables and the external inputs, without having to go through extensive recalculations of the optimal control strategies. Secondly, the present study confirmed that a sensitivity analysis of the model to parameter variations and a sensitivity analysis of an optimal control problem including the same model might yield different results. Though the process model has an undisputed role in optimal control, it is the balancing of various objectives such as costs, revenues, and penalties that determine the optimal control strategies. Finally, the intermediate variables in the solution of the optimal control problem such as the Hamiltonian and the costate trajectories, were found to be instrumental for a better understanding of the role of the model and model parameters in the determination of the optimal control strategies.

Application of the sensitivity analysis to an optimal control problem in greenhouse crop production led to the following conclusions:

1. The constraint on the humidity strongly influences the performance of optimal greenhouse climate management. Because these constraints were included as a first step to deal with adverse effects of high relative humidity on the quality of the crop, future research on greenhouse climate management should focus on a proper assessment of these effects in terms of quantitative models or modified climate constraints.

2. In optimal greenhouse climate management, the dynamics of crop growth play a dominant role and require accurate models of crop growth and development.
3. The relatively small performance sensitivity to changes in the heat and mass capacities of the greenhouse air indicate that the response time of the greenhouse climate is not a limiting factor for economic optimal control.
4. The outside climate conditions such as solar radiation, carbon dioxide concentration, and humidity, and to a lesser extent the temperature, are important in greenhouse climate management. Consequently their accurate measurement and prediction is required.

6.4.4 Time-Scale Decomposition

Following the work of Van Henten (1994), Van Henten and Bontsema (2009) and the methodology description of Chapter 4, this section presents the results of the time-scale decomposition of the optimal greenhouse climate management problem.

6.4.4.1 Materials and Methods

In this example a growing period of 50 days was considered. Two-minute measurements of outdoor climate conditions, including solar radiation, wind speed, temperature and humidity, obtained during a growing experiment in early 1992, were used as external inputs during the solution (Van Henten, 1994).

Three optimization runs were done. First, using the systems equations in Equations 6.1 through 6.4, the performance measure of Equation 6.14, and the resulting Hamiltonian of Equation 6.15, the full problem was solved to yield trajectories for all state variables W, $C_{CO_2,a}$, $C_{H_2O,a}$, and T_a^C, associated costates λ_W, $\lambda_{CO_2,a}$, $\lambda_{H_2O,a}$, and $\lambda_{T_a^C}$, as well as a value of the performance measure J. These results were used to evaluate the decomposition.

Second, the so-called slow subproblem was solved. In this subproblem the dynamics of the slow state variable W was explicitly taken into account, but for the fast state variables $C_{CO_2,a}$, $C_{H_2O,a}$, and T_a^C the dynamic model Equations 6.2 through 6.4 were reduced to their quasi-steady-state form. As a performance measure, again Equation 6.14 was used. This slowsubproblem yielded trajectories of $\bar{W}$, $\bar{C}_{CO_2,a}$, $\bar{C}_{H_2O,a}$, and $\bar{T}_a^c$, their associated costates $\bar{\lambda}_W$, $\bar{\lambda}_{CO_2,a}$, $\bar{\lambda}_{H_2O,a}$, and $\bar{\lambda}_{T_a^C}$, as well as the control inputs $\bar{u}_{CO_2}$, $\bar{u}_q$, and $\bar{u}_v$. The bar indicates the slow subproblem.

Third, using the trajectories of $\bar{W}$ and $\bar{\lambda}_W$, the so-called fast subproblem was solved. This optimization problem took explicitly into account the dynamic model equations of the fast state variables in Equations 6.2 through 6.4. For this optimization problem the modified performance measure was derived in accordance with the methodology described in Van Henten (1994), Van Henten and Bontsema (2009), and Chapter 4.

The decomposition was evaluated using the following criteria:

- The optimal control trajectories of the fast subproblem should approximate the optimal control trajectories of the original full optimal control problem, i.e., $\tilde{\mathbf{u}}^*(t) \cong \mathbf{u}^*(t)$.
- The evolution of the slow state and costate variables simulated with the optimal control trajectories of the fast subproblem should approximate the state trajectories of the full control problem, i.e., $\mathbf{x}(\tilde{\mathbf{u}}^*) \cong \mathbf{x}(\mathbf{u}^*)$ and $\boldsymbol{\lambda}(\tilde{\mathbf{u}}^*) \cong \boldsymbol{\lambda}(\mathbf{u}^*)$.
- The evolution of the fast state and costate variables calculated in the fast subproblem should approximate the evolution of the fast state variables calculated in the full control problem, i.e., $\tilde{\mathbf{x}}^f(t) \cong \mathbf{x}^f(t)$ and $\tilde{\boldsymbol{\lambda}}^f(t) \cong \boldsymbol{\lambda}^f(t)$.
- The performance of the optimal control system using the solution of the fast subproblem $\tilde{\mathbf{u}}(t)$ should approximate the performance of the full optimal control system, i.e., $J(\tilde{\mathbf{u}}) \cong J(\mathbf{u})$.

6.4.4.2 Results

In Figure 6.10a it is shown that the crop growth trajectory obtained in the slow sub-problem agrees well with the evolution of the crop state obtained in the solution of the full control problem, though an underestimation does occur. However, if the full system dynamics were simulated using the optimal controls of the fast subproblem, a more accurate approximation was obtained. The same can be observed for the slow costate trajectories (see Figure 6.10b). Figure 6.11 shows the rapid fluctuations of the solar radiation over two days during the 50-day production cycle. Figure 6.12 presents, for the same two days, the optimal state trajectories of the carbon dioxide concentration in the greenhouse.

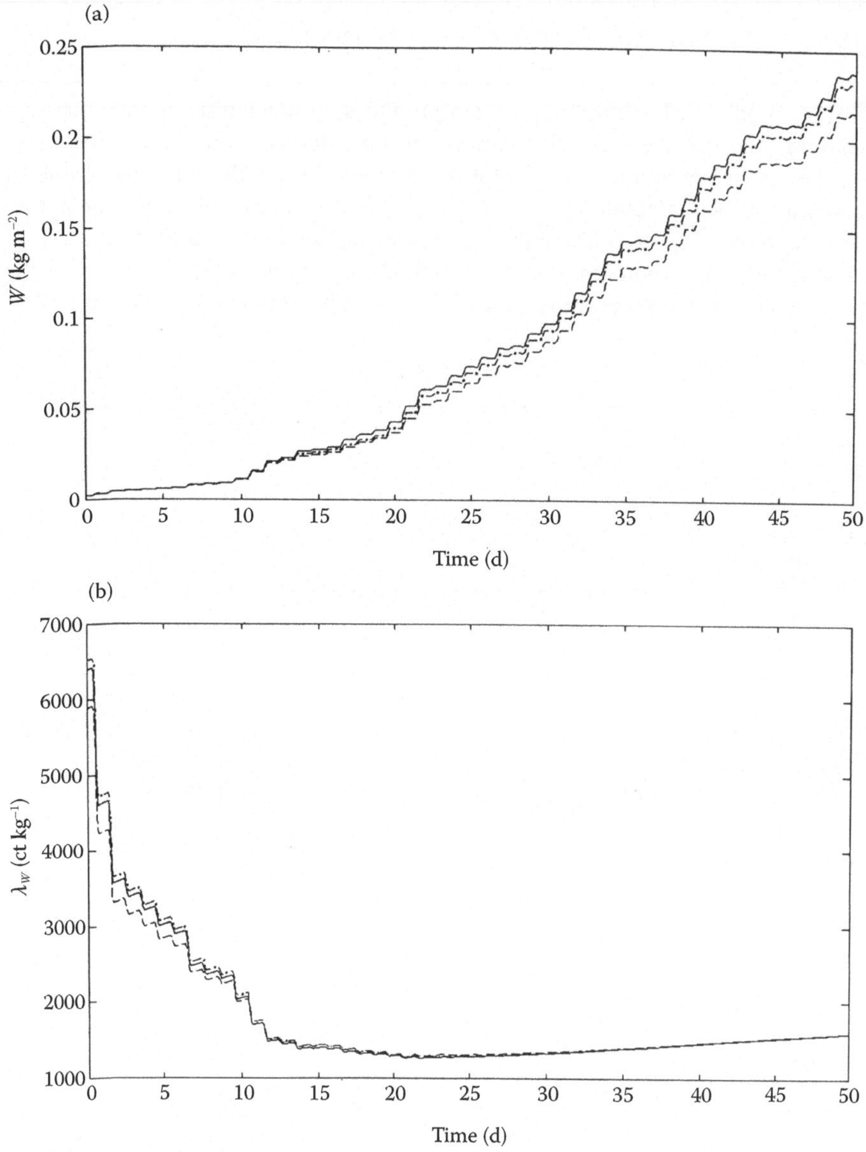

FIGURE 6.10 State trajectories (a) and costate trajectories (b) obtained in the solution of the full control problem (—) and the slow subproblem (---) and a simulation of the full system dynamics using the optimal control trajectories of the fast subproblem (– •).

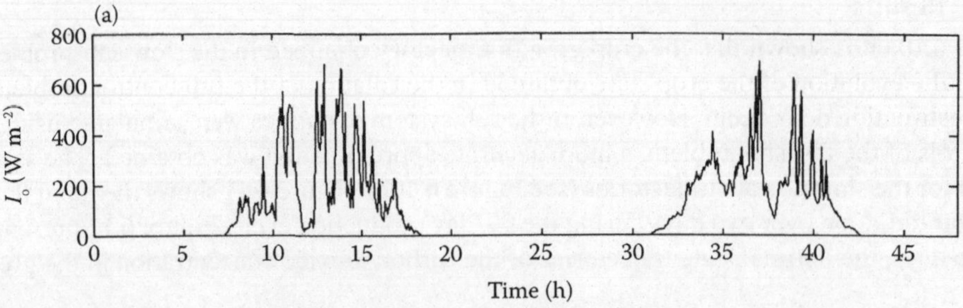

FIGURE 6.11 Solar radiation during two days of the 50-day production cycle.

It is clearly illustrated that the slow subproblem produces an inaccurate approximation of the fast system dynamics, because the rapid fluctuations in the external inputs are much faster than the dynamics of the fast subsystem (compare Figure 6.12a and b). On the other hand, the same figure shows that the trajectories calculated in the fast subproblem accurately describe the dynamics of the fast subsystem (compare Figure 6.12a and c). The same can be observed for the fast costate trajectories associated with the carbon dioxide concentration (see Figure 6.13) as well as for the state and costate trajectories of the temperature and humidity in the greenhouse (not shown).

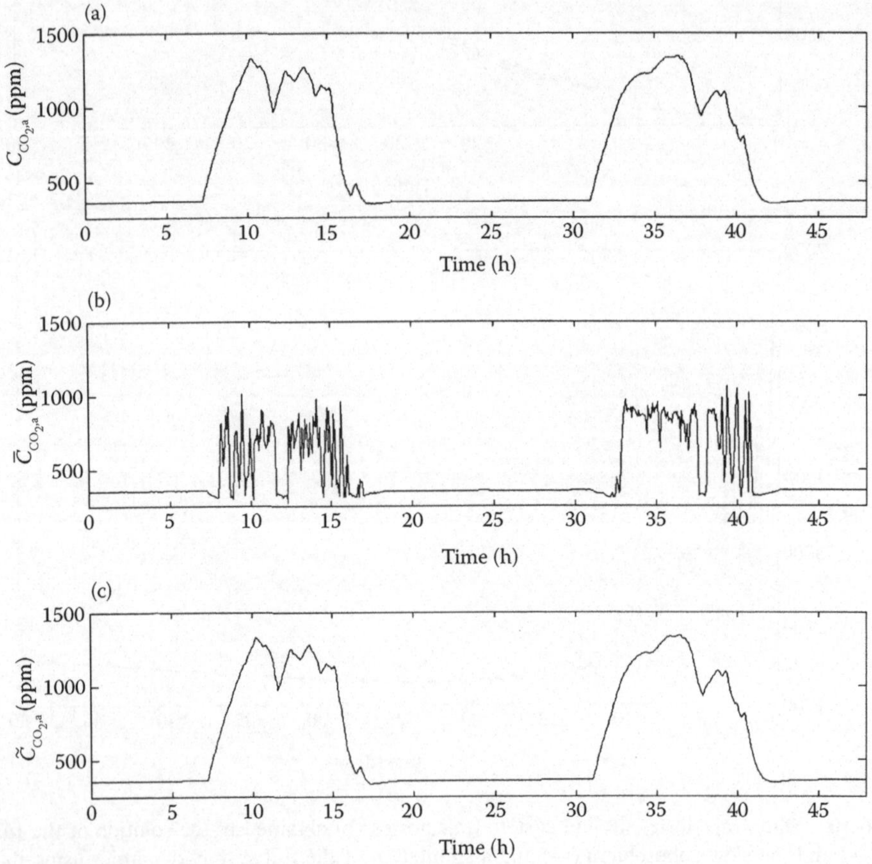

FIGURE 6.12 Optimal carbon dioxide concentration obtained in the solution of (a) the full problem, (b) the slow subproblem, and (c) the fast subproblem.

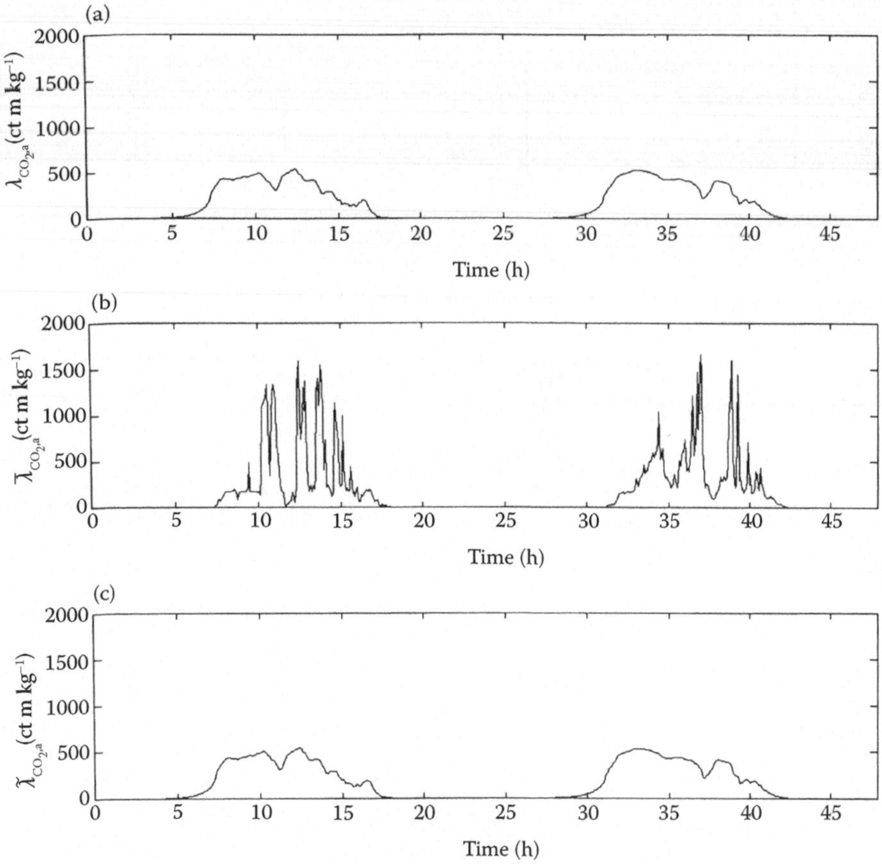

FIGURE 6.13 Costate of carbon dioxide concentration obtained after solving (a) the full problem, (b) the slow subproblem, and (c) the fast subproblem.

Figure 6.14 shows the trajectories of the carbon dioxide supply rate. Again, the solution of the slow subproblem is not an accurate approximation of the optimal control of the full problem, but the solution of the fast subproblem is. Similar observations can be made for the control trajectories of energy supply and ventilation rate (not shown).

Finally, evaluation of the performance using the control inputs calculated in the fast subproblem, yielded only a 2% reduction compared with the performance obtained in the solution of the full optimal control problem. So, an accurate approximation was obtained for this optimal control problem. This result is obtained because as shown in Figure 6.10 the slow system dynamics indeed do not respond to a large extent to the rapid fluctuations in the fast system dynamics, external inputs and control inputs, which was the main assumption on which this decomposition was based.

It was just shown that an optimal control problem arising in greenhouse climate management during one lettuce production cycle can be hierarchically decomposed into two subproblems: one dealing with the slow crop growth dynamics and one dealing with the relatively fast greenhouse climate dynamics. But what is the implication of this result for greenhouse climate management in horticultural practice?

Figure 6.15 shows in a schematic diagram the procedure a grower uses nowadays to control crop growth and production by means of climate conditioning. Depending on the current state of the crop, the grower decides on the setpoints of the greenhouse climate variables such as air temperature, humidity and carbon dioxide concentration. These setpoints are usually not defined as fixed

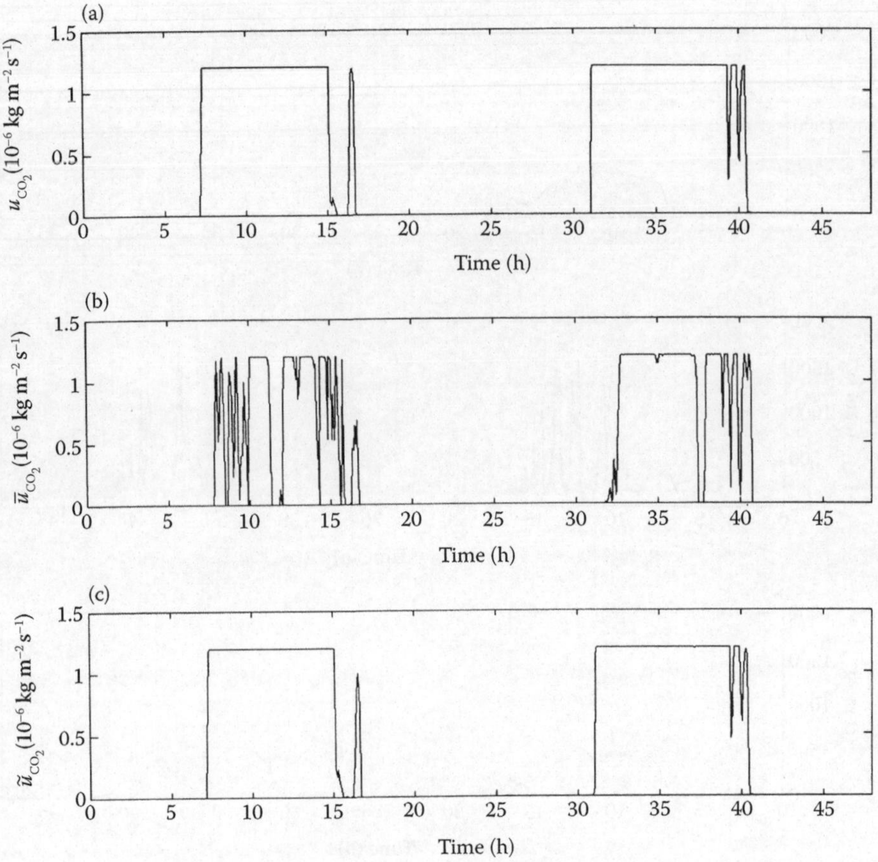

FIGURE 6.14 Optimal carbon dioxide supply rate obtained in the solution of (a) the full problem, (b) the slow subproblem, and (c) the fast subproblem.

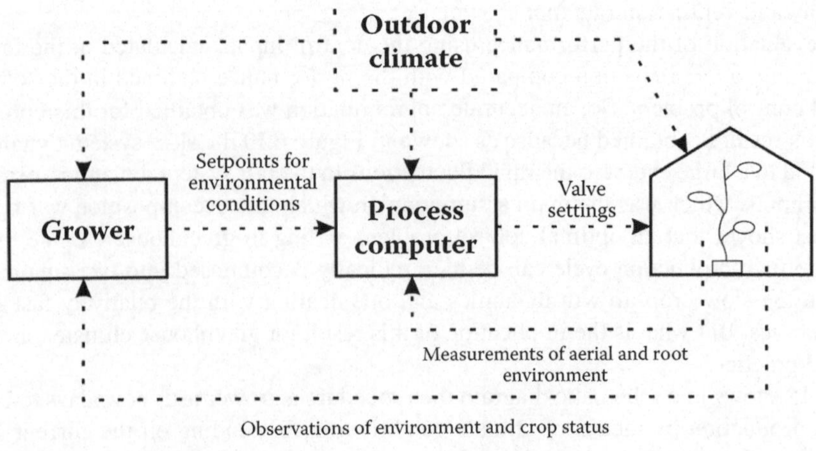

FIGURE 6.15 A schematic diagram of the climate control procedure in current horticultural practice.

values. Following rules defined by the grower based on experience, they may change during actual operation of the climate conditioning equipment in response to changes in the outside climate conditions. Such adaptations of the setpoints include, for instance, a solar radiation dependent change of the air temperature setpoint, and radiation and ventilation dependent adaptation of the carbon dioxide setpoint. The grower may also put bounds on the ventilator's aperture and the temperature of the heating pipes. A minimum temperature of the heating pipes is often used to assure circulation of air within the canopy. All together in modern greenhouse climate control computer systems, a large number of parameters (>150) need to be specified by the grower. Once the grower has decided on the settings of all these parameters, the greenhouse climate computer aims to achieve the desired climate in the greenhouse using measurements of the indoor climate and standard PI feedback control techniques. But there is a second indirect feedback loop from the crop growth process to the grower, in which during the growing season the grower may decide to modify the settings on the control computer based on observations of the actual state of the crop and detected or expected indoor and outdoor climate conditions.

Optimal control techniques can improve the performance of a crop production cycle, as was shown by Van Henten, Bontsema, and Van Straten (1997). But there is the challenging issue of implementing an optimal control system given modeling errors, rapidly fluctuating disturbances that are hard to predict but have a strong impact on the economic performance of the system, and finally, large differences in dynamic response times. It is at this point where the above shown decomposition may play a role. Figure 6.16 shows a hierarchical scheme for greenhouse climate control based on the previously described and validated decomposition of optimal greenhouse climate management. The hierarchical control scheme contains two control loops, an outer loop controlling the (slow) crop growth dynamics and an inner loop controlling the (fast) greenhouse climate dynamics.

Using a long-term weather prediction (e.g., long-term averages), a prediction of the auction price and a measurement of the initial state of the crop, the slow subproblem is solved for the outer control loop. Due to modeling errors and errors in the predictions of the weather and the auction price, the actual state and costate may deviate from the pre-calculated trajectories and state feedback is therefore required. Repeated solution of the control problem using new information about the state of the crop, the auction price and the weather is hence needed. Because we are dealing with a fixed final time problem, the recurrent solution will consider a slowly decreasing time interval as t_0 slowly approaches the harvest time t_f. Recalculation should take place once a week, i.e., the time-scale of the slow crop growth dynamics.

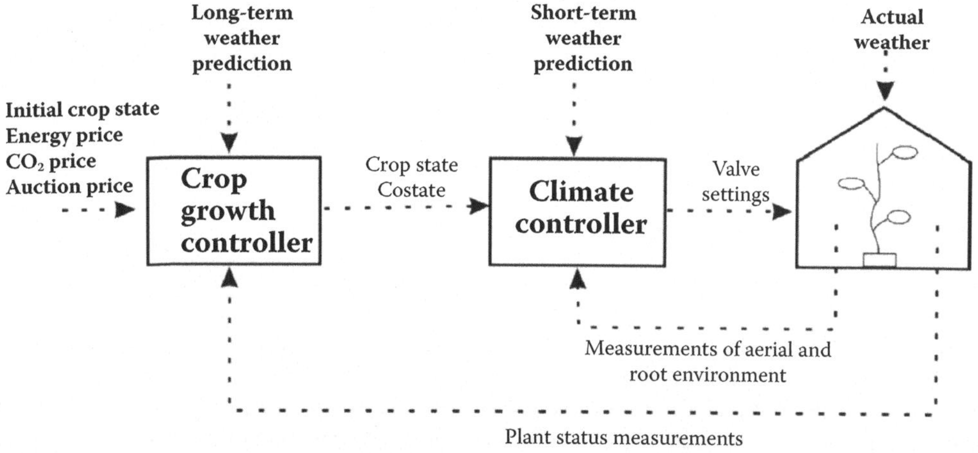

FIGURE 6.16 Optimal greenhouse climate management: a new hierarchical concept.

Using the state and costate trajectories calculated in the outer loop and a short-term weather prediction, the fast subproblem is solved for the inner control loop to control the greenhouse climate dynamics. This can be achieved in a receding horizon optimal control framework, as was demonstrated by Tap (2000). For that purpose the performance criterion of the fast subproblem has the favorable property that it does not have to be optimized over a whole growing period, but only over that time interval during which a control action affects the greenhouse climate. In horticultural practice this time interval will be in the order of one hour or less.

It is interesting to observe that the schemes shown in Figures 6.15 and 6.16 have a strong resemblance, both showing a hierarchical decomposition. However, in the latter the grower's expert knowledge is at least partially replaced by a quantitative model of crop growth and production.

Hierarchical control schemes for greenhouse climate management have been proposed before by, for instance, Udink ten Cate, Bot, and Van Dixhoorn (1978), Tantau (1991), and Challa and Van Straten (1991). They were all built on a proven scheme due to Richalet et al. (1978) that found common application in industrial practice. These schemes used setpoint optimization techniques to improve system performance at the highest level, neglecting the fast system dynamics, and relied on setpoint tracking techniques for the lowest control level in which economic performance was not explicitly considered. The interesting feature of the hierarchical structure presented in this chapter is the fact that at each level, i.e., both at the highest and the lowest control level, an economic performance criterion is used that has a direct relationship with the main objective of economic optimal greenhouse climate management defined at the highest level.

6.4.4.3 Concluding Remarks

In this chapter it has been shown that based on differences in response times, an optimal control problem in greenhouse climate management can be decomposed into two subproblems, one dealing with the slow system dynamics concerning crop growth and evolution, and one dealing with the faster greenhouse climate dynamics. For this particular problem, the decomposition was found to be sufficiently accurate. To generalize this concept, further research is required aiming at deriving a formal proof of this decomposition.

Based on this decomposition, a hierarchical scheme for greenhouse climate management has been proposed. This scheme is characterized by the fact that:

1. At each control level, control of the dynamic process responses is emphasized.
2. At each control level, a performance criterion is used that has a clear and direct relationship with the main objective of economic optimal greenhouse climate management defined at the highest level.
3. The relation between the control levels is defined in terms of state and costate trajectories, with the costate trajectories expressing the economic value of achieving the reference state trajectory at the lower level.
4. It is easy to implement in a receding horizon optimal control framework.

6.5 CONCLUDING REMARKS

This chapter has shed some light onto the advantages and characteristics of optimal greenhouse climate management.

Solutions of the optimal control problem are sometimes straightforward and clear. However, sometimes, results are counterintuitive. Section 6.4.1 has produced some insight into the particular characteristics of the optimal control problem. This offers a basis for further understanding of the optimization problem considered. Additionally, it can be used to instruct growers on how to improve the performance of their greenhouse climate management and, moreover, this insight can be used to convince growers of the advantages of model-based optimal greenhouse climate management.

Section 6.4.2 showed that optimal climate management may yield a better performance than current climate control practices. It is this potential advantage that should inspire researchers to proceed along this route and to improve climate control focusing on more efficient and effective use of resources using the optimal control paradigm. Still, in this section some limitations were also discussed, of which modeling errors and uncertainty in the auction prices as well as the outdoor weather are the most important ones. Some of these issues are considered in later chapters of this book.

Further to Section 6.4.1, Section 6.4.3 produced more insight into the characteristics of the optimal problem considered. It was shown that humidity constraint plays a dominant role in this optimization problem. This is not unexpected, because climate control in temperate climates like The Netherlands is largely based on consideration about the humidity level in the greenhouse and its consequences for both quantity and quality of the produced crop. Further research is needed to quantify effects of humidity on the crop status as well as to more accurately assess the possibly time varying bounds on the indoor climate as used in the optimal control problem.

In literature many hierarchical schemes have been proposed essentially to deal with the differences in response times in the greenhouse crop production system. As shown in Section 6.4.4, exploiting the differences in response times offers the opportunity to decompose the optimal control problem, yet maintaining a properly and consistently defined performance criterion at all control levels. This approach can be used to put optimal control to work in practice, as demonstrated in Chapter 7.

REFERENCES

Bailey, B.J., and Z.S. Chalabi. 1994. Improving the cost effectiveness of greenhouse climate control. *Computers and Electronics in Agriculture* 10: 203–214.

Brown, B.G., R.W. Katz, and A.H. Murphy. 1984. Time series models to simulate and forecast wind speed and wind power. *Journal of Climate and Applied Meteorology* 23: 1184–1195.

Challa, H., and G. Van Straten. 1991. Reflections about optimal climate control in greenhouse cultivation. In *Mathematical and Control Applications in Agriculture and Horticulture*, ed. Y. Hashimoto and W. Day, 1:13–18, IFAC Workshop Series.

Courtin, P., and J. Rootenberg. 1971. Performance index sensitivity of optimal control systems. *IEEE Transactions on Automatic Control* AC-16: 275–277.

Critten, D.L. 1991. Optimization of CO_2 concentration in greenhouses: A modelling analysis for a lettuce crop. *Journal of Agricultural Engineering Research* 48: 261–271.

Evers, A.H. 1979. *Sensitivity analysis of optimal control problems*, PhD diss., Technical University, Enschede, The Netherlands.

Evers, A.H. 1980. Sensitivity analysis in dynamic optimization. *Journal of Optimization Theory and Applications* 32: 17–37.

Gill, P.E., W. Murray, and M.H. Wright. 1981. *Practical Optimization*. New York: Academic Press, Inc.

Huang, Z., and Z.S. Chalabi. 1995. Use of time-series analysis to model and forecast wind speed. *Journal of Wind Engineering and Industrial Aerodynamics* 56: 311–322.

Ioslovich, I., I. Seginer, P.-O. Gutman, and M. Borshchevsky. 1995. Sub-optimal CO_2 enrichment of greenhouses. *Journal of Agricultural Engineering Research* 60: 117–136.

Kirk, D.E. 1970. *Optimal Control Theory*. Englewood Cliffs, New Jersey: Prentice-Hall.

Pierre, D.A. 1969. *Optimization Theory with Applications*. New York: John Wiley and Sons Inc.

Press, W.H., B.P. Flannery, S.A. Teukolsky, and W.T. Vetterling. 1986. *Numerical Recipes*. Cambridge, UK: Cambridge University Press.

Richalet, S., A. Rault, J.L. Testud, and J. Papon. 1978. Model predictive heuristic control: application to industrial process. *Automatica* 14: 413–428.

Schapendonk, A.H. C.M., and P. Gaastra. 1984. A simulation study on CO_2 concentration in protected cultivation. *Scientia Horticulturae* 23: 217–229.

Seginer, I., 1991. Optimal greenhouse temperature trajectories for a multi state variable tomato model. In *Mathematical and Control Applications in Agriculture and Horticulture*, ed. Y. Hashimoto and W. Day, 1: 73–79. IFAC Workshop Series.

Tantau, H.J. 1991. Optimal control for plant production in greenhouses. In *Mathematical and Control Applications in Agriculture and Horticulture*, ed. Y. Hashimoto and W. Day, 1: 1–6. IFAC Workshop Series.

Tap, R.F. 2000. *Economics-based optimal control of greenhouse tomato crop production*, PhD dissertation, Wageningen Agricultural University, Wageningen, The Netherlands.

Tap, R.F., L.G. Van Willigenburg, and G. Van Straten. 1996. Experimental results of receding horizon optimal control of greenhouse climate. *Acta Horticulturae* 406: 229–238.

Tap, R.F., L.G. Van Willigenburg, G. Van Straten, and E.J. Van Henten. 1993. Optimal control of greenhouse climate: computation of the influence of fast and slow dynamics. In *Proceedings of the 12th IFAC World Congress*, ed. G.C. Goodwin, R.J. Evans, J.B. Cruz Jr, and U. Jaaksoo, 321–324. Sydney, Australia: The Institution of Engineers.

Udink ten Cate, A.J., G.P.A. Bot, and J.J. Dixhoorn. 1978. Computer control of greenhouse climates. *Acta Horticulturae* 87: 265–272.

Van Henten, E.J. 1994. *Greenhouse climate management: an optimal control approach*. Ph.D dissertation, Wageningen Agricultural University, Wageningen, The Netherlands.

Van Henten, E.J. 2003. Sensitivity analysis of an optimal control problem in greenhouse climate management. *Biosystems Engineering* 85: 355–364.

Van Henten, E.J., and J. Bontsema. 1991. Optimal control of greenhouse climate. In *Mathematical and Control Applications in Agriculture*, ed. Y. Hashimoto and W. Day, 1: 27–32. IFAC Workshop Series.

Van Henten, E.J., and J. Bontsema. 1992. Singular perturbation methods applied to a variational problem in greenhouse climate control. In *Proceedings of the 31st IEEE Congress on Decision and Control*, ed. T. Basar, 3068–3069. New York: IEEE Control Systems Society.

Van Henten, E.J., and J. Bontsema. 1996. Greenhouse climate control: a two time-scale approach. *Acta Horticulturae* 406: 213–219.

Van Henten, E.J., and J. Bontsema. 2009. Time scale decomposition of an optimal control problem in greenhouse climate management. *Control Engineering Practice* 17: 88–96.

Van Henten, E.J., and G. Van Straten. 1994. Sensitivity analysis of a dynamic growth model of lettuce. *Journal of Agricultural Engineering Research* 59: 19–31.

Van Henten, E.J., J. Bontsema, and G. Van Straten. 1997. Improving the efficiency of greenhouse climate control: an optimal control approach. *Netherlands Journal of Agricultural Science* 45: 109–125.

Van Meurs, W. Th. M. 1980. The climate control computer system at the IMAG, Wageningen. *Acta Horticulturae* 106: 77–83.

Van Meurs, W. Th. M., and E.J. Van Henten. 1994. An experiment on the optimization of CO_2 in greenhouse climate control. *Acta Horticulturae* 366: 201–208.

Yang, T.H., and E. Polak. 1993. Moving horizon control of non-linear systems with input saturations, disturbances and plant uncertainty. *International Journal of Control* 58: 875–903.

7 An Experimental Application: Tomato*

7.1 INTRODUCTION

In the previous chapters, the foundation is laid for the optimal control of greenhouse cultivation on the basis of the theory of optimal control. The solution of open-loop control was simplified by distinguishing two time scales. Simulation results showed the necessary optimal control actions under a priori known but realistic weather fluctuations. Important insights were obtained regarding the potential benefits of optimal control that integrates greenhouse dynamics and crop models, expressed in direct profit to the grower.

To arrive at a real application, two additional issues must be solved. The first has to do with the variability in the weather. The use of realistic weather patterns, as done in the previous chapter, is important to study the effect of weather dynamics on the control actions, but it is not sufficient for online control. The premise of open-loop control computation, namely that the weather is known over the full optimization horizon, is obviously violated in a real online situation. It should be noted that the issue is not to suppress the influence of the weather, which is the normal situation in control, but rather to exploit the weather as much as possible. Hence, a solution for this problem is needed. The second issue is that open-loop simulations assume the model and model parameters to be correct and that the initial conditions and exogenous variables are known with great precision. In practice, deviations are unavoidable.

The answer to uncertainty is feedback. The two–time-scale decomposition as described in the previous chapter is transformed into a *receding horizon optimal controller*, with a goal function expressed in the same economic terms as before but over a shorter horizon. The long-term effects are incorporated via the slow costates, as discussed in Chapter 4. Feedback is provided by correcting the predicted state by observations. The weather is incorporated by short-term weather forecasts. The majority of the contents of this chapter originated from Tap (2000).

The crop chosen here as a case is *tomato*. This is a generative crop that poses larger challenges to the model as compared with a crop like lettuce. Moreover, it is of larger economic interest. A distinctive feature is that in the productive stage, the harvest is more or less continuous. The crop model is a three-state model, with the assimilates and the fruit and leaf biomass as states. In addition, a virtual development state is used to control the harvest. The *greenhouse climate model* is a relatively simple lumped parameter model, with the greenhouse air temperature, a virtual soil temperature, the CO_2 concentration, and the water vapor concentration in the air as states. In addition, the heating pipe temperature dynamics is modeled. Much effort is put into the experimental calibration and validation of the models.

Although in the previous chapter the evaluation of the optimal control was done by simulations only, here a comparative experiment is described showing the difference between the conventional control and the receding horizon optimal control over the period of one month. The assessment of seasonal effects is done by using the calibrated models in simulation that mimics the online control.

* Notation: Chapter-specific symbols are defined in the text and are listed together with parameter values in Tables 7.2 and 7.3.

7.2 TOMATO MODEL

The tomato state–space model has been developed by Tap (2000) on the basis of the work of De Koning (1994). It is called a big leaf–big fruit model because no distinction is made between leaf or fruit number. It is formulated here in a slightly more generic form than the original version. The model describes the evolvement of leaf and fruit biomass after anthesis of the first fruit. The model has three principle states: nonstructural biomass, that is, assimilates, which can be viewed as an assimilate buffer (W_B), and structural biomass in leaves (W_L) and fruits (W_F), as shown in Figure 7.1.

At first, we describe the mass balances for the vegetative structural parts, that is, the sum of leaves, stems, and roots (W_V), and the generative parts (fruit biomass W_F). Later, instead of working with the sum of all vegetative parts, the leaves only are taken as the states. The basic mass balances are as follows.

Assimilates:

$$\frac{dW_B}{dt} = P - G_V - \theta_V G_V - G_F - \theta_F G_F - R_{B,V} - R_{B,F}. \qquad (\mathrm{kg[dw]\ m^{-2}[gh]\ s^{-1}}) \qquad (7.1)$$

rate of change of assimilates ($\dot{W}_B$) =

+ Production of assimilates by photosynthesis (P)
– Conversion of assimilates to vegetative biomass by growth (G_V)
– Use of assimilates as energy to drive vegetative growth (fraction θ_V of G_V)
– Conversion of assimilates to generative (fruit) biomass by growth (G_F)
– Use of assimilates as energy to drive generative (fruit) growth (fraction θ_F of G_F)
– Drain of assimilates for maintenance of vegetative parts ($R_{B,V}$)
– Drain of assimilates for maintenance of generative parts ($R_{B,F}$)

Vegetative parts:

$$\frac{dW_V}{dt} = G_V - R_{V,V} - H_L, \qquad (\mathrm{kg[dw]\ m^{-2}[gh]\ s^{-1}}) \qquad (7.2)$$

rate of change of leaf biomass ($\dot{W}_V$) =

+ Conversion of assimilates to leaf biomass by growth (G_V)
– Use of biomass for maintenance when there is a lack of assimilates ($R_{V,V}$)
– Leaf picking rate (H_L)

Generative parts (fruits):

$$\frac{dW_F}{dt} = G_F - R_{F,F} - H_F, \qquad (\mathrm{kg[dw]\ m^{-2}[gh]\ s^{-1}}) \qquad (7.3)$$

rate of change of fruit biomass ($\dot{W}_F$) =

+ Conversion of assimilates to fruit biomass by growth (G_F)
– Use of biomass for maintenance when there is a lack of assimilates ($R_{F,F}$)
– Fruit harvest rate (H_F)

where all biomasses and rates are expressed in dry weight per unit greenhouse projected area.

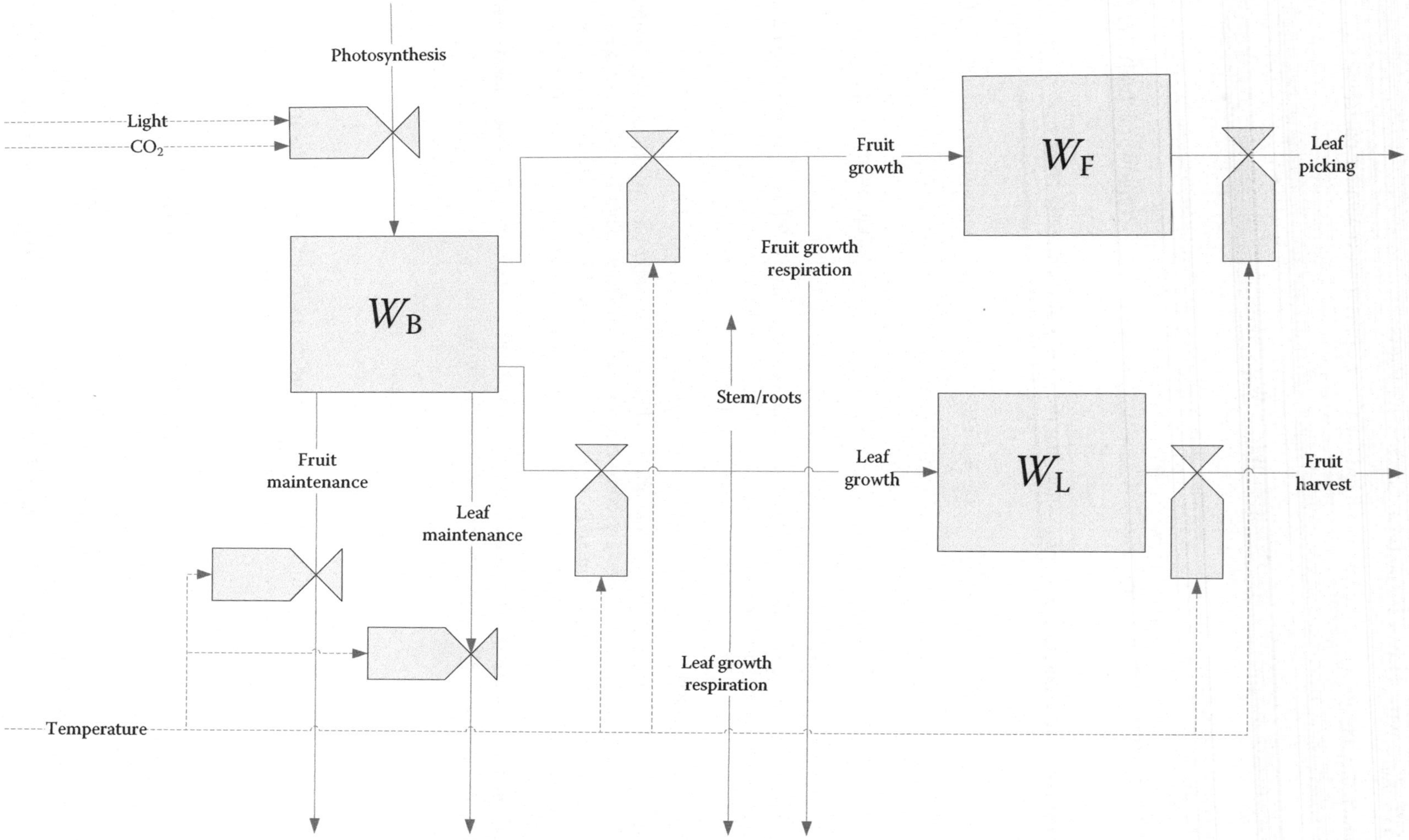

FIGURE 7.1 Big fruit–big leaf model. The rectangles represent the biomass states (W_B assimilate buffer, W_L leaf biomass, and W_F fruit biomass). Solid lines represent carbon mass flows, dashed lines information flows. Internal information loops are left out for readability.

7.2.1 Working with Leaves Instead of Generative Parts

In the model, it is assumed that the leaf is a fixed fraction z of the total vegetative parts:

$$z = \frac{W_L}{W_L + W_{stem} + W_{roots}}. \qquad (-) \quad (7.4)$$

Consequently,

$$W_V = W_L + W_{stem} + W_{roots} = \frac{1}{z} W_L, \qquad (kg[dw]\ m^{-2}[gh]) \quad (7.5)$$

and provided that the initial time is chosen such that the fixed ratio is already obeyed,

$$G_V = \frac{1}{z} G_L, \qquad (kg[dw]\ m^{-2}[gh]\ s^{-1}) \quad (7.6)$$

$$R_V = \frac{1}{z} R_L, \qquad (kg[dw]\ m^{-2}[gh]\ s^{-1}) \quad (7.7)$$

where G_L and R_L are the leaf growth and leaf maintenance, respectively. It has to be noted that leaf picking disturbs the fixed ratio. This effect is ignored in the model.

7.2.2 Assimilate Pool

Assimilates are being produced by photosynthesis. The gross canopy photosynthesis rate in dry matter per unit area is P. Assimilates are converted to leaf and fruits. This is commonly known as *growth*. Leaf and fruits have a "demand" for assimilates, which will be honored if there are sufficient assimilates available. In addition to the leaves, there are other plant parts, such as stem and roots.

If we denote the unit area growth demand of leaves and fruits as G_L^{dem} and G_F^{dem}, then the amount of assimilates turned into structural matter by the actual growth is

$$h\{\cdot\}\left(\frac{G_L^{dem}}{z} + G_F^{dem}\right), \qquad (kg[dw]\ m^{-2}[gh]\ s^{-1}) \quad (7.8)$$

where $h\{\cdot\}$ is a smooth switching function that has a value of one if the assimilates are plenty and goes to zero when the assimilates are depleted. Hence, the actual leaf and fruit growth rates are given by

$$G_L = h\{\cdot\} G_L^{dem} \qquad (kg[dw]\ m^{-2}[gh]\ s^{-1}) \quad (7.9)$$

and

$$G_F = h\{\cdot\} G_F^{dem}, \qquad (kg[dw]\ m^{-2}[gh]\ s^{-1}) \quad (7.10)$$

respectively. This shows that actual growth comes to a halt when the buffer is empty. The curly brackets in $h\{\cdot\}$ are used here to remind us that h is not a constant parameter but a function of the relative abundance of the assimilates (to be explained later).

To realize structural growth, energy is needed, which is also drawn from the assimilate pool. If we denote the additional amount of assimilates needed for one unit of structural fruit and vegetative parts by θ_F and θ_V, respectively, the actual drawing from the assimilate pool for growth, including growth respiration, is

$$h\{\cdot\}\left(\left(1+\theta_V\right)\frac{G_L^{dem}}{z}+\left(1+\theta_F\right)G_F^{dem}\right). \qquad (kg[dw]\ m^{-2}[gh]\ s^{-1}) \quad (7.11)$$

The terms $(1 + \theta_F)$ and $(1 + \theta_V)$ are also known as the fruit assimilate requirement ratio and the vegetative assimilate requirement ratio, respectively.

Even if the plant does not grow, it requires energy for maintenance. This is obtained by respiring some of the assimilates. Unlike growth, respiration will continue even if the assimilate buffer is empty. In that case, some of the structural biomass is reallocated to be used as maintenance energy. The total drawing of assimilates from the assimilation pool for respiration is

$$h\{\cdot\}\left(\frac{R_L}{z}+R_F\right), \qquad (kg[dw]\ m^{-2}[gh]\ s^{-1}) \quad (7.12)$$

where R_F and R_L are the respiration needs of fruits and leaves, and, as before, $h\{\cdot\}$ is a smooth function that goes to 0 when the assimilates get depleted. For convenience, the same function is used as for growth. This is not a severe limitation because the exact form of $h\{\cdot\}$ is not so important. Equation 7.12 implies that $R_{B,V}$ and $R_{B,F}$ in Equation 7.1 are replaced by

$$R_{B,V}=h\{\cdot\}\frac{R_L}{z} \qquad (kg[dw]\ m^{-2}[gh]\ s^{-1}) \quad (7.13)$$

and

$$R_{B,F}=h\{\cdot\}R_F, \qquad (kg[dw]\ m^{-2}[gh]\ s^{-1}) \quad (7.14)$$

respectively. If the buffer is empty, the respiration needs will be satisfied by conversion of structural biomass. Denoting by W_B the total assimilates in the canopy, expressed as dry weight per unit area, the following mass balance holds,

$$\frac{dW_B}{dt}=P-h\{\cdot\}\left(\frac{\left(1+\theta_V\right)}{z}G_L^{dem}+\left(1+\theta_F\right)G_F^{dem}\right)-h\{\cdot\}\left(\frac{R_L}{z}+R_F\right). \qquad (kg[dw]\ m^{-2}[gh]\ s^{-1}) \quad (7.15)$$

To see more clearly what happens, we can split this into

when W_B is plenty ($h = 1$): $\dfrac{dW_B}{dt}=P-\dfrac{\left(1+\theta_V\right)}{z}G_L^{dem}-\left(1+\theta_F\right)G_F^{dem}-\dfrac{R_L}{z}-R_F;$

when W_B is near zero ($h = 0$): $\dfrac{dW_B}{dt}=P.$

It is perhaps useful to analyze the role of the function $h\{\cdot\}$ a little further. It serves as a kind of switching function to indicate the change in operation mode of the plant when the assimilates change from an abundant state to a depleted state. Typical shapes of switching functions as a function of W_B are shown in Figure 7.2. The sharp discrete switch (curve 1) leads to jittering of the model around the switching value of W_B. This can be seen because during the day, when $P > 0$ and $h = 0$, W_B will increase, but as soon as it reaches the switching level, that is, if h jumps to one, the growth and respiration terms will drag W_B down again. Hence, even with constant photosynthesis and assimilate demands, there is no steady state. By smoothing the function, for instance, by an S-shape form (curve 2), there is a steady state, which is given by the value of W_B belonging to

$$h^{ss}\{\cdot\} = \frac{P^{ss}}{\frac{(1+\theta_V)}{z} G_L^{dem,ss} + (1+\theta_F) G_F^{dem,ss} + \frac{R_L^{ss}}{z} + R_F^{ss}}. \quad (-) \quad (7.16)$$

Tap (2000) has used the saturated curve given by

$$h = 1 - e^{-p_h W_B}, \quad (-) \quad (7.17)$$

with p_h a (positive) parameter (curve 3).

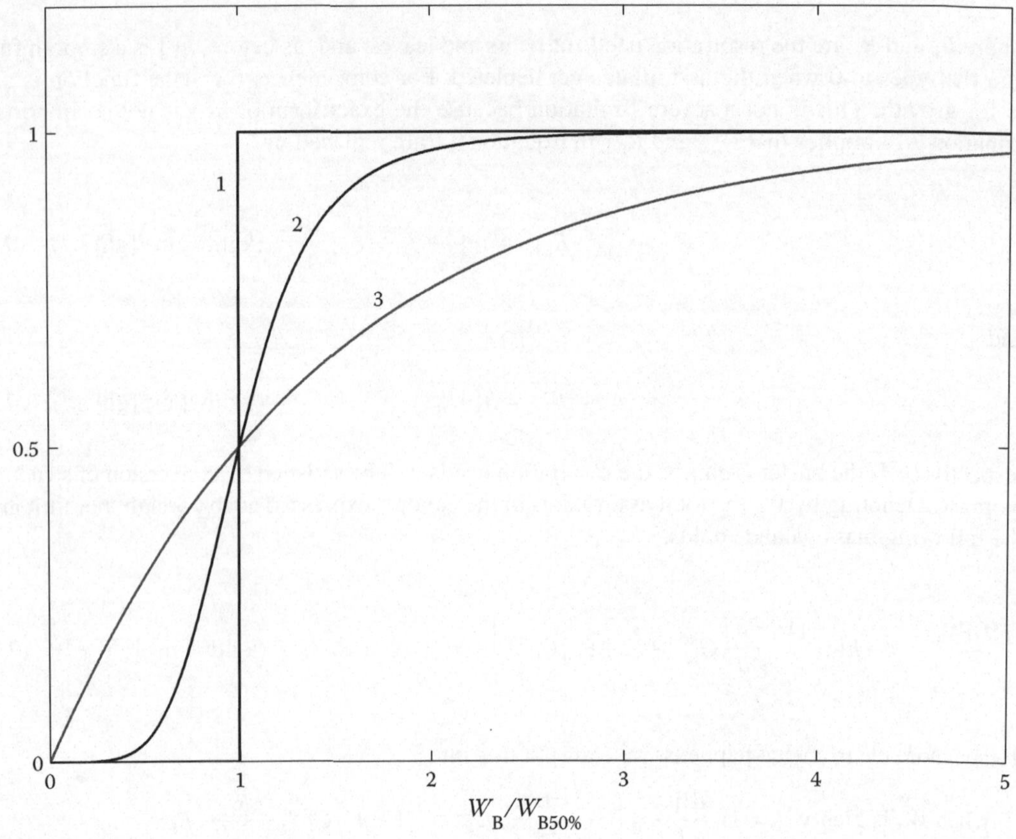

FIGURE 7.2 Examples of switching functions (step, exponential, and S-shape).

7.2.3 Leaf and Fruit Biomass

The leaf growth is equal to the amount of assimilates converted to structural leaf biomass in the canopy and is given by $h\{\cdot\}G_L^{dem}$. The model does not incorporate an extra state for stem and roots, but by the factor z it is assumed that each increment in leaf will be accompanied by a proportional increment in stem and root. If there are no sufficient assimilates, growth simply stops, but maintenance respiration is assumed to happen at all times. Normally, assimilates are used for maintenance, but in lack of assimilates, maintenance in the model goes at the expense of structural parts, that is, the leaves and the fruits. The mass balance for the canopy leaf biomass per unit area, W_L, reads

$$\frac{dW_L}{dt} = h\{\}G_L^{dem} - (1 - h\{\})R_L - H_L. \qquad (\text{kg[dw] m}^{-2}\text{[gh] s}^{-1}) \quad (7.18)$$

The term H_L is a formal term to express the fact that leaves are pruned from time to time.

Similarly, for the fruits, the following mass balance holds

$$\frac{dW_F}{dt} = h\{\cdot\}G_F^{dem} - \left(1 - h\{\cdot\}\right)R_F - H_F. \qquad (\text{kg[dw] m}^{-2}\text{[gh] s}^{-1}) \quad (7.19)$$

We see that

when W_B is plenty ($h = 1$):

$$\frac{dW_L}{dt} = G_L^{dem} - H_L$$

$$\frac{dW_F}{dt} = G_F^{dem} - H_F;$$

when W_B is near zero ($h = 0$):

$$\frac{dW_L}{dt} = -R_L - H_L$$

$$\frac{dW_F}{dt} = R_F - H_F.$$

7.2.4 Losses

To know the amount of CO_2 returned to the greenhouse environment, it is relevant to list all loss terms here. By adding up all biomass forms, we obtain the following:

in case of abundant W_B: $$\frac{dW_{tot}}{dt} = P - \frac{\theta_V}{z}G_L^{dem} - \theta_F G_F^{dem} - \frac{1}{z}R_L - R_F - \frac{1}{z}H_L - H_F;$$

in case of depleted W_B: $$\frac{dW_{tot}}{dt} = P - \frac{1}{z}R_L - R_F - \frac{1}{z}H_L - H_F.$$

The fixed ratio between leaves and roots–shoot requires the rather artificial condition that whenever leaves are pruned, an equivalent proportion of stem and root is taken out. The harvest obviously is not returned as CO_2 to the greenhouse, but the respiratory terms are, so the total amount respired per unit time R is given by

$$R = h\{\cdot\}\left(\frac{\theta_V}{z}G_L^{dem} + \theta_F G_F^{dem}\right) + \frac{R_L}{z} + R_F. \qquad (\text{kg[dw] m}^{-2}\text{[gh] s}^{-1}) \quad (7.20)$$

Equation (7.20) expresses that the total respiration is obviously the sum of the total growth respiration—in case there is growth—and the total maintenance respiration.

7.2.5 Constitutive Relations

To complete the model, equations are needed that link the rate of change terms to the states of the model and the external inputs.

7.2.5.1 Photosynthesis

The gross photosynthesis depends on the photosynthetic activate radiation (PAR) (I^{PAR}) and on the CO_2 concentration (C_{CO_2}), both with a saturation curve. With young plants, the photosynthesis is roughly proportional to the leaf area per unit greenhouse area (LAI), whereas the photosynthesis becomes practically independent of the standing biomass when the canopy closes. As the LAI is not a separate state in the model, a reduction function is used that depends on the leaf biomass:

$$f_m\{\cdot\} = \frac{(W_L/p_m)^m}{1+(W_L/p_m)^m}. \qquad (-) \quad (7.21)$$

The factor f_m can be seen as a kind of "maturity" factor. It is near zero when the plant is young and will approach one as the canopy closes and the plant gets more mature. The parameter p_m is the leaf dry weight where the maturity factor is 0.5 (kg[dw] m^{-2}[gh]), and m is a dimensionless parameter.

The overall photosynthesis rate is given by

$$P = P^{max}\frac{I^{PAR}}{I^{PAR}+K_I}\frac{C_{CO_2}}{C_{CO_2}+K_C}f_m\{\cdot\}, \qquad (\text{kg[dw] m}^{-2}\text{[gh] s}^{-1}) \quad (7.22)$$

with parameters P^{max} as the maximum gross canopy photosynthetic rate (kg[dw] m^{-2}[gh] s^{-1}), K_I as the half saturation PAR light intensity (W[PAR] m^{-2}), and K_C as the half saturation CO_2 concentration in the air (kg[CO_2] m^{-3}). The PAR light intensity at the crop level I^{PAR} is simply set proportional to the global radiation

$$I^{PAR} = f_{PAR/I}\tau_r I_o, \qquad (\text{W[PAR] m}^{-2}) \quad (7.23)$$

where $f_{PAR/I}$ is the fraction PAR of the global radiation, τ_r is the transmittance of the roof, and I_o is the global radiation (W m^{-2}[gh]).

7.2.5.2 Growth Demand

The growth demand of the leaves and the fruits is assumed to be roughly proportional to the total biomass. An attempt has been made, however, to encapsulate the fact that in the early stages of bearing fruits the size distribution is toward smaller fruits, the relative growth rate tends to be larger. This is expressed in a dimensionless development factor $f_D\{\cdot\}$ to be explained later. Growth of fruits is also assumed to be temperature dependent, which is modeled with a Q_{10} relation. The growth demand equation therefore is

$$G_F^{dem} = k_{GF}^{ref} f_{TG}(T) f_D\{\cdot\} W_F, \qquad (\text{kg[dw] m}^{-2}\text{[gh] s}^{-1}) \quad (7.24)$$

with

$$f_{TG}(T) = Q_{10G}^{\frac{T-T_G^{ref}}{10}}. \qquad (-) \quad (7.25)$$

The parameter k_{GF}^{ref} is the reference growth rate coefficient of fruits (s^{-1}), that is, the growth rate at the reference temperature.

The growth demand of leaves and hence of all vegetative parts is modeled as

$$G_L^{dem} = f_{L/F}(T)k_{GF}^{ref}f_{TG}(T)f_D\{\cdot\}W_L. \qquad (\text{kg[dw] m}^{-2}\text{[gh] s}^{-1}) \quad (7.26)$$

In this formulation, the same growth rate coefficient as for the fruits is applied to the leaves, but it is modified by a temperature-dependent ratio that expresses by which factor the specific leaf growth rate is larger or smaller than the specific fruit growth rate. This ratio plays an important role in controlling the vegetative and generative growth relative to each other. It is expressed as

$$f_{L/F}(T) = f_{L/F}^{ref}e^{\nu_2\left(T-T_{L/F}^{ref}\right)}. \qquad (-) \quad (7.27)$$

Parameter $T_{L/F}^{ref}$ is the pivot temperature. With a negative ν_2, increasing the temperature above the pivot temperature means relatively more generative growth. Parameter $f_{L/F}^{ref}$ is a dimensionless constant expressing how much larger the growth demand of the leaves is as compared with that of the fruits at the pivot temperature.

7.2.5.3 Maintenance Respiration

The maintenance respiration demand of the leaves is modeled as a simple first-order expression of the biomass,

$$R_L = k_{RL}^{ref}f_{TR}(T)W_L, \qquad (\text{kg[dw] m}^{-2}\text{[gh] s}^{-1}) \quad (7.28)$$

where k_{RL}^{ref} is the respiration rate at the reference temperature T_{ref}, and $f_{TR}(T)$ is a function of the temperature, modeled as a Q_{10} relation, that is,

$$f_{TR}(T) = Q_{10R}^{\frac{T-T_R^{ref}}{10}}. \qquad (-) \quad (7.29)$$

The temperature here should be, strictly speaking, the leaf temperature, but as this is not a state in the model, the air temperature is used instead.

The same holds, by analogy, for the fruits:

$$R_F = k_{RF}^{ref}f_{TR}(T)W_F, \qquad (\text{kg[dw] m}^{-2}\text{[gh] s}^{-1}) \quad (7.30)$$

where, for simplicity, the same temperature dependency is assumed.

Refer to Section 9.4.3 for a discussion on modeling maintenance and growth respiration.

7.2.5.4 Development State

The correction factor for the fruit growth rate in Equation 7.24 is given by

$$f_D = \frac{c_{f1} - c_{f2}D}{c_{f1} - c_{f2}}. \qquad (-) \quad (7.31)$$

Here, D is the development stage for the plant, defined in the next paragraph. Note that f_D is equal to 1 if $D = 1$. With the parameters given by Tap (2000), the correction factor stays near 1, even at low values of the development stage.

The development stage is an artificial state variable that measures in a way the temperature history passed since anthesis. It could, in principle, also be used to predict the onset of fruits, although that is not done in the proposed model. The basis for the formula derived as approximation by Tap (2000) goes back to the research of De Koning (1994), who provided a detailed analysis of growth of individual fruits, which obviously cannot be incorporated in the big fruit–big leaf model. The overall crop development stage is modeled as an additional state variable, according to the empirical expression, originally presented as

$$\frac{\mathrm{d}D}{\mathrm{d}t} = c_{\mathrm{d1}} + c_{\mathrm{d2}} \ln\left(\frac{T}{c_{\mathrm{d3}}}\right) - c_{\mathrm{d4}}\left(t - t_{\mathrm{anthesis}}\right) - k_{\mathrm{H}}. \qquad (\mathrm{s}^{-1}) \quad (7.32)$$

Here, the constants c_{d1}, c_{d2}, c_{d3}, and c_{d4} are parameters deduced from the underlying empirical development models. There is a term that reduces the development stage proportional to the time since anthesis, which expresses a kind of aging of the plant or, put differently, expresses that the vigor of the plant gets less when the plant gets older. Parameter c_{d3} is a pivot temperature (293 K, 20°C) so that when $T > c_{d3}$, the development rate grows faster than at the pivot temperature and slower when $T < c_{d3}$. The term k_H (s^{-1}) is the development rate effect due to harvest as discussed in the next section. It is included to express that the average plant development reduces when the ripest fruits are harvested.

The disadvantage of the formulation of Equation 7.32 is the explicit dependency on time, which is not desirable. The term $t - t_{anthesis}$ is actually a state variable that expresses the crop age since anthesis. Therefore, it is better to introduce a new state variable, called e_D here. The state equation for e_D is

$$\frac{\mathrm{d}e_{\mathrm{D}}}{\mathrm{d}t} = \begin{cases} 0 & t < t_{\mathrm{anthesis}} \\ 1 & t \geq t_{\mathrm{anthesis}} \end{cases}, \qquad (-) \quad (7.33)$$

where $t_{anthesis}$ should somehow be linked to the development state. Lack of this relation restricts the validity of the model to stages in the crop development since fruit set begins.

7.2.5.5 Harvest Rate

In practice, growers prune the leaves of lower internodes, and of course fruits that are ripe will be picked. Strictly speaking, the harvest is a controllable input. In stand-alone simulations for assessment of the potential options for harvest, it makes sense to model the harvest process. To this end, it can be assumed that harvest starts when some fruits are ripe. This is set to appear when D reaches the value one. A continuous harvest is achieved by regulating the harvest so that D stays equal to one. Hence,

$$k_{\mathrm{H}} = c_{\mathrm{H}}(D)\left(c_{\mathrm{d1}} + c_{\mathrm{d2}} \ln\left(\frac{T}{c_{\mathrm{d3}}}\right) - c_{\mathrm{d4}} e_{\mathrm{D}}\right), \qquad (\mathrm{s}^{-1}) \quad (7.34)$$

where the control function $c_H(D)$,

$$c_H(D) = \mathcal{U}(D - 1), \qquad (-) \quad (7.35)$$

is the unit Heaviside step function, that is, if $D < 1$, $c_H(D) = 0$, and if $D \geq 1$, $c_H(D) = 1$. In practical operational ranges, when harvest has started, it will continue and D remains one so that the goal to control k_H is easily achieved by discarding the development stage state equation entirely. In other words, as soon as fruit picking begins, D remains one, and k_H is computed from

$$k_H = c_{d1} + c_{d2} \ln\left(\frac{T}{c_{d3}}\right) - c_{d4} e_D. \tag{7.36}$$

The harvest rate itself is in fact a control, or it may be known from the data. In stand-alone simulations, an assumption has to be made to relate the harvest rate to the states of the model. In that case, it is simply set proportional to the biomass, that is,

$$H_F = k_{HF} W_F, \quad (\text{kg[dw] m}^{-2}\text{[gh] s}^{-1}) \tag{7.37}$$

$$H_L = k_{HL} W_L. \quad (\text{kg[dw] m}^{-2}\text{[gh] s}^{-1}) \tag{7.38}$$

The rate coefficients k_{HF} and k_{HL} are not constants but are made functions of the development stage effect due to harvest,

$$k_{HF} = c_{yF} k_H, \quad (\text{s}^{-1}) \tag{7.39}$$

$$k_{HL} = c_{yL} k_H, \quad (\text{s}^{-1}) \tag{7.40}$$

with c_{yF} and c_{yL} as the dimensionless harvest parameters. In view of Equation 7.34, this implies that as the crop matures, the relative harvest rate reduces.

Although leaf pruning has no direct economic benefit, it is important, as the presence of leaf biomass in this model entails more respiration losses.

It should be noticed that the total harvested fruit amount can be easily computed from

$$S_{HF}(t) = \int_{t_0}^{t_f} k_{HF} W_F \, dt, \quad (\text{kg[dw] m}^{-2}\text{[gh] s}^{-1}) \tag{7.41}$$

or, alternatively, from the auxiliary state equation,

$$\frac{dS_{HF}}{dt} = k_{HF} W_F. \quad (\text{kg[dw] m}^{-2}\text{[gh] s}^{-1}) \tag{7.42}$$

As this state does not feed back to the crop behavior, it is a nonessential state for the dynamics (see Section 7.4).

7.3 GREENHOUSE CLIMATE MODEL

As in the case of lettuce discussed in the previous chapter, it is important to discuss the choices of the system boundaries and the consequences of this choice. Figure 7.3 shows two common solutions. It shows the greenhouse compartment as an information flow diagram, with separate blocks for the pipe system and the ventilation system. In the top picture, the boundaries are drawn such that the heat input from the heating pipes and the ventilation flux are considered as independent inputs.

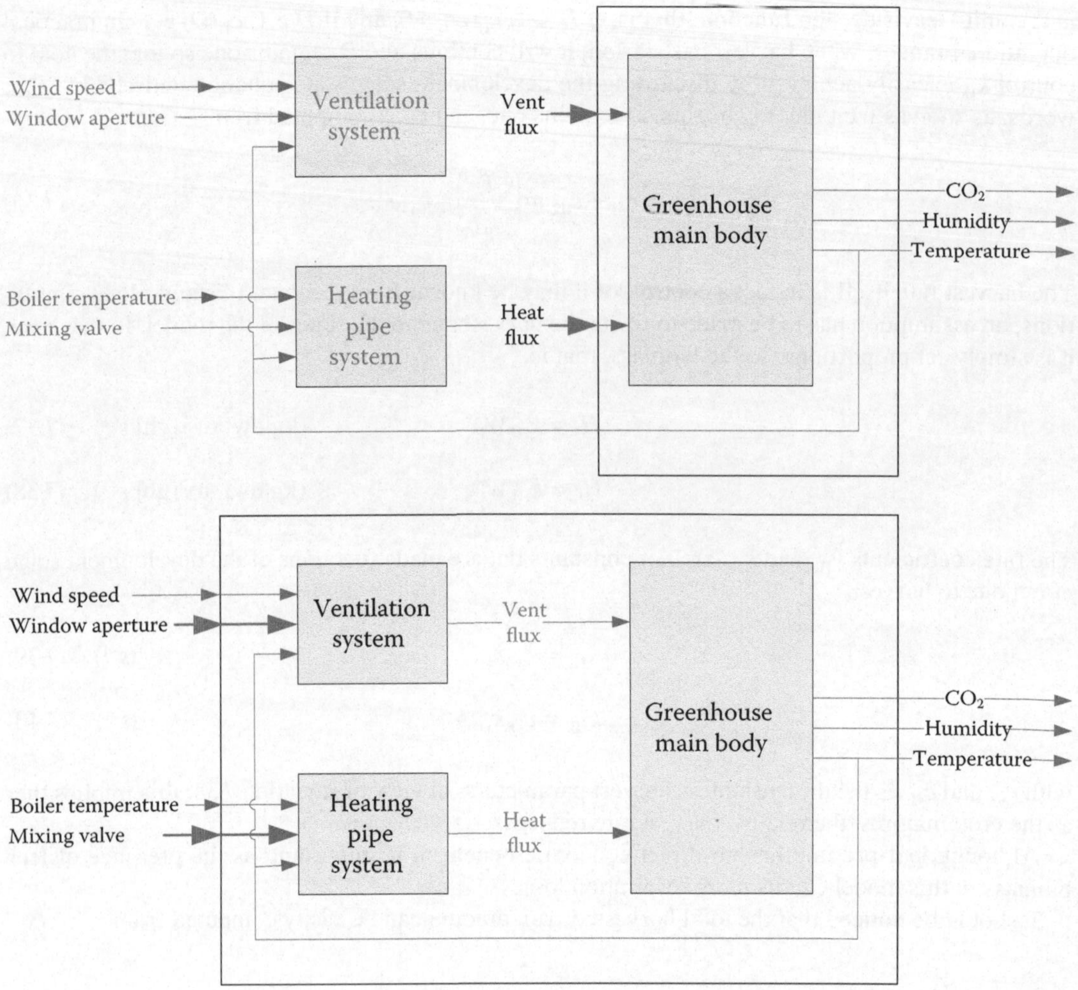

FIGURE 7.3 Different choices for the system boundaries. Top: excluding the pipe and ventilator systems. Bottom: with integrated models for heating pipe and ventilation dynamics.

Alternatively, the boundaries can be drawn in such a way that the heating pipe and the ventilation subsystems are an integral part of the total system. In that case, the control inputs are the mixing valve for the heating and the window aperture for the ventilation. Additional (disturbance) inputs are the boiler temperature, the greenhouse temperature, and the wind speed. A separate model is needed to describe the pipe temperature dynamics. Similarly, a model is needed to express the ventilation flux as a function of wind speed, buoyancy, and window aperture.

It is clear that the first option is simpler to use but is more remote from practice. Moreover, it can be seen that there is a difficulty because the ventilation and the heat fluxes depend on the greenhouse temperature and are therefore not truly freely controllable inputs. On the other hand, if it were possible to build a fast controller that would be able to realize the demanded heating flux, one could in the frame of optimal control refrain from separate modeling of the heating pipe system. It would, however, require an accurate online measurement of the transferred heat. This is not easy to realize. A compromise would be to measure and to control the pipe temperature, in which case for the optimal control it would not be necessary to model the piping system. Instead, the pipe temperature now becomes the system input, and any fluctuations in boiler temperature are dealt with by the local controller. Another aspect is that it simply might not be possible to supply the demanded heat under all circumstances. This occurs when actuators get saturated. Moreover, a heat input in the optimal control could become negative, but it is clear that the heating pipe system cannot provide the required cooling in that case. Hence, there are constraints that must be taken into account when the heat input is taken as a variable to be optimized. These are the reasons to model the pipe system separately. This model can then also be used to study the effects of a local pipe temperature controller on the optimization result.

Similar arguments apply to the ventilation flux, with the additional difficulty that there are to date no good solutions to obtain a direct measurement of the ventilation flux, which would be a requirement if one would wish to use a local ventilation controller. In addition, constraints play an important role because the ventilation flux capacity depends quite a lot on the wind speed. So, particularly on quiet days, the required ventilation may not be realized. A disadvantage clearly is that it is now necessary to have the wind speed as external disturbance.

For all these reasons, in this study, it was decided to model not only the main greenhouse compartment but also the heating pipe system and the ventilation. We come back to this point in Section 7.3.3.

7.3.1 Heat Balances

Greenhouse compartment (terms toward greenhouse compartment are positive, loss terms negative)

$$K_g \frac{dT_g}{dt} = q_{o\_g}^{rad} - q_{g\_o}^{vent} - q_{g\_o}^{cond} - q_{g\_s} + q_{p\_g} + q_{g\_r}^{cons} - q_{g\_c}^{trans}. \qquad (\text{W m}^{-2}[\text{gh}]) \quad (7.43)$$

In Equation 7.43, we chose to express all flux variables in flows per square meter projected ground area of the greenhouse. The advantage of this is that in this form it is easy to analyze the various contributions in heat and mass balances because they are all expressed in the same units. The disadvantage is that some care must be taken to convert fluxes that conventionally are taken with respect to local surface areas into the correct per unit greenhouse area units.

K_g is the (virtual) heat capacity of the greenhouse compartment (J m^{-2}[gh] K^{-1}). If it would be determined solely by the air heat capacity, it could be computed from

$$K_g = \rho_a c_{p,a} \frac{V_g}{A_g}, \qquad (\text{J m}^{-2}[\text{gh}]\ \text{k}^{-1}) \quad (7.44)$$

but this value should be seen as a lower limit because the crop and the greenhouse materials such as gutters and the like also contribute to the heat capacity. The greenhouse compartment thus is a lumped compartment, combining radiation-receiving materials (except the heating pipes) and greenhouse air. The second-order effect of radiation received by materials, which is then transferred to the air, is ignored in this simple model (for a more elaborate physical model, see Chapter 8). To clearly distinguish this lumped model from models where the greenhouse compartment is separated in various components, such as in Chapter 8, the subscript g is used to remind the reader that we are talking about the lumped greenhouse compartment and not about just the air (which is denoted by subscript "a"). The rate of change of the (sensible) heat content in the greenhouse compartment (W m^{-2}[gh]) is the sum of various contributions, specified as follows:

- Solar radiation absorbed

$$q_{o\_g}^{rad} = \eta_g I_o, \qquad (\text{W m}^{-2}[\text{gh}]) \quad (7.45)$$

where I_o is the global radiation and η_g is the fraction contributing to heat gain of the greenhouse compartment (air and solids). In the model, the heat absorbed by the heating pipe system is considered separately and is equal to

$$\eta_p = \frac{A_p}{A_g}\eta, \qquad (-) \quad (7.46)$$

where the term A_p/A_g amounts to the total (visual) pipe surface area per unit greenhouse projected area. The fraction η_g therefore can be written as

$$\eta_g = \eta - \eta_p = \left(1 - \frac{A_p}{A_g}\right)\eta. \qquad (\mathrm{W\ m^{-2}[gh]}) \quad (7.47)$$

It should be noted that the η used here will be higher than the heat absorption coefficient of Chapter 6 because in Chapter 6, the factor is also needed to account for the temperature mitigating effect of evapotranspiration, which is separated out here by the last term of Equation 7.43. As an approximation, it is simply set equal to the transmittance of the roof

$$\eta = \tau_r. \qquad (-) \quad (7.48)$$

- Heat loss by ventilation

$$q_{g\_o}^{vent} = u_v \rho_a c_{p,a}\left(T_g - T_o\right), \qquad (\mathrm{W\ m^{-2}[gh]}) \quad (7.49)$$

where u_v is the volumetric ventilation flow rate per unit greenhouse area (m^3[air] m^{-2}[gh] s^{-1}). With the standard symbol convention, $u_v \equiv \frac{1}{A_g} F_{g\_o}^{vent}$. The notation u_v is used here to allow comparison with the model presented in Chapter 6, where the ventilation flow is treated as a control input (for more details, see Section 7.3.3). Remark that the ratio u_v/h is the refreshment rate (s^{-1}) of the greenhouse (with the effective height $h = V_g/A_g$). The ventilation flow rate is related to the wind speed and the windward ($0 \le u_v^{Apwsd} \le 1$) and lee side ($0 \le u_v^{Aplsd} \le 1$) relative window openings according to

$$u_v = \left(\frac{p_{v1} u_v^{Aplsd}}{1 + p_{v2} u_v^{Aplsd}} + p_{v3} + p_{v4} u_v^{Apwsd}\right) v + p_{v5}, \qquad (\mathrm{m^3[air]\ m^{-2}[gh])\ s^{-1}}) \quad (7.50)$$

where p_{vj}, $j = 1, \ldots, 5$ are parameters, and v is the wind speed (De Jong, 1990; Van Henten, 1994). The superscripts "Aplsd" and "Apwsd" are used to distinguish the relative apertures u_v^{Aplsd} and u_v^{Apwsd} as control inputs from the resulting airflow u_v that serves as control input in Chapter 6 (for more details, see Section 7.3.3). As normally the windows are opened first on the leeward side and then on the windward side, it is convenient to introduce a single control variable $0 \le u_v^{Ap} \le 2$, from which u_v^{Aplsd} and u_v^{Apwsd} can be computed according to

$$u_v^{Apwsd} = \frac{u_v^{Ap} - 1 + \left|u_v^{Ap} - 1\right|}{2}, \qquad (-) \quad (7.51)$$

$$u_v^{Aplsd} = u_v^{Ap} - u_v^{Apwsd}. \qquad (-) \quad (7.52)$$

In Equation 7.49, the latent heat transport associated with the moisture content difference of the exchanged air, is ignored.

- Heat loss by conduction through the roof and walls,

$$q_{g\_o}^{cond} = U_{g\_o} \frac{A_{g\_o}}{A_g} \left(T_g - T_o \right), \qquad (W\ m^{-2}[gh]) \quad (7.53)$$

where U_{g_o} (W m^{-2} K^{-1}) is the overall heat transfer coefficient of the cover, defined as the sum of roof and walls, and the factor A_{g_o}/A_g is the total area of the cover per unit greenhouse projected area.

- Heat exchange with the soil

$$q_{g\_s} = U_{g\_s} \frac{A_{g\_s}}{A_g} \left(T_g - T_s \right), \qquad (W\ m^{-2}[gh]) \quad (7.54)$$

where U_{g_s} is the overall heat transfer coefficient toward the soil, and A_{g_s}/A_g is the ratio of the effective contact area of the soil relative to the ground area. This ratio will be close to one. The soil temperature T_s is another state variable of the model (Section 7.3.1.1).

- Heat supplied via the pipe system

$$q_{p\_g} = U_{p\_g} \frac{A_{p\_g}}{A_g} \left(T_p - T_g \right), \qquad (W\ m^{-2}[gh]) \quad (7.55)$$

where U_{p_g} is the overall heat transfer coefficient from the water in the heating pipes to the greenhouse air, and A_{p_g}/A_p is the ratio of the contact area of the pipes relative to the ground area. The pipe temperature T_p is modeled with its own heat balance and is therefore a state variable as well (Section 7.3.1.2).

The overall heat transfer coefficient depends on a mix of forced and free convection. It is modeled by

$$U_{p\_g} = p_{p1} \sqrt{p_{p2} + \sqrt{\left| T_g - T_p \right|}}, \qquad (W\ K^{-1}\ m^{-2}) \quad (7.56)$$

where p_{p1} and p_{p2} are parameters.

- Part of the heat released to the greenhouse compartment due to condensation of moisture on walls and roof,

$$q_{g\_r}^{cons} = \Lambda(1-\kappa)\varphi_{H_2O\_g\_r}^{cons}, \qquad (W\ m^{-2}[gh]) \quad (7.57)$$

where Λ (J kg^{-1}) is the evaporation heat of water, and $\varphi_{H_2O\_g\_r}^{cons}$ (kg[H_2O] m[gh]$^{-2}$ s^{-1}) is the condensation flux, as specified later. The factor $1 - \kappa$ expresses which part of the condensation heat is favoring the greenhouse air. It means that a fraction κ is lost to the environment. The factor κ is the ratio of inner resistance to total resistance toward transport of heat via the greenhouse cover.

- Heat withdrawn from the greenhouse by the canopy for evapotranspiration

$$q_{g\_c}^{trans} = \Lambda E_c, \qquad \text{(W m}^{-2}\text{[gh])} \quad (7.58)$$

where E_c is the evapotranspiration rate of the canopy (kg[H_2O] m^{-2}[gh] s^{-1}), and Λ is the heat of evaporation of water (J kg^{-1}). The modeling of E_c is described in Section 7.3.2.

7.3.1.1 Soil

A lumped soil compartment is considered as a first-order approximation of exchange of heat with the ground in the greenhouse. In the model, it is assumed that the soil does not receive direct heat by radiation (full soil coverage).

$$K_s \frac{dT_s}{dt} = q_{g\_s} - q_{s\_ss}. \qquad \text{(W m}^{-2}\text{[gh])} \quad (7.59)$$

The rate of change of the heat contents in the soil, with virtual heat capacity K_s (J m^{-2}[gh] K^{-1}), is equal to the heat loss from the greenhouse air (q_{g_s}) minus the heat loss from soil to deeper soil:

$$q_{s\_ss} = U_{s\_ss}(T_s - T_{ss}), \qquad \text{(W m}^{-2}\text{[gh])} \quad (7.60)$$

where U_{s_ss} is the overall heat transfer coefficient toward the soil, and T_{ss} is the subsoil temperature, assumed to be a constant parameter.

7.3.1.2 Heating Pipe System

$$K_p \frac{dT_p}{dt} = q_{boil\_p} - q_{p\_g} + q_{o\_p}^{rad}. \qquad \text{(W m}^{-2}\text{[gh])} \quad (7.61)$$

The rate of change of the heat in the pipes is given by the left-hand side. Note that the pipe heat capacity refers, as before, to a unit ground area, that is,

$$K_p = \rho_{H_2O} c_{p,H_2O} \frac{V_p}{A_g}. \qquad \text{(J K}^{-1}\text{ m}^{-2}\text{ [gh])} \quad (7.62)$$

In Equation 7.61, the pipe temperature is a lumped variable representing the effective mean temperature over the length of the pipe, irrespective of the presence of a gradient. The rate of change is equal to the heat supply to the pipe system, q_{boil_p}, which is equivalent to the difference in heat content of water entering and leaving the pipe system per unit ground area (W m[gh]$^{-2}$):

$$q_{\text{boil\_p}} = q_{\text{in\_p}} - q_{\text{p\_out}} = \frac{1}{A_g} \rho_{\text{H}_2\text{O}} c_{\text{p,H}_2\text{O}} F_p \left(T_{\text{p\_in}} - T_{\text{p\_out}} \right), \quad (\text{W m}^{-2}[\text{gh}]) \quad (7.63)$$

minus the loss of heat to the greenhouse $q_{\text{p\_g}}$, Equation 7.55, plus some gain due to radiation toward the pipes $q_{\text{o\_p}}^{\text{rad}}$. The latter is introduced because it was observed that even without heat input from the boiler, the water in the pipes gets hotter during daytime. The received radiation is modeled by

$$q_{\text{o\_p}}^{\text{rad}} = \eta_p I_o, \quad (\text{W m}^{-2}[\text{gh}]) \quad (7.64)$$

where η_p is given by Equation 7.46.

Note that the total heat absorbed by the complete greenhouse is given by

$$q^{\text{rad}} = q_{\text{o\_g}}^{\text{rad}} + q_{\text{o\_p}}^{\text{rad}} = \eta I_o. \quad (\text{W m}^{-2}[\text{gh}]) \quad (7.65)$$

The heat supplied to the pipe system depends on the layout of the system. In the experimental setup, the layout was as given in Figure 7.4. In the system, it is assumed that the flow from the boiler (F_{boil}) and the flow of the circulation pump (F_p) are known fixed variables. Also the temperature of the boiler T_{boil} is considered to be known and fixed. The heating valve is a three-way valve, defined by its opening u_q^{vp}. Position 0 (0%) means no supply to the heating pipe circuit with full recirculation back to the boiler via the bypass, position 1 (100%) means that all hot water is entering the heating pipe circuit. Using an overall heat balance for this system and the mean temperature,

$$T_p = \frac{T_{\text{p\_in}} + T_{\text{p\_out}}}{2}, \quad (\text{K}) \quad (7.66)$$

it can be shown that the heat energy supply rate from the boiler to the lumped pipe system is given by

$$u_{\text{boil\_p}} = \frac{1}{A_g} \rho_w c_{pw} F_p \frac{2u_q^{\text{vp}}}{\left(2\frac{F_p}{F_{\text{boil}}} - u_q^{\text{vp}} \right)} \left(T_{\text{boil}} - T_p \right). \quad (\text{W m}^{-2}[\text{gh}]) \quad (7.67)$$

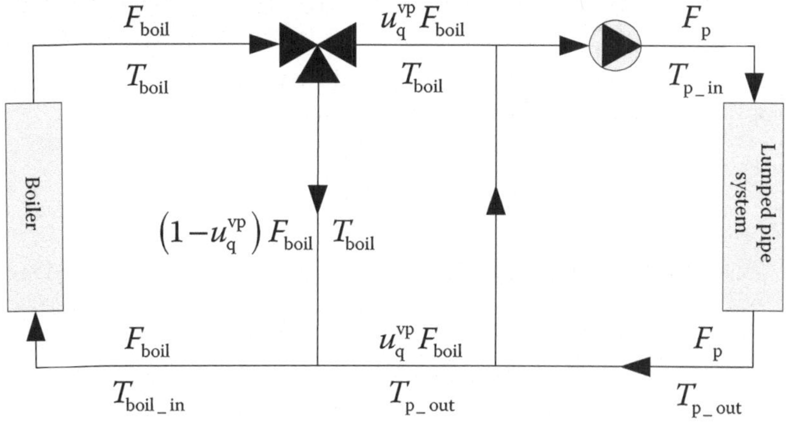

FIGURE 7.4 Heating pipe system as used in the experiment.

Note that the relationship between the valve position u_q^{vp} (control variable) and the actual heat input to the greenhouse is nonlinear due to this construction.

An alternative to the model above is to assume the existence of a well-working pipe temperature controller. In that case, the heat transfer to the greenhouse would still be given by Equation 7.55, but then the pipe temperature setpoint is the control input (if the controller is really fast, it can be set equal to the actual pipe temperature), and any disturbances in boiler temperature or flows would be accommodated by the controller. On the other hand, at the expense of two flow sensors (needed anyway if one wishes to know the realized heat input) and a thermometer to measure the boiler temperature (usually already there), it is easy to correct a computed control for variations in these quantities. Once again, using the pipe model automatically takes into account constraints, which is not guaranteed if the pipe temperature is taken as the free manipulative variable, as it is quite conceivable that under circumstance the commanded pipe temperature cannot be realized.

Ideally, the parameterization of the lumped pipe model above has to take account of the longitudinal temperature gradient in the pipe system. This would require a distributed heating pipe model, for instance, the one described by De Zwart (1996). One would then have to relate the lumped parameters to the parameters of the distributed model by equating the heat transfer from both models. This is possible for stationary conditions, but the dynamics will still be somewhat different. Another option is to minimize the difference in the Bode diagram of the lumped and the distributed model over the frequency range of interest.

7.3.2 Mass Balances

7.3.2.1 Water Vapor in the Greenhouse Air

$$\frac{V_g}{A_g}\frac{dC_{H_2O}}{dt} = E_c - \varphi^{vent}_{H_2O,g\_o} - \varphi^{cons}_{H_2O,g\_r}. \qquad (kg[H_2O]\ m^{-2}[gh]\ s^{-1}) \quad (7.68)$$

The rate of change of the concentration of vapor in the air times the capacity, that is, the volume V_g (m^3) per unit greenhouse ground area A_g (m^2[gh]) equals the evapotranspiration E_c (kg[H_2O] m^{-2}[gh] s^{-1}) minus the loss with ventilation $\varphi^{vent}_{H_2O,g\_o}$ minus the loss due to condensation $\varphi^{cons}_{H_2O,g\_r}$.

The evapotranspiration is modeled with a general Penman–Monteith form (cf. Jolliet and Bailey, 1992):

$$E_c = \alpha_c \eta_g I_o + \beta_c \Delta p_{H_2O}\{\cdot\}, \qquad (kg[H_2O]\ m^{-2}[gh]\ s^{-1}) \quad (7.69)$$

where $\Delta p_{H_2O}\{\cdot\}$ is the vapor pressure deficit, that is, the difference between the saturation vapor pressure at the canopy temperature T_c and the actual vapor pressure in the air:

$$\Delta p_{H_2O} = p^{sat}_{H_2O}(T_c) - p_{H_2O}. \qquad (Pa) \quad (7.70)$$

The saturation pressure $p^{sat}_{H_2O}$ at temperature T (K) is an empirical relation as defined in Section 8.B.1,

$$p^{sat}_{H_2O}(T) = c_{s1}\exp\left(\frac{c_{s2}T^C}{c_{s3}+T^C}\right), \qquad (Pa) \quad (7.71)$$

with $T^C = T - 273$, the temperature in degrees Celsius. Because the canopy temperature is not a state in this simple model, its temperature is set equal to the greenhouse air temperature, that is,

$$T_c = T_g. \qquad \text{(K)} \quad (7.72)$$

Equation 7.69 is valid for full coverage. To correct for stages of crop growth where full coverage has not been reached, here we simply multiply by the crop maturity factor (Equation 7.21), that is,

$$E_c = f_m\{\cdot\}\left(\alpha_c \eta_g I_o + \beta_c \Delta p_{H_2O}\{\cdot\}\right). \qquad (\text{kg[H}_2\text{O] m}^{-2}\text{[gh] s}^{-1}) \quad (7.73)$$

The parameters α_c and β_c depend slightly on temperature, radiation intensity, and CO_2 concentration. This is ignored here. A more elaborate evapotranspiration model is presented in Chapter 8.

It should be noted that Equation 7.73 merely expresses how the evaporation depends on the short wave radiation, that is, the heat load to the crop is not expressed explicitly. Instead, in the model, the total heat load entering the greenhouse first is used to increase the temperature, and part of it is taken back due to the evapotranspiration (see Equation 7.43). This is different from more elaborate models that have separate states for the canopy temperature, such as the model in Chapter 8.

The ventilation loss is given by

$$\varphi_{H_2O,g\_o}^{vent} = u_v\left(C_{H_2O} - C_{H_2O,o}\right). \qquad (\text{kg[H}_2\text{O] m}^{-2}\text{[gh] s}^{-1}) \quad (7.74)$$

Condensation only occurs when the moisture concentration in the air is higher than the saturation concentration belonging to the roof temperature. The driving force for moisture transport can be expressed in terms of the vapor concentration (kg[H_2O] m^{-3}), the humidity mixing ratio (kg[H_2O] kg^{-1}[dry air]), or the vapor pressure (Pa). In view of the similarity with crop evapotranspiration, we choose to express the condensation as function of the vapor pressure surplus:

$$\varphi_{H_2O,g\_r}^{cons} = k_{g\_r}\frac{A_r}{A_g}\left(p_{H_2O} - p_{H_2O}^{sat}(T_r)\right), \qquad (\text{kg[H}_2\text{O] m}^{-2}\text{[gh] s}^{-1}) \quad (7.75)$$

where A_r is the total surface of the cover, that is, roof plus walls. Condensation only occurs when there is a positive driving force toward the cold surface, which can be expressed in a single formula without if statements as

$$\varphi_{H_2O,g\_r}^{cons} = k_{g\_r}\frac{A_r}{A_g}\frac{p_{H_2O} - p_{H_2O}^{sat}(T_r) + \left|p_{H_2O} - p_{H_2O}^{sat}(T_r)\right|}{2}. \qquad (\text{kg[H}_2\text{O] m}^{-2}\text{[gh] s}^{-1}) \quad (7.76)$$

The mass transfer rate is computed from

$$k_{g\_r} = c_{m1}\left|\frac{T_g - T_r}{c_{T_{g\_r}}}\right|^{c_{m2}}, \qquad (\text{kg[H}_2\text{O] m}^{-2}\text{ s}^{-1}\text{ Pa}^{-1}) \quad (7.77)$$

with c_{m1} and c_{m2} as parameters (Monteith and Unsworth, 1990). The constant $c_{T_{g\_r}}$ in the denominator is equal to 1 K and is added to make the equation dimensionally correct.

The roof (cover) temperature is a weighted mean between outside and inside air temperature,

$$T_r = (1-\kappa)T_g + \kappa T_o, \qquad \text{(K)} \quad (7.78)$$

with $0 \le \kappa \le 1$ as given before.

7.3.2.2 Carbon Dioxide in the Greenhouse Air

$$\frac{V_g}{A_g}\frac{dC_{CO_2}}{dt} = -\eta_{CO_2/dw}P + \eta_{CO_2/dw}R - \varphi^{vent}_{CO_2,g\_o} + u_{CO_2}. \qquad (kg[CO_2]\ m^{-2}[gh]\ s^{-1}) \quad (7.79)$$

The total mass of CO_2 in the greenhouse is $V_g C_{CO_2}$, where V_g (m^3) is the greenhouse air volume, and C_{CO_2} (kg[CO_2] m^{-3}) is the concentration. The rate of change of the CO_2 mass (kg) per unit greenhouse ground area A_g equals the amount taken up by photosynthesis, $\eta_{CO_2/dw}P$, where $\eta_{CO_2/dw}$ (kg[CO_2) kg^{-1}[dw]) is the amount of CO_2 needed to form one unit of biomass plus the amount returned by respiration $\eta_{CO_2/dw}R$ minus the loss by ventilation ($\varphi^{vent}_{CO_2,g\_o}$) plus the supply ($u_{CO_2}$). P and R (kg[dw] m^{-2}[gh] s^{-1}) are the photosynthesis and the total crop respiration as described in Sections 7.2.4 and 7.2.5.

The ventilation loss is given by

$$\varphi^{vent}_{CO_2,g\_o} = u_v\left(C_{CO_2} - C_{CO_2\_o}\right). \qquad (kg[CO_2]\ m^{-2}[gh]\ s^{-1}) \quad (7.80)$$

The supply can be conveniently expressed in relative valve opening $u^{vp}_{CO_2}$ (in practice often realized via the duty cycle of a pulse-width modulated valve)

$$u_{CO_2} \equiv \varphi_{CO_2,in\_g} = u^{vp}_{CO_2}\varphi^{max}_{CO_2,in\_g}. \qquad (kg[CO_2]\ m^{-2}[gh]\ s^{-1}) \quad (7.81)$$

7.3.3 Comparison of Lumped Model with Control Input by Actuators or by Fluxes

In the model above, the heating valve position, the window aperture, and the CO_2 valve position are handled as the control input. These are therefore directly referring to the actual actuators physically present in the greenhouse. In Chapter 6 and in many studies in the literature, the control problem is simplified to deal with the heat flux, the ventilation flux, and the CO_2 flux (flows per unit greenhouse projected area) as controls. The relationship is depicted in Figure 7.5. In fact, in the lumped model of Chapter 6, the boundary is shifted more inward. It can be seen that the heat input and the ventilation rate are not simply determined by the valve position or window aperture alone but also depend on the temperature of the greenhouse air and on external factors, such as the boiler temperature and the wind speed. As argued before, this means that treating the fluxes as input has the risk that a flux demanded by the optimal controller cannot be delivered in practice. It is possible to incorporate this in the optimal control, but in that case, we end up with state and external input-dependent bounds, which is more complicated. On the other hand, the flux approach is easier to implement and to understand because it does not require extensive modeling of the heat and ventilation system. As a final remark, it has to be noted that because of the heat capacity of the pipe system, the net heat delivered to the greenhouse $u_q = q_{p\_g} - q^{rad}_{o\_g}$ is not at all times equal to the net heat supplied by the boiler $u_q = q_{boil\_p}$, as

$$q_{p\_g} - q^{rad}_{o\_p} = q_{boil\_p} - K_p\frac{dT_p}{dt}, \qquad (W\ m^{-2}[gh]) \quad (7.82)$$

so that, strictly speaking, it matters how u_q in the flux lumped model is evaluated. This is the reason to put u_q between brackets in Figure 7.5. However, when the costs are evaluated by taking the integrals over time of the lumped heat supply to the greenhouse as evaluated by the left-hand side

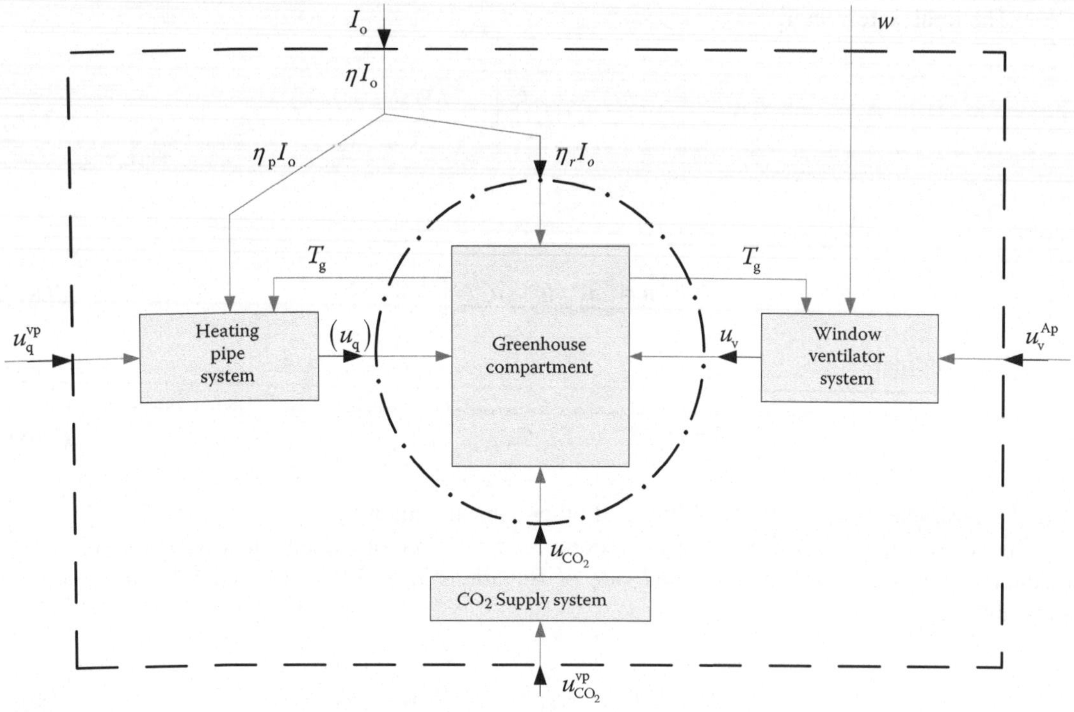

FIGURE 7.5 Information flow comparison between lumped systems with actuators as control input and lumped systems with fluxes as input (u_q).

of Equation 7.82 or the actual heat supply withdrawn from the boiler q_{boil_p} as defined by Equation 7.67, the contribution of the d/dt term in the right-hand side of Equation 7.82 vanishes so that the two alternative terms will lead to the same energy costs in the optimization.

7.4 STATE–SPACE FORM OF THE COMPLETE GREENHOUSE-CROP MODEL

Together, Equations 7.1 through 7.81 form a set of ordinary differential equations that can easily be cast in standard state–space form

$$\dot{\mathbf{x}} = \mathbf{f}(\mathbf{x}, \mathbf{u}, \mathbf{d}, \mathbf{p})$$
$$\mathbf{y} = \mathbf{g}(\mathbf{x}, \mathbf{u}, \mathbf{d}, \mathbf{p}) \tag{7.83}$$

by defining

The crop state vector,*

$$\mathbf{x}_c = [W_B, W_L, W_F, D]^T, \tag{7.84}$$

The greenhouse state vector,

$$\mathbf{x}_g = \left[T_g, T_s, T_p, C_{H_2O}, C_{CO_2}\right]^T, \tag{7.85}$$

* Strictly speaking, also e_D, Eq. 7.33.

The total state vector,

$$\mathbf{x} = \begin{bmatrix} \mathbf{x}_c \\ \mathbf{x}_g \end{bmatrix}, \tag{7.86}$$

The control inputs,

$$\mathbf{u} = \left[u_q^{vp},\, u_v^{Ap},\, u_{CO_2}^{vp}\right]^T, \tag{7.87}$$

The external inputs,

$$\mathbf{d} = [I_o,\, T_o,\, v,\, T_{ss},\, C_{H_2O\_o},\, C_{CO_2\_o}]^T, \tag{7.88}$$

and the parameter vector **p** that combines all other constant quantities.

The vector-valued function **f** can be decomposed in a vector-valued function $\mathbf{f}_c$ for the crop, essentially consisting of the right-hand side of Equations 7.15, 7.18, 7.19, and 7.32, with proper substitutions,

$$\dot{\mathbf{x}}_c = \mathbf{f}_c(\mathbf{x}_c,\, \mathbf{u},\, [\mathbf{d} \quad \mathbf{x}_g]^T,\, \mathbf{p}), \tag{7.89}$$

and a vector-valued function $\mathbf{f}_g$, essentially consisting of the right-hand sides of Equations 7.43, 7.59, 7.61, 7.68, and 7.79, with proper substitutions,

$$\dot{\mathbf{x}}_g = \mathbf{f}_g(\mathbf{x}_g,\, \mathbf{u},\, [\mathbf{d} \quad \mathbf{x}_c]^T,\, \mathbf{p}). \tag{7.90}$$

The outputs (variables of interest) can be chosen freely, but if additional integrated values are needed, such as the harvested fruit biomass as defined in 7.42 or the total energy consumption, then the system must be expanded with additional integrators to compute these nonessential states, which is easy to do. These additional states are nonessential because there is no feedback to the system itself. Figure 7.6 illustrates the information flow in the model, with separate representations for the greenhouse subsystem and the crop subsystem. Note that the states of the greenhouse act as inputs to the crop—they can be viewed as disturbance or as control inputs—and that the states of the crop act as disturbance inputs to the greenhouse. The equations for output variables and integrated outputs can be modified at any time, without affecting the underlying state dynamics of the system. An important subset of the outputs are the equations that relate the commonly measured variables temperature, relative humidity (RH) (%), and CO_2 (ppm) to the states of the system:

$$\mathbf{y} = \begin{pmatrix} T \\ \mathrm{RH} \\ C_{CO_2}^{ppm} \end{pmatrix} = \begin{pmatrix} g_1(T) \\ g_2(C_{H_2O},\, T) \\ g_3(C_{CO_2}) \end{pmatrix}, \tag{7.91}$$

where $g_1(T) = T$, g_2 is a function that relates RH to vapor concentration and temperature (see Section 8.B.2), and g_3 is a conversion function from kg[CO_2] m^{-3} to ppm.

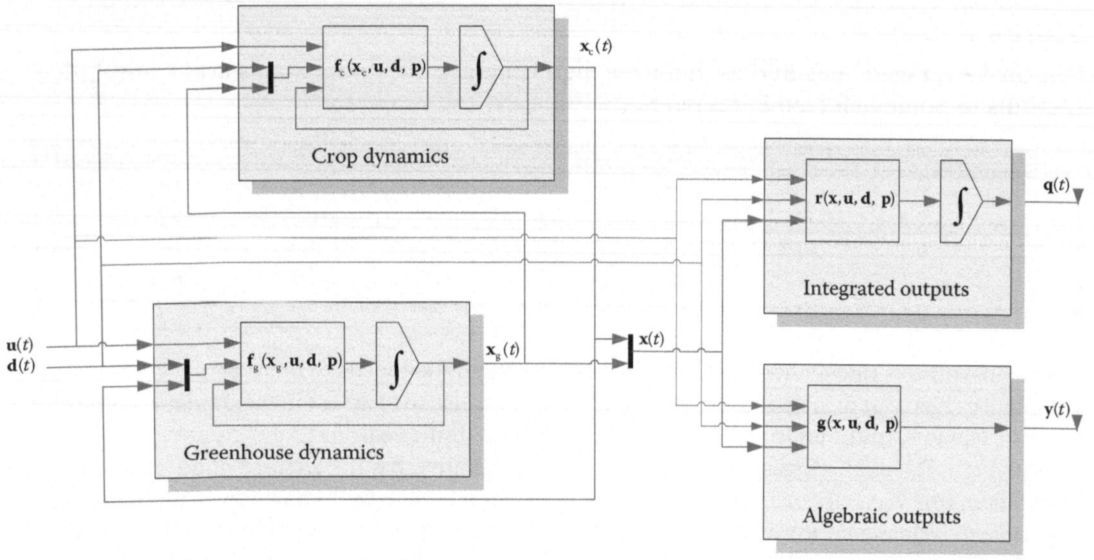

FIGURE 7.6 Information flow in state–space form of the greenhouse-crop system together with outputs and integrated outputs.

7.5 CALIBRATION AND MODEL RESULTS

Before the model can be used for control, it is necessary to calibrate it by matching the model outputs with observations. Some parameters are known from prior knowledge or independent measurements, and others need to be adjusted to the actual situation. A too large number is not feasible, so it makes sense to apply a systematic procedure for parameter selection for calibration. Here, a sensitivity analysis was used to identify unknown but sensitive parameters that were later used for parameter estimation.

7.5.1 Calibration of the Big Leaf–Big Fruit Model

The big leaf–big fruit model is in fact a reduced model of more elaborate descriptions of tomato growth, which takes into account the development of separate internodes and fruit classes (De Koning, 1994). Although a number of parameters could be obtained from the underlying investigations, there are still a number left for calibration.

A sensitivity analysis was performed with the first-order sensitivity system (see, e.g., Frank, 1978). Together with information on the reliability of parameters from the literature, this led to a preliminary selection of parameters for which further analysis would be worthwhile.

Table 7.1 gives the mean sensitivities for eight selected parameters, evaluated as the mean of the absolute values of the relative sensitivities over a cultivation period. They are expressed as percentage change in the variable of interest per percentage change in the parameter. The analysis was done using a preliminary calibrated model on data from an experimental greenhouse.

The parameters c_{d1} and c_{d3} are related to the development stage (Equation 7.32) together with c_{d2} and c_{d4}, which have a similar effect and are not shown for that reason. All other parameters have no effect on the development stage because it only depends on the temperature. However, as the development rate determines the moment where harvesting starts, its parameters also affect the other states. The reference temperature $T_{L/F}^{ref}$ in the relation that determines the vegetative/generative distribution (Equation 7.27) also has a large effect on the fruit and leaf biomass (W_F and W_L, respectively). The same holds for ν_2 and $f_{L/F}^{ref}$. It is also interesting to discuss briefly the role of P^{max} (Equation 7.22). One percent increase over the full period leads to a substantial relative increase in leaf and fruits on the plant, on average. The increase is more than proportional because increasing

TABLE 7.1
Seasonal Averaged Sensitivities, Ignoring Sign Changes, of Model States and Cumulative Outputs to Some Selected Parameters (%/%)

	c_{yL}	c_{yF}	$f_{L/F}^{ref}$	P^{max}	p_m	c_{d1}	c_{d3}	$T_{L/F}^{ref}$
W_F	0.4	0.1	3.4	3.8	4.2	1.0	1.1	10.9
W_L	0.1	0.1	2.1	2.4	3.1	0.7	0.8	6.8
D	0.0	0.0	0.0	0.0	0.0	1.0	1.2	0.0
S_{WF}	0.2	0.7	1.9	1.6	1.2	7.7	8.9	6.1
S_{WL}	0.1	0.7	1.2	1.0	0.9	7.9	9.2	4.0

the photosynthesis rate without a similar increase in respiration leads to more gain. The cumulative fruit and leaf harvest also increase, but less, which is because the harvest in the model is a first-order process, leaving fruits and leaves on the plant at the end of the season.

Although Table 7.1 suggests that the vegetative/generative parameters are good candidates for calibration, this was not done because they were considered to be well known from the independent experiments. Moreover, it was questionable whether the temperature conditions in the greenhouse would have a sufficiently wide range to get reliable results. The excitation of inputs in crop growth experiments is an issue that needs further discussion in relation to optimal control. There is a danger that the optimal control exploits regions of the model—if that is favorable to the goal function—for which the model has not been properly calibrated. This should of course be avoided, for example, by introducing constraints.

Ultimately, the following parameters were selected for calibration:

- The maximum photosynthesis rate P^{max} as it directly determines the biomass level.
- The parameter p_m that determines the reduction factor in early stages of plant development when the leaf area index has not reached its maximum (Equation 7.21).
- The parameter $f_{L/F}^{ref}$ that determines the vegetative/generative ratio (Equation 7.27).
- The proportionality factors of the harvest c_{yF} and c_{yL} (Equations 7.39 and 7.40).

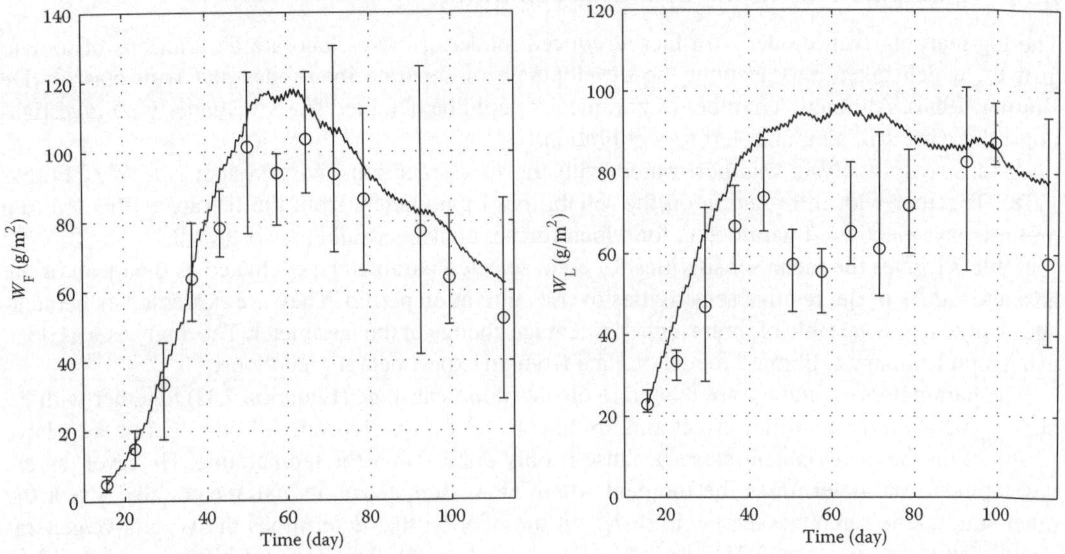

FIGURE 7.7 Validation of the big fruit–big leaf model on independent experimental data. Circles: average biomass (g[dw] m^{-2}[gh]) over five or six replicas. Error bars show upper and lower limits of observed values. Solid line: model simulation.

It should be noted that for the purpose of calibration it is probably more accurate to treat the actually observed harvest as input, but in an optimal control setting, it is necessary to have a harvest model because otherwise it is not possible to compute the benefit. With the given formulation, the effect of the harvest model on the crop biomass is that with higher the temperature, more fruits and leaves can be harvested because conversion of assimilates into structural biomass is faster, at least temporarily over a period that the assimilate buffer can support it.

The calibration was performed on experimental data described by Heuvelink (1995). The model was calibrated on the simultaneous fit of experiments 7 and 8. A validation was done on the data from experiment 10. This experiment lasted from early August until the end of November. Consequently, the global radiation was decreasing, on average, during this period. The temperature setpoint in the greenhouse was 18°C, but especially during August, early September, and early October, temperatures during the day were considerably higher, in the order of magnitude of 25°C. CO_2 was occasionally dosed, but concentrations were not much higher than the target value of 300 ppm. The solar radiation was gradually decreasing toward the end of the period, with lower spells around days 50–60 and 75–85 after the first appearance of the fruits. Figure 7.7 shows the result of experiment and simulation. In the experiment, the assessment of fruit and leaf dry weight was done by slaughtering five to six plants. The bars show the upper and lower limit of these data, thus clearly indicating that crop experiments are subject to large experimental uncertainty. The big fruit–big

TABLE 7.2
Nominal Crop Model Parameters, in Order of Appearance in the Text

Symbol	Value	Units	Meaning
z	0.6081	–	Fraction leaf of total vegetative mass
θ_F	0.2	–	Surplus assimilate requirement factor per unit fruit increment
θ_V	0.23	–	Surplus assimilate requirement factor per unit vegetative increment
p_h	2.70×10^{-3}	$m^2\ kg^{-1}$[dw]	Parameter of switching function
p_m	1.78×10^{-2}	$kg\ m^{-2}$	Parameter in maturity factor
m	2.511	–	Parameter in maturity factor
p_{max}	2.25×10^{-6}	kg[dw] m^{-2}[gh] s^{-1}	Maximum gross canopy photosynthesis rate
K_I	577	$W\ m^{-2}$	Monod constant for PAR
K_C	0.211	$kg[CO_2]\ m^{-3}$	Monod constant for CO_2
$f_{PAR/I}$	0.475	–	PAR fraction of global radiation
τ_r	0.7	–	Transmittance of the roof
k_{GF}^{ref}	3.81941×10^{-6}	s^{-1}	Reference fruit growth rate coefficient
T_{GF}^{ref}	20	°C	Reference temperature
Q_{10G}	1.6	–	Temperature function parameter growth
$f_{L/F}^{ref}$	1.3774	–	Reference leaf–fruit partitioning factor
ν_2	–0.168	K^{-1}	Fruit–leaf partitioning parameter
$T_{L/F}^{ref}$	19	°C	Fruit–leaf partitioning reference temperature
k_{RL}^{ref}	2.89×10^{-7}	s^{-1}	Maintenance respiration coefficient leaf
Q_{10R}	2	–	Temperature function parameter respiration
T_R^{ref}	25	°C	Reference temperature for respiration
k_{RF}^{ref}	1.16×10^{-7}	s^{-1}	Maintenance respiration coefficient leaf
c_{f1}	4.6296×10^{-6}	s^{-1}	Parameter in fruit growth function
c_{f2}	8.1019×10^{-7}	s^{-1}	Parameter in fruit growth function
c_{d1}	2.13×10^{-7}	s^{-1}	Parameter in development rate function
c_{d2}	2.47×10^{-7}	s^{-1}	Parameter in development rate function
c_{d3}	20	°C	Parameter in development rate function
c_{d4}	7.50×10^{-11}	s^{-2}	Parameter in development rate function
c_{yF}	1.636	–	Parameter in harvest function (fruit)
c_{yL}	0.4805	–	Parameter in harvest function (leaf)

leaf model performs quite well in reproducing the observed fruit weights, but from the leaf weight, it is clear that the model is not perfect. Especially in the period around day 60, the true leaf weight is quite a bit lower than expected by the model. This may be due to the attempt to model the leaf picking instead of using the real events driven by the grower's decision. A full list of the crop parameters is given in Table 7.2.

7.5.2 Calibration of the Heating Pipe and Greenhouse Climate Model

In a similar fashion, the heating pipe model was calibrated using the observed greenhouse air temperatures as inputs. After calibration of the pipe model, using the greenhouse data as inputs, the greenhouse model was calibrated using the observed heating pipe temperatures as inputs. Details about the parameter selection procedure as well as the covariance matrix of the parameter estimates can be found in the study of Tap (2000). An example of a calibration result for one particular day is shown in Figure 7.8. Despite the lumped nature of the model, which is certainly a limitation, quite good results are obtained.

When the model is validated by applying it to independent data in the same period of the year, quite good results are obtained, as shown in Figure 7.9. The largest deviations are observed with the calculated RH. The model is quite sensitive to the parameters of the crop evaporation model (Equation 7.73). Moreover, any temperature deviations will also be visible in the RH fit. In comparing real data to model results, it should be kept in mind that local CO_2, moisture content, and temperature differences, which can be quite substantial, are not captured in a lumped model. A full list of the greenhouse parameters is given in Table 7.3.

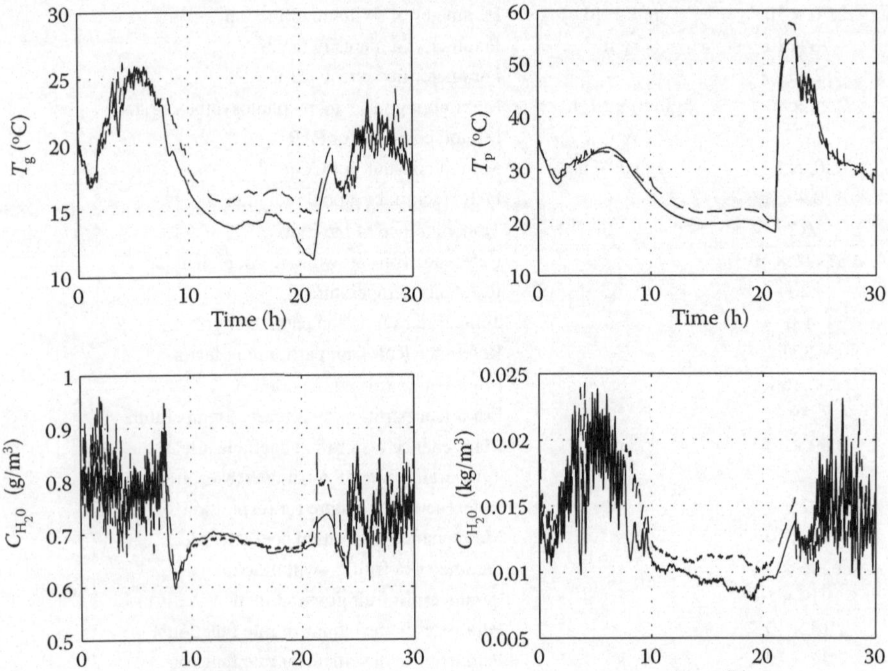

FIGURE 7.8 Example of calibration result for the heating pipe submodel and greenhouse climate model. Left-top: greenhouse air temperature; right-top: heating pipe temperature; lower left: CO_2 concentration; lower right: water vapor concentration; dashed: measured; solid: simulated.

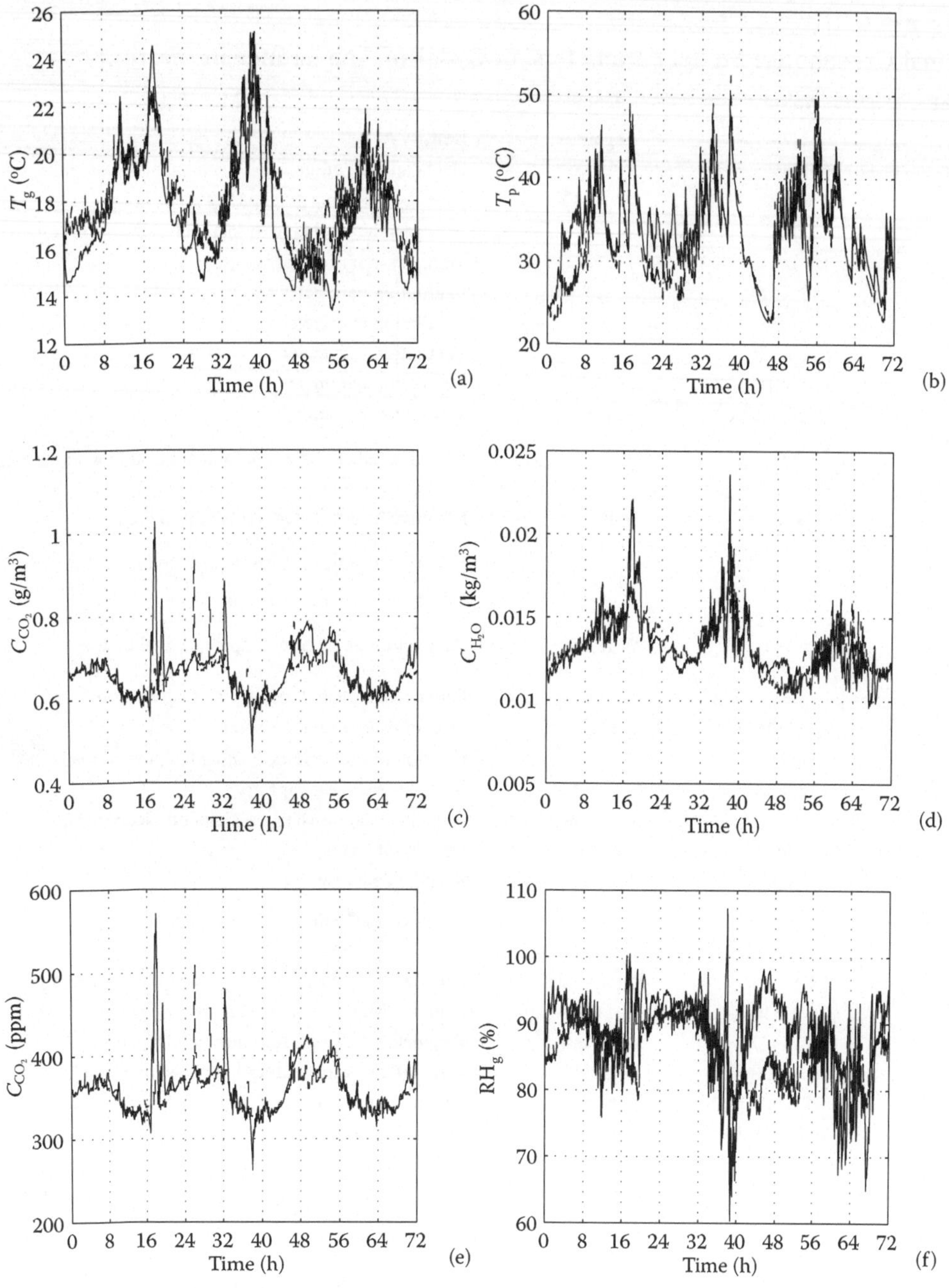

FIGURE 7.9 Greenhouse validation result for three consecutive days: (a) air temperature, (b) heating pipe temperature, (c) CO_2 concentration, (d) water vapor concentration, (e) CO_2 in ppm, and (f) RH. Dashed: measured; solid: simulated.

TABLE 7.3
Nominal Greenhouse Model Parameters, in Order of Appearance in the Text

Symbol	Value	Units	Meaning
ρ_a	1.29	kg m^{-3}	Density of air
c_{pa}	1010	J K^{-1} kg^{-1}	Heat capacity of air
K_g	32000	J m^{-2}[gh] K^{-1}	Greenhouse effective capacity
η	0.7	–	Absorbed heat relative to total net radiation energy received
τ_r	0.7		Transmissivity of greenhouse cover
p_{v1}	7.17×10^{-5}	%$^{-1}$	Ventilation parameter
p_{v2}	0.0156	%$^{-1}$	Ventilation parameter
p_{v3}	2.71×10^{-5}	–	Ventilation parameter
p_{v4}	6.32×10^{-5}	%$^{-1}$	Ventilation parameter
p_{v5}	7.40×10^{-5}	m s^{-1}	Ventilation parameter
$U_{g\_o} \frac{A_{g\_o}}{A_g}$	7.9	W K^{-1} m^{-2}	Effective heat transfer coefficient through wall per unit greenhouse area
U_{g_s}	5.75	W K^{-1} m^{-2}	Effective heat transfer coefficient to the soil
$U_{p\_g} \frac{A_{p\_g}}{A_g}$			Effective heat transfer from pipes to air per unit greenhouse area
$p_{p1} \frac{A_{p\_g}}{A_g}$	1.524	W K^{-1} m^{-2}	Parameter in heat transfer pipe to greenhouse
p_{p2}	3	K$^{0.5}$	Parameter in heat transfer pipe to greenhouse
Λ	$2.26 \times 10^{+6}$	J kg^{-1}	Heat of evaporation of water
κ	0.75	–	Fraction of condensation latent heat lost to outdoors
K_s	120000	J m^{-2}[gh] K^{-1}	Soil effective heat capacity
U_{s_ss}	2	W K^{-1} m^{-2}	Effective heat transfer coefficient to deeper soil
ρ_{H_2O}	998	kg m^{-3}	Density of water
c_{p,H_2O}	4180	J K^{-1} kg^{-1}	Specific heat of water
$\frac{V_p}{A_g}$	0.004	m^3 m^{-2}[gh]	Heating pipe volume per unit greenhouse area
α_c	2.60×10^{-7}	kg[H_2O] J^{-1}	Penman–Monteith factor for crop evapotranspiration
β_c	2.30×10^{-8}	kg[H_2O] m^{-2} s^{-1} Pa^{-1}	Penman–Monteith factor for crop evapotranspiration
c_{s1}	0.18407	Pa °C^{-3}	Parameter in vapor saturation pressure formula
c_{s2}	0.97838	Pa °C^{-2}	Parameter in vapor saturation pressure formula
c_{s3}	51.492	Pa °C^{-1}	Parameter in vapor saturation pressure formula
F_p	4.00×10^{-4}	m^3 s^{-1}	Pipe pump flow
T_{boil}	80	°C	Boiler temperature
c_{m1}	6.25×10^{-12}	kg[H_2O] m^{-2} s^{-1} Pa^{-1}	Condensation mass transfer parameter (pressure deficit base)
c_{m2}	0.33	–	Condensation mass transfer parameter
c_{T_gr}	1	K	Dimension parameter in condensation mass transfer
$\eta_{CO_2/dw}$	1.4667	kg[CO_2] kg^{-1}[dw]	Ratio CO_2 per unit dry weight
$\varphi^{max}_{CO_2,in\_g}$	2.10×10^{-6}	kg[CO_2] m^{-2}[gh] s^{-1}	Maximum CO_2 injection flux

7.5.3 Conclusions about the Models

The efforts to calibrate and validate the greenhouse and crop models finally result in a set of useful dynamic models in state–space form that can be used for dynamic optimization and optimal control. Rather than trying to make the models better, at the expense of larger data needs or increased complication, it is better to first see how sensitive to model errors the ultimate problem is for which the models are designed in the first place—in this case, optimal control. If the problem is robust against model errors, improvements do not pay off. If not, methods for online model adaptation may also provide an answer.

7.6 OPEN-LOOP OPTIMIZATION

As a first step in the design of an optimal greenhouse cultivation controller, open-loop calculations are performed. The methodology is summarized in Section 7.6.2. The goal of the open-loop computation is in fact to obtain estimates of the time evolution of the costates of the slow variables because these are needed in the ultimate feedback control to account for the long-term effects, as outlined in Chapter 5. The open-loop calculations also provide insight into the expected optimal time evolution of the states under nominal weather conditions and of the expected seasonal costs and benefits. Moreover, the results can indicate bottlenecks in the system, in particular when constraints are frequently violated.

7.6.1 Problem to Be Solved

The problem to be solved is to minimize the cost function,

$$J = -c_{\mathrm{F}} S_{\mathrm{HF}} + \int_{t_0}^{t_{\mathrm{f}}} \left(c_{\mathrm{CO_2}} u_{\mathrm{CO_2}} + c_{\mathrm{q}} u_{\mathrm{q}} + P_{\mathrm{CO_2}} + P_{\mathrm{H_2O}} + P_{\mathrm{T}} \right) \mathrm{d}t, \qquad (€\ \mathrm{m}^{-2}[\mathrm{gh}]) \quad (7.92)$$

subject to the state Equation (7.83) and the initial conditions,

$$\mathbf{x}(t_{\mathrm{o}}) = \mathbf{x}_{\mathrm{o}}, \quad (7.93)$$

and control constraints,

$$\begin{aligned} &0 \le u_{\mathrm{v}}^{\mathrm{Ap}} \le 2 \\ &0 \le u_{\mathrm{q}}^{\mathrm{vp}} \le 1 \\ &0 \le u_{\mathrm{CO_2}}^{\mathrm{vp}} \le 1. \end{aligned} \quad (7.94)$$

In Equation 7.92, the harvested fruit biomass S_{HF} has a market price c_{F}, and the benefit from selling the crop is hence a negative contribution to the costs. The first two terms under the integral are the costs of CO_2 injection (with CO_2 market price $c_{\mathrm{CO_2}}$) and the costs of heating, where it is assumed that the heat delivered to the greenhouse can be valued with market price c_{q}. The variable u_q and $u_{\mathrm{CO_2}}$ are related to the true controls (value positions $u_{\mathrm{q}}^{\mathrm{vp}}$ and $u_{\mathrm{CO_2}}^{\mathrm{vp}}$) via the relations given in Sections 7.3.1 and 7.3.2, respectively. Instead of choosing hard state constraints to express the desire to stay away from unmodeled detrimental effects of climate variables on the crop, here penalties are used that value constraint violations in virtual cost terms. The penalties have the general form

$$P_j = \begin{cases} \alpha_j (x_j^{\mathrm{lb}} - x_j) \\ 0 \\ \alpha_j (x_j^{\mathrm{ub}} - x_j) \end{cases} \qquad j = \left\{ \mathrm{CO_2} \quad \mathrm{H_2O} \quad \mathrm{T} \right\},$$

$$x_{\mathrm{CO_2}} = C_{\mathrm{CO_2}},\ x_{\mathrm{H_2O}} = C_{\mathrm{H_2O}},\ x_{\mathrm{T}} = T_{\mathrm{g}}. \qquad (€\ \mathrm{s}^{-1}\ \mathrm{m}^{-2}[\mathrm{gh}]) \quad (7.95)$$

Note that Equation 7.92 can also be written as

$$J = \int_{t_0}^{t_f} \left(-c_F \frac{dS_{HF}}{dt} + c_{CO_2} u_{CO_2} + c_q u_q + P_{CO_2} + P_{H_2O} + P_T \right) dt, \qquad (€\ m^{-2}[gh]) \quad (7.96)$$

thus showing the harvest rate as a (negative) running cost.

To keep the problem numerically tangible, on the scale of the open-loop seasonal optimization, it was assumed that the greenhouse part of the state equations was in pseudoequilibrium. Hence, Equation 7.90 changes into an algebraic relation

$$\mathbf{0} = \mathbf{f}_g(\mathbf{x}_g, \mathbf{u}, \begin{bmatrix} \mathbf{d} & \mathbf{x}_c \end{bmatrix}^T, \mathbf{p}). \quad (7.97)$$

This is an implicit formula from which, at any given value of the external inputs, the crop states and the control inputs of the actual greenhouse variables can be computed. This assumption implies that on the scale of the seasonal optimization it is assumed that the greenhouse responds to changes by reaching a new steady state rapidly.

7.6.2 Method

- The problem was solved for a cultivation period from March 1 until October 31. The period before March 1 did not have fruits yet, and is not considered.
- The choice was made to take as the nominal weather the hourly average of real data, collected during a greenhouse control experiment over the cultivation period. Data are for Wageningen, 1995.
- The pseudostatic greenhouse Equation 7.97 was solved numerically by a zero finding procedure.
- In the calculation of the pseudostatic greenhouse, the optimal seasonal values of the slow state variables $\mathbf{x}_c^*$ are needed. Instead, values computed by simulation using the actual controls as applied in the experiment were used. This is more accurate. However, this is a rather special case because normally the controls are not known yet. It could be done here, thanks to the fact that an experiment had already been done.
- Penalties were used on the basis of Equation 7.95, with constraint values and parameters set as shown in Table 7.4.
- The optimization problem was solved with the sequential search method by Seginer and Sher (1993). This amounted to finding 3 (controls) × 245 (days) × 24 (hours) numbers that together minimize the goal function, in an iterative fashion.

TABLE 7.4
Soft Constraints and Penalty Function Parameters

	T_g Night (°C)	T_g Day (°C)	C_{CO_2} (ppm)	RH (%)
Lower bound	15	17	–	65
Upper bound	25	25	1000	90
Slope	5×10^{-6} Dfl °C^{-1}	5×10^{-6} Dfl °C^{-1}	1×10^{-8} Dfl ppm^{-1}	1.17 Dfl %$^{-1}$

7.6.3 Results

The open-loop calculations were done afterward with a true hourly averaged weather pattern. The resulting state trajectories will thus depend on the weather as it occurred. In a hierarchical setup, one could conceive to use the calculated state trajectories as setpoints, but because the future weather will surely deviate from the weather used for the optimization, this is not a good policy. The most important outcomes of the procedure are the costates for the slow state variables, in particular the costates for the leaf and fruit biomass, because those will be used in the receding horizon controller (Van Straten, Van Willigenburg, and Tap, 2002). Figure 7.10 shows the time evolution of the slow states together with the costates for leaves and fruits, respectively. As the costate represents the sensitivity of the goal function to a marginal change in the corresponding state and as the goal function in Equation 7.92 is formulated as costs, a positive value means that a change in the state, without affecting the other states, would lead to higher costs and vice versa. It can be seen that over the period shown, where the plant is carrying fruits, it is always beneficial to produce more fruits if that would be possible without additional costs. In reality, as the graphs are valid for the optimum, such action would entail additional costs that would exactly balance the gain. The value of additional fruits decreases and finally becomes zero at the end of the optimization period, meaning that an additional unit of fruit at the end does not contribute to the goal function value anymore, which makes sense. The same holds for the leaves at the end. In the beginning of a fruit carrying plant, further investments in leaves is not profitable (positive costate in the beginning). In other words, putting more efforts in leaves would lead to higher costs; that is, an increment in leaves has a negative marginal price. Near day 80, however, an increment in leaves leads to less costs; that is, in that period, leaf increments bring more money (negative costate value in cost terms) and hence have a positive marginal price. It makes sense to invest in leaves in that period to guarantee the ability of the crop to produce fruits later on.

Figure 7.11 allows the analysis of the time evolution of the long-term criterion and its components. All values are expressed in monetary units, here Dfl m^{-2} (price level 1995; 1 € = Dfl 2.20). Note that the scales are different for each subplot. On the benefit side (negative costs) is the income of about 28 Dfl m^{-2} from the harvested fruits (*a*). Among the direct costs, the CO_2 injection costs (less than 1 Dfl m^{-2} (*b*)) are negligible as compared with the heating costs of 12–13 Dfl m^{-2}. The true net costs (J_{real}; Dfl m^{-2}) are determined by the difference between the CO_2 and heating costs ($b + c$) and the fruit income (*a*) and amount to about –28 + 13 = –15 Dfl m^{-2}, in other words, a net-positive income. However, this is only realized if nothing happens to the crop because of violation of the constraints. The penalty for CO_2 (*d*) is negligible, but the temperature (*e*) and humidity penalties (*f*) make significant contributions to the overall long-term criterion (*i*). During summer, the temperature penalty is quite high because, during daytime, it is often physically impossible to keep the temperature below the upper temperature boundary. In summer also the RH penalty has its highest values. Hence, the grower pays an insurance premium to himself of about 13 Dfl m^{-2} for a too high temperature and puts aside about 6 Dfl m^{-2} for a too high humidity level. If nothing happens, indeed he can add those amounts to (*i*), thus giving 4–19 ≈ 15 Dfl m^{-2} real profit, which is of course the same as the real direct costs (*g*). It should be kept in mind that the true economic costs of greenhouse operation are also determined by other factors, such as labor and capital costs.

Despite the limited direct value of the state trajectories obtained in the open-loop computation, it is interesting and instructive to look at the state trajectories and the necessary controls in some detail. The discussion will be done on the basis of fourteen-day excerpts from the data for four selected periods: March 1–14, April 20–May 4, June 15–28, and September 17–30, with weather data for 1995.

Figure 7.12 (top) shows the outside temperature and radiation in early March. The temperature in this period was quite low, below 10°C, and so was the global radiation. The middle panel shows the pipe temperature that resulted from the optimal heating valve position as computed by the optimization. During the night, in the absence of the solar heat input, the heating valve is often fully open,

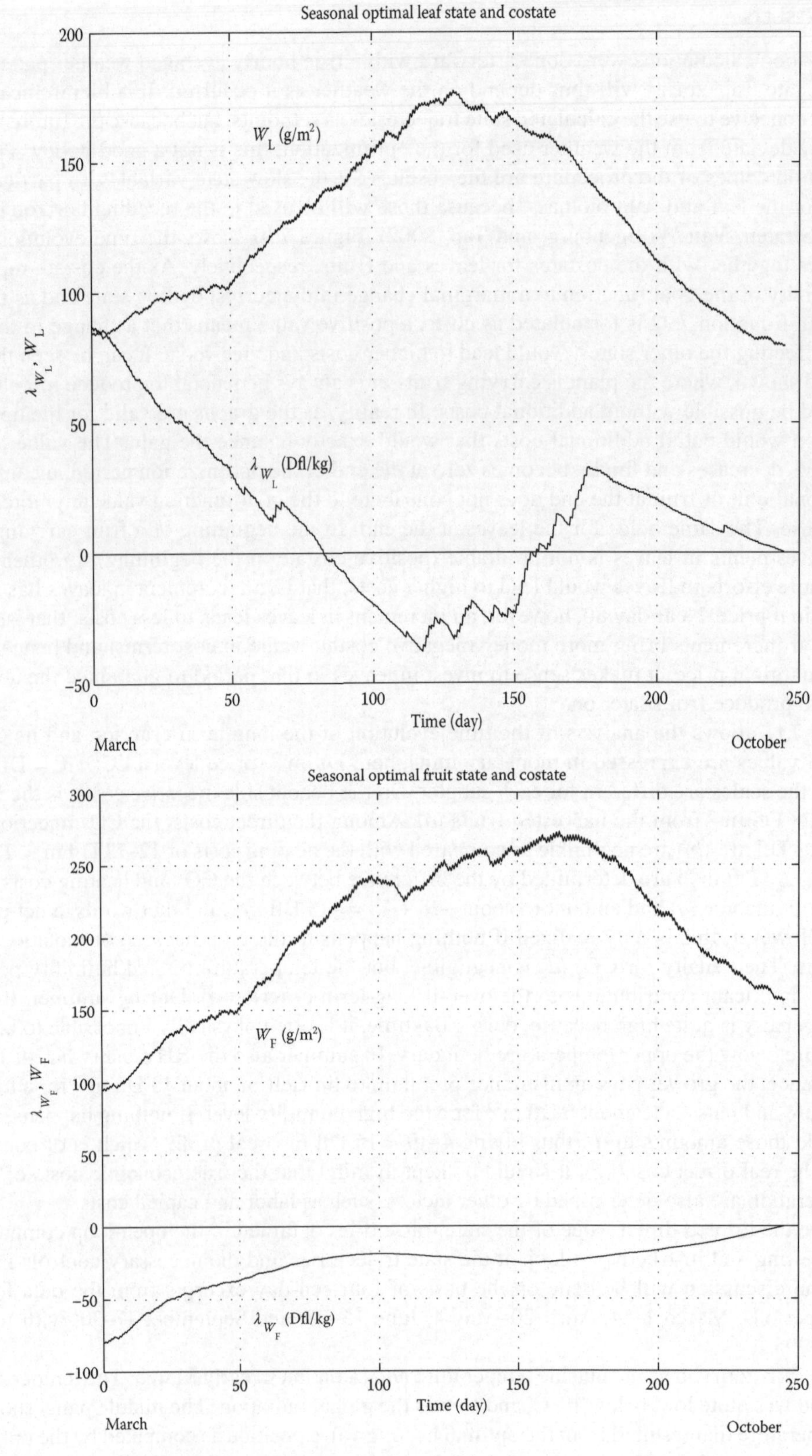

FIGURE 7.10 Slow states and costates. Top: leaves; bottom: fruits.

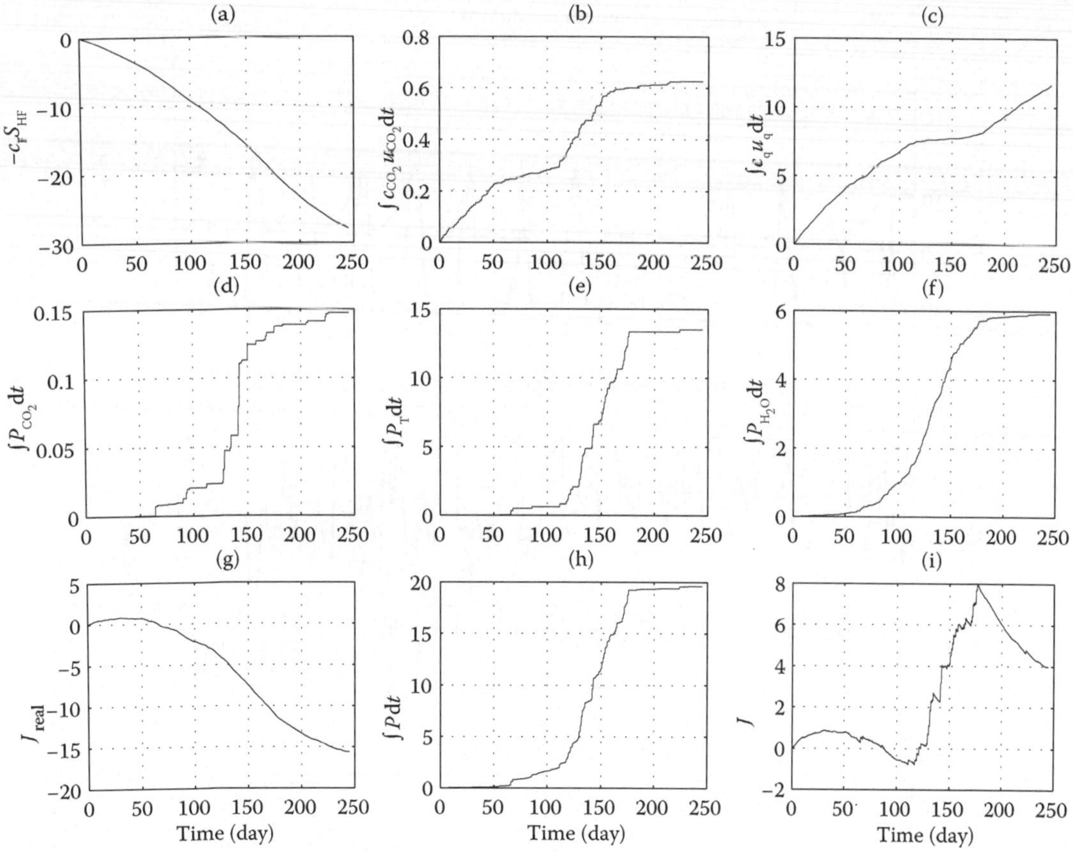

FIGURE 7.11 Trajectories of the criterion function (i), the cash costs (g), the sum of the penalty costs (h), and the cost function components (a through f) for the open-loop optimization from March 1 until October 31 (weather 1995). Units: Dfl m^{-2}.

and the pipe temperature reaches the maximum achievable temperature at the given boiler temperature setting. During the day, the greenhouse is heated as well, but less. The resulting greenhouse temperature reaches a plateau of about 24°C, whereas at night the temperature is allowed to drop to its lower bound. This shows that the optimization in a way confirms common practice, where during the day a higher temperature is set than at night. However, it is also clear that the range suggested by the optimal control is wider than what is currently used by growers (typically between 16 and 17 at night and 22 during the day). It is also interesting to point to the behavior of the optimal control at day 5. Because there is almost no light, the onset of the heating is delayed, and the optimal greenhouse temperature is lower than usual.

Figure 7.13 (top) shows the assimilate buffer. It is plotted together with the solar radiation to show that the buildup of assimilates is of course strongly correlated with radiation. This is clear at day 5, where no assimilates are formed because of the lack of radiation. Figure 7.13 also shows that the assimilate buffer pattern has a slight lag, meaning that the assimilate load to the plant reaches a peak in the afternoon. The middle panel shows that during the day there is still so much evapotranspiration that the RH reaches its allowed maximum of 90%. Despite this, it is not necessary to open the windows during the day because the higher temperatures allow higher moisture contents in the air without violating the RH constraint. Conversely, at night, when temperatures drop, it is sometimes required to ventilate. The required window opening varies among similar days because

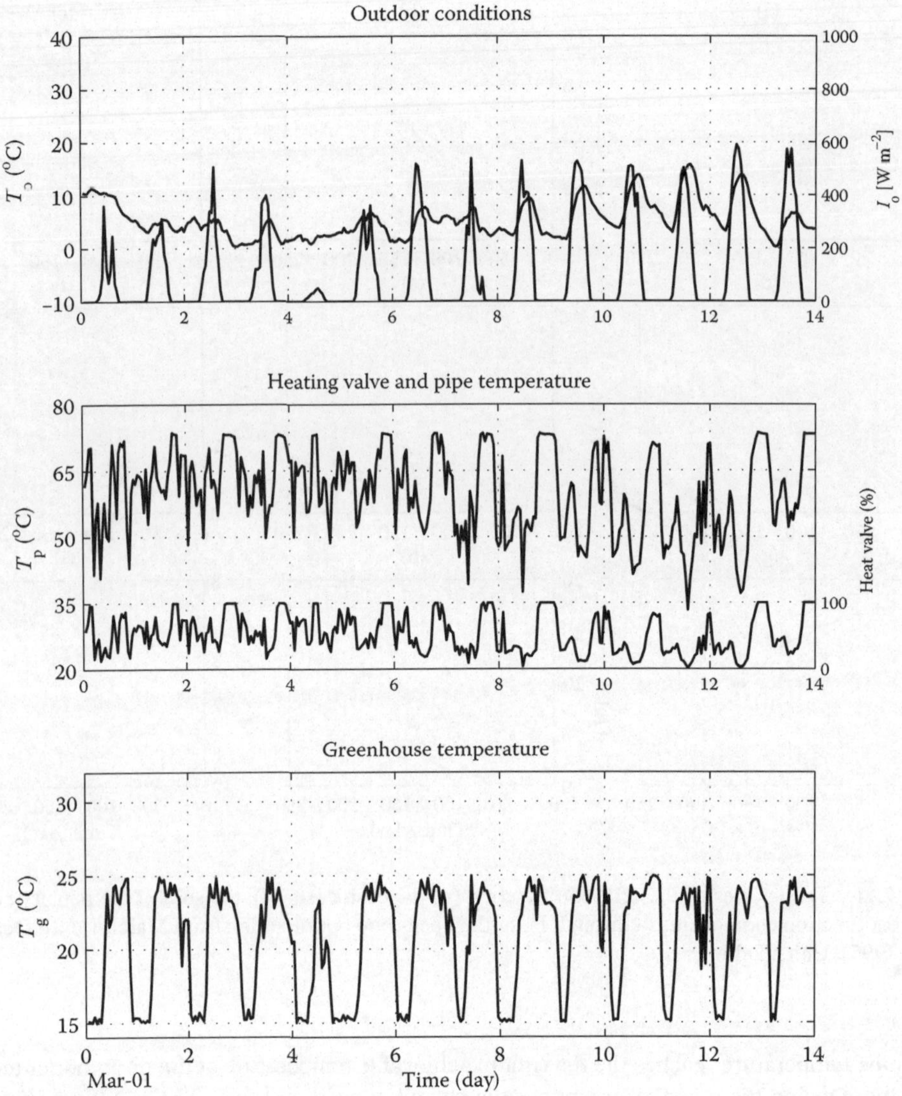

FIGURE 7.12 Open-loop optimal control results (a), March 1–14.

of differences in wind speed (not shown). CO_2 dosages and resulting CO_2 concentrations are shown in the lower panel. The patterns are not very clear, but it can be seen that the higher CO_2 concentrations often occur during night as a consequence of the respiration assumed in the model. During the day, this stored CO_2 is first consumed before any additional CO_2 is given. One would perhaps have expected more CO_2 dosage, for instance, on day 10, but apparently the rise in assimilates is already large so that extra CO_2 dosage does not pay off. It should be remembered that in the optimization, CO_2 has a price. This is different from a situation where CO_2 from flue gas is injected in the system because this CO_2 is essentially free as long as the generated heat can be used.

Figures 7.14 and 7.15 show the situation at the end of April. Outdoor temperatures have gone up, reaching a day maximum of 20°C on day 55 (Figure 7.14, top). Also the solar radiation is higher, peaking to about 700 W m^{-2}. As a consequence, the optimal heating pipe temperature goes down

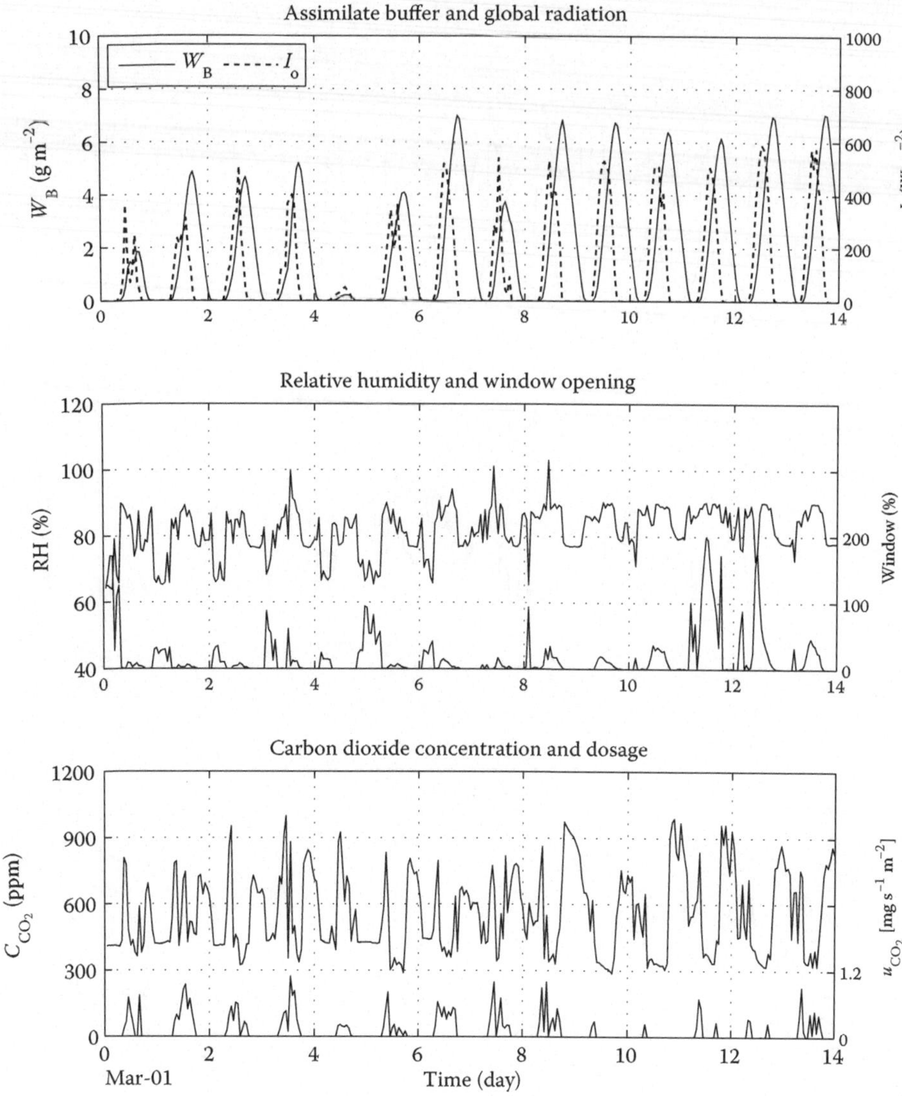

FIGURE 7.13 Open-loop optimal control results (b), March 1–14.

(Figure 7.14, middle), with a rise again in the colder spell in the middle of the period. Interestingly enough, the required greenhouse temperatures during the day are lower than those in March. Note that the period of elevated temperature within a day is longer than the daylight period. This is because an elevated temperature is favorable for converting assimilates into biomass as long as there are assimilates (Figure 7.15, top), even if the assimilate biomass is low (e.g., on day 58). One might think that it could be more favorable to have the greenhouse temperature high, but at a lower level, for an ever longer period, to make sure that the assimilates are just exhausted when the new day begins. The reason that this does not happen probably has to do with the maintenance losses, which are also temperature driven. Another factor in the model is that the distribution of fruit and leaves is also temperature driven, while over the season, the marginal values of leaf and fruit vary with respect to each other. Figure 7.15 (middle) shows that the RH is always at its upper bound, requiring

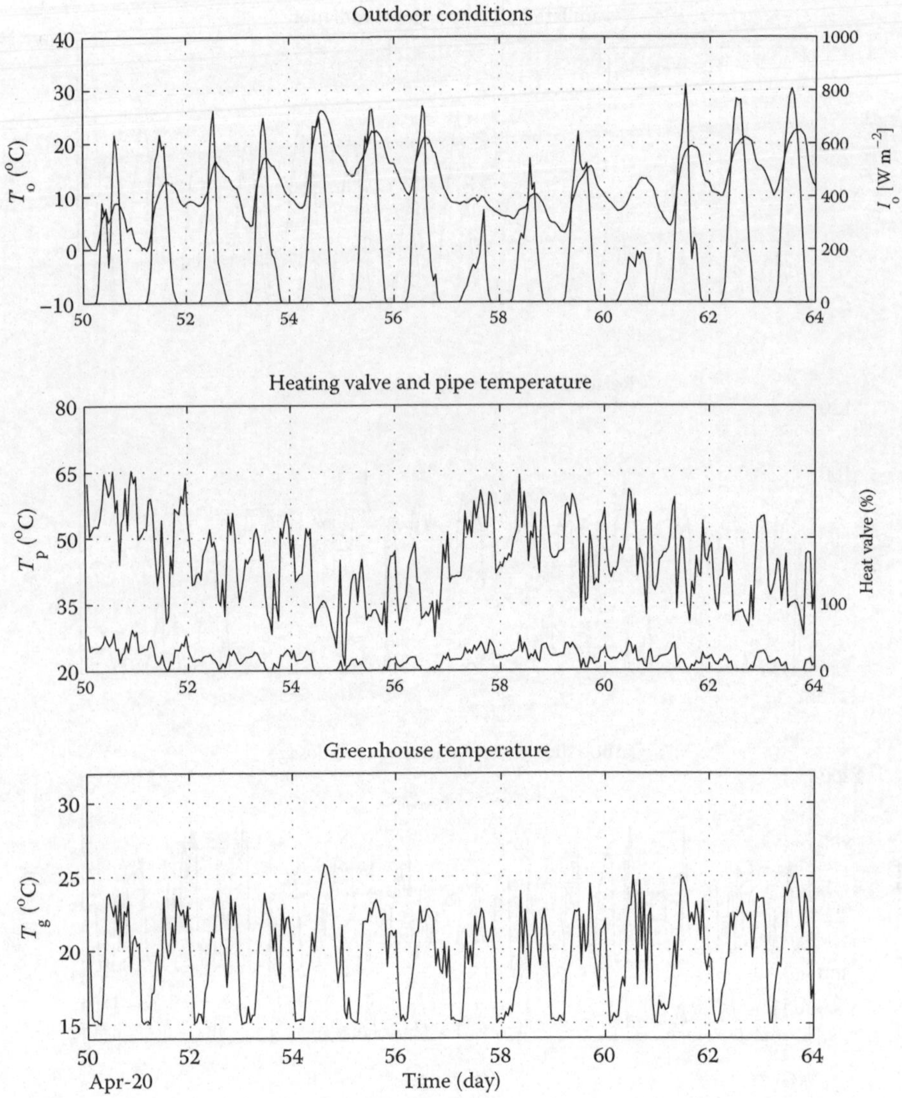

FIGURE 7.14 Open-loop optimal control results (a), April 20–May 4.

ventilation, and it is also clear from the lower panel that the need for ventilation suppresses the CO_2 concentrations in the greenhouse. Dosage of CO_2 then does not pay off because all injected CO_2 would disappear via the open windows.

In Figures 7.16 and 7.17, the patterns are given for summer conditions starting at June 15. In the beginning of the period, vary bad weather occurred. Unlike similar conditions as in the beginning of the season (March, Figure 7.12), daytime greenhouse temperatures are not raised to 24°C but are settled at much lower values (lower panel). This might have to do with the marginal leaf value. In this period, investment in leaves is not profitable, and consequently the temperature can stay low when the outside conditions are unfavorable. There is still some heating but quite low (middle panel). It is interesting to note that the optimization yields a pipe temperature that is still elevated. This may seem to look similar to current practice of "minimum pipe" temperatures used

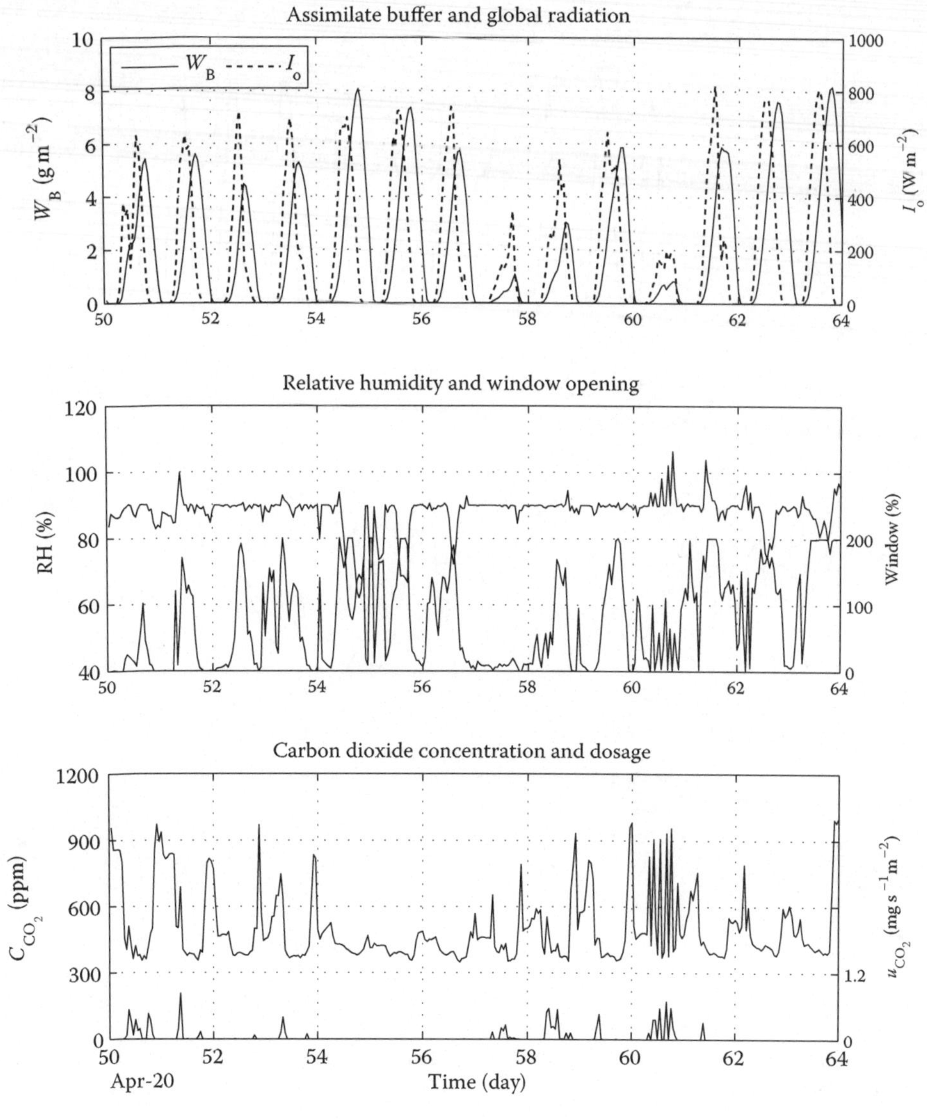

FIGURE 7.15 Open-loop optimal control results (b), April 20–May 4.

by growers "to keep the crop activated." Hence, the optimal control seems to give a justification for this empirical behavior. It should not be forgotten, though, that reasons that are not in the model, such as keeping the root zone at higher temperature, might equally well be a valid reason. As long as this is not described by scientifically based models, it will not be possible to demystify this practice.

The larger standing crop biomass creates more evapotranspiration when the sun is shining (Figure 7.17), and it often occurs that ventilation by opening the windows is not enough to keep the RH at its allowed maximum. Values larger than 100% suggest that the condensation in the model may not be sufficiently fast. On bright days, occasionally CO_2 dosage is applied (Figure 7.17, lower panel), even on days where the windows are fully open. On such days, there are low wind speeds, thus rendering the ventilation losses still relatively low, as testified by the high humidity. The dosage

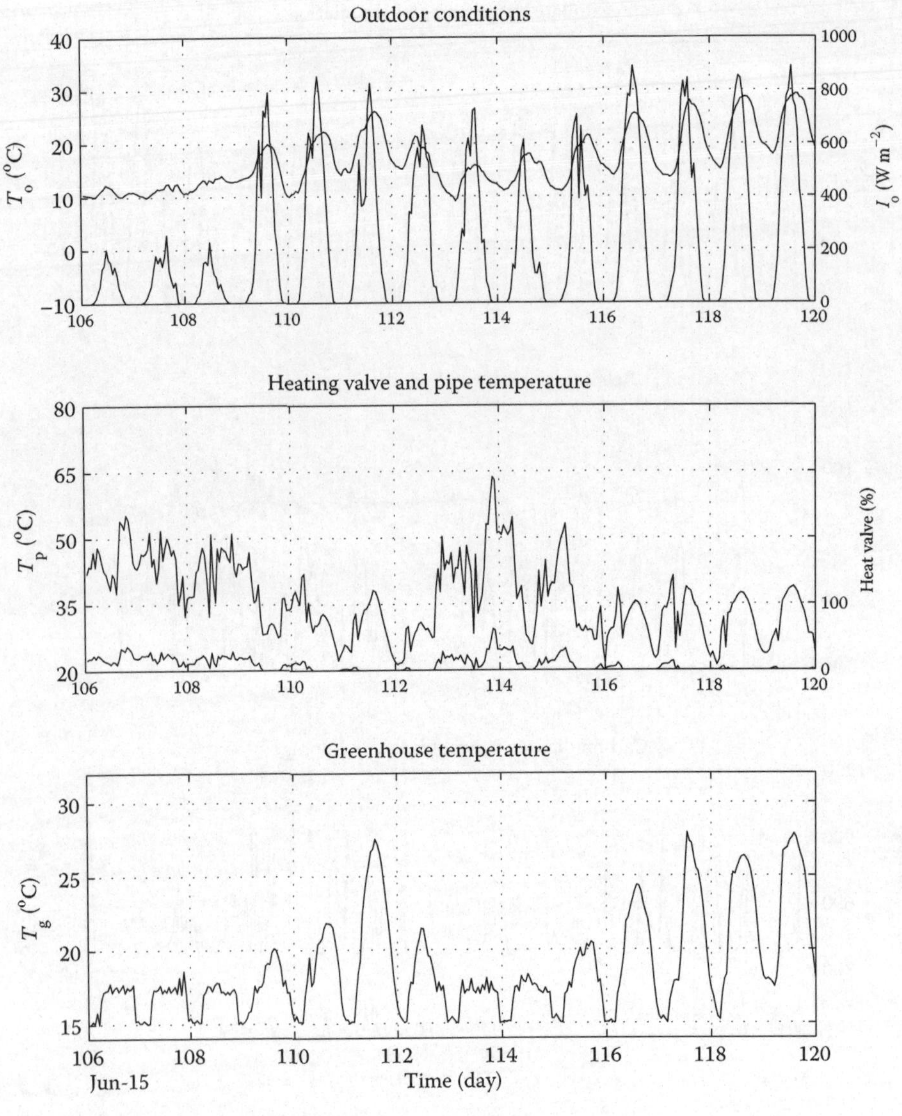

FIGURE 7.16 Open-loop optimal control results (a), June 15–28.

of CO_2 is hardly visible in the CO_2 concentrations themselves because all dosed CO_2 is immediately taken up by the crop.

Finally, Figures 7.18 and 7.19 present the situation at the end of September, which is the end of the optimization period. Outdoor conditions are characteristic for late summer and early autumn, with fair levels of outdoor temperature and solar radiation (Figure 7.18, top). Still heating is applied (middle panel), as it is profitable to have elevated temperatures again (lower panel). At the end of the season, the marginal value of the leaves that became positive again should be steered to zero to direct as much assimilates as possible to the fruits. In the model, the production of fruits is favored over leaves if the temperature is high (Equation 7.27), which explains the elevated day temperatures. Figure 7.19 (middle) shows that the ventilation is sufficient to keep the RH at its maximum bound, but the RH is high all the time. CO_2 concentrations can be high, particularly if they rose directly

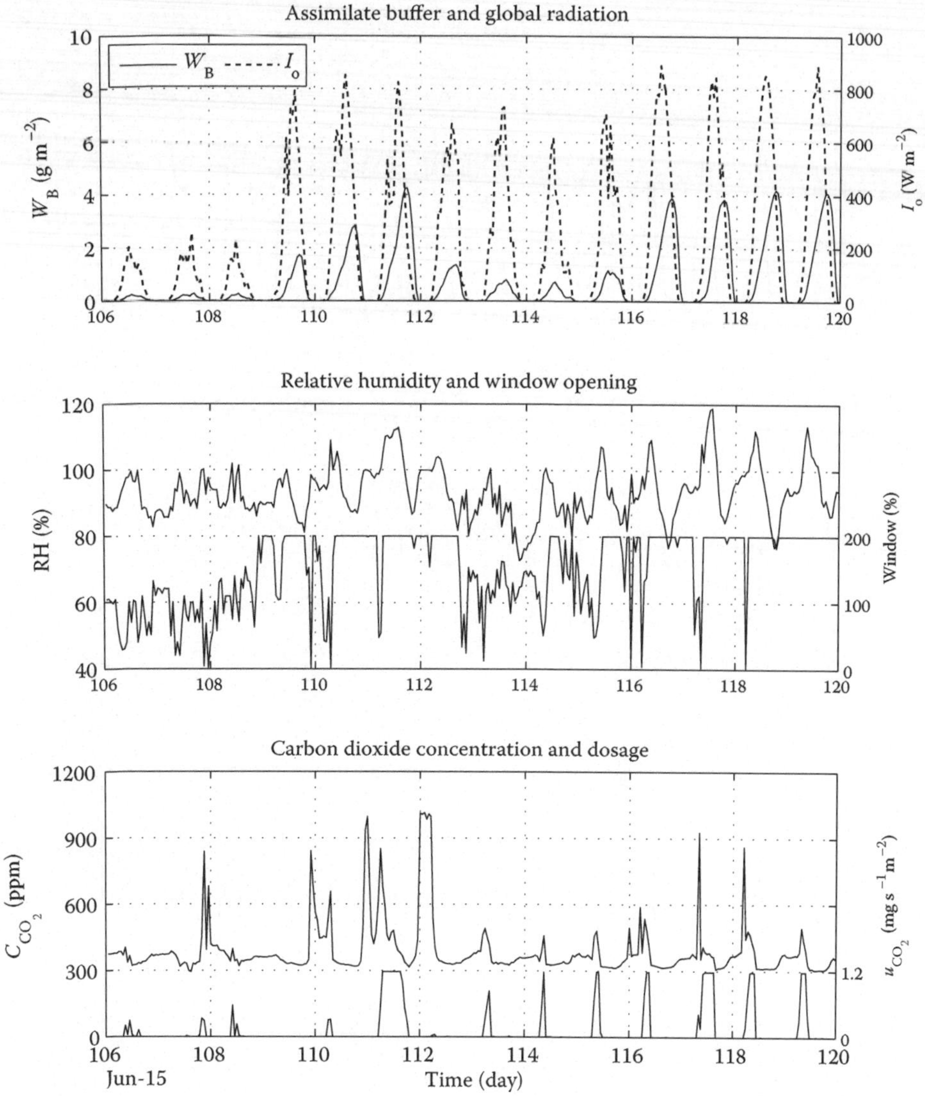

FIGURE 7.17 Open-loop optimal control results (b), June 15–28.

after sunset because of respiration. Once high, they stay high when the overall ventilation rate is low. No dosage of CO_2 takes place in this period (Figure 7.19, bottom).

Looking at all the figures together, it is interesting to note that according to the model, the control is chosen in such a way that there is no carryover of assimilates from one day to the other because assimilates are always depleted before the next sunrise. One could say that the optimization assures that the source–sink ratio averaged over a day is about 1. It also suggests that with this model, there is no point in compensating low temperatures a few days earlier by higher temperatures later.

7.6.4 Recapitulation of the Open-Loop Step

In the previous section, it has been shown how the open-loop optimization leads to trajectories of controls, greenhouse, and crop states over the season. The calculation was performed afterward,

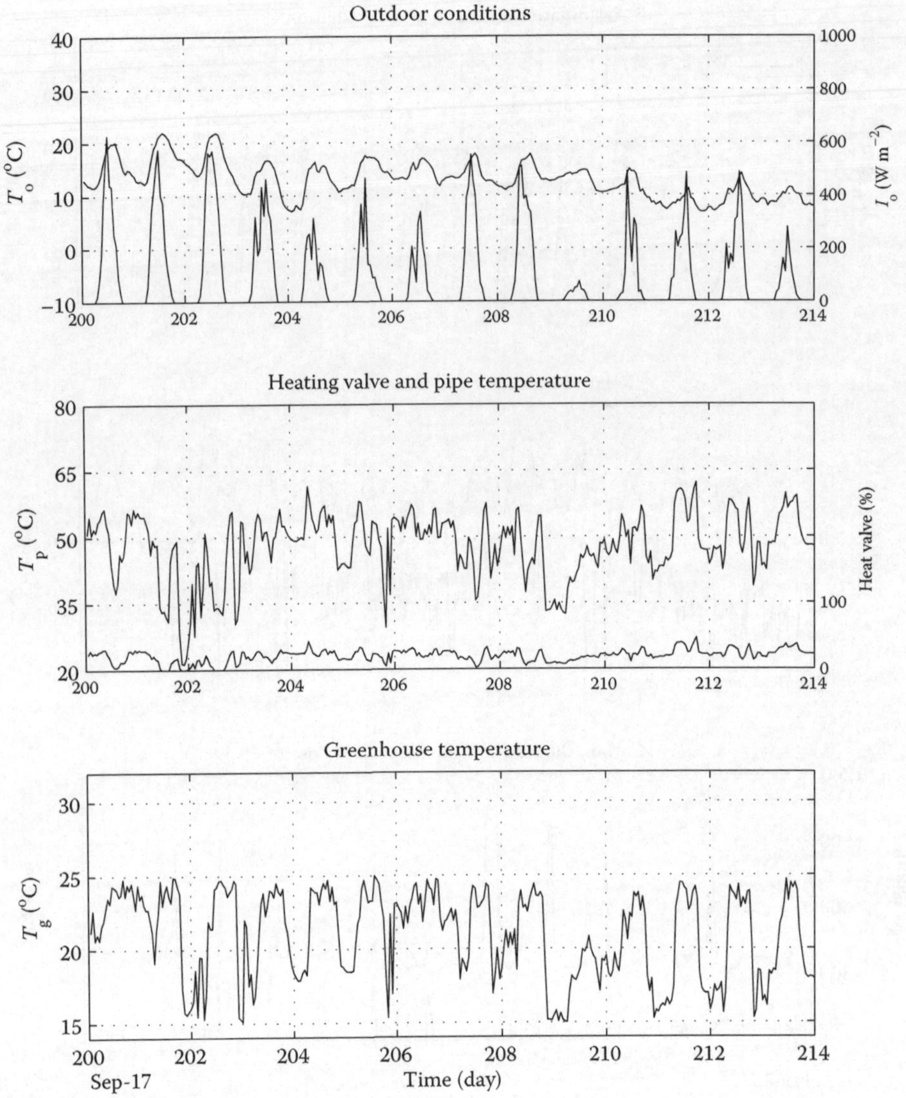

FIGURE 7.18 Open-loop optimal control results (a), September 17–30.

using the actual weather. In a realistic situation, of course, the weather is not yet known. One possibility to tackle this is to use a nominal weather file. It is not immediately clear how this should be constructed. Using a single realization of real weather has the advantage that all high frequent information (on the scale of half an hour or so) is maintained so that the estimates of the achievable goal function value are as realistic as possible for that particular weather pattern. Ideally, one would need a large set of realistic weather patterns and compute the optimal controls, states and costates, as well as the goal function for each to obtain a statistical distribution. It would show what the effect of the weather is on achievable goals under the assumption of optimal control.

However, even if this information would be available, it is not yet obvious how to use it to build an online optimal controller. The results above support the view outlined in Chapter 5 that it is not a good idea to use the states obtained from optimization under realistic weather as setpoints. As explained in that chapter, following the two–time-scale decomposition principle, the slow costates

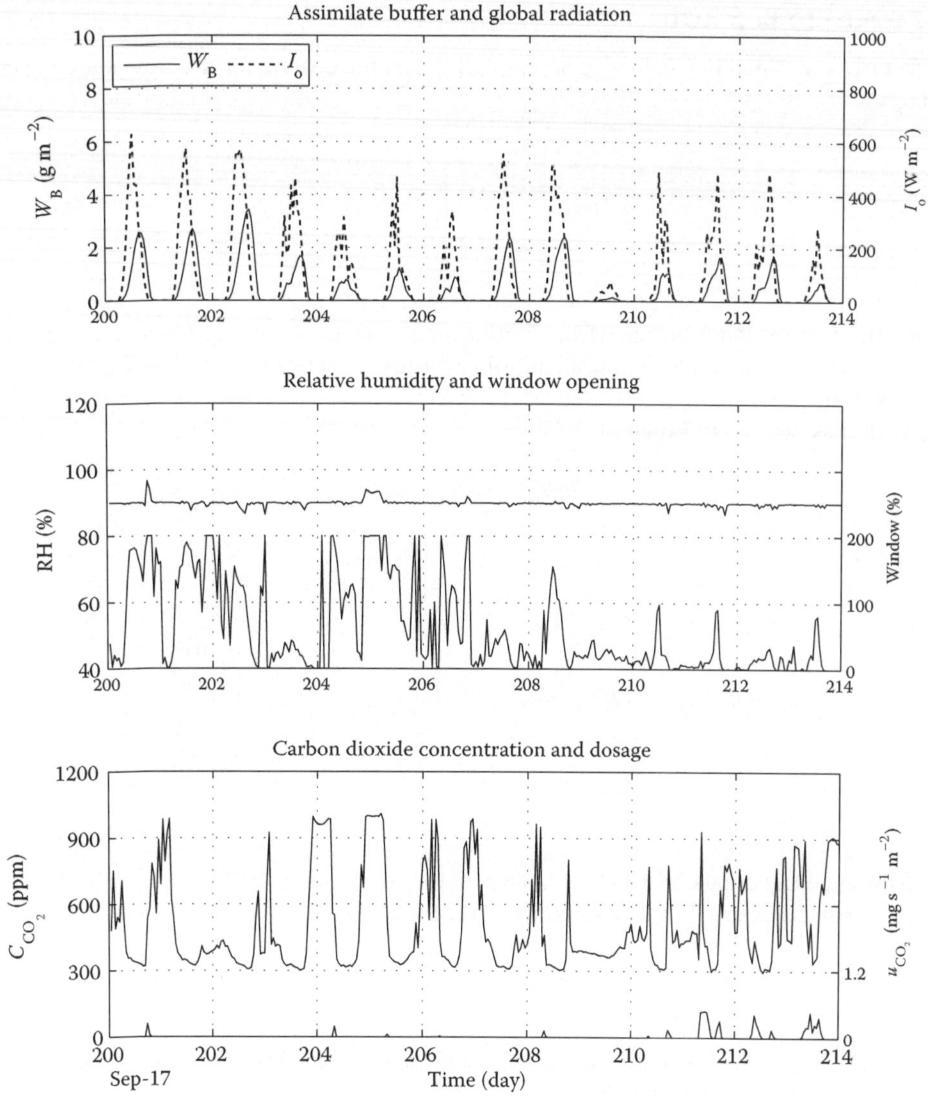

FIGURE 7.19 Open-loop optimal control results (b), September 17–30.

are better candidates to convey long-term information to the short term. In the next section, this is worked out for the real application described in this chapter.

7.7 TWO-TIME-SCALE RECEDING HORIZON OPTIMAL CONTROLLER (RHOC)

According to the procedure outlined in Chapter 5, the next step is to develop an online optimal controller. Once again, an optimal control problem is solved, but now for the short term with a limited horizon and using offline information from a seasonal optimization. The calculated control is implemented, and at the next sampling instant, the actual state is measured and used as a new initial value. At the next control interval, the game is repeated, whereby the horizon recedes toward the future. As explained, this approach is known as a receding horizon optimal controller (RHOC). As in this case the long-term information originates from a time-scale decomposition, we sometimes refer to this procedure as the two-step receding horizon optimal control.

7.7.1 Problem to Be Solved

The procedure is outlined in Chapter 5, Section 3.1.2. Let the costate for the slow state variables be denoted by

$$\boldsymbol{\lambda}^{s*}(t) = \begin{pmatrix} \lambda^*_{WF}(t) \\ \lambda^*_{WL}(t) \end{pmatrix}, \quad t_0 < t < t_f, \qquad (€\ kg^{-1}[dw]) \quad (7.98)$$

known for the full seasonal horizon from initial time t_0 to final time t_f. Let the current time be denoted by t_s. Then, to obtain a feedback controller for the short term, the following optimal control problem is solved:

Find, with $\mathbf{u}$ as defined in Equation 7.87,

$$\mathbf{u}^*(t), \quad t_s < t < t_s + h, \qquad (7.99)$$

that minimizes

$$J = \int_{t_s}^{t_s+h} \left(\left(-c_F \frac{dS_{HF}}{dt} + c_c u_c + c_q u_q + P_{CO_2} + P_{H_2O} + P_T \right) + \begin{pmatrix} \lambda^*_{WF} & \lambda^*_{WL} \end{pmatrix} \begin{pmatrix} \frac{dW^*_F}{dt} \\ \frac{dW^*_L}{dt} \end{pmatrix} \right) dt, \qquad (€\ m^2[gh]) \quad (7.100)$$

subject to the state Equations 7.89 and 7.90, the penalties of Equation 7.95, and using as disturbances the current expected external input trajectories over the horizon,

$$\hat{\mathbf{d}}(t), \quad t_s < t < t_s + h, \qquad (7.101)$$

with $\mathbf{d}$ as defined in Equation 7.88. The initial conditions at t_s for the greenhouse states are reconstructed from the observations, that is,

$$\mathbf{x}_g(t_s) = \mathbf{f}\left(\mathbf{x}_g^{obs}(t_s)\right), \qquad (7.102)$$

whereas the initial conditions for the slow crop states follow from the slow subproblem,

$$\mathbf{x}_c(t_s) = \mathbf{x}^*_c(t_s). \qquad (7.103)$$

Equation 7.100 specifies that in the short term, the goal function is the same economics-based goal function as used for the seasonal optimization but enhanced with a contribution that is proportional to the rate of change of the crop variables, with the instantaneous costates as proportionality factor. Thus, the costates act as shadow prices to value the investment in fruit dry weight (before harvest) and in leaf dry weight. This assures that the optimization takes into account the long-term effect and prevents the selling out of resources on the short term.

Once the optimal trajectory over the horizon has been found, only the initial value of the control trajectory at time t_s, $\mathbf{u}_s^*(t_s)$ is applied to the plant, during the control interval T, that is,

$$\mathbf{u}(t) = \mathbf{u}_s^*(t_s), \quad t_s < t < t_s + T. \tag{7.104}$$

At the next control interval $t_s + T$, the optimization is repeated using the reconstructed greenhouse states from the new observations at $t_s + T$ as initial conditions.

7.7.2 Method

- The two–time-scale receding horizon controller was implemented in a real greenhouse experiment in Wageningen, over the period August 3 until October 31, 1995. The crop already had fruits.
- At the time of the experiment, the seasonal costates for the slow variables were not yet available. Therefore, a fixed price was used of 20 Dfl kg^{-1}[dw], equal for leaves and fruits, that is, $\lambda_{WF} = -0.02$ Dfl g^{-1}[dw] and $\lambda_{WL} = -0.02$ Dfl g^{-1}[dw]. A zero marginal value was set for the buffer costate because the assimilate buffer is only an intermediate variable of which the optimal trajectory will be driven by the benefits from an increase in leaf and fruits.
- The optimal control problem was solved online, using a sample interval of 60 s, that is, every minute an optimal control problem was solved, and the control computed was imposed on the greenhouse and held during the control interval.
- Instead of using the crop variables from the slow subproblem, online-simulated values are used. These are more accurate as they are computed with the real weather as it appeared rather than with the assumed long-term nominal weather. With this choice, the crop costates are the only information arising from the long-term optimization that is conveyed to the short term.
- A simple calculation based on inversion of Equation 7.91 was used to reconstruct the actual greenhouse states.
- The horizon for the optimization was set at one hour. This is motivated by a separate analysis of the effect of the horizon on the optimization result.
- The weather forecast over the control horizon was a "lazy man" forecast, meaning that the actually observed climate conditions at the sample interval are supposed to remain the same over the full prediction horizon. The prediction horizon was set the same as the optimization horizon.
- The penalties were the same as in Table 7.2, except for the maximum of the RH, which was set at 95%.
- Initial conditions were taken from the actual situation on August 3, 1995, including estimates of biomass of leaf and fruit.
- The solution is constrained by the real-time restriction that a solution should be available at the end of each one-minute control interval. This means that the optimization is forced to stop searching, even if no convergence to the optimum is yet obtained. This should be kept in mind in interpreting the results. The solution found is used as the initial guess for the next time interval.

7.7.3 Results

Data have been stored for each of the days during the experiments. These include at any control instant with one-minute interval the actual solar radiation, outdoor temperature, and wind speed, the computed optimal control sequences in steps of one minute over the coming hour, the evolution of the modified goal function and goal function components in steps of one minute over the coming hour, and the resulting predicted greenhouse conditions over the coming hour. In addition, experimental data for the controls and climate variables in the greenhouse are available.

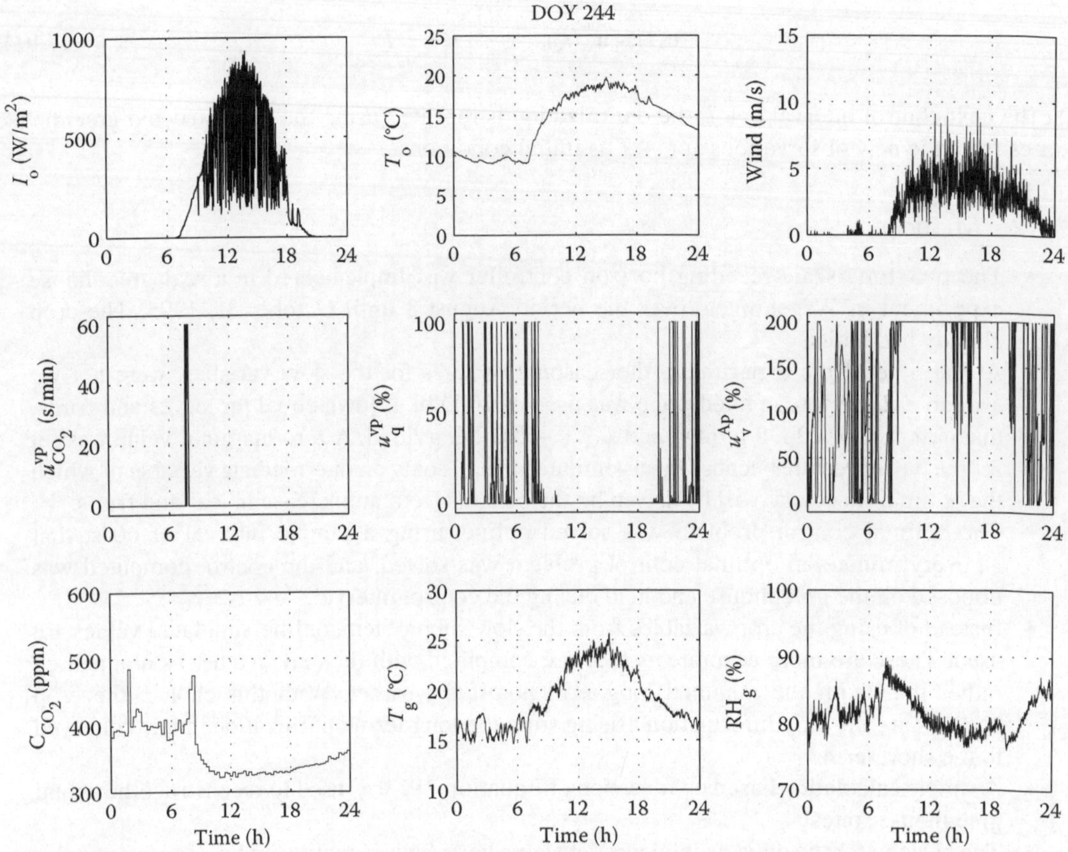

FIGURE 7.20 Closed-loop results for September 1, 1995. Top row: external inputs; middle row: controls; lower row: outputs.

In general, the measured controls show short delays as compared with the calculated ones because in reality it takes some time to adjust the valves and windows. In the plots below, only the calculated controls are shown because on this scale the difference with the realized ones is negligible. Of each sequence of sixty computed control actions evaluated at any time *t*, only the first is presented because these are actually implemented to the real greenhouse. Likewise, with the state variables, not the predicted state over the coming hour, but the observed states at the control instant are presented.

To illustrate the behavior of the receding horizon controller, two days are shown in more detail here. Figure 7.20 shows the pattern at September 1, 1995 (day of year 244). This day is characterized by a cold night, followed by a bright day and elevated day temperatures. During the night, the heating valve (u_q^{vp}) is opened a number of times to satisfy the lower temperature constraint. Notice the almost bang-bang nature of the heating. During the day, no heating is needed as the temperature rises naturally, and the RH remains below its lower bound. The windows are opened during the day to prevent too high temperatures that would only lead to respiration losses. Because of the open windows, there is no point in dosing CO_2, except during a short period in the morning, when the windows are still closed. Notice that during the night, with the windows partly closed, CO_2 rises because of the respiration, whereas the open windows during the day ensure sufficient exchange with the outside conditions to prevent the CO_2 from dropping very low.

Another example is September 23, 1995 (day of year 266, Figure 7.21). This day is characterized by a mild night and cloudy weather during the day, with overall low radiation. Not much heating

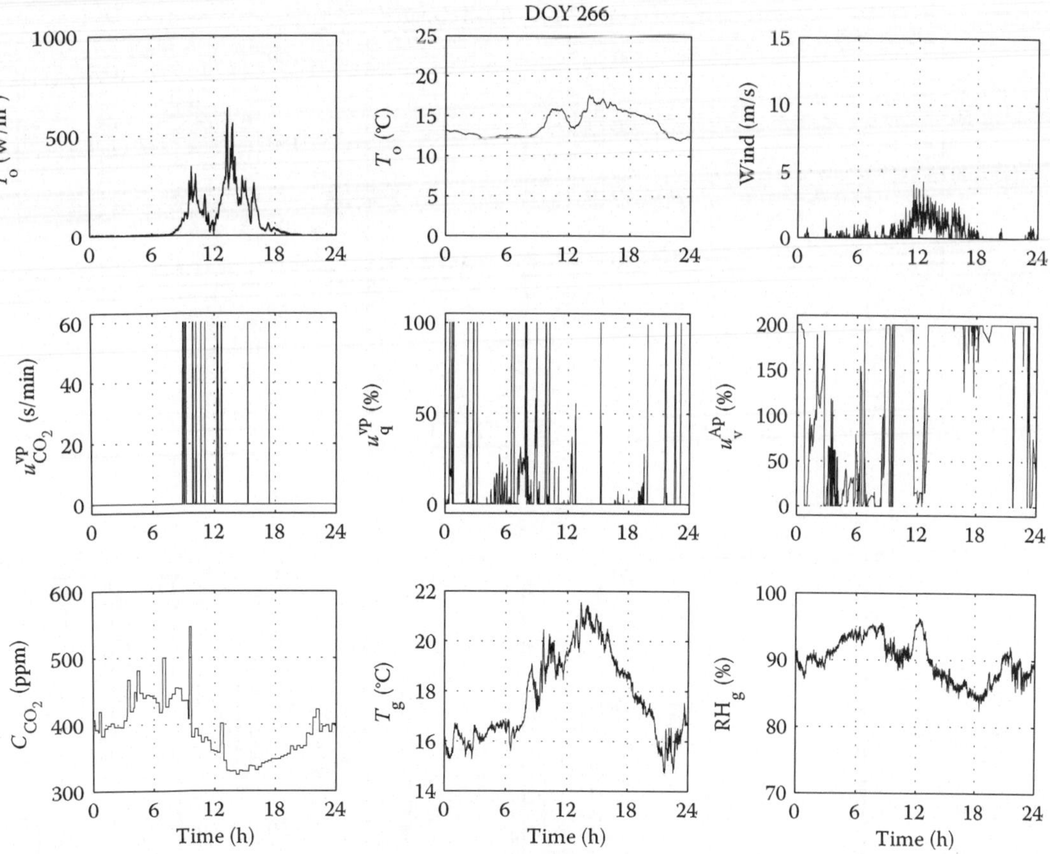

FIGURE 7.21 Closed-loop results for September 23, 1995. Top row: external inputs; middle row: controls; lower row: outputs.

is needed during the night to stay above the lower temperature limit. The moisture content reaches the upper bound in the morning, leading to opening of the windows. When radiation peaks at about midday, windows are closed and CO_2 is dosed, until the moisture contents hit the 95% bound, after which ventilation is needed to get rid of the surplus moisture. Again, it is clearly seen that during the night, CO_2 rises, even if the windows are open in the evening hours. The ventilation effect of opening the windows at the end of the day is limited because of the low wind speed.

7.8 EVALUATION OF OPTIMAL CONTROL

7.8.1 Sensitivity of RHOC to Modeling Errors

During the experiment, the actual measurements are used to update the states to serve as initial conditions for the coming horizon. To investigate the influence of modeling errors, a simulation was done as if the receding horizon optimal controller was online, but without an update from the observations. Any differences between reality and model prediction then show up as differences in calculated controls.

Figure 7.22 compares the simulated and observed states together with the differences between the controls. Comparing the experimental and simulation results reveals that during the night in simulation, the controller can keep the temperature much closer to its lower boundary value. Looking at the heating valve (u_q^{vp}) and window controls (u_v^{Ap}), it appears that the overall pattern is the same

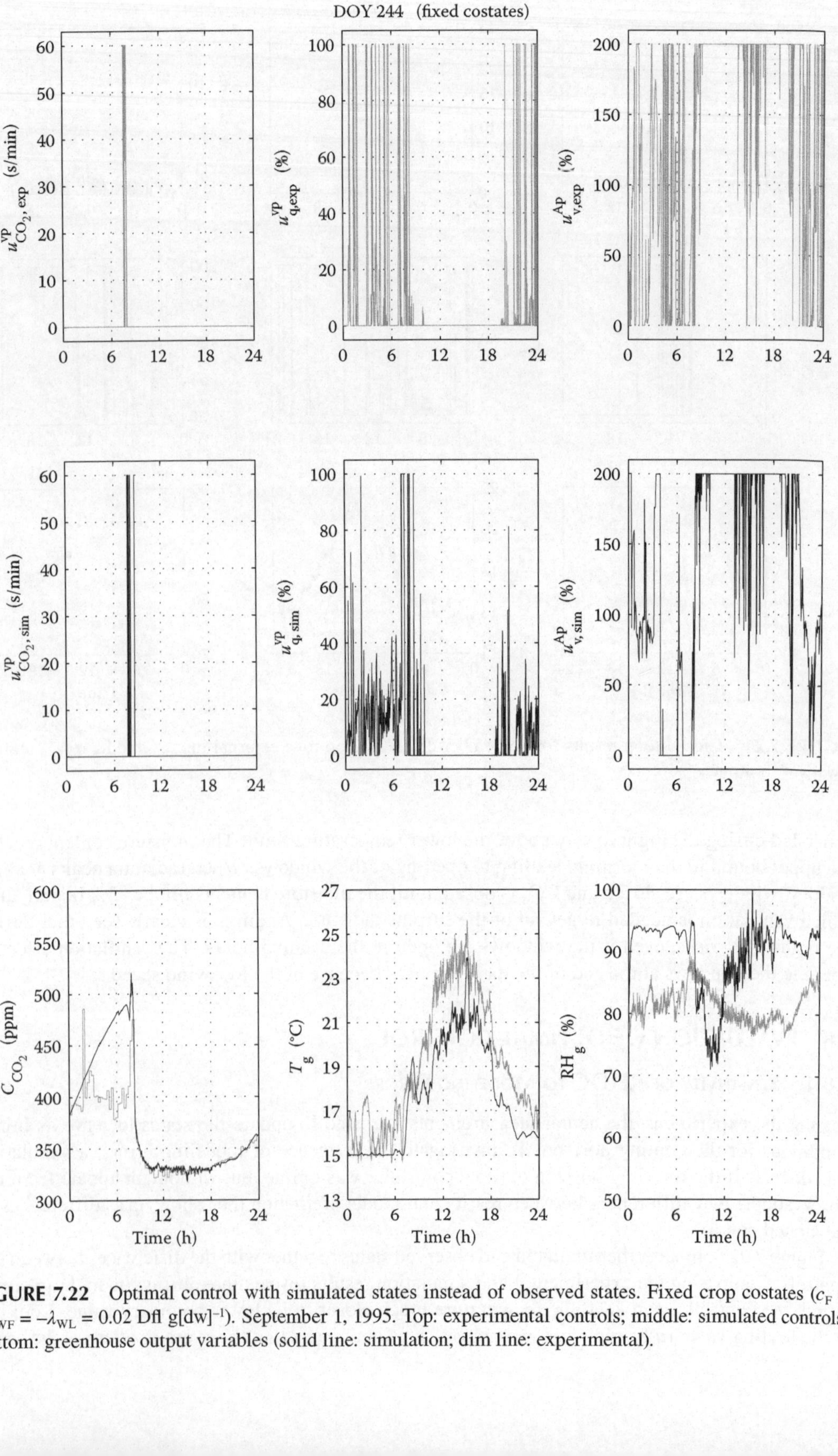

FIGURE 7.22 Optimal control with simulated states instead of observed states. Fixed crop costates ($c_F = -\lambda_{WF} = -\lambda_{WL} = 0.02$ Dfl g[dw]$^{-1}$). September 1, 1995. Top: experimental controls; middle: simulated controls; bottom: greenhouse output variables (solid line: simulation; dim line: experimental).

in both cases, but in simulation the controls have less large excursions. This may be an indication of the effects of modeling errors, mainly in the ventilation part, although it might partly be also because in simulation there is no need to force the optimization routine to stop searching after the control interval elapses. The ventilation part of the model does not include free convection; thus, implicitly it assumes that for zero wind speed, there is no ventilation. An indication for this is that in the preceding night hours, when the wind speed is low, the simulated CO_2 concentration is quite a bit higher than the measured one. It also supports the observation that the simulated RH during that period is higher than that in the real system.

The effect on the control patterns of replacing real observations with simulated values appears to be rather limited. Some suboptimality may be expected if predictions are steadily wrong. On the other hand, comparing one-step-ahead predictions with real observations should give clues to (automatic) adjustment of models. Overall, the RHOC does not show signs of instability and therefore looks rather robust against modeling errors.

7.8.2 Sensitivity of Slow Costates to the Nominal Weather

Unfortunately, no analysis was made on the sensitivity of the costate patterns to the weather. Figure 7.10 suggests that the slow costates have less high-frequency variation than the weather itself or than the optimal state trajectories. Consequently, they are more suitable to convey long-term information to the short-term optimization than those rapidly fluctuating signals. Ideally, it would be required to calculate the costate trajectories for nominal weather and apply these in the optimization. During the season, updates can be made to reflect the actual past.

7.8.3 Sensitivity of RHOC to Slow Costates

Because after the experiment the real costate trajectories for the slow state variables were known, it was possible to compute afterward the sensitivity of the short-term receding horizon controller to the slow costates. In Figure 7.10, it can be seen that at September 1 the value of the fruit weight costate (λ_{WF}) is about –0.02 Dfl g[dw]$^{-1}$, which is the same as used in the experiments and which is also equal to the value of the harvested fruits. The costate associated to the leaf biomass (λ_{WL}) is about 0.027 Dfl g[dw]$^{-1}$, meaning that the marginal costs increase if more leaves are produced at that time. At September 1, being near to the end of the season, investing in more leaves does not pay off. However, in the experiment, a fixed value of –0.02 Dfl g[dw]$^{-1}$ was used, which is just the opposite. Thus, the controls in the experiment tend to favor generative growth more than is the case in the simulation with the correct costates. This can be seen in Figure 7.23, which shows that with the correct leaf costate, the temperature is made considerably higher. At higher temperature in the model the allocation of assimilates is more directed to fruits, which makes sense if investment in leaves is no longer necessary. A higher temperature is achieved by invoking less ventilation, as can be seen by comparing the window openings. The action is not hampered by humidity constraints as higher temperatures also help decrease the RH. Because there are less ventilation losses, the optimal control algorithm tries to take advantage of this situation by opening the CO_2 valve more often, although the effect on the CO_2 content remains quite small.

7.8.4 Sensitivity of RHOC to Weather Forecast and Prediction Horizon

The control solution depends on the horizon length and, consequently, on the accuracy of the weather forecast over the horizon length. In the tomato case of this chapter, this was not investigated explicitly. The long-term effects of crop growth are encapsulated via the marginal value of the biomass increase, according to Equation 7.100. Therefore, one would perhaps expect that the short-term receding horizon controller does not have to look over longer time horizons than that in agreement with the physical time constants of the system. These can be evaluated by looking at the

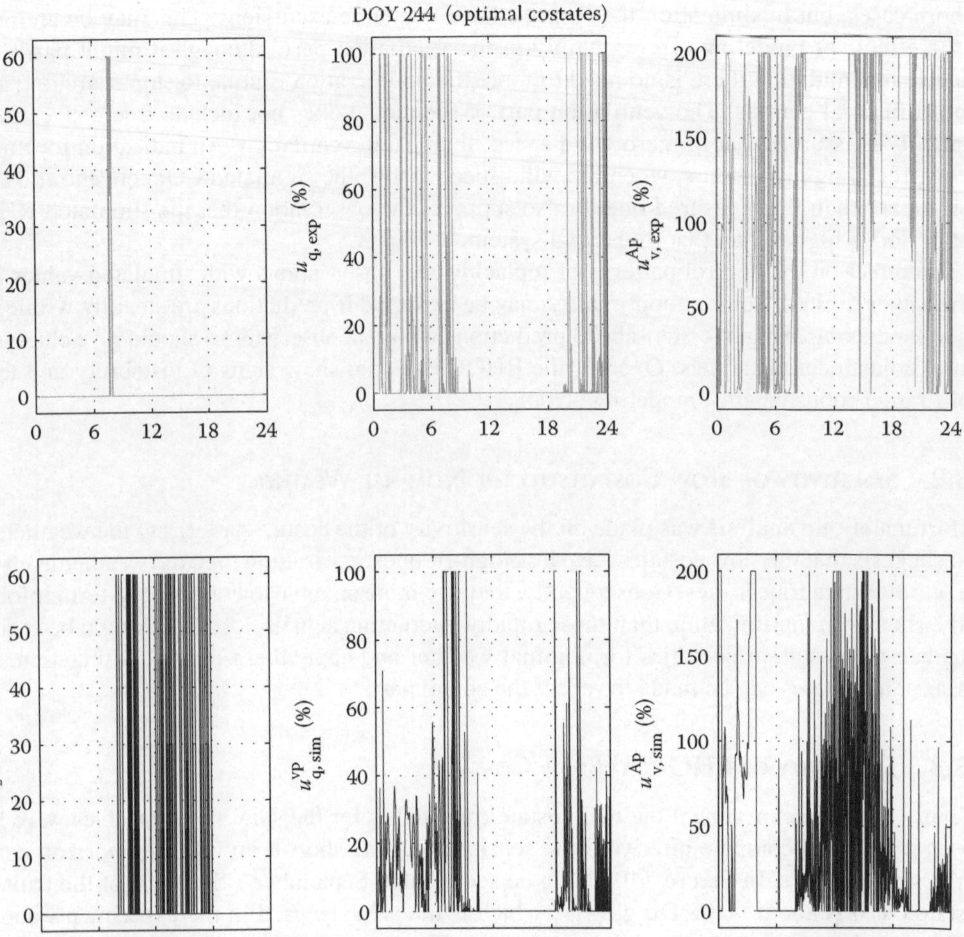

FIGURE 7.23 Optimal control with posterior costates instead of fixed costates. September 1, 1995. Top: experimental controls with fixed costate; middle: simulated controls with true costate; bottom: greenhouse output variables (solid line: simulation; dim line: experimental).

eigenvalues of the linearized physical model and appear to be on the order of, say, half an hour. This supports the choice for a horizon of one hour.

Another indication is obtained from a study reported by Tap, Van Willigenburg, and Van Straten (1996b). The crop in that study was lettuce, but a goal function of the same shape as in Equation 7.100 was used. The receding horizon controller as described in this chapter was simulated over 24 hours for several selected days during the season. It was observed how the goal function value achieved for a particular day changed with the prediction horizon for four different cases. An example of an outcome is shown in Figure 7.24. The following situations have been compared:

1. Open-loop optimal control over 24 hours with perfect weather. This results in the reference value, that is, the best theoretically achievable value, represented by the straight line in Figure 7.24.
2. Perfect weather prediction. The difference with the open-loop calculation is that the control can only take into account the forecast over the selected horizon length. It is seen that losses occur if the horizon is very short (45% loss with a horizon of one minute, about 20% loss with a horizon of 20 minutes), but as the horizon length gets more than one hour, almost no further losses occur, suggesting that knowing the weather over a longer time horizon has little influence on the immediate decisions to be taken. Conversely, it can be said that knowing the weather in some detail over the coming few hours can significantly improve the economic result.
3. Lazy man weather prediction. That is, at any control interval of one minute, it is assumed that the weather over the coming control horizon remains the same as currently observed. The constant pattern is updated every minute, but unlike a perfect weather prediction, there is no time variation in the pattern. It can be seen that there is a loss with respect to the

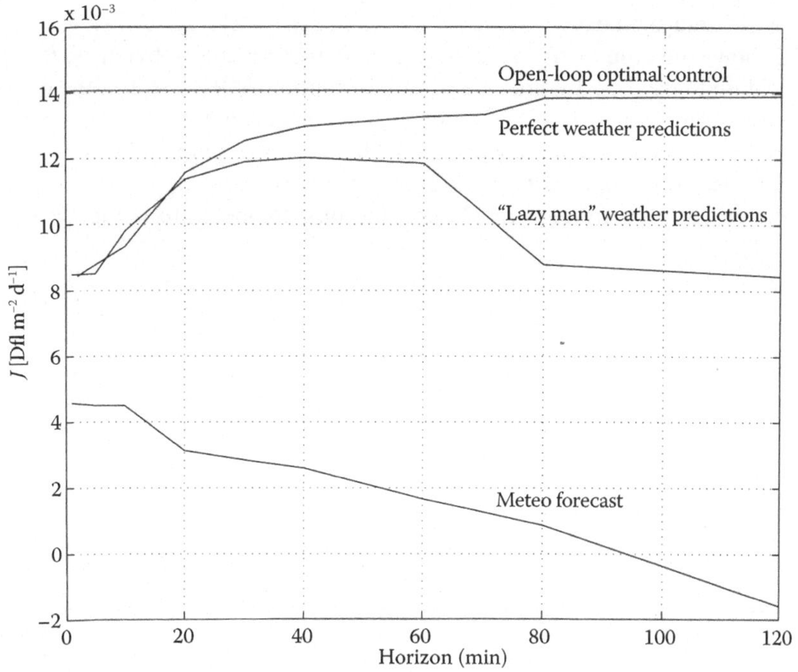

FIGURE 7.24 Effect of the optimization horizon and type of weather forecast on the criterion function. Lettuce model (adapted from Tap, Van Willigenburg, and Van Straten, 1996b).

perfect weather, but it remains limited up to a horizon of one hour. With longer horizons, the loss becomes larger again.

4. Also a comparison was made with a commercially available consumer prediction of hourly radiation over the coming 24 hours, which are made available at 7:00 and 11:00 A.M. Because these forecasts are intended for the general public and pertain to a large area, deviations from the local weather are large at any time, and consequently there are considerable losses.

Clearly, cases 1 and 2 cannot be realized in practice. However, forecasts that are currently available commercially for the greenhouse industry in The Netherlands become more and more detailed and can therefore be used with advantage, provided they are specific for local conditions. Estimates for local conditions can be obtained by combining regional forecasts with local measurements via Kalman filters, as shown by Doeswijk and Keesman (2005). The study also shows that if such forecasts are lacking, still very good results can be obtained by using local observations together with the lazy man prediction. The choice of the one hour horizon in the tomato experiment was motivated by the result described above.

7.9 ASSESSMENT OF ECONOMIC RESULT AS COMPARED WITH CONVENTIONAL CONTROL

The optimal control algorithm has been tested and compared with a conventional controller in two experiments in 1994 and 1995 (Tap, 2000; Tap, Van Willigenburg, and Van Straten, 1996a). The experiments were done in two compartments of a real greenhouse with tomatoes. One compartment was controlled by a conventional climate computer and the other by the receding horizon optimal control algorithm. The first experiment from 1994 is mainly used to overcome practical implementation problems. In the second experiment in 1995, both compartments were controlled by a conventional controller until day 214. From then on, the optimal controller has taken over control in the optimal control compartment.

Figure 7.25 shows the crop measurements in both compartments. Obviously, the differences in growth and yield up to day 214 are not caused by the control. After that day, the optimal and conventional fruit weight and harvested fruit weight stay almost equal to each other, except for slightly more leaves in the optimal control compartment. This can be explained by the lower greenhouse temperature obtained in the optimally controlled greenhouse, causing a more vegetative crop. In general, in the optimally controlled greenhouse, the pipe temperature is higher and the optimal

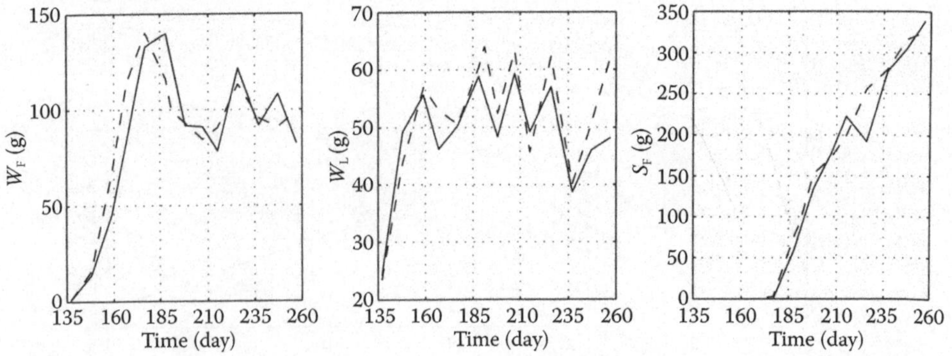

FIGURE 7.25 Comparison of fruit and leaf biomass and fruit yield between conventionally controlled compartment (solid line) and optimally controlled compartment (active from day 214, dashed line). Measured data from May 17 until September 14, 1995.

window opening is larger than the conventional ones. Together, this results in an RH that is about 5–10% lower, despite the fact that the conventional controller was given an upper RH boundary setting of 90% and the optimal controller one of 95%. It shows that the optimal controller, with the present settings, values the maintaining of the upper RH boundary more than the conventional controller, resulting in a lower optimal RH at the expense of a higher heat input. As to the control actions, these are more smooth in the conventional greenhouse than in the optimally controlled one. The latter tends to bang-bang control because no costs were imposed on control movements. It means that windows are opened further, with in theory a potential risk of cold spells. In practice, however, no visual differences in crop health were observed, with the crop in the optimal compartment looking more vigorous. The experiments clearly indicated that the control of the RH strongly influences the overall behavior of the controller. To stick to the upper RH boundary, the optimal pipe temperature is higher and the optimal windows are opened farther than is the case in the conventional case, resulting in a lower CO_2 concentration and temperature.

7.9.1 Simulated Comparison

The experiment has not run long enough in comparison with the time constants of the crop to expect any significant differences in crop behavior between conventional and optimal control. Moreover, the experiment is performed only once, so considering the variability in plants, the results are not statistically reliable. Therefore, to demonstrate any differences by experiments, experiments must last longer, and several repetitions are needed.

To get a good impression of the expected difference over a whole season, simulation is a good tool. In contrast to practice, the crop is deterministic, thus eliminating any stochastical differences; in addition, reliable energy consumption evaluations are much easier to perform in simulation. Although simulation is subject to model inaccuracies, a simulated comparison is still useful as both controllers control the same greenhouse-crop model. The absolute value of the simulated harvest and energy consumption may differ from the true ones, but the relative performance of the different controllers is expected to be affected less.

7.9.1.1 Initial Conditions

Because the optimal controller computations are lengthy (at the time of the experiment, it was about a quarter of real time), computation over a whole season is infeasible. Therefore, the comparison is performed for four characteristic days throughout the year 1995 (March 2, May 22, August 13, and October 31), using the observed weather. Initial states are taken equal and are obtained from the seasonal optimization for the crop states and from a run for the preceding day with a conventional controller for the greenhouse states. It should be noted that as the conventional controller behaves differently from the optimal one, the initial values obtained this way are suboptimal for the optimal case.

7.9.1.2 Matching the Humidity Constraint Violation

As can be learned from the experimental results, the conventional controller takes the constraints much less seriously than the optimal controller. This creates a different playing ground. To make a fair comparison, it is necessary to grant the optimal controller the same optimization space as the conventional one. As shown before, this is particularly relevant for the humidity constraint. The task is performed by adapting the humidity penalty in such a way that it is the same in the conventional and optimal case.

The procedure used to match the humidity constraint violation between the two controllers is based on the following reasoning. Because the conventional controller is widespread in practice, it has been assumed that the humidity trajectories it realizes are acceptable to the growers, despite the violation of the bounds. In addition, it has been assumed that it does not matter when exactly the constraint violations occur. The humidity penalty is defined in Equation 7.95, where the slope parameter α represents the costs in money units of exceeding these boundaries by 1% during one second. By an iterative procedure, the slope parameter in the optimal controller is selected in such a way that on average the penalty equals the conventional one. The details of this procedure were described by Tap (2000).

The matching α, named α^*, is quite sensitive to the chosen day and season. Ideally, one would have to compute it for the whole season. Here, simply, the value is taken that matches the humidity penalties averaged over the four selected days (0.5×10^{-7} Dfl m^{-2} s^{-1} (%RH)$^{-1}$).

7.9.1.3 Humidity Penalty and Heat Input

The slope parameter determines to what extent the humidity boundaries are maintained. As the slope increases, the heat input will increase, but it turns out that this strongly depends on the weather of the considered day. The total heat input for the four days increases approximately by 11% when the slope is increasing from $\alpha = 0$ to the matching one $\alpha = \alpha^*$. This indicates that 11% of the optimal energy consumption is related to dehumidification. The actual percentage for a specific day can be larger or smaller. This value is of the same order of magnitude as the annual values De Halleux and Gauthier (1998) found in their simulation experiments under Quebec climatic conditions using different ventilation regimes (12.6% for on–off control, and 18.4% for proportional control).

7.9.1.4 Results

Figure 7.26 summarizes the results as averaged daily values of the criterion function and its components expressed in income terms (Equation 7.100 with an opposite sign), both for the optimal and the conventional controller averaged over the four simulated days.

First, it can be seen that the humidity penalty matching procedure was successful because the penalties are practically equal for both greenhouses (oval in Figure 7.26). It has to be noted, however, that a considerable difference can still occur during individual days because it is not possible to find a single matching α that equates the humidity penalty on every individual day. On average, in the optimally controlled greenhouse fewer fruits are harvested (5%), but then the heat demand is 13% lower. Hence, per unit weight of harvested fruit, there is an 8% lower energy consumption; that is, the optimally controlled greenhouse has an 8% higher energy efficiency than the conventional one. None of the controllers violate the CO_2 bounds. The cost of CO_2 dosage, the CO_2 penalty, and the marginal value of the costate are combined in the "other" category because they are generally low. The temperature constraints are violated more often in the conventional greenhouse. The higher temperature penalty represents a higher risk of damage to the crop because of too high or too low temperatures. The optimal controller takes the penalties more seriously, probably at the expense of some costs. If the "optimal" temperature penalty parameter was set such that the penalty would be equal to the conventional one, by a similar procedure as for the humidity constraint, the comparison would have been more in favor of the optimal controller.

There are quite remarkable differences in the long-term investment terms of the goal function (the last bracketed term in the integral of Equation 7.100). In the conventional controller, the (future) benefits of both fruits and leaves are negative. Because they are not taken into account in the conventional controller, there is a tendency to neglect investment in future production capacity. In practice, it is the task of the grower to adjust the settings to maintain a good balance between vegetative and generative growth, but this is not always successful. It is one of the advantages of the two–time-scale receding horizon optimal controller that this is automatically achieved, provided that the distribution model (especially Equation 7.27) is an accurate description of reality.

Together, the fruit harvest, the heating costs, and the CO_2 costs constitute the cash flow over the day. Over the four chosen days, the conventional controller has a 3% higher immediate cash flow. This is possible because on the selected days the conventional controller cashes costs invested earlier in fruits and invests less in future harvest. The optimal controller is, however, achieving a better goal function value ($-J$), as expected, because in the minimization, not only is the cash flow

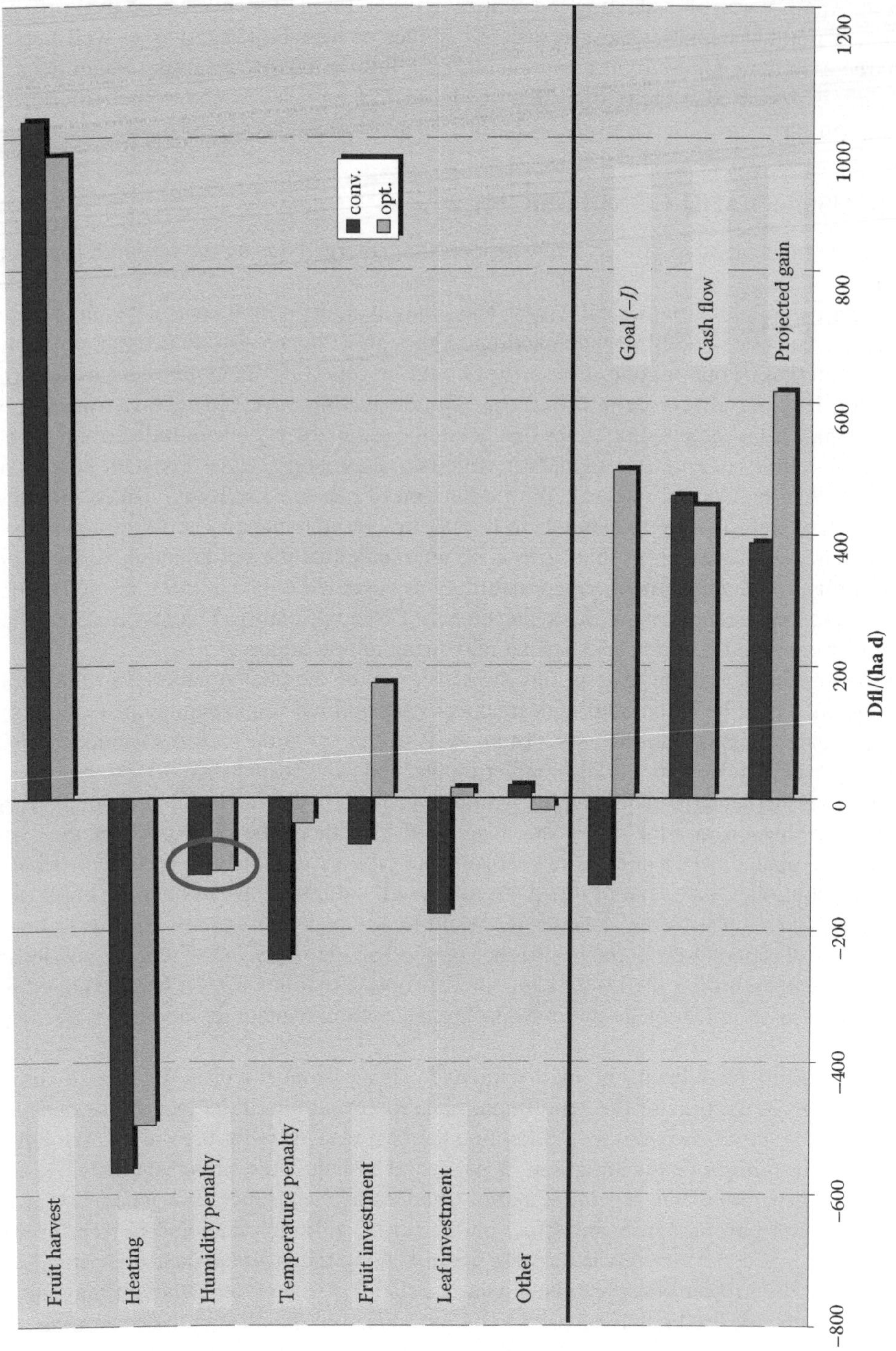

FIGURE 7.26 Comparison between conventionally and optimally controlled greenhouse. Averages over four selected days distributed over the cultivation period.

taken into account but also the risk of damage to the plant because of violation of the bounds and the expected future profit because of investments in leaf and fruit. The value of $-J$ shows that on average for these four days, the conventional controller suffers a loss, whereas the optimal control gains a profit. Another way of evaluating the controllers is by taking the investment in future fruits into account. If nothing happens, these fruits will sooner or later be harvested as well and then generate a true cash flow. The sum of instantaneous cash flow and expected future benefits, coined *projected gain*, is presented at the bottom line of Figure 7.26 and clearly shows the superiority of the optimal controller.

7.10 DISCUSSION AND CONCLUSIONS

Before concluding, it is good to discuss a few aspects that emerged during the research reported in this chapter.

First, there are a couple of numerical issues. The solution of an optimal control problem at every control instant is computationally very demanding. Although with a prediction horizon of one hour and a control interval of one minute, trajectories could be found in the experiment as described in this chapter, it is not always certain that the solution really converged in that time interval. Moreover, in simulation, computation times that get in the same order of magnitude as real time are a problem. A possibility to reduce computation time is to assume piecewise constant control over a longer period than the control interval. With a horizon of one hour and, say, fifteen minutes of piecewise constant control, only four numbers per control variable have to be found to minimize the cost function. Another aspect is that there is no guarantee that the optimization is not trapped into local minima. In the experiment, this possibility has never led to instabilities or odd solutions, but local minima may be hampering the achievement of true optimality. The fact that significant improvements were found is perhaps somewhat reassuring to practitioners.

On the theoretical side, more insight into the sensitivity of the goal function J to the various influential factors would help come up with practical near-optimal implementations or with solutions that are easier to explain to the user. An analysis of the variation of J to variation in model parameters (e.g., see Van Henten, 2003), weather inputs, and slow costates would also be welcome to see how much effort needs to be put into model improvement. In the current application, it has not been investigated how sensitive the slow costates are for the weather. One can expect that the costate patterns, when computed over a number of realizations of the weather, have a certain distribution. In a "nominal" controller, the mean of this distribution will naturally have to be used, but it would be good to know how much worse or better the result in any real year can be. It needs to be said that optimal control cannot prevent the occurrence of good or bad years, but within the possibilities offered by the weather, it does the best it can. Another option that has not yet been explored is to repeat the long-term optimization a few times during the season to adapt the nominal trajectory to the actual past.

The current results have been obtained without feedback from the crop. In lack of suitable online crop sensors at the time of the experiment, the crop evolution of the model was open loop during the receding horizon experiment. Sampling data came too late for use online. Crop monitoring is expected to improve the situation, in particular to adjust the important distribution of assimilates over generative and vegetative parts. Changes in the temperature sensitivity of this process are likely to have a significant effect on the result, although it should be kept in mind that in a conventional controller this is the case as well. As in the classical controller, it will still be necessary that the grower observes visual characteristics of the crop. In the optimal controller, deviations from a desired behavior need to be introduced by modifying the constraints. The advantage of optimal control is that within this new playground, a new optimum is automatically found.

It was assumed in the application that prices of commodities and yield market prices were constant throughout the optimization period. It is known that this is not the case in normal practice.

In case of substantial price movements it may be necessary to recompute the long-term optimization. However, unlike classical control, it is very easy to include price information at any time. The optimal controller will allow adjustment to the actual situation in an automatic fashion and in an optimal way. The correct formulation of the goal function is crucial in this process, but in doing so it stimulates clear and crisp thinking about what the grower really wants to achieve.

On the basis of this comprehensive real experiment, as well as the experimentally supported simulations presented in this chapter, the following can be concluded:

- The receding horizon controller appears to be quite robust against modeling errors. Modeling errors will lead to less than optimal performance but do not seem to give rise to controller instability.
- The results suggest that it does make sense to exploit the actual weather. A lazy man local forecast already gives good results, and some further benefits can be gained by using local short-term weather forecasts. Gains in energy efficiency in the order of magnitude of 10–15% seem achievable, just on the basis of control, without investments in equipment.
- Humidity control contributes significantly to the costs of greenhouse operation (10–20%). The use of penalty functions in the goal function provides for a mechanism to satisfy the constraints in a much better way than with classical control. Without precautions, this may lead to higher energy costs in exchange for much lower risk. Carefully choosing the humidity constraints and penalty parameters is therefore important. There is a delicate balance between the risk accepted and the economic result that can be achieved. More insight into the risks of stretching the bounds is an important premise to achieve further energy savings. This is true as well in a conventional controller setting.
- The receding horizon optimal controller as developed here incorporates in an elegant way the long-term (crop) effects in the short-term control. The result is a variable climate, which is much more in agreement with a natural situation. Crop quality was at least as good as with conventional control.
- The experiment described in this chapter has provided the proof of principle that optimal control is feasible in practice.

REFERENCES

De Halleux, D., and L. Gauthier. 1998. Energy consumption due to dehumidification of greenhouses under northern latitudes. *Journal of Agricultural Engineering Research* 69 (1): 35–42.

De Jong, T. 1990. *Natural Ventilation of Large Multi-span Greenhouses*. Wageningen: Wageningen Agricultural University.

De Koning, A.N.M. 1994. *Development and Dry Matter Distribution in Glasshouse Tomato: A Quantitative Approach*. Wageningen: Wageningen Agricultural University.

De Zwart, H.F. 1996. *Analyzing Energy-Saving Options in Greenhouse Cultivation Using a Simulation Model*. Wageningen: Agricultural University Wageningen.

Doeswijk, T.G., and K.J. Keesman. 2005. Adaptive weather forecasting using local meteorological information. *Biosystems Engineering* 91 (4): 421–431.

Frank, P.M. 1978. *Introduction to System Sensitivity Theory*. New York: Academic Press.

Heuvelink, E. 1995. Growth, development and yield of a tomato crop: Periodic destructive measurements in a greenhouse. *Scientia Horticulturae* 61 (1–2): 77–99.

Jolliet, O., and B.J. Bailey. 1992. The effect of climate on tomato transpiration in greenhouses: Measurements and models comparison. *Agricultural and Forest Meteorology* 58 (1–2): 43–62.

Monteith, J.L., and M.H. Unsworth. 1990. *Principles of Environmental Physics*. 2nd ed. London: Edward Arnold.

Seginer, I., and A. Sher. 1993. Optimal greenhouse temperature trajectories for a multi-state-variable tomato model. In *The Computerized Greenhouse*, ed. Y. Hashimoto, G.P.A. Bot, W. Day, H.-J. Tantau, and H. Nonami. San Diego: Academic Press, 153–174.

Tap, R.F. 2000. *Economics-Based Optimal Control of Greenhouse Tomato Crop Production, Systems and Control*. Wageningen, The Netherlands: Wageningen Agricultural University.

Tap, R.F., L.G. Van Willigenburg, and G. Van Straten. 1996a. Experimental results of receding horizon optimal control of greenhouse climate. *Acta Horticulturae* 406: 229–238.

Tap, R.F., L.G. Van Willigenburg, and G. Van Straten. 1996b. Receding horizon optimal control of greenhouse climate based on the lazy man weather prediction. In *Proceedings of the 13th IFAC World Congress, San Francisco, CA, USA*, ed. J.J. Gerler, J.B. Cruz, and M. Peshkin. Elsevier Science. Vol. B, 387–392.

Van Henten, E.J. 1994. *Greenhouse Climate Management: An Optimal Control Approach*. Wageningen: Agricultural University Wageningen.

Van Henten, E.J. 2003. Sensitivity analysis of an optimal control problem in greenhouse climate management. *Biosystems Engineering* 85 (3): 355–364.

Van Straten, G., L.G. Van Willigenburg, and R.F. Tap. 2002. The significance of crop co-states for receding horizon optimal control of greenhouse climate. *Control Engineering Practice* 10 (6): 625–632.

8 An Advanced Application: The Solar Greenhouse

8.1 INTRODUCTION

The flexibility of the receding horizon optimal control is illustrated in this chapter, where the methodology is applied to a novel advanced greenhouse design, called the solar greenhouse. In this greenhouse, surplus heat collected during the summer is stored in an aquifer, using a heat exchanger, and retrieved in the winter using a heat pump. In addition, ventilation, when it is needed, is done with heat recovery. Moreover, materials with better heat insulation properties are used for the cover.

In contrast to the time-scale decomposition approaches in the previous chapters, a different route is followed here (Van Ooteghem, 2007). There is no specific crop; the crop model is a simple one-state biomass model, using the net photosynthesis. Differences between crops can be expressed by different model parameter choices. The long-term effects are circumvented by assuming that any biomass produced sooner or later will lead to income. This amounts to the assumption that a unit of biomass increase has the same value at any time during the cultivation. Unwanted developmental effects are avoided by constraints on temperature and by the temperature integral concept. The long-term effects of the aquifer are tackled by calculating an averaged annual time pattern for the energy content, and requiring that the controller remain in the neighborhood of this pattern.

Elaborate models for photosynthesis and crop evapotranspiration are used to encapsulate the larger operational ranges to be expected from the optimal control. Also the *greenhouse climate model* is an elaborate model. The states are: CO_2 concentration, water concentration, air temperatures in compartments above and below a thermal screen, and temperatures of the roof, the crop, and the soil. Additional models describe the temperatures of upper and lower heating pipes, the cooling net, the heat pump, and the energy content of the aquifer. Exchanges by convection, conduction, and radiation are described separately.

Part of the model is calibrated against *experimental data* for a conventional greenhouse. The parameters for the new elements such as heat exchanger and heat pump are derived by design.

To obtain control solutions, a restriction is first introduced on the degrees of freedom for the control actions, avoiding actions that are unlikely to be optimal, such as heating and cooling at the same time. To avoid local minima, a grid search is performed to get a crude optimal trajectory, and finally a gradient method is used for fine-tuning. These modifications are significant to reduce the computation time, which is a prerequisite for on-line applications.

Results have been obtained by year-round simulations, thus allowing the assessment of the savings of the solar greenhouse as compared with conventional greenhouses without the solar elements. Note that, in this case, it was assumed that the conventional greenhouse was also controlled by a receding horizon optimal controller. Effects of the temperature integral and effects of the weather forecasts on the expected benefits (i.e., before a control action is invoked) and the real benefits (i.e., with the realized control) are investigated, leading to more insight into the aspects that are relevant to the economic result.

8.2 DESCRIPTION OF THE SOLAR GREENHOUSE CONCEPT

An accurate model of the controlled system is necessary for the successful application of optimal control. Based on this model and a mathematical description of the control objectives, the optimal controller finds the best solution. In practice, the successful application of optimal control depends critically on the quality of the model. Van Henten (1994) and Tap (2000) found that parts of the greenhouse behavior were not well described by their models. This negatively affects the performance of the optimal control.

For the receding horizon optimal control concept used in Section 8.7.2, a state-space description of the system is needed. The model should be sufficiently small with respect to the number of differential equations, controls, and external inputs to limit the on-line computational load. On the other hand, it should be sufficiently accurate.

In this section a dynamic model for the solar greenhouse is developed. With a few small modifications this model can be turned into a model for the conventional greenhouse.

A conventional greenhouse is heated by a boiler, which in The Netherlands is also used to provide CO_2 for crop growth. The roof has a high transmission of solar radiation, but poor heat-insulating properties. The greenhouse can be cooled by opening the windows, which also provides a means to decrease humidity.

In the solar greenhouse design, the heat insulation and the transmission of solar radiation are maximized. A warm- and a cold-water aquifer* layer are used to store and retrieve the surplus solar energy. At times of heat demand, the greenhouse can be heated with little energy input with a heat pump and warm aquifer water. At times of heat surplus, the greenhouse can be cooled with a heat exchanger and cold aquifer water, while energy is harvested for use at times of heat demand. In contrast to common greenhouses, the CO_2 supply in the solar greenhouse concept is detached from the boiler, thus avoiding the need to use the boiler at times of CO_2 demand. It is assumed that the CO_2 can be acquired from a power plant.† Ventilation with heat recovery is used to dehumidify the greenhouse at times of heat demand.

The solar greenhouse has the following changes compared with a conventional greenhouse:

- *Improved insulation value and improved light transmission cover*: To minimize heat loss to outdoor air and maximize the input of solar radiation. This will result in a higher crop yield and lower energy consumption.
- *Ventilation with heat recovery*: If ventilation is needed for high humidity but not for cooling, the sensible heat loss can be partially recovered by exchanging the air through a heat exchanger. The outdoor air is preheated by the indoor greenhouse air, while the humidity content is decreased. Latent heat that is vented out is lost. If ventilation is needed to prevent high humidity and high temperature, the windows are used—as in normal greenhouse practice.
- *Aquifer*: A long-term storage of water in the lower soil layers. The aquifer has a cold ($T_{aq}^{C,cold} = 10°C$) and a warm ($T_{aq}^{C,warm} = 16°C$) side. When the greenhouse is cooled, cold water is taken from the cold aquifer side, heat is extracted from the greenhouse with the heat exchanger, and the resulting warm water is stored in the warm aquifer side. When the greenhouse is heated, warm water is taken from the warm aquifer side, heat is supplied to the greenhouse with the heat pump, and the resulting cold water is stored in the cold aquifer side.
- *Heat exchanger*: Heat can be extracted from the greenhouse by a heat exchanger. The heat exchanger is used to cool water in the finned upper cooling net pipes with water from the

* An aquifer is a formation of water-bearing sand material in the soil that can contain and transmit water. Wells can be drilled into the aquifers and water can be pumped into and out of the water layers.

† It is possible in The Netherlands to retrieve pure CO_2. Shell Pernis/OCAP currently supplies about 200 growers with CO_2, thus reducing the CO_2 emission by 170 kilotons CO_2 per year. This saves the growers 95 million m^3 gas.

cold aquifer side. The cooling net extracts energy from the greenhouse. Water from the cold aquifer side is heated to a temperature above T_{aq}^{warm}. Water with a temperature T_{aq}^{warm} is stored in the warm aquifer side.

- *Heat pump*: Heat can be supplied to the greenhouse by a heat pump. The heat pump is used to heat water in the lower heating net with water from the warm aquifer side. The heat pump can attain a heating temperature of about 33°C. The lower heating net supplies energy to the greenhouse. Water from the warm aquifer side is cooled to a temperature below T_{aq}^{cold}. Water with a temperature T_{aq}^{cold} is stored in the cold aquifer side.
- *Boiler*: Used for additional heating if the heat pump cannot supply enough heat.
- *Carbon dioxide supply*: Separate CO_2 supply, because CO_2 is no longer supplied by the boiler.
- *Gas motor or electric drive*: Used to run the heat pump; the exhaust gas can be used to give additional heat.

The gas motor might be replaced by a windmill that supplies electricity (sustainable instead of fossil energy). In the ideal set-up, the boiler is only needed as a backup. The heating and storage devices have to be controlled to optimize the heat use. This will ensure appropriate production and quality and low energy consumption.

The model of the conventional greenhouse used in this research is developed based on the model by Heesen (1997), who exploited the research by Bot (1983), De Jong (1990), De Zwart (1996), and Van Henten (1994). This conventional greenhouse model has been modified to include a thermal screen and a double glass cover. For the solar greenhouse it has been extended with the solar greenhouse elements described above, which give new possibilities for heating, cooling, and dehumidification. The greenhouse model uses the crop model for the exchange of heat, CO_2, and water with the crop. This section gives a complete and detailed description of the greenhouse model.

The outline of the succeeding sections is as follows. The system is described in Section 8.3, where an overview is given of the system with its states, control inputs, and external inputs. The state equations mainly contain terms that describe the exchange of heat, water, and CO_2. In Section 8.4, these exchange terms are worked out. In Section 8.5 the crop biophysics are described. In Section 8.6 the model is calibrated and validated to investigate its accuracy and suitability for optimal control purposes. In Section 8.7 the optimal control method is explained and the results are presented and discussed.

8.3 SYSTEM DESCRIPTION

The greenhouse configuration is described in Section 8.3.1. In Section 8.3.2, all assumptions made in this model are described. Section 8.4 gives the states, control and external inputs, and the state equations that govern the system behavior.

8.3.1 Greenhouse Configuration

The greenhouse configuration is given in Figure 8.1. The greenhouse is a Venlo greenhouse with a north–south orientation. A Venlo greenhouse is a multi-span greenhouse.

A heating system consisting of a boiler, a condenser, and a heat pump can be used to heat the greenhouse. The lower heating net can be heated with the boiler to a temperature of 90°C and with the heat pump to a temperature of about 33°C. The upper heating net is heated by the condenser to a temperature of 45°C. The condenser is heated by the flue gas of the boiler. The heating system is described in Sections 8.4.6.1 and 8.4.6.3.

A cooling system consisting of a heat exchanger can be used to cool the greenhouse. The upper cooling net can be cooled with the heat exchanger to a temperature of about 10°C. The cooling system is described in Section 8.4.6.4.

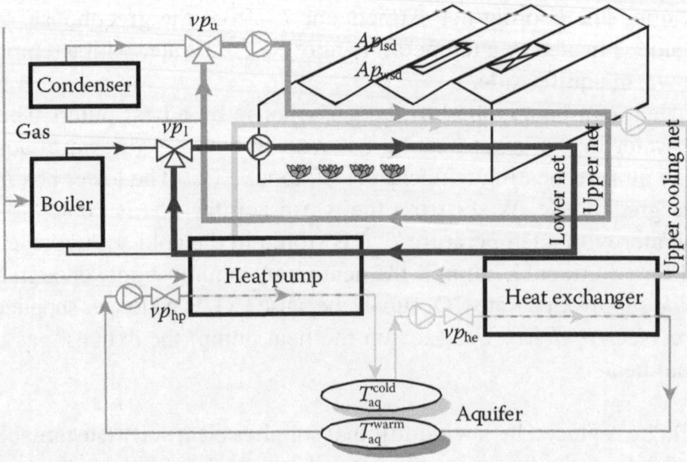

FIGURE 8.1 Greenhouse configuration.

The heat pump and the heat exchanger operate in conjunction with an aquifer. A warm- and a cold-water aquifer layer are used to store and retrieve the surplus solar energy. The warm-water layer has a temperature of $T_{aq}^{C,warm} = 16°C$ and the cold-water layer has a temperature of $T_{aq}^{C,cold} = 10°C$. The warm water is used by the heat pump to heat the greenhouse. The cold water is used by the heat exchanger to cool the greenhouse.

A thermal screen can be closed during the night to reduce the heat loss to the environment if the temperature of the outdoor air is low. The thermal screen is operated based on rules used in common practice, which are described in Section 8.4.4.

Ventilation by opening windows can be used to cool the greenhouse and to decrease humidity. At times of heat demand, the humidity can be decreased by using ventilation with heat recovery. The sensible heat that is normally lost during ventilation through windows is partially recovered by exchanging the air through a heat exchanger. The ventilation model is described in Section 8.4.5.

The roof has a double-layer zigzag cover, which has a high insulation value and light transmission. This decreases heat loss to the environment and increases radiation in the greenhouse. To minimize fossil energy consumption, no lighting is used.

The control input trajectories consist of actuator settings, such as window apertures and valve positions of, for instance, the boiler. For the heat and mass transport, the following elements are taken into account: air (above and below the screen), crop, heating and cooling net pipes, roof, screen, and soil. These elements are modeled as lumped parameter models, which are assumed internally homogeneous. The soil and the roof are divided into two layers/parts.

8.3.2 Assumptions

The following assumptions are made:

- The greenhouse has a north–south orientation.
- Each span in the multispan greenhouse has the same layout with respect to the configuration of the heating and the cooling net, the thermal screen, and its size.
- The lower heating net is below the canopy and the upper heating and cooling net are above the canopy, but below the screen.
- All outdoor weather conditions are not influenced by the greenhouse climate conditions.
- All compounds (crop, roof glass, upper soil layer, lower and upper heating net, upper cooling net, screen, aquifer, etc.) and gasses (greenhouse air above and below the screen) are

homogeneous—they have a uniform temperature. The air in the greenhouse above and below the screen is perfectly mixed (with respect to CO_2 and H_2O concentration).

- When the screen is fully opened ($Cl_{sc} = 0$), the temperature and the CO_2 and H_2O concentrations can be averaged (proportional with the heat capacity and volume of the air above and below the screen). This is necessary to avoid numerical problems in the integration (see Section 8.4.4.3). The screen is impermeable for all gasses (H_2O, CO_2, and air). The screen transmits part of the solar radiation. The exchange of heat, CO_2, and H_2O through the screen opening can be described by a simple air-exchange rate.
- The heating nets and the cooling net can be described as a number of loops of pipes with a specific length and diameter. The temperature of the water in the lower and upper heating net and the cooling net can be described by simplified equations. In these equations, it is assumed that one temperature can be used to describe the energy content of the net. This temperature depends on the incoming temperature, from which the outgoing temperature can be directly computed with the heat-exchange terms (Van Ooteghem, 2007).
- One soil layer can be used to approximate the temperature of the upper soil layer T_s (Van Ooteghem, 2007).
- The double-glass zigzag roof cover consists of two layers of glass with air in between. The temperature of the outdoor side of the roof T_{ro} can be directly computed from the temperature of the indoor side of the roof T_{ri} and their heat exchange terms (Van Ooteghem, 2007).
- The heat pump is a compression heat pump. It is assumed that the heat transfer between the heat pump and the lower net has no dynamics (direct transfer of heat).
- The heat exchanger is a countercurrent heat exchanger. It is assumed that the heat transfer between the heat exchanger and the upper cooling net has no dynamics (direct transfer of heat).
- The aquifer has an infinite amount of warm and cold water available. The loading and unloading of the aquifer buffers is limited by government demands, which indirectly corrects for the fact that the buffers are not infinite.
- Water that condensates on the indoor side of the roof, on the screen and on the upper cooling net pipes is directly removed and therefore not available for evaporation.
- When ventilation with heat recovery is used, a fixed fraction η_{vhr} of the sensible heat is recovered.
- The CO_2 assimilation by the crop is instantaneously converted to biomass.
- The boiler runs on (natural) gas.
- The CO_2 supply in the solar greenhouse is assumed to be detached from the boiler. It is assumed that the CO_2 can be acquired from a power plant.

8.4 THE SOLAR GREENHOUSE MODEL

The greenhouse model is written in state-space form

$$\dot{\mathbf{x}} = \mathbf{f}(t, \mathbf{x}, \mathbf{u}, \mathbf{d})$$

where t is time, $\mathbf{x}$ are the states, $\mathbf{u}$ are the control inputs, $\mathbf{d}$ are the external inputs, and $\mathbf{f}$ is a nonlinear function. This function is integrated by using a Runge-Kutta fourth-order integration algorithm (Press et al., 1986) to obtain the states.

The model description given here is based on the model described by Heesen (1997), which in turn is based on the research by Bot (1983), De Jong (1990), De Zwart (1996), and Van Henten (1994). This model has been extended with a thermal screen, a double-glass cover, and the so-called solar greenhouse elements: heat pump, heat exchanger, ventilation with heat recovery, and a cooling net to describe the solar greenhouse behavior. The main external input is the weather.

The state equations have been formed based on the laws of conservation of enthalpy and matter. The dynamic behavior of the states is described using first-order differential equations, which match the state-space description of the systems. The notational conventions for the model used in this section are given in Table 8.1.

A description of the states **x**, the control inputs **u**, and the external inputs **d** is given in Table 8.2. The state variables **x**, the external inputs **d**, and the control inputs **u** are shown in Figure 8.2. In this figure, the boxes for the state variables **x** are bold, for the external inputs **d** are dashed, and for the control inputs **u** are dotted or dash-dotted. The dotted and the dash-dotted boxes are used to distinguish between the control inputs that are set by the optimal control and the control inputs that are directly derived from external inputs or from other control inputs.

The screen condition c_{sc} is either zero or one, where $c_{sc} = 0$ indicates that the screen is fully opened and $c_{sc} = 1$ indicates that the screen is (possibly partly) closed. The screen condition is a discrete switch that can be interpreted as an external input d, since it only depends on the outdoor shortwave solar radiation I_o and the temperature T_o of the outdoor air (see Section 8.4.4.1).

TABLE 8.1
Notational Conventions

Symbol	Description	Unit	Symbol	Description	Unit
			Variables		
Φ	Mass flow rate	kg s^{-1}	T	Temperature	K
F	Volume rate	m^3 s^{-1}	V	Volume	m^3
A	Surface area	m^2	$\rho \cdot c_p \cdot V$	Heat capacity	J K^{-1}
Q	Heat exchange	W			
			Subscripts		
a	Greenhouse air below screen		out	Going out of the system	
as	Greenhouse air above screen		rad	Shortwave radiation	
c	Crop		ri	Roof indoor side	
CO_2	Carbon dioxide		ro	Roof outdoor side	
he	Heat exchanger		s	Upper soil layer	
hp	Heat pump		ss	Subsoil layer	
H_2O	Water		sc	Screen	
in	Going into the system		sk	Sky	
l	Lower heating net		u	Upper heating net	
o	Outdoor		uc	Upper cooling net	
			Superscripts		
*	Optimal control		ppm	In unit (ppm)	
★	Estimated reference		rad	Shortwave radiation	
C	In unit (°C)		ref	Reference	
cold	Cold side aquifer		sat	Saturation	
cons	Condensation		sp	Setpoint	
lwv	Longwave radiation		T	Temperature	
max	Maximum		TI	Temperature integral	
mbar	In unit (mbar)		trans	Transpiration	
min	Minimum		warm	Warm side aquifer	
mol	In unit (mol)		wb	Wet bulb	

TABLE 8.2
States, Control Inputs, and External Inputs

Symbol	Description	Unit
	States x	
$C_{CO_2,a}$, $C_{CO_2,as}$	CO_2 concentration indoor air below/above screen	kg[CO_2] m^{-3}
$C_{H_2O,a}$, $C_{H_2O,as}$	H_2O concentration indoor air below/above screen	kg[H_2O] m^{-3}
T_a, T_{as}	Temperature indoor air below/above screen	K
T_c	Temperature crop	K
T_{ri}	Temperature roof indoor side	K
T_s	Temperature soil (upper layer)	K
T_l, T_u	Temperature lower/upper heating net	K
T_{uc}	Temperature upper cooling net	K
T_{sc}	Temperature thermal screen	K
S_T	Temperature integral	K day
B	Total biomass (fresh weight)	kg[fw] m^{-2}[gh]
E_{aq}	Aquifer energy content	J m^{-2}[gh]
	Control Inputs u	
vp_{CO_2}	Valve position CO_2 supply	[0,1]
Ap_{lsd}, Ap_{wsd}	Window aperture lee-side/windward-side	[0,1]
Cl_{sc}	Thermal screen closure	[0,1]
op_{vhr}	Option ventilation heat recovery	{0,1}
vp_l, vp_u	Valve position lower/upper net	[0,1]
vp_{he}	Valve position heat exchanger	[0,1]
vp_{hp}	Valve position heat pump	[0,1]
	External Inputs d	
I_o	Outdoor shortwave solar radiation	W m^{-2}[gh]
v_o	Outdoor wind speed	m s^{-1}
T_o	Temperature outdoor air	K
T_o^{wb} †	Temperature wet bulb	K
T_{sk}	Temperature sky	K
$C_{CO_2,o}$	CO_2 concentration outdoor air	kg[CO_2] m^{-3}

† H_2O concentration outdoor air $C_{H_2O,o}$ can be computed from the temperatures T_o and T_o^{wb} (see Appendix B).

The state equations are:

Carbon dioxide concentration indoor air below the screen

$$\frac{dC_{CO_2,a}}{dt} = \begin{cases} \dfrac{\Phi_{CO_2,in\_a} - \Phi_{CO_2,a\_c} - \Phi_{CO_2,a\_as}}{V_a} & \text{if } c_{sc} = 1 \\ \dfrac{\Phi_{CO_2,in\_a} - \Phi_{CO_2,a\_c} - \Phi_{CO_2,as\_o}}{V_a + V_{as}} & \text{if } c_{sc} = 0 \end{cases} \qquad (\text{kg}[CO_2]\ m^{-3}\ s^{-1}) \quad (8.1)$$

Carbon dioxide concentration indoor air above the screen

$$\frac{dC_{CO_2,as}}{dt} = \begin{cases} \dfrac{\Phi_{CO_2,a\_as} - \Phi_{CO_2,as\_o}}{V_{as}} & \text{if } c_{sc} = 1 \\ \dfrac{\Phi C_{CO_2,a}}{dt} & \text{if } c_{sc} = 0 \end{cases} \qquad (\text{kg}[CO_2]\ m^{-3}\ s^{-1}) \quad (8.2)$$

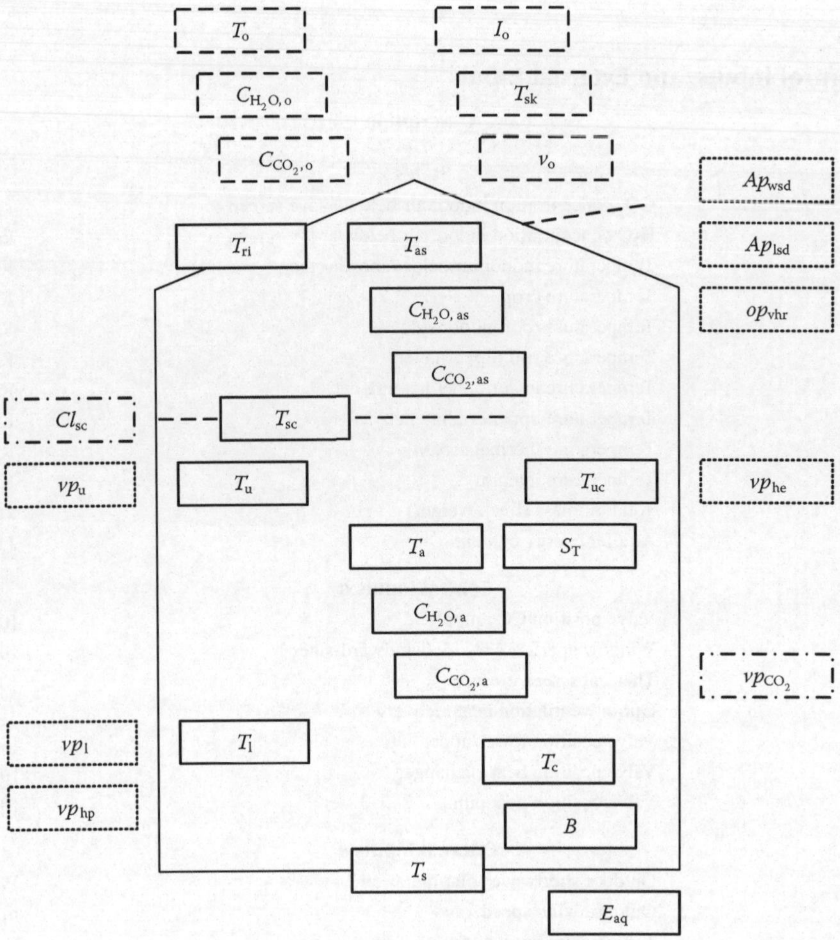

FIGURE 8.2 States x (solid box), external inputs d (dashed box) and control inputs u (dotted and dash-dotted boxes) in the solar greenhouse.

Water concentration indoor air below the screen

$$\frac{dC_{H_2O,a}}{dt} = \begin{cases} \dfrac{\Phi_{H_2O,c\_a} - \Phi_{H_2O,a\_uc} - \Phi_{H_2O,a\_sc} - \Phi_{H_2O,a\_as}}{V_a} & \text{if } c_{sc} = 1 \\ \dfrac{\Phi_{H_2O,c\_a} - \Phi_{H_2O,a\_uc} - \Phi_{H_2O,a\_sc} - \Phi_{H_2O,as\_ri} - \Phi_{H_2O,as\_o} - \Phi_{H_2O,as\_sc}}{V_a + V_{as}} & \text{if } c_{sc} = 0 \end{cases} \quad (\text{kg[H}_2\text{O] m}^{-3}\,\text{s}^{-1}) \quad (8.3)$$

Water concentration indoor air above the screen

$$\frac{dC_{H_2O,as}}{dt} = \begin{cases} \dfrac{\Phi_{H_2O,a\_as} - \Phi_{H_2O,as\_ri} - \Phi_{H_2O,as\_o} - \Phi_{H_2O,as\_sc}}{V_{as}} & \text{if } c_{sc} = 1 \\ \dfrac{\Phi C_{H_2O,a}}{dt} & \text{if } c_{sc} = 0 \end{cases} \qquad (\text{kg[H}_2\text{O] m}^{-3}\,\text{s}^{-1}) \quad (8.4)$$

Temperature indoor air below the screen

$$\frac{dT_a}{dt} = \begin{cases} \dfrac{Q_{l\_a} + Q_{u\_a} + Q_{uc\_a} + Q_{sc\_a} - Q_{a\_c} - Q_{a\_s} - Q_{a\_as}}{\rho_a \cdot c_{p,a} \cdot V_a} & \text{if } c_{sc} = 1 \\ \dfrac{Q_{l\_a} + Q_{u\_a} + Q_{uc\_a} + Q_{sc\_a} - Q_{a\_c} - Q_{a\_s} - Q_{as\_o} - Q_{as\_ri} + Q_{sc\_as}}{\rho_a \cdot c_{p,a} \cdot V_a + \rho_a \cdot c_{p,a} \cdot V_{as}} & \text{if } c_{sc} = 0 \end{cases} \qquad (\text{K s}^{-1}) \quad (8.5)$$

Temperature indoor air above the screen

$$\frac{dT_{as}}{dt} = \begin{cases} \dfrac{Q_{a\_as} + Q_{sc\_as} - Q_{as\_ri} - Q_{as\_o}}{\rho_{as} \cdot c_{p,a} \cdot V_{as}} & \text{if } c_{sc} = 1 \\ \dfrac{dT_a}{dt} & \text{if } c_{sc} = 0 \end{cases} \qquad (\text{K s}^{-1}) \quad (8.6)$$

Temperature crop

$$\frac{dT_c}{dt} = \frac{Q_{rad\_c} + Q_{a\_c} + Q_{l\_c} + Q_{ri\_c} + Q_{s\_c} + Q_{u\_c} + Q_{uc\_c} - Q_{c\_a}^{trans} - Q_{c\_sc}}{\rho_c \cdot c_{p,c} \cdot V_c} \qquad (\text{K s}^{-1}) \quad (8.7)$$

Temperature soil

$$\frac{dT_s}{dt} = \frac{Q_{rad\_c} + Q_{a\_s} + Q_{l\_s} + Q_{uc\_s} + Q_{u\_s} - Q_{s\_c} - Q_{s\_ri} - Q_{s\_ss} - Q_{s\_sc}}{\left(0.7\rho_s \cdot c_{p,s} + 0.2\rho_{H_2O} \cdot c_{p,H_2O} + 0.1\rho_a \cdot c_{p,a}\right) \cdot V_s} \qquad (\text{K s}^{-1}) \quad (8.8)$$

Temperature lower net

$$\frac{dT_l}{dt} = \frac{Q_{in\_l} - Q_{l\_out} + Q_{rad\_l} - Q_{l\_a} - Q_{l\_c} - Q_{l\_ri} - Q_{l\_s} - Q_{l\_sc}}{\rho_{H_2O} \cdot c_{p,H_2O} \cdot V_l} \quad (K\ s^{-1}) \quad (8.9)$$

Temperature upper net

$$\frac{dT_u}{dt} = \frac{Q_{in\_u} - Q_{u\_out} + Q_{rad\_u} - Q_{u\_a} - Q_{u\_c} - Q_{u\_ri} - Q_{u\_s} - Q_{u\_sc}}{\rho_{H_2O} \cdot c_{p,H_2O} \cdot V_u} \quad (K\ s^{-1}) \quad (8.10)$$

Temperature upper cooling net

$$\frac{dT_{uc}}{dt} = \frac{Q_{in\_uc} - Q_{uc\_out} + Q_{rad\_uc} + Q_{a\_uc}^{cons} - Q_{uc\_a} - Q_{uc\_c} - Q_{uc\_ri} - Q_{uc\_s} - Q_{uc\_sc}}{\rho_{H_2O} \cdot c_{p,H_2O} \cdot V_{uc}} \quad (K\ s^{-1}) \quad (8.11)$$

Temperature indoor side of the roof (double or single glass cover)

$$\frac{dT_{ri}}{dt} = \begin{cases} \dfrac{Q_{rad\_ri} + Q_{as\_ri} + Q_{as\_ri}^{cons} + Q_{l\_ri} + Q_{s\_ri} + Q_{sc\_ri} + Q_{u\_ri} + Q_{us\_ri} - Q_{ri\_c} - Q_{ri\_ro} - Q_{ri\_ro}^{lwv}}{\rho_r \cdot c_{p,r} \cdot V_r} & \text{double} \\[2ex] \dfrac{Q_{rad\_ri} + Q_{as\_ri} + Q_{as\_ri}^{cons} + Q_{l\_ri} + Q_{s\_ri} + Q_{sc\_ri} + Q_{u\_ri} + Q_{uc\_ri} - Q_{ri\_c} - Q_{ro\_o} - Q_{ro\_sk}}{\rho_r \cdot c_{p,r} \cdot V_r} & \text{single} \end{cases} \quad (K\ s^{-1}) \quad (8.12)$$

Temperature screen

$$\frac{dT_{sc}}{dt} = \frac{Q_{rad\_sc} + Q_{c\_sc} + Q_{l\_sc} + Q_{s\_sc} + Q_{u\_sc} + Q_{us\_sc} + Q_{a\_sc}^{cons} + Q_{as\_sc}^{cons} - Q_{sc\_a} - Q_{sc\_as} - Q_{sc\_ri}}{\rho_{sc} \cdot c_{p,sc} \cdot V_{sc}} \quad (K\ s^{-1}) \quad (8.13)$$

Temperature integral temperature indoor air below the screen (more details in Section 8.5.3)

$$\frac{dS_T}{dt} = \frac{T_a - T_a^{ref}}{n_{secs}} \quad \text{(K day s}^{-1}\text{)} \quad (8.14)$$

Total biomass (fresh weight)

$$\frac{dB}{dt} = \frac{f_{CO_2,B} \cdot \Phi_{CO_2,a\_c}}{A_s} \quad \text{(kg[fw] m}^{-2}\text{[gh] s}^{-1}\text{)} \quad (8.15)$$

Aquifer energy content

$$\frac{dE_{aq}}{dt} = \frac{Q_{he} - Q_c}{A_s} \quad \text{(J m}^{-2}\text{[gh] s}^{-1}\text{)} \quad (8.16)$$

Table 8.3 gives the sizes of greenhouse components, while Table 8.4 gives some parameter values. In the subsequent sections, the carbon dioxide, water, and heat exchanges in the greenhouse are described.

8.4.1 Carbon Dioxide Model

The differential equations for the carbon dioxide concentrations of the indoor air below and above the screen ($C_{CO_2,a}$, $C_{CO_2,as}$ (kg[CO_2] m^{-3})) and the total biomass (B (kg[fw] m^{-2}[gh])) are given in Section 8.4. The carbon dioxide mass flow rates are described in the subsequent sections. The carbon dioxide concentrations, the biomass, and the carbon dioxide mass flows are shown in Figure 8.3.

All carbon dioxide concentrations $C_{CO_2,x}$ are here expressed in the SI unit (kg[CO_2] m^{-3}[air]). Often the carbon dioxide concentration $C_{CO_2,x}^{ppm}$ is used, which is expressed in (μmol[CO_2] mol^{-1}[air]), because this is the unit used in practice. This concentration can be computed with $C_{CO_2,x}^{ppm} = \frac{C_{CO_2,x}}{f^{ppm}}$ (see Equation 8.135).

TABLE 8.3
Sizes

Component	Diameter, Thickness, or Width (m)	Length (m)	Quantity (–)
Lower heating net	$d_l = 0.051$	$l_l = 2\,(l_s - 2 + 0.75) = 109.5$	$n_l = 2.5$
Upper heating net	$d_u = 0.028$	$l_u = 2\,(l_s - 2 + 0.75) = 109.5$	$n_u = 2.5$
Upper cooling net	$d_{uc} = 0.051$	$l_{uc} = 2\,(l_s - 2 + 0.75) = 109.5$	$n_{uc} = 2.5$
Roof (glass)	$d_r = 0.004$	$d_{ra} = 0.01$ (air cavity)	
Soil layer	$d_s = 0.65$	$dx_s = 1.247$ (centers)	
Leaf	$l_f = 0.035$		$LAI = 3$
Screen	$d_{sc} = 0.0005$		
Greenhouse span	$w_s = 4$	$l_s = 56$	$n_s = 14$
Gutter, ridge	$h_g = 4.5$	$h_r = h_g + 0.5\,w_s \cdot \tan(\gamma)$	
Window	$w_{win} = 2.25$	$h_{win} = 1$	$n_{win} = n_s \cdot n_{wins} = 14 \cdot 12 = 168$

TABLE 8.4
Parameter Values

Material		Density	Specific Heat Capacity	Area	Volume	Heat Capacity
		ρ $\left(\frac{kg}{m^3}\right)$	c_p $\left(\frac{J}{kg \cdot K}\right)$	A (m²)	V (m³)	$\rho \cdot c_p \cdot V$ $\cdot 10^6 \left(\frac{J}{K}\right)$
Water	Lower heating net	$\rho_{H_2O} = 998$	$c_{p,H_2O} = 4186$	$A_l = n_s \cdot n_l \cdot \pi \cdot d_l \cdot l_l \approx 614.05$	$V_l = A_l \cdot \frac{1}{4} \cdot d_l \approx 7.83$	32.71
	Upper heating net			$A_u = n_s \cdot n_u \cdot \pi \cdot d_u \cdot l_u \approx 337.12$	$V_u = A_u \cdot \frac{1}{4} \cdot d_u \approx 2.36$	9.86
	Upper cooling net			$A_{uc} = 4n_s \cdot n_{uc} \cdot \pi \cdot d_{uc}\ l_{uc} \approx 2456.19$	$V_{uc} = \frac{A_{uc}}{4} \cdot \frac{1}{4} \cdot d_{uc} \approx 7.83$	32.71
Air	Below screen	$\rho_a = 1.29 \cdot \frac{T_0}{T_a}$	$c_{p,a} = 1000$		$V_a = A_s \cdot h_g \approx 14112$	16.68
	Above screen	$\rho_{as} = 1.29 \cdot \frac{T_0}{T_{as}}$			$V_{as} = A_s \cdot \frac{1}{2}(h_n - h_g) \approx 1267.03$	1.50
Sand		$\rho_s = 1600$	$c_{p,s} = 800$			
Soil	70% sand, 20% water, 10% air			$A_s = n_s \cdot l_s \cdot w_s \approx 3136$	$V_s = A_s \cdot d_s \approx 2038.40$	3529.78
Crop		$\rho_c = 700$	$c_{p,c} = 3500$	$A_c = 2LAI \cdot A_s \approx 18816$	$V_c = LAI \cdot A_s \cdot d_c \approx 28.22$	69.14
Roof (glass)		$\rho_r = 2500$	$c_{pr} = 840$	$A_r = \frac{1}{\cos(\gamma)} \cdot A_s \approx 3382.28$	$V_r = A_r \cdot d_r \approx 13.53$	28.41
Screen		$\rho_{sc} = 1500$	$c_{p,sc} = 1500$	$A_{sc} = A_s \approx 3136$	$V_{sc} = A_{sc} \cdot d_{sc} \approx 1.57$	3.53

Note: Where necessary, a temperature of 25°C is used to calculate the values in this table. Here n_s, l_s and w_s is the number of spans, length and width per span; n is number of heating or cooling pipes per span, d is the outer diameter of the pipe and l is the length of one loop of pipe; $2LAI$ is the leaf area index, 2 for two sides of the leaf; 4 indicates the surface area amplification for the finned cooling pipe compared with a normal pipe and $\gamma = 22° \cdot \frac{\pi}{180°}$ (rad) is the angle of the roof with the horizontal plane.

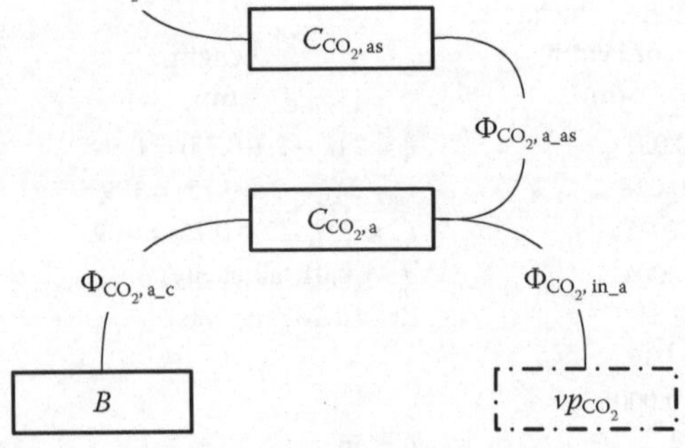

FIGURE 8.3 States x (solid box), external input d (dashed box), and control input u (dash-dotted box) in the carbon dioxide model.

8.4.1.1 Carbon Dioxide Supply

In the solar greenhouse case, the carbon dioxide supply is independent of boiler operation, which means that the maximum CO_2 supply $\Phi_{CO_2}^{max}$ is a design parameter.

The mass flow rate of carbon dioxide Φ_{CO_2,in_a} supplied to the indoor air is described by

$$\Phi_{CO_2,in\_a} = vp_{CO_2} \cdot \Phi_{CO_2}^{max} \qquad (kg[CO_2]\ s^{-1}) \quad (8.17)$$

in which $\Phi_{CO_2}^{max} = 5 \times 10^{-5} A_s$ kg[CO_2] s^{-1} is the maximum mass flow rate CO_2 supply, where $vp_{CO_2} \in [0,1]$ is the valve position carbon dioxide supply (control input).

In a conventional greenhouse in The Netherlands, carbon dioxide is a side product of energy supply by the boiler. The carbon dioxide supply by the boiler is therefore limited by the amount of carbon dioxide Φ_{CO_2,in_a}^{max} produced by the boiler, so

$$\Phi_{CO_2,in\_a} = \min\left(vp_{CO_2} \cdot \Phi_{CO_2}^{max}, \Phi_{CO_2,in\_a}^{max}\right) \qquad (kg[CO_2]\ s^{-1}) \quad (8.18)$$

in which

$$\Phi_{CO_2,in\_a}^{max} = F_{gas} \cdot f_{CO_2\_gas} \qquad (kg[CO_2]\ s^{-1}) \quad (8.19)$$

in which the conversion factor $f_{CO_2\_gas}$ = 1.78 kg[CO_2] m^{-3}[gas], where F_{gas} (m^3[gas] s^{-1}) is the gas flow needed by the boiler.

The gas flow F_{gas} needed by the boiler is defined by

$$F_{gas} = \frac{Q_{boil}}{\eta_{boil} \cdot H_u} \qquad (m^3[gas]s^{-1}) \quad (8.20)$$

in which the efficiency of the boiler η_{boil} = 0.95 and the (high) combustion value of gas H_u = 35.17×10^6 J m^{-3}[gas], where Q_{boil} (W) (Equation 8.78) is the energy supply by the boiler for heat supply. The high combustion value of gas is the amount of energy available from its complete combustion, including condensation of water vapor that results from the combustion.

8.4.1.2 Photosynthesis and Respiration

The mass flow rate of carbon dioxide Φ_{CO_2,a_c} from the indoor air to the canopy (the net photosynthesis rate of the canopy) is described by

$$\Phi_{CO_2,a\_c} = P_{cg} - r_c \qquad (kg[CO_2]\ s^{-1}) \quad (8.21)$$

in which the gross assimilation rate of the canopy P_{cg} is given by

$$P_{cg} = A_s \cdot P_g \qquad (kg[CO_2]\ s^{-1}) \quad (8.22)$$

and the dark respiration rate of the canopy r_c is given by

$$r_c = A_s \cdot r_D \qquad (kg[CO_2]\ s^{-1}) \quad (8.23)$$

where P_g (kg[CO_2] m^{-2}[gh] s^{-1}) is the gross assimilation rate of the canopy, r_D (kg[CO_2] m^{-2}[gh] s^{-1}) is the dark respiration rate of the canopy, and A_s (m^2[gh]) is the surface area of the soil. Several different models can be used to compute these rates (Van Ooteghem, 2007). A new photosynthesis model was formed based on existing models in literature (see Section 8.5.2.1), which is used to compute the gross assimilation rate P_g and the dark respiration rate r_D of the canopy.

This carbon dioxide is used to produce biomass. It is assumed that the CO_2 assimilation by the crop is instantaneously converted to biomass. For the conversion from the consumed CO_2 to the fresh weight biomass increase rate, the conversion factor $f_{CO_2,B}$ is used

$$f_{CO_2,B} = \frac{1}{1-\frac{p_w}{100}} \cdot \frac{c_f \cdot c_{cs}}{ASRQ} \qquad (kg[fw]\ kg^{-1}[CO_2]) \quad (8.24)$$

in which the percentage water in total fresh weight biomass p_w = 94%, the fraction of produced biomass material for dry weight c_f = 1, the conversion factor from CO_2 to CH_2O (fraction of molar masses) $c_{cs} = \frac{30}{44}$ kg[CH_2O] kg[CO_2]$^{-1}$, and the conversion factor from dry weight to CH_2O (glucose requirement) ASRQ = 1.2 kg[CH_2O] kg[dw]$^{-1}$.

For the long-term temperature effects on crop development, temperature integration is used. This is described in Section 8.5.3.

8.4.1.3 Carbon Dioxide Transport due to Ventilation

The mass flow rate of carbon dioxide Φ_{CO_2,as_o} from the indoor to the outdoor air is described by

$$\Phi_{CO_2,as\_o} = F_{as\_o} \cdot (C_{CO_2,as} - C_{CO_2,o}) \qquad (kg[CO_2]\ s^{-1}) \quad (8.25)$$

where F_{as_o} (m^3 s^{-1}) is the ventilation flow (Equation 8.58), $C_{CO_2,as}$ (kg m^{-3}) is the carbon dioxide concentration of indoor air above the screen and $C_{CO_2,o}$ (kg m^{-3}) is the carbon dioxide concentration of outdoor air.

8.4.1.4 Carbon Dioxide Transport past the Screen

The mass flow rate of carbon dioxide Φ_{CO_2,a_as} from the indoor air below the screen to the indoor air above the screen is described by

$$\Phi_{CO_2,a\_as} = F_{a\_as} \cdot \left(C_{CO_2,a} - C_{CO_2,as}\right) \qquad (kg[CO_2]\ s^{-1}) \quad (8.26)$$

where F_{a_as} (m^3 s^{-1}) is the volume flow of air from below the screen to above the screen (Equation 8.53), $C_{CO_2,a}$ (kg m^{-3}) is the carbon dioxide concentration of indoor air below the screen, and $C_{CO_2,as}$ (kg m^{-3}) is the carbon dioxide concentration of indoor air above the screen.

8.4.2 Water Vapor Model

The differential equations for the water vapor concentrations of the indoor air below and above the screen ($C_{H_2O,a}$, $C_{H_2O,as}$ (kg[H_2O] m^{-3})) are given in Section 8.4. The water vapor mass flow rates are described in the subsequent sections. The water vapor concentrations and the water vapor mass flows are shown in Figure 8.4.

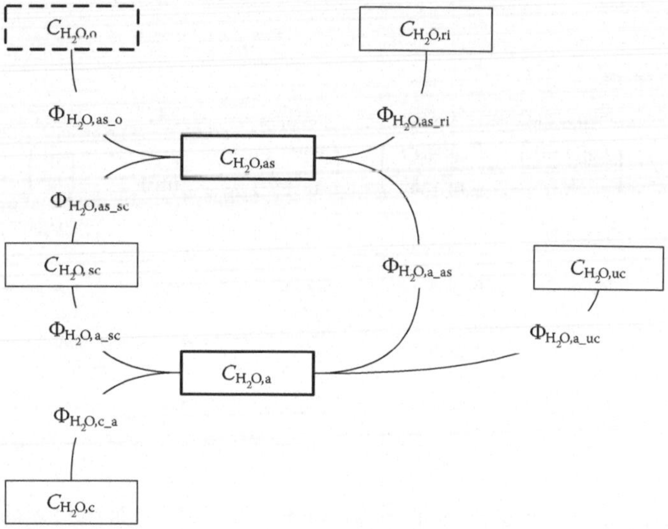

FIGURE 8.4 States x (solid box) and external input d (dashed box) in the water vapor model.

8.4.2.1 Canopy Transpiration

The canopy transpiration is determined based on the thesis of Stanghellini (1987). The mass flow rate of water vapor Φ_{H_2O,c_a} from the canopy to the indoor air due to transpiration (kg[H_2O] s^{-1}) is given in Section 8.5.1.

8.4.2.2 Condensation of Water

Water will condense on a surface when the concentration of water vapor in the air is higher than the saturation concentration of water vapor of the surface. The saturation concentration of water vapor of the surface depends on the surface temperature and the humidity of the air. It is assumed that water that condenses is directly removed, and is therefore not available for evaporation.

The mass flow rate of water vapor Φ_{H_2O,A_B} from compartment A that condensates on surface B is

$$\Phi_{H_2O,A\_B} = \max(A_B \cdot k_{H_2O,A\_B} \cdot (C_{H_2O,A} - C^{sat}_{H_2O,B}), 0) \qquad (\text{kg[H}_2\text{O] s}^{-1}) \quad (8.27)$$

where A_B (m^2) is the surface area of surface B, k_{H_2O,A_B} (m s^{-1}) is the mass transfer coefficient of water vapor from the air in compartment A to surface B, $C^{sat}_{H_2O,B}$ (kg[H_2O] m^{-3}[air]) is the saturation concentration of water vapor at the temperature of surface B (see Appendix B.1), and $C_{H_2O,A}$ (kg[H_2O] m^{-3}[air]) is the concentration of water vapor at the temperature of the air in compartment A. If $C_{H_2O,A} \leq C^{sat}_{H_2O,B}$, then $\Phi_{H_2O,A\_B} = 0$ (no condensation).

The mass transfer coefficient of water vapor k_{H_2O,A_B} from the air in compartment A to the surface B is defined by Bot (1983)

$$k_{H_2O,A\_B} = \frac{\alpha_{A\_B}}{\rho_A \cdot c_{p,a} \cdot Le^{\frac{2}{3}}} \qquad (\text{m s}^{-1}) \quad (8.28)$$

where α_{A_B} (W m^{-2} K^{-1}) is the heat transfer coefficient from the air in compartment A to the surface B (see Table 8.5), ρ_A (kg m^{-3}) is the density of air in compartment A (temperature dependent),

TABLE 8.5
Condensation of Water

	Φ_{H_2O,A_B} $\left(\frac{kg[H_2O]}{s}\right)$	$C_{H_2O,A}$ $\left(\frac{kg[H_2O]}{m^3[air]}\right)$	$C^{sat}_{H_2O,B}$ $\left(\frac{kg[H_2O]}{m^3[air]}\right)$	A_B (m^2)	k_{H_2O,A_B} $\left(\frac{m}{s}\right)$	α_{A_B} $\left(\frac{W}{m^2K}\right)$	ρ_A $\left(\frac{kg}{m^3}\right)$
Indoor air above screen to roof indoor side	Φ_{H_2O,as_ri}	$C_{H_2O,as}$	$C^{sat}_{H_2O,ri}$	A_r	k_{H_2O,as_ri}	α_{as_ri}	ρ_{as}
Indoor air below screen to upper cooling net	Φ_{H_2O,a_uc}	$C_{H_2O,a}$	$C^{sat}_{H_2O,uc}$	A_{uc}	k_{H_2O,a_uc}	α_{uc_a}	ρ_a
Indoor air below screen to screen	Φ_{H_2O,a_sc}	$C_{H_2O,a}$	$C^{sat}_{H_2O,sc}$	A_{sc}	k_{H_2O,a_sc}	α_{a_sc}	ρ_a
Indoor air above screen to screen	Φ_{H_2O,as_sc}	$C_{H_2O,as}$	$C^{sat}_{H_2O,sc}$	A_{sc}	k_{H_2O,as_sc}	α_{as_sc}	ρ_{as}

$c_{p,a}$ (J kg^{-1} K^{-1}) is the specific heat capacity of air and $Le = 0.89$ (–) is the Lewis number for water vapor in air.

In Table 8.5, the condensation of water is given for all surfaces in the greenhouse to which it applies, based on the Equations 8.27 and 8.28. The equations and the values for the surfaces and the heat transfer coefficients α_{A_B} are given in Tables 8.4 and 8.6.

8.4.2.3 Water Vapor Transport due to Ventilation

The mass flow rate of water vapor Φ_{H_2O,as_o} from the indoor air above the screen to the outdoor air is described by

$$\Phi_{H_2O,as\_o} = F_{as\_o} \cdot (C_{H_2O,as} - C_{H_2O,o}) \qquad (kg[H_2O]\ s^{-1}) \quad (8.29)$$

where F_{as_o} (m^3 s^{-1}) is the ventilation flow (Equation 8.58), $C_{H_2O,as}$ (kg m^{-3}) is the water concentration of indoor air above the screen, and $C_{H_2O,o}$ (kg m^{-3}) is the water concentration of outdoor air.

8.4.2.4 Water Vapor Transport past the Screen

The mass flow rate of water vapor Φ_{H_2O,a_as} from the indoor air below the screen to the indoor air above the screen is described by

$$\Phi_{H_2O,a\_as} = F_{a\_as} \cdot (C_{H_2O,a} - C_{H_2O,as}) \qquad (kg[H_2O]\ s^{-1}) \quad (8.30)$$

where F_{a_as} (m^3 s^{-1}) is the volume flow of air from below the screen to above the screen (Equation 8.53), $C_{H_2O,a}$ (kg m^{-3}) is the water concentration of indoor air below the screen and $C_{H_2O,as}$ (kg m^{-3}) is the water concentration of indoor air above the screen.

8.4.3 Thermal Model

The differential equations for the temperatures of the roof, the indoor air below, and above the screen, the crop, the soil (upper layer), the lower and the upper heating net, the upper cooling net, and the thermal screen (T_{ri}, T_a, T_{as}, T_c, T_s, T_l, T_u, T_{uc}, T_{sc} (K)) are given in Section 8.4. The heat transfer terms are described in the subsequent sections. The temperatures and the heat transfer terms are shown in Figure 8.5. The control inputs vp_l, vp_u, vp_{hp}, and vp_{he} in the thermal model, and the heat transfer terms corresponding to these control inputs Q_{in_l}, Q_{l_out}, Q_{in_u}, Q_{u_out}, Q_{in_uc}, Q_{uc_out}, Q_{he}, and Q_{hp} are not incorporated in this figure.

TABLE 8.6
Convection

	Q_{A_B} (W)	T_A (K)	T_B (K)	A_{A_B} (m²)	α_{A_B} $\left(\frac{W}{m^2K}\right)$	Source Reference
Indoor air below screen to canopy	Q_{a_c}	T_a	T_c	A_c	$\alpha_{a\_c} = \frac{\rho_a \cdot c_{p,a}}{R_{heat,b}}$	z
Indoor air below screen to soil	Q_{a_s}	T_a	T_s	A_s	$\alpha_{a\_s} = \begin{cases} 1.7\sqrt[3]{\lvert T_a - T_s \rvert} \text{ if } T_a < T_s \\ 1.3\sqrt[4]{\lvert T_a - T_s \rvert} \text{ if } T_a \geq T_s \end{cases}$	z
Indoor air above screen to roof indoor side	Q_{as_ri}	T_{as}	T_{ri}	A_r	$\alpha_{as\_ri} = 3\sqrt[3]{\lvert T_{as} - T_{ri} \rvert}$	s
Lower net to indoor air below screen	Q_{l_a}	T_l	T_a	A_l	$\alpha_{l\_a} = 1.28\sqrt[4]{\frac{\lvert T_l - T_a \rvert}{d_l}}$	z
Roof outdoor side to outdoor air	Q_{ro_o}	T_{ro}	T_o	A_r	$\alpha_{ro\_o} = \begin{cases} 2.8 + 1.2v_o \text{ if } v_o < 4 \\ 2.5\, v_o^{0.8} \text{ if } v_o \geq 4 \end{cases}$	b
Screen to indoor air below screen	Q_{sc_a}	T_{sc}	T_a	A_{sc}	$\alpha_{a\_sc} = Cl_{sc} \cdot 3\sqrt[3]{\lvert T_a - T_{sc} \rvert}$	s
Screen to indoor air above screen	Q_{sc_as}	T_{sc}	T_{as}	A_{sc}	$\alpha_{as\_sc} = Cl_{sc} \cdot 3\sqrt[3]{\lvert T_{as} - T_{sc} \rvert}$	s
Upper net to indoor air below screen	Q_{u_a}	T_u	T_a	A_u	$\alpha_{u\_a} = 1.28\sqrt[4]{\frac{\lvert T_u - T_a \rvert}{d_u}}$	z
Upper cooling net to indoor air below screen	Q_{uc_a}	T_{uc}	T_a	A_{uc}	$\alpha_{uc\_a} = 1.28\sqrt[4]{\frac{\lvert T_{uc} - T_a \rvert}{d_{uc}}}$	z

Source: From [z]De Zwart (1996); [s]Stoffers (1989); [b]Bot (1983).

Note: Where $R_{heat,b}$ is the boundary layer resistance to convective heat transfer, v_o is the wind speed and Cl_{sc} is the screen closure.

8.4.3.1 Convection

Convection is the heat transfer between a solid and a gas or fluid material. Convection is also part of the ventilation process and the heat exchange past the screen. The heat transfer Q_{A_B} from A to B is described by the equation (Newton's law of cooling)

$$Q_{A\_B} = A_{A\_B} \cdot \alpha_{A\_B} \cdot (T_A - T_B) \qquad \text{(W)} \quad (8.31)$$

where A_{A_B} (m²) is the surface area for heat transfer, α_{A_B} (W m^{-2} K^{-1}) is the heat transfer coefficient between A and B, and T_A and T_B (K) are the temperatures of A and B.

In Table 8.6, the convective heat transfer is given for all surfaces in the greenhouse to which it applies, based on Equation 8.31. The equations and the values for the surfaces can be found in Table 8.4.

Convection, Heat Exchange past the Screen

The heat exchange Q_{a_as} between indoor air below the screen and indoor air above the screen is defined by

$$Q_{a\_as} = \overline{\rho_a} \cdot c_{p,a} \cdot F_{a\_as} \cdot (T_a - T_{as}) \qquad \text{(W)} \quad (8.32)$$

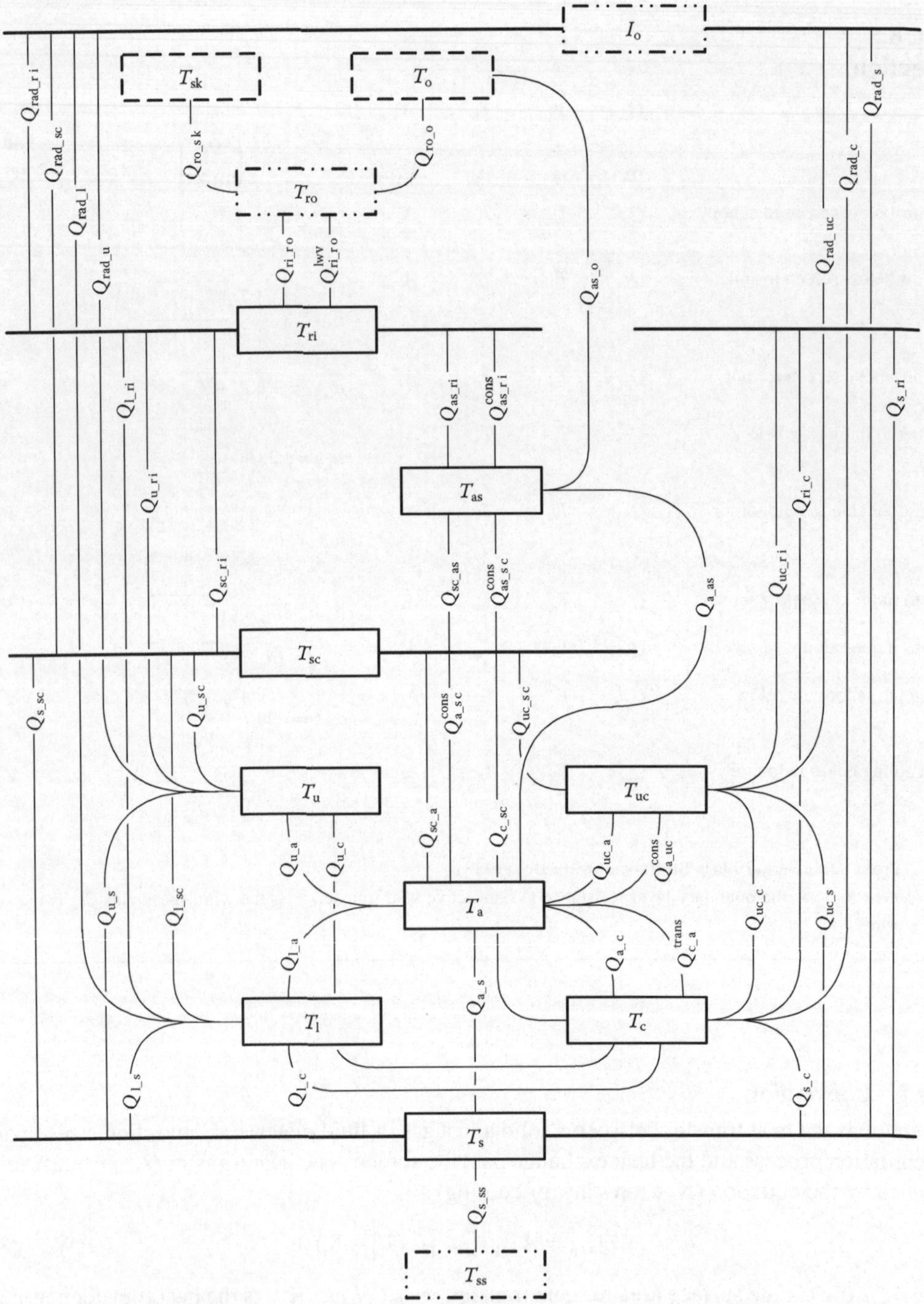

FIGURE 8.5 States x (solid box) and external inputs d (dashed and dash-dotted boxes) in the thermal model.

where $\overline{\rho_a}$ (kg m^{-3}) is the average density of air below and above the screen, $c_{p,a}$ (J kg^{-1} K^{-1}) is the specific heat capacity of air, F_{a_as} (m^3 s^{-1}) is the volume flow of air from below the screen to above the screen (Equation 8.53), and T_a and T_{as} (K) are the temperatures of the air below and above the screen.

Convection, Heat Exchange through Ventilation

The heat exchange Q_{as_o} between indoor and outdoor air by natural ventilation is defined by

$$Q_{as\_o} = (1 - op_{vhr} \cdot \eta_{vhr}) \cdot \rho_{as} \cdot c_{p,a} \cdot F_{as\_o} \cdot (T_{as} - T_o) \qquad \text{(W)} \quad (8.33)$$

where ρ_{as} (kg m^{-3}) is the density of air above the screen, $c_{p,a}$ (J kg^{-1} K^{-1}) is the specific heat capacity of air, F_{as_o} (m^3 s^{-1}) is the ventilation flow (Equation 8.58), and T_{as} and T_o (K) are the temperatures of the air above the screen and the outdoor air.

The option ventilation with heat recovery $op_{vhr} = 1$ indicates that ventilation with heat recovery is used. The option $op_{vhr} = 0$ applies to normal ventilation. When ventilation with heat recovery is used, the outdoor air is preheated by the indoor greenhouse air with a heat exchanger. It is assumed that a fixed fraction of the sensible heat is recovered. The efficiency factor for the ventilation with heat recovery $\eta_{vhr} = 0.9$, which means that 90% of the sensible heat is recovered. Latent heat that is vented out is lost. Ventilation with heat recovery is used when the greenhouse is heated; otherwise, normal ventilation is used:

$$\begin{aligned} op_{vhr} &= 0 \qquad \forall vp_{hp} = 0 \ \& \ vp_l = 0 \\ op_{vhr} &= 1 \qquad \forall vp_{hp} > 0 \ \mid \ vp_l > 0 \end{aligned} \qquad (-) \quad (8.34)$$

where $vp_{hp} \in [0, 1]$ and $vp_l \in [0, 1]$ (–) are the valve positions of the heat pump and the lower net (boiler) (both control inputs).

8.4.3.2 Longwave Radiation Absorption

Radiation absorption is the heat transfer due to radiation between two materials. For longwave radiation absorption, the heat transfer Q_{A_B} from object A to object B is described by the equation (Stefan–Boltzmann)

$$Q_{A\_B} = A_{A\_B} \cdot E_A \cdot E_B \cdot \chi_{A\_B} \cdot \sigma \cdot (T_A^4 - T_B^4) \qquad \text{(W)} \quad (8.35)$$

where A_{A_B} (m^2) is the surface area for heat transfer, E_A and E_B (–) are the emission coefficients for A and B, χ_{A_B} (–) is the view factor from A to B, $\sigma = 5.67051 \times 10^{-8}$ W m^{-2} K^{-4} is the Stefan–Boltzmann constant and T_A and T_B (K) are the temperatures of A and B. The view factors are derived in Van Ooteghem (2007) based on the work by Sparrow and Cess (1970) and Van Strien (1988). The values of the emission coefficients can be found in Table 8.7.

In Table 8.7, the heat transfer through longwave radiation is given for all surfaces in the greenhouse to which it applies, based on Equation 8.35. The equations and the values for the surfaces can be found in Table 8.4.

The transmittance of longwave radiation τ_{c_ll} by the canopy is computed based on the extinction coefficient for longwave radiation by the canopy $k_{c\_ll} = 0.64$ and the leaf area index LAI (m^2[leaf] m^{-2}[gh])

$$\tau_{c\_ll} = e^{-k_{c\_ll} \cdot LAI} \qquad (-) \quad (8.36)$$

8.4.3.3 Shortwave Radiation Absorption

The shortwave radiation (also called global or solar radiation) is defined as the radiation received directly or indirectly from the sun by an horizontal plane at the Earth's surface, integrated over all

TABLE 8.7
Longwave Radiation

	Q_{A_B} (W)	T_A (K)	T_B (K)	A_{A_B} (m²)	E_A (–)	E_B (–)	χ_{A_B} (–)
Lower net to roof indoor side	Q_{l_ri}	T_l	T_{ri}	A_l	E_l	E_{ri}	$\chi_{l\_ri} = 0.5\,(1 - Cl_{sc})\cdot\tau_{c\_ll}$
Lower net to screen	Q_{l_sc}	T_l	T_{sc}	A_l	E_l	E_{sc}	$\chi_{l\_sc} = 0.5\,Cl_{sc}\cdot\tau_{c\_ll}$
Lower net to soil	Q_{l_s}	T_l	T_s	A_l	E_l	E_s	$\chi_{l\_s} = 0.5$
Lower net to canopy	Q_{l_c}	T_l	T_c	A_l	E_l	E_c	$\chi_{l\_c} = 1 - \chi_{l\_s} - \chi_{l\_ri} - \chi_{l\_sc}$
Upper net to roof indoor side	Q_{u_ri}	T_u	T_{ri}	A_u	E_u	E_{ri}	$\chi_{u\_ri} = 0.5\,(1 - Cl_{sc})$
Upper net to screen	Q_{u_sc}	T_u	T_{sc}	A_u	E_u	E_{sc}	$\chi_{u\_sc} = 0.5\,Cl_{sc}$
Upper net to soil	Q_{u_s}	T_u	T_s	A_u	E_u	E_s	$\chi_{u\_s} = 0.5\,\tau_{c\_ll}$
Upper net to canopy	Q_{u_c}	T_u	T_c	A_u	E_u	E_c	$\chi_{u\_c} = 1 - \chi_{u\_ri} - \chi_{u\_sc} - \chi_{u\_s}$
Upper cooling net to roof indoor side	Q_{uc_ri}	T_{uc}	T_{ri}	A_{uc}	E_{uc}	E_{ri}	$\chi_{uc\_ri} = 0.5\,(1 - Cl_{sc})$
Upper cooling net to screen	Q_{uc_sc}	T_{uc}	T_{sc}	A_{uc}	E_{uc}	E_{sc}	$\chi_{uc\_sc} = 0.5\,Cl_{sc}$
Upper cooling net to soil	Q_{uc_s}	T_{uc}	T_s	A_{uc}	E_{uc}	E_s	$\chi_{uc\_s} = 0.5\,\tau_{c\_ll}$
Upper cooling net to canopy	Q_{uc_c}	T_{uc}	T_c	A_{uc}	E_{uc}	E_c	$\chi_{uc\_c} = 1 - \chi_{uc\_ri} - \chi_{uc\_sc} - \chi_{uc\_s}$
Soil to canopy	Q_{s_c}	T_s	T_c	A_s	E_s	E_c	$\chi_{s\_c} = (1 - \tau_{c\_ll})\cdot(1 - \chi_{s\_l})$
Soil to roof indoor side	Q_{s_ri}	T_s	T_{ri}	A_s	E_s	E_{ri}	$\chi_{s\_ri} = (1 - Cl_{sc})\cdot(1-\chi_{s\_c} - \chi_{s\_l} - \chi_{s\_u} - \chi_{s\_uc})$
Soil to screen	Q_{s_sc}	T_s	T_{sc}	A_s	E_s	E_{sc}	$\chi_{s\_sc} = Cl_{sc}\cdot(1-\chi_{s\_c} - \chi_{s\_l} - \chi_{s\_u} - \chi_{s\_uc})$
Roof outdoor side to sky	Q_{ro_sk}	T_{ro}	T_{sk}	A_r	E_{ro}	E_{sk}	$\chi_{ro\_sk} = 1\cdot\frac{A_s}{A_r} = \cos(\gamma)$
Roof indoor side to roof outdoor side	$Q^{lwv}_{ri\_ro}$	T_{ri}	T_{ro}	A_r	E_{ri}	E_{ro}	$\chi_{ri\_ro} = 1$
Roof indoor side to canopy	Q_{ri_c}	T_{ri}	T_c	A_r	E_{ri}	E_c	$\chi_{ri\_c} = (1-Cl_{sc})\cdot(1-\tau_{c\_ll})\cdot\left(\chi_{ro\_sk} - 0.5\frac{A_u}{A_r} - 0.5\frac{A_{uc}}{A_r}\right)$
Screen to roof indoor side	Q_{sc_ri}	T_{sc}	T_{ri}	A_s	E_{sc}	E_{ri}	$\chi_{sc\_ri} = Cl_{sc}$
Canopy to screen	Q_{c_sc}	T_c	T_{sc}	A_s	E_c	E_{sc}	$\chi_{c\_sc} = Cl_{sc}\cdot(1-\tau_{c\_ll})\cdot\left(1-0.5\frac{A_u}{A_{sc}} - 0.5\frac{A_{uc}}{A_{sc}}\right)$

Note: $Cl_{sc} \in [0,1]$ is the thermal screen closure, τ_{c_ll} (–) is the transmittance of longwave radiation by the canopy and γ (rad) is the angle of the roof with the horizontal plane. The emission coefficients values are: $E_l = E_u = E_{uc} = E_{ri} = E_{ro} = 0.95$, $E_s = 0.7$, $E_{sc} = 0.9$, $E_{sk} = 1$, and $E_c = 1 - \tau_{c\_ll}$ (–). Some intermediate variables: $\chi_{s\_l} = \frac{A_l}{A_s}\cdot\chi_{l\_s}$, $\chi_{s\_u} = \frac{A_u}{A_s}\cdot\chi_{u\_s}$, $\chi_{s\_uc} = \frac{A_{uc}}{A_s}\cdot\chi_{uc\_s}$.

wavelengths in the shortwave interval. In good approximation, the sun can be considered a black body with a surface temperature of about 5700 K. It consists of two components: the direct and the diffuse radiation.

For shortwave radiation absorption, the heat Q_{rad_A} absorbed by object A is described by the equation

$$Q_{rad\_A} = A_A \cdot \eta_{A\_Is} \cdot I_o \quad \text{(W)} \quad (8.37)$$

where A_A (m^2) is the surface area for heat transfer, η_{A_Is} (–) is the shortwave radiation absorption coefficient for A, and I_o (W m^{-2}[gh]) is the outdoor shortwave solar radiation. The absorption coefficient depends on the transmitted shortwave radiation, which depends on the position of the object A in the greenhouse. Also, corrections are made for the transmittance of shortwave radiation by the roof τ_{r_Is} and by the screen τ_{sc_Is} (–).

In Table 8.8, the heat transfer through shortwave radiation is given for all surfaces in the greenhouse to which it applies, based on Equation 8.37. The equations and the values for the surfaces can be found in Table 8.4.

The transmittance τ_{r_Is} of shortwave radiation by the roof (–) is given by

$$\tau_{r\_Is} = f_{dif} \cdot \tau_{difR} + (1 - f_{dif}) \cdot \tau_{dirR} \quad (-) \quad (8.38)$$

with transmittance of diffuse radiation by the roof $\tau_{difR} = 0.78$, where τ_{dirR} (–) is the transmittance of direct radiation by the roof and f_{dif} (–) is the fraction diffuse radiation in shortwave radiation (see Appendix A).

The transmittance τ_{dirR} of the roof for direct PAR radiation is determined from transmissivity table by De Zwart (1996), here approximated by a function:

$$\tau_{dirR} = \begin{cases} 0.85\left(1 - e^{-0.083 \cdot \frac{360}{2\pi} \cdot \beta_{sun}}\right) & \text{single glass} \\ 0 & \text{if } sun_{up} = 0 \end{cases} \quad (-) \quad (8.39)$$

TABLE 8.8
Shortwave Radiation

	Q_{rad_A} (W)	A_A (m^2)	η_{A_Is} (–)
Roof indoor side	Q_{rad_ri}	A_r	$\eta_{ri\_Is} = 0.02\sqrt{\tau_{r\_Is}}$
Roof outdoor side	Q_{rad_ro}	A_r	$\eta_{ro\_Is} = 0.02$
Screen	Q_{rad_sc}	A_{sc}	$\eta_{sc\_Is} = \tau_{r\_Is} \cdot Cl_{sc} \cdot (1 - \beta_{sc\_Is} - \tau_{sc\_Is0})$
Upper net	Q_{rad_u}	A_u	$\eta_{u\_Is} = \tau_{r\_Is} \cdot \tau_{sc\_Is} \cdot 0.5\,(1 - \beta_{u\_Is})$
Upper cooling net	Q_{rad_uc}	A_{uc}	$\eta_{uc\_Is} = \tau_{r\_Is} \cdot \tau_{sc\_Is} \cdot 0.5\,(1 - \beta_{uc\_Is})$
Lower net	Q_{rad_l}	A_l	$\eta_{l\_Is} = \tau_{r\_Is} \cdot \tau_{sc\_Is} \cdot (\tau_{c\_Il} + (1 - \tau_{c\_Il}) \cdot \tau_{c\_Is}) \cdot 0.5\,(1 - \beta_{l\_Is})$
Soil	Q_{rad_s}	A_s	$\eta_{s\_Is} = \tau_{r\_Is} \cdot \tau_{sc\_Is} \cdot (\tau_{c\_Il} + (1 - \tau_{c\_Il}) \cdot \tau_{c\_Is}) \cdot 0.5\,(1 - \beta_{s\_Is})$
Canopy	Q_{rad_c}	A_s	$\eta_{c\_Is} = \tau_{r\_Is} \cdot \tau_{sc\_Is} \cdot (1 + \tau_{c\_Is} \cdot \beta_{s\_Is}) \cdot (1 - \tau_{c\_Is} - \beta_{c\_Is})$

Note: $Cl_{sc} \in [0,1]$ is the thermal screen closure. The shortwave radiation coefficients for reflection by the lower, upper, and upper cooling net, soil, screen and canopy (dense stand) are: $\beta_{l\_Is} = \beta_{u\_Is} = \beta_{uc\_Is} = 0.83$, $\beta_{s\_Is} = 0.58$, $\beta_{sc\_Is} = 0.05$, and $\beta_{c\_Is\infty} = 0.12$ (–). For the pipes 0.5 indicates that only half the pipe surface is seen by the shortwave radiation. τ_{c_Is} and τ_{c_Il} (–) are the transmittance of shortwave and longwave radiation by the canopy. The term $\tau_{c\_Il} + (1 - \tau_{c\_Il}) \cdot \tau_{c\_Is}$ is used for the total transmittance of the canopy for radiation. Note that for Q_{rad_c} the surface area A_s of the soil is used, not the surface area A_c of the crop. This is due to the definition of the absorption coefficient η_{c_Is} by Stanghellini (1987). For the roof the transmittance $\sqrt{\tau_{r\_Is}}$ is used because it is a single layer of glass, while τ_{r_Is} holds for a double glass cover. If the roof has a single glass cover, $\eta_{ro\_Is} = 0$.

where β_{sun} is the elevation of the sun. For the zigzag roof it is assumed that the transmittance of the roof for direct PAR radiation τ_{dirR} is as high as with a single glass roof.

The transmittance τ_{sc_Is} of shortwave radiation by the screen (–) is given by

$$\tau_{sc\_Is} = (1 - Cl_{sc}) + \tau_{sc\_Is0} \cdot Cl_{sc} \qquad (-) \quad (8.40)$$

in which the transmittance of the fully closed screen $\tau_{sc\_Is0} = 0.8$. This gives the transmittance $\tau_{sc\_Is} = 0.8$ if the screen is fully closed ($Cl_{sc} = 1$) and $\tau_{sc\_Is} = 1$ if the screen is fully opened ($Cl_{sc} = 0$).

The shortwave radiation absorption coefficient η_{A_Is} for A describes the part of the shortwave radiation that is absorbed by object A. In general, there is a term for the fraction going past the screen and a term correcting for reflection

$$\eta_{A\_Is} = 1 - \beta_{A\_Is} - \tau_{A\_Is} \qquad (-) \quad (8.41)$$

where β_{A_Is} (–) is the shortwave radiation coefficient for reflection by the object A. It is assumed that all shortwave radiation not reflected or transmitted by the object A is absorbed.

The reflection coefficient β_{c_Is} for shortwave radiation by the canopy is given by Stanghellini (1987) as

$$\beta_{c\_Is} = (1 - \tau_{c\_Il}) \cdot \beta_{c\_Is\infty} \qquad (-) \quad (8.42)$$

where $\beta_{c\_Is\infty}$ (–) is the shortwave radiation coefficient for reflection by the canopy for a dense stand and τ_{c_Il} (–) (Equation 8.36) is the transmittance of longwave radiation by the canopy, which is used here as a measure for the permeability of the canopy.

The transmittance τ_{c_Is} of shortwave radiation by the canopy is computed based on the extinction coefficient $k_{c\_Is} = 0.48$ for shortwave radiation by the canopy and the leaf area index LAI (m^2[leaf] m^{-2}[gh])

$$\tau_{c\_Is} = e^{-k_{c\_Is} \cdot LAI} \qquad (-) \quad (8.43)$$

8.4.3.4 Conduction

For conduction, the heat transfer Q_{A_B} between the locations A and B in a homogeneous medium is described by the equation

$$Q_{A\_B} = A_{A\_B} \cdot \frac{\lambda}{d} \cdot (T_A - T_B) \qquad (W) \quad (8.44)$$

where A_{A_B} (m^2) is the surface area for heat transfer, λ (W m^{-1} K^{-1}) is the thermal conductivity of the homogeneous medium, d (m) is the distance between the locations A and B and T_A and T_B (K) are the temperatures of A and B.

In Table 8.9, the heat transfer through conduction is given for conduction through soil and air based on Equation 8.44. The equations and the values for the surfaces can be found in Table 8.4.

TABLE 8.9
Conduction

	Q_{A_B} (W)	A_{A_B} (m^2)	λ $\left(\frac{W}{m\,K}\right)$	d (m)	T_A (K)	T_B (K)
Upper soil layer to subsoil layer	Q_{s_ss}	A_s	$\lambda_s = 0.86$	dx_s	T_s	T_{ss}
Roof indoor side to roof outdoor side	Q_{ri_ro}	A_r	$\lambda_a = 24 \times 10^{-3}$	d_{ra}	T_{ri}	T_{ro}

The temperature T_{ss} of the subsoil layer is a function of the day number day_{NR} [1365].

$$T_{ss} = T_0 + 15 + 2.5\sin(1.72 \times 10^{-2}\,(day_{NR} - 140)) \qquad (\text{K}) \quad (8.45)$$

Instead of six soil layers (De Zwart, 1996), only one layer is used. This defines the value for the distance dx_s between the center of the upper soil layer and the subsoil layer in Table 8.9, and the distance d_s that defines the volume V_s in Equation 8.8. Values for these distances have been derived in Van Ooteghem (2007) from the model by De Zwart (1996). The estimated values are $d_s = 0.65$ m and $dx_{s\{0,7\}} = 1.247$ m.

8.4.3.5 Latent Heat Exchange

Latent heat exchange is due to a change in the level of "free" energy of water. It is not directly sensed as an increase or decrease in temperature. As the surrounding environment loses heat, water condenses, and it changes from a higher to a lower state of "free" energy, whereby latent heat is released to the environment. As the surrounding environment is heated, water evaporates, and it changes from a lower to a higher state of "free" energy, whereby latent heat is absorbed from the environment. The same equation is also used for crop transpiration.

The heat transfer $Q^{cons}_{A\_B}$ between air and a surface A and B is described by the equation

$$Q^{cons}_{A\_B} = \Lambda \cdot \Phi_{H_2O,A\_B} \qquad (\text{W}) \quad (8.46)$$

where $\Lambda = 2.26 \times 10^6$ J kg^{-1} is the heat of evaporation of water and Φ_{H_2O,A_B} (kg s^{-1}) (see Table 8.6) is the mass flow rate of water vapor from air to surface or vice versa.

In Table 8.10, the heat transfer through latent heat exchange is given, based on Equation 8.46.

8.4.4 Modeling the Screen

A thermal screen is used to decrease heat loss during cold periods with little solar radiation, in particular at night. Thermal screens are either open or closed, and are not designed for frequent operation during the day. In principle, the discrete screen condition $c_{sc} \in \{0,1\}$ is a control variable that could be part of the optimal control. However, because strong dynamic operation is undesirable, and the rules for opening or closure are rather straightforward, it was decided to assume that the screens are operated in a similar fashion as in commercial greenhouses. This solution can be seen as a local control solution, that need not be the concern of the optimal controller. The adopted rules for the screen closure are described in Section 8.4.4.1.

In the calculations it is assumed that the screen can be opened or closed within 3 min. If the screen is closed, temperature, CO_2, and H_2O concentrations above and below the screen will differ.

TABLE 8.10
Latent Heat Exchange

	$Q^{cons}_{A\_B}$ (W)	Φ_{H_2O,A_B} $\left(\frac{\text{kg}[H_2O]}{\text{s}}\right)$
Condensation of water from indoor air above screen on roof indoor side	$Q^{cons}_{as\_ri}$	Φ_{H_2O,as_ri}
Condensation of water from indoor air below screen on upper cooling net	$Q^{cons}_{a\_uc}$	Φ_{H_2O,a_uc}
Condensation of water from indoor air below screen on screen	$Q^{cons}_{a\_sc}$	Φ_{H_2O,a_sc}
Condensation of water from indoor air above screen on screen	$Q^{cons}_{as\_sc}$	Φ_{H_2O,as_sc}
Transpiration of water from canopy to indoor air below screen	$Q^{trans}_{c\_a}$	Φ_{H_2O,c_a}

Therefore, separate state variables are used to describe these temperatures and concentrations above and below the screen (see Section 8.4). When the screen is open, it is assumed that temperatures and concentrations are averaged proportional to the heat capacities above and below the screen (Section 8.4.4.3).

It is assumed that the screen material is impermeable for all gases (air, H_2O, and CO_2). If the screen is closed, only part of the solar radiation is transmitted through the screen. The transmittance by the screen is given in Equation 8.40.

When the screen is closed, a small crack opening of 3% is left to carry off water, meaning that some exchange of air, H_2O, and CO_2 is still possible. This air exchange is described by a simple air exchange rate (Section 8.4.4.2).

8.4.4.1 Screen Closure

The screen crack of 3% gives a screen closure Cl_{sc} (0,1) defined by

$$Cl_{sc} = 0.97c_{sc} \qquad (-) \quad (8.47)$$

where $c_{sc} \in \{0,1\}$ is the screen condition.

The control variable for screen closure is the screen condition $c_{sc} \in \{0,1\}$ with a value zero if the screen is closed, and one if the screen is open. The screen condition depends only on the external inputs d: the outdoor shortwave solar radiation I_o, and the temperature of the outdoor air T_o. The screen condition is a logical (Boolean) combination of separate screen conditions for radiation c_{sc}^{rad} and temperature c_{sc}^{T}.

$$c_{sc} = c_{sc}^{rad} \,\&\, c_{sc}^{T} \qquad (-) \quad (8.48)$$

which means that both conditions must be one for the screen to close.

The screen condition c_{sc}^{rad} for radiation is switched if the outdoor shortwave solar radiation I_o (W m^{-2}[gh]) enters the region between the two parameters I_o^{min} and I_o^{max}, according to

$$c_{sc}^{rad}\left(t_k + t_{s,u}\right) = \begin{cases} 0 & \text{if } \; c_{sc}^{rad}\left(t_k\right) = 1 \,\&\, I_0 \geq I_0^{max} \\ 1 & \text{if } \; c_{sc}^{rad}\left(t_k\right) = 0 \,\&\, I_0 \leq I_0^{min} \\ c_{sc}^{rad}\left(t_k\right) & \text{if } \; \left(c_{sc}^{rad}\left(t_k\right) = 0 \,\&\, I_0 > I_0^{min}\right) \Big| \\ & \quad\;\; \left(c_{sc}^{rad}\left(t_k\right) = 1 \,\&\, I_0 < I_0^{max}\right) \end{cases} \qquad (-) \quad (8.49)$$

in which $I_0^{min} = 20$ W m^{-2}[gh], $I_0^{max} = 60$ W m^{-2}[gh], and $t_{s,u} = 30$ min is the time interval for the control inputs. In this way, a region is created in which the screen condition c_{sc}^{rad} does not change. This is done to prevent the screen from opening and closing more than once in response to small variations in the radiation at sunrise and sunset. The screen condition is initialized at initial time t_0 with

$$c_{sc}^{rad}\left(t_0\right) = \begin{cases} 0 & \text{if } \; I_0 > I_0^{min} \\ 1 & \text{if } \; I_0 \leq I_0^{min} \end{cases} \qquad (-) \quad (8.50)$$

A similar construct is used for the screen condition c_{sc}^{T} for temperature, according to

$$c_{sc}^{T}\left(t_k + t_{s,u}\right) = \begin{cases} 0 & \text{if} \quad c_{sc}^{T}\left(t_k\right) = 1 \;\&\; T_0 \geq T_0^{max} \\ 1 & \text{if} \quad c_{sc}^{T}\left(t_k\right) = 0 \;\&\; T_0 \leq T_0^{min} \\ c_{sc}^{T}\left(t_k\right) & \text{if} \quad \left(c_{sc}^{T}\left(t_k\right) = 0 \;\&\; T_0 > T_0^{min}\right) \Big| \\ & \quad\quad \left(c_{sc}^{T}\left(t_k\right) = 1 \;\&\; T_0 < T_0^{max}\right) \end{cases} \qquad (-) \quad (8.51)$$

in which $T_0^{min} = T_0 + 8$ K and $T_0^{max} = T_0 + 12$ K, where T_0 = 273.15 K. As with c_{sc}^{rad}, this creates a region in which the screen condition c_{sc}^{T} does not change. The screen condition is initialized at initial time t_0 with

$$c_{sc}^{T}\left(t_0\right) = \begin{cases} 0 & \text{if} \quad T_0 > T_0^{min} \\ 1 & \text{if} \quad T_0 \leq T_0^{min} \end{cases} \qquad (-) \quad (8.52)$$

8.4.4.2 Volume Flow Air past the Screen

The convective air exchange through the screen material and along the crack opening is caused by the temperature difference and the pressure difference above and below the screen. A physical model of the air exchange exists (De Zwart, 1996). It describes the air exchange rate past the screen as a function of the temperature difference and the pressure difference below and above the screen. In the simulation, these equations give numerical problems due to the fast dynamics of this air exchange. It is therefore described by a simplified equation, in which it is assumed that the air exchange speed between the air below and above the screen (past the screen) is constant. It is assumed that the screen is impermeable for all gases (H_2O, CO_2, and air), so there is no exchange through the screen material.

The volume flow F_{a_as} of air from below the screen to above the screen is given by

$$F_{a\_as} = v_{a\_as} \cdot A_{sc} \cdot (1 - Cl_{sc}) \qquad (m^3\ s^{-1}) \quad (8.53)$$

in which the air exchange rate v_{a_as} between the air below and above the screen is given by

$$v_{a\_as} = 0.05 \qquad (m\ s^{-1}) \quad (8.54)$$

where A_{sc} (m^2) is the surface area of the screen and $Cl_{sc} \in [0,1]$ is the thermal screen closure.

8.4.4.3 Temperatures and Concentrations of CO_2 and H_2O When the Screen is Open

When screens are opened or closed, the system structure changes: The differential equations for temperature, CO_2 concentration, and H_2O concentration above and below the screen are different. This can be seen in the corresponding state equations (Equations 8.1 through 8.6). When the screens are open the concentrations and temperatures above and below the screen are equal, while when the screens are closed the concentrations and temperatures differ.

When the screen is opened, a very rapid exchange of air, and therefore heat, CO_2, and H_2O occurs. This exchange causes the system to become very stiff, which leads to numerical integration problems (large derivatives, small step size for integration).

To avoid numerical integration problems the temperatures and concentrations above and below the screen are averaged when the screen is opened. To average, in the temperature the heat capacity $\rho \cdot c_p \cdot V$ is used, and in the concentrations the volume V is used. If $c_{sc} = 0$ (screen is opened), this results in

$$\left.\begin{aligned} T_a(t_k + t_{s,u}) \\ T_{as}(t_k + t_{s,u}) \end{aligned}\right\} = \frac{T_a(t_k) \cdot \rho_a \cdot c_{p,a} \cdot V_a + T_{as}(t_k) \cdot \rho_{as} \cdot c_{p,a} \cdot V_{as}}{\rho_a \cdot c_{p,a} \cdot V_a + \rho_{as} \cdot c_{p,a} \cdot V_{as}} \quad \text{(K)} \quad (8.55)$$

$$\left.\begin{aligned} C_{CO_2,a}(t_k + t_{s,u}) \\ C_{CO_2,as}(t_k + t_{s,u}) \end{aligned}\right\} = \frac{C_{CO_2,a}(t_k) \cdot V_a + C_{CO_2,as}(t_k) \cdot V_{as}}{V_a + V_{as}} \quad \text{(kg[CO}_2\text{] m}^{-3}\text{)} \quad (8.56)$$

$$\left.\begin{aligned} C_{H_2O,a}(t_k + t_{s,u}) \\ C_{H_2O,as}(t_k + t_{s,u}) \end{aligned}\right\} = \frac{C_{H_2O,a}(t_k) \cdot V_a + C_{H_2O,as}(t_k) \cdot V_{as}}{V_a + V_{as}} \quad \text{(kg[H}_2\text{O] m}^{-3}\text{)} \quad (8.57)$$

which means that the new values (after the screen has opened) are computed based on the values on the previous control time interval (before the screen was opened). These computations are based on steady state assumptions: Before the screen is opened, the temperatures and CO_2 and H_2O concentrations above and below the screen are different, and after the screen is opened, they are equal.

8.4.5 Modeling Ventilation

Greenhouse ventilation is performed through the opening of windows. Furthermore, there is always a certain amount of ventilation due to leakage. The air exchange through the windows is caused by the temperature difference between the indoor and outdoor air and by the outdoor wind speed. The equations given in this section are taken from the thesis of De Jong (1990).

Ventilation causes the exchange of carbon dioxide, water vapor exchange, and heat. These are described in Sections 8.4.1.3, 8.4.2.3, and 8.4.3.1.

8.4.5.1 Volume Flow of Air through Windows and Leakage

The ventilation flow F_{as_o} is the sum of the ventilation through the windows and the ventilation due to leakage

$$F_{as\_o} = F_{leak} + F_{win} \quad \text{(m}^3\text{ s}^{-1}\text{)} \quad (8.58)$$

where F_{leak} (m^3 s^{-1}) is the ventilation due to leakage of the greenhouse construction and F_{win} (m^3 s^{-1}) is the ventilation through the windows.

The ventilation flow F_{leak} due to leakage of the greenhouse construction is described by

$$F_{leak} = A_s \cdot (8.3 \times 10^{-5} + 3.5 \times 10^{-5}\, v_o \cdot f_a) \quad \text{(m}^3\text{ s}^{-1}\text{)} \quad (8.59)$$

where A_s (m^2) is the surface area of the soil, v_o (m s^{-1}) is the outdoor wind speed and $f_a = 1$ is the infiltration factor.

The ventilation flow F_{win} through the windows is described by

$$F_{win} = \sqrt{F_{wind}^2 + \left(F_{\Delta T\_lsd} + F_{\Delta T\_wsd}\right)^2} \qquad (m^3\ s^{-1}) \quad (8.60)$$

where F_{wind} ($m^3\ s^{-1}$) is the wind-induced component, and $F_{\Delta T\_lsd}$ and $F_{\Delta T\_wsd}$ ($m^3\ s^{-1}$) are the components determined by the temperature difference between indoor and outdoor air, subdivided into lee- and windward-side.

The wind-induced component F_{wind} is described by

$$F_{wind} = n_{win} \cdot (f_{lsd} + f_{wsd}) \cdot v_o \cdot w_{win} \cdot h_{win} \qquad (m^3\ s^{-1}) \quad (8.61)$$

where f_{lsd} and f_{wsd} (–) are the ventilation functions for the lee- and the windward-side, w_{win} and h_{win} (m) are the width and the height of the window and n_{win} (–) is the total number of windows per side of the greenhouse span.

The ventilation functions f_{lsd} and f_{wsd} for the lee- and the windward-side are defined by

$$f_{lsd} = c_{lsd3} \cdot Ap_{lsd}^{\ 3} - c_{lsd2} \cdot Ap_{lsd}^{\ 2} + c_{lsd1} \cdot Ap_{lsd} + c_{lsd0} \qquad (–) \quad (8.62)$$

$$f_{wsd} = c_{wsd3} \cdot Ap_{wsd}^{\ 3} - c_{wsd2} \cdot Ap_{wsd}^{\ 2} + c_{wsd1} \cdot Ap_{wsd} + c_{wsd0} \qquad (–) \quad (8.63)$$

where $Ap_{lsd} \in [0,1]$ and $Ap_{wsd} \in [0,1]$ are the window aperture of the lee- and the windward-side (control inputs), and the ventilation function coefficients for the lee- and the windward-side (–) are given by: $c_{lsd0} = 1.26730 \times 10^{-4}$, $c_{lsd1} = 4.71176 \times 10^{-2}$, $c_{lsd2} = 3.77646 \times 10^{-2}$, $c_{lsd3} = 1.21774 \times 10^{-2}$, $c_{wsd0} = 5.91362 \times 10^{-4}$, $c_{wsd1} = 7.08828 \times 10^{-2}$, $c_{wsd2} = 4.48621 \times 10^{-2}$, and $c_{wsd3} = 2.42856 \times 10^{-2}$.

The components $F_{\Delta T\_lsd}$ and $F_{\Delta T\_wsd}$ determined by the temperature difference between indoor and outdoor air for lee- and windward-side are defined by

$$F_{\Delta T\_lsd} = n_{win} \cdot \frac{1}{3} c_w \cdot w_{win} \cdot \sqrt{g \cdot \frac{|T_o - T_{as}|}{T_o}} \cdot L_{win\_lsd}^{\ 1.5} \qquad (m^3\ s^{-1}) \quad (8.64)$$

$$F_{\Delta T\_wsd} = n_{win} \cdot \frac{1}{3} c_w \cdot w_{win} \cdot \sqrt{g \cdot \frac{|T_o - T_{as}|}{T_o}} \cdot L_{win\_wsd}^{\ 1.5} \qquad (m^3\ s^{-1}) \quad (8.65)$$

where w_{win} (m) is the window width, $c_w = 0.6$ (–) is the discharge coefficient through the windows, $g = 9.81\ m\ s^{-2}$ is gravity, L_{win_lsd} and L_{win_wsd} (m) are the lengths of the vertical projection of the windows opening on the lee- and the windward-side, and T_{as} and T_o (K) are the temperatures of the air above the screen and the outdoor air.

The lengths L_{win_lsd} and L_{win_wsd} of the vertical projection of the windows opening on the lee- and the windward-side are computed by

$$L_{win\_lsd} = 2\ h_{win} \cdot \cos(\gamma - 0.5\ Ap_{lsd} \cdot \Theta^{max}) \cdot \sin(0.5\ Ap_{lsd} \cdot \Theta^{max}) \qquad (m) \quad (8.66)$$

$$L_{win\_wsd} = 2\, h_{win} \cdot \cos(\gamma - 0.5\, Ap_{wsd} \cdot \Theta^{max}) \cdot \sin(0.5\, Ap_{wsd}\, \Theta^{max}) \quad \text{(m)} \quad (8.67)$$

where h_{win} (m) is the window height, γ (rad) is the angle of the roof with the horizontal plane and $\Theta^{max} = 40° \cdot \frac{\pi}{180°}$ (rad) is the maximum angle of the window aperture.

8.4.6 Modeling the Heating and the Cooling System

This part of the model is often left out by assuming local controllers. This option was not chosen here, due to physical constraints it cannot be guaranteed that the assumed setpoints can be realized, which would make the optimal control suboptimal.

The heating system consists of a boiler, a condenser, and a heat pump (see Figure 8.6). The boiler can be used to heat the lower net to a temperature of about 90°C. The condenser is heated by the flue gas of the boiler. It can be used to heat the upper net to a temperature of about 45°C. The heat pump can be used to heat the lower net to a temperature of about 33°C. The heating system with the boiler, the condenser, and the heat pump is described in Sections 8.4.6.1 and 8.4.6.3.

The cooling system consists of a heat exchanger (see Figure 8.7). The heat exchanger can be used to cool the upper cooling net to a temperature of about 10°C. The cooling system is described in Section 8.4.6.4.

The heat pump and the heat exchanger operate in conjunction with an aquifer. The aquifer has a warm ($T_{aq}^{C,warm} = 16°C$) and a cold ($T_{aq}^{C,cold} = 10°C$) side. The warm water is used by the heat pump to heat the greenhouse. The cold water is used by the heat exchanger to cool the greenhouse.

In the dimensions (length of pipes, number of loops) of the lower and upper heating net and the cooling net it is assumed that the layout is the same in every greenhouse span. The net can then be described by a number of loops of pipes with a specific length and diameter. The flow entering a net is equal to the flow leaving the net, and for each net there is one water temperature entering the net and one water temperature leaving the net.

8.4.6.1 Heating System Boiler and Condenser

The boiler is used to heat the lower net to a temperature of about 90°C. The flue gas of the boiler is used to heat a condenser, which heats the upper net to a temperature of about 45°C. The heating system is shown in Figure 8.6. The input of the lower net is taken from the boiler. The output of the lower net is the input of the heat pump. The output of the heat pump can be partly led through the

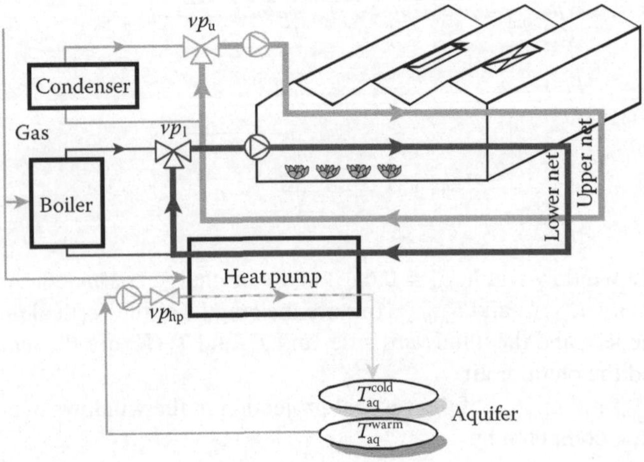

FIGURE 8.6 Heating with boiler, condenser, and heat pump.

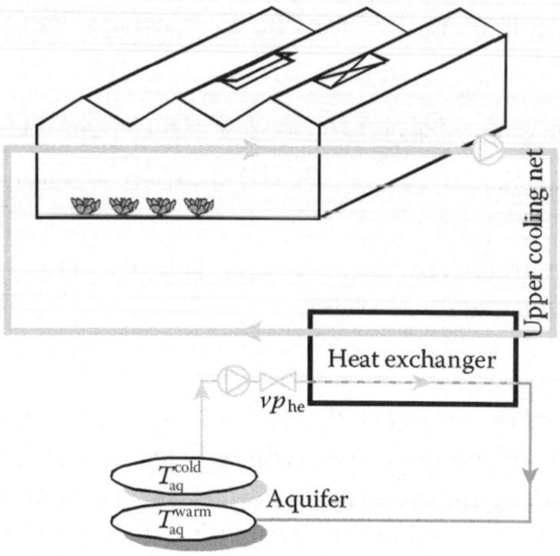

FIGURE 8.7 Cooling with heat exchanger.

boiler and partly through the lower net bypass. If the valve position of the lower net $vp_l = 0$, then no water is led to the boiler, and if the valve position of the lower net $vp_l = 1$, then no water is led through the lower net bypass. The input of the upper net is taken from the condenser. The output of the upper net can be partly led through the condenser and partly through the upper net bypass. If the valve position of the upper net $vp_u = 0$, then no water is led to the condenser, and if the valve position of the upper net $vp_u = 1$, then no water is led through the upper net bypass.

Lower Net

The input of the lower net (where it enters the greenhouse) is taken from the boiler and the lower net bypass. The output of the lower net (where it leaves the greenhouse) is the input of the heat pump. The energy transport terms Q_{in_l} and Q_{l_out} due to the water flow into and out of the lower net are defined by

$$Q_{in\_l} = \rho_{H_2O} \cdot c_{p,H_2O} \cdot F_l^{max} \cdot T_{in\_l} \qquad (W) \quad (8.68)$$

$$Q_{l\_out} = \rho_{H_2O} \cdot c_{p,H_2O} \cdot F_l^{max} \cdot T_{l\_out} \qquad (W) \quad (8.69)$$

where ρ_{H_2O} kg m^{-3} is the density of water, c_{p,H_2O} · (J kg^{-1} K^{-1}) is the specific heat capacity of water, $F_l^{max} = 1.62 \times 10^{-3}$ m^3 s^{-1} is the maximum pump flow rate of water into the lower net and T_{in_l} and T_{l_out} (K) are the water temperatures entering and leaving the lower net.

The water temperature T_{in_l} entering the lower net is defined by

$$T_{in\_l} = vp_l \cdot T_{boil} + (1 - vp_l) \cdot T_{l\_bypass} \qquad (K) \quad (8.70)$$

in which the boiler water temperature $T_{boil} = T_0 + 90$ K, where $vp_l \in [0,1]$ is the valve position of the lower net (control input). The water temperature T_{l_bypass} through the lower net bypass is equal to the water temperature T_{hp_out} leaving the heat pump

$$T_{l\_bypass} = T_{hp\_out} \qquad (K) \quad (8.71)$$

The water temperature T_{l_out} leaving the lower net is given by (Van Ooteghem, 2007)

$$T_{l\_out} = T_l + \frac{Q_{rad\_l} - Q_{l\_a} - Q_{l\_c} - Q_{l\_ri} - Q_{l\_s} - Q_{l\_sc}}{\rho_{H_2O} \cdot c_{p,H_2O} \cdot F_l^{max}} \quad (K) \quad (8.72)$$

where T_l (K) is the lower net water temperature, ρ_{H_2O} (kg m^{-3}) is the density of water, and c_{p,H_2O} (J kg^{-1} K^{-1}) is the specific heat capacity of water.

Upper Net

The input of the upper net (where it enters the greenhouse) is taken from the condenser and the upper net bypass. The output of the upper net (where it leaves the greenhouse) can be partly led through the condenser and partly through the upper net bypass. The energy transport terms Q_{in_u} and Q_{u_out} due to the water flow into and out of the upper net are defined by

$$Q_{in\_u} = \rho_{H_2O} \cdot c_{p,H_2O} \cdot F_u^{max} \cdot T_{in\_u} \quad (W) \quad (8.73)$$

$$Q_{u\_out} = \rho_{H_2O} \cdot c_{p,H_2O} \cdot F_u^{max} \cdot T_{u\_out} \quad (W) \quad (8.74)$$

where ρ_{H_2O} (kg m^{-3}) is the density of water, c_{p,H_2O} (J kg^{-1} K^{-1}) is the specific heat capacity of water, $F_l^{max} = 1.08 \times 10^{-3}\,m^3\,s^{-1}$ is the maximum pump flow rate of water into the upper net and T_{in_u} and T_{u_out} (K) are the water temperatures entering and leaving the upper net.

The water temperature T_{in_u} entering the upper net is defined by

$$T_{in\_u} = vp_u \cdot T_{cond} + (1 - vp_u) \cdot T_{u\_bypass} \quad (K) \quad (8.75)$$

in which the condenser water temperature $T_{cond} = T_0 + 45$ K, where $vp_u \in [0,1]$ is the valve position of the upper net (control input). The water temperature T_{u_bypass} through the upper net bypass is equal to the water temperature T_{u_out} leaving the upper net

$$T_{u\_bypass} = T_{u\_out} \quad (K) \quad (8.76)$$

The water temperature T_{u_out} leaving the upper net is given by (Van Ooteghem, 2007)

$$T_{u\_out} = T_u + \frac{Q_{rad\_u} - Q_{u\_a} - Q_{u\_c} - Q_{u\_ri} - Q_{u\_s} - Q_{u\_sc}}{\rho_{H_2O} \cdot c_{p,H_2O} \cdot F_u^{max}} \quad (K) \quad (8.77)$$

where T_u (K) is the upper net water temperature, ρ_{H_2O} (kg m^{-3}) is the density of upper net water, and c_{p,H_2O} (J kg^{-1} K^{-1}) is the specific heat capacity of water.

Boiler

The energy supply Q_{boil} by the boiler is defined by

$$Q_{boil} = \rho_{H_2O} \cdot c_{p,H_2O} \cdot F_{boil} \cdot (T_{boil} - T_{in\_boil}) \quad (W) \quad (8.78)$$

where ρ_{H_2O} (kg m^{-3}) is the density of water, c_{p,H_2O} (J kg^{-1} K^{-1}) is the specific heat capacity of water, F_{boil} (m^3 s^{-1}) is the flow rate of water leaving the boiler, and T_{boil} and T_{in_boil} (K) are the water temperatures in the boiler and entering the boiler. The water temperature T_{boil} is either set to a fixed value ($T_{boil} = T_0 + 90$ K), or follows from data.

The flow rate of water F_{boil} leaving the boiler is determined by

$$F_{boil} = vp_l \cdot F_l^{max} \qquad (m^3[H_2O]\ s^{-1}) \quad (8.79)$$

where F_l^{max} ($m^3[H_2O]\ s^{-1}$) is the maximum pump flow rate into the lower net and $vp_l \in [0,1]$ is the valve position of the lower net (control input).

The water temperature T_{in_boil} entering the boiler is equal to the water temperature in the lower net bypass T_{l_bypass}

$$T_{in\_boil} = T_{l\_bypass} \qquad (K) \quad (8.80)$$

8.4.6.2 The Aquifer

An aquifer is a formation of water-bearing sand material in the soil that can contain and transmit water. Wells can be drilled into the aquifers and water can be pumped into and out of the water layers. The heat pump and the heat exchanger operate in conjunction with an aquifer. The aquifer has a warm ($T_{aq}^{C,warm} = 16°C$) and a cold ($T_{aq}^{C,cold} = 10°C$) side. The warm water is used by the heat pump to heat the greenhouse and the cold water is used by the heat exchanger to cool the greenhouse.

It is assumed that the aquifer has an infinite amount of warm and cold water available. The loading and unloading of the aquifer buffers is limited by government demands, because the aquifer has to be energy neutral year-round. This indirectly corrects for the fact that the buffers are not infinite. These issues are worked out in the optimal control in Section 8.7.1.1.

8.4.6.3 Heating System Heat Pump

A gasfired heat pump is used to heat the lower net to a temperature of about 33°C. The heat pump heats the water in the lower net with water from the warm side of the aquifer. The cold water obtained in this process is led to the cold side of the aquifer. The heating system is shown in Figure 8.6. It is assumed that the heat transfer between the heat pump and the lower net has no dynamics (direct transfer of heat).

The lower net equations are given in Section 8.4.6.1. The heat pump equations are taken from Van Ooteghem (2007). The configuration and the energy transport of the heat pump is given in Figure 8.8. In this figure, the following water temperatures (K) are shown: T_{in_hp} and T_{hp_out} flowing into and out of the heat pump (greenhouse side), T_{aq}^{warm} and $T_{hp\_aq}^{cold}$ of the warm and the cold side of the aquifer (aquifer side) and T_{hs} and T_{cs} of the warm and the cold side of the heat pump (inside the heat pump). Water temperature $T_{hp\_aq}^{cold}$ should be lower than the desired water temperature T_{aq}^{cold}, so that the

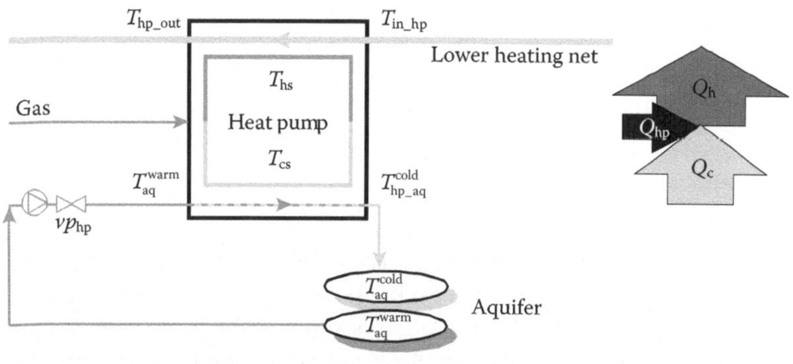

FIGURE 8.8 Configuration and energy transport heat pump.

water temperature T_{aq}^{cold} can be achieved by mixing with water with temperature T_{aq}^{warm}. This has to be solved locally (outside the system boundary considered here).

The water temperature T_{aq}^{warm} of the warm side of the aquifer (aquifer side) is known, as well as the water temperature T_{in_hp} flowing into the heat pump (greenhouse side). With the valve position vp_{hp} of the heat pump (control input), the water temperature T_{hp_out} flowing out of the heat pump (greenhouse side), the water temperatures T_{hs} and T_{cs} (inside the heat pump), the water temperature $T_{hp\_aq}^{cold}$ (aquifer side), the energy transport terms Q_h, and Q_c, and the energy used by the heat pump Q_{hp} can be computed.

Heat Pump

The water temperature T_{in_hp} entering the heat pump is equal to the water temperature T_{l_out} leaving the lower net

$$T_{in\_hp} = T_{l\_out} \qquad \text{(K)} \quad (8.81)$$

The water temperature T_{hp_out} (K) leaving the heat pump can be partly led through the boiler and partly through the lower net bypass. If the valve position of the heat pump $vp_{hp} = 0$, then no water is led through the heat pump (the heat pump is off). It is assumed here that a compression heat pump is used. The operation of a compression heat pump is based on the compression and evaporation of a fluid. The fluid evaporates when thermal energy Q_c is added from the warm side of the aquifer. The fluid condenses when thermal energy Q_h is subtracted by the lower net. The energy used by the heat pump to drive this process Q_{hp} is given by

$$Q_{hp} = Q_h - Q_c \qquad \text{(W)} \quad (8.82)$$

This amount of energy determines the coefficient of performance COP of the heat pump

$$COP = \frac{Q_h}{Q_h - Q_c} \qquad \text{(W)} \quad (8.83)$$

The energy transport Q_h due to the water flow on the lower net side and the energy transport Q_c due to the water flow on the aquifer side are given by (see Figure 8.8)

$$Q_h = \rho_{H_2O} \cdot c_{p,H_2O} \cdot F_l^{max} \cdot \left(T_{hp\_out} - T_{in\_hp}\right) \qquad \text{(W)} \quad (8.84)$$

$$Q_c = \rho_{H_2O} \cdot c_{p,H_2O} \cdot vp_{hp} \cdot F_{hp}^{max} \cdot \left(T_{aq}^{warm} - T_{hp\_aq}^{cold}\right) \qquad \text{(W)} \quad (8.85)$$

where ρ_{H_2O} (kg m^{-3}) is the density of water, c_{p,H_2O} (J kg^{-1} K^{-1}) is the specific heat capacity of water, F_l^{max} and $F_{hp}^{max} = 1.5\times10^{-3}(\text{m}^3\ \text{s}^{-1})$ are the maximum pump flow rates of water into the lower net and through the heat pump and $vp_{hp} \in [0,1]$ is the valve position of the heat pump (control input).

The resulting temperature $T_{hp\_aq}^{cold}$ of the aquifer water cooled by the heat pump is given by (Van Ooteghem, 2007)

$$T_{\mathrm{hp\_aq}}^{\mathrm{cold}} = \frac{\begin{array}{c}\eta_{\mathrm{hp}} \cdot (c_1 - 1) \cdot (h_1 \cdot T_{\mathrm{hp\_out}} - T_{\mathrm{in\_hp}}) \cdot vp_{\mathrm{hp}} \cdot F_{\mathrm{hp}}^{\max} \cdot T_{\mathrm{aq}}^{\mathrm{warm}} \\ + F_1^{\max} \cdot (T_{\mathrm{hp\_out}} - T_{\mathrm{in\_hp}}) \\ \cdot \left((1-\eta_{\mathrm{hp}}) \cdot (c_1 - 1) \cdot (h_1 \cdot T_{\mathrm{hp\_out}} - T_{\mathrm{in\_hp}}) + T_{\mathrm{aq}}^{\mathrm{warm}} \cdot (h_1 - 1)\right)\end{array}}{\begin{array}{c}\eta_{\mathrm{hp}} \cdot (c_1 - 1) \cdot (h_1 \cdot T_{\mathrm{hp\_out}} - T_{\mathrm{in\_hp}}) \cdot vp_{\mathrm{hp}} \cdot F_{\mathrm{hp}}^{\max} \\ + F_1^{\max} \cdot (T_{\mathrm{hp\_out}} - T_{\mathrm{in\_hp}}) \cdot c_1 \cdot (h_1 - 1)\end{array}} \quad \text{(K)} \quad (8.86)$$

in which $h_1 = \mathrm{e}^{\frac{k_{\mathrm{hp}} \cdot A_{\mathrm{hp}}}{\rho_{\mathrm{H_2O}} \cdot c_{p,\mathrm{H_2O}} \cdot F_1^{\max}}}$ and $c_1 = \mathrm{e}^{\frac{k_{\mathrm{hp}} \cdot A_{\mathrm{hp}}}{\rho_{\mathrm{H_2O}} \cdot c_{p,\mathrm{H_2O}} \cdot vp_{\mathrm{hp}} F_{\mathrm{hp}}^{\max}}}$, where $k_{\mathrm{hp}} = 250$ W m^{-2} K^{-1} is the heat pump heat transfer coefficient, $A_{\mathrm{hp}} = 500$ m^2 is the heat pump surface for heat transfer, and $\eta_{\mathrm{hp}} = 0.4$ (–) is the efficiency of the heat pump.

The water temperature $T_{\mathrm{hp\_out}}$ leaving the heat pump is given by (Van Ooteghem, 2007)

$$T_{hp\_out} = \frac{1}{12 \cdot a_{\mathrm{T}}} \cdot \left(-3b_{\mathrm{T}} - \sqrt{3} p_{\mathrm{T7}} + \left(\frac{3}{p_{\mathrm{T6}} \cdot p_{\mathrm{T7}}} \cdot \left(2 \cdot \left(30 b_{\mathrm{T}}^2 - 8 a_{\mathrm{T}} \cdot c_{\mathrm{T}} - a_{\mathrm{T}} \cdot p_{\mathrm{T6}}\right) \cdot p_{\mathrm{T6}} \cdot p_{\mathrm{T7}} + 8 a_{\mathrm{T}} \cdot \left(3 b_{\mathrm{T}} \cdot d_{\mathrm{T}} - 12 a_{\mathrm{T}} \cdot e_{\mathrm{T}} - c_{\mathrm{T}}^2\right) \cdot p_{\mathrm{T7}} + 6\sqrt{3} \cdot \left(8 a_{\mathrm{T}}^2 \cdot d_{\mathrm{T}} + b_{\mathrm{T}}^3 - 4 a_{\mathrm{T}} \cdot b_{\mathrm{T}} \cdot c_{\mathrm{T}}\right) \cdot p_{\mathrm{T6}} \right) \right)^{\frac{1}{2}} \right) \quad \text{(K)} \quad (8.87)$$

where a_{T}, b_{T}, c_{T}, d_{T} and e_{T}, are parameters of a fourth-order equation and p_{T1}, p_{T2}, p_{T3}, p_{T4}, p_{T5}, and p_{T6} are parameter combinations (Van Ooteghem, 2007).

The minimum valve position vp_{hp}—determined by the heat pump characteristics—is $vp_{\mathrm{hp}}^{\min} = 0.57$. Below this value the temperature $T_{\mathrm{hp\_aq}}^{\mathrm{cold}} \leq T_0 (T_0 = 273.15\ \mathrm{K})$. Because the optimal control will compute the value for the valve position vp_{hp}^* between 0 and 1, the valve position is scaled between the minimum valve position $vp_{\mathrm{hp}}^{\min}$ and the maximum value of 1

$$vp_{\mathrm{hp}} = \begin{cases} \left(1 - vp_{\mathrm{hp}}^{\min}\right) \cdot vp_{\mathrm{hp}}^* + vp_{\mathrm{hp}}^{\min} & \text{if } vp_{\mathrm{hp}}^* = 0 \\ 0 & \text{if } vp_{\mathrm{hp}}^* = 0 \end{cases} \quad (–) \quad (8.88)$$

where vp_{hp}^* is the valve position computed by the optimal control.

There are some restrictions as to the operation of the heat pump. The heat pump cannot be operated:

- If the temperature $T_{\mathrm{in\_hp}}$ is too high ($T_{\mathrm{in\_hp}} > T_{\mathrm{in\_hp}}^{\max}$), because the heat pump cannot increase this temperature any further. With the chosen heat pump characteristics this temperature $T_{\mathrm{in\_hp}}^{\max} = T_0 + 30.1$ K.

- If the valve position vp_{hp}^{*} of the heat pump is so low that the temperature $T_{hp\_aq}^{cold} \leq T_0 (vp_{hp}^{*} < vp_{hp}^{min})$, because this would mean that the aquifer water would freeze. This is avoided by using Equation 8.88.
- If the resulting water temperature $T_{hp\_aq}^{cold}$ for the cold side of the aquifer is higher than the desired temperature T_{aq}^{cold}, the desired temperature cannot be reached by mixing with water with temperature T_{aq}^{warm}. If this occurs, the valve position vp_{hp}^{*} is decreased by 0.1, making the water run slower and decreasing the temperature $T_{hp\_aq}^{cold}$. The valve position is decreased further until $T_{hp\_aq}^{cold} < T_{aq}^{cold}$.
- The heat pump is turned off ($vp_{hp} = 0$) if any of the temperature differences between the incoming and outgoing water from the lower net or the aquifer and the condensation and the evaporation temperature T_{hs} and T_{cs} (K) is lower than or equal to zero, which would mean that the heat transfer would take place in the opposite direction. It is also turned off if $T_{hp\_aq}^{cold} \leq T_0$, because then the aquifer water would freeze.

8.4.6.4 Cooling System Heat Exchanger

The heat exchanger is used to cool the upper cooling net to a temperature of about 10°C. The heat exchanger cools the water in the upper cooling net with water from the cold side of the aquifer. The warm water obtained in this process is led to the warm side of the aquifer. The cooling system is shown in Figure 8.7. It is assumed that the heat transfer between the heat exchanger and the upper cooling net has no dynamics (direct transfer of heat).

The upper cooling net equations are given in Section 8.4.6.4. The heat exchanger equations are taken from Van Ooteghem (2007). The configuration and the energy transport of the heat exchanger is given in Figure 8.9. In this figure, the following water temperatures (K) are shown: T_{in_he} and T_{he_out} flowing into and out of the heat exchanger (greenhouse side) and T_{aq}^{cold} and $T_{he\_aq}^{warm}$ of the cold and the warm side of the aquifer (aquifer side). Water temperature T_{aq}^{warm} should be higher than the desired water temperature T_{aq}^{warm}, so that the water temperature T_{aq}^{warm} can be achieved by mixing with water with temperature T_{aq}^{cold}. This has to be solved locally (outside the system boundary considered here).

The water temperature T_{aq}^{cold} of the cold side of the aquifer (aquifer side) is known, as well as the water temperature T_{in_he} flowing into the heat exchanger (greenhouse side). With the valve position vp_{he} of the heat exchanger (control input), the water temperature T_{he_out} flowing out of the heat exchanger (greenhouse side), the water temperature $T_{he\_aq}^{warm}$ (aquifer side) and the energy transport term Q_{he} can be computed.

Upper Cooling Net

The input of the upper cooling net is taken from the heat exchanger. The output of the upper cooling net is the input of the heat exchanger. The energy transport terms Q_{in_uc} and Q_{uc_out} due to the water flow into and out of the upper cooling net are defined by

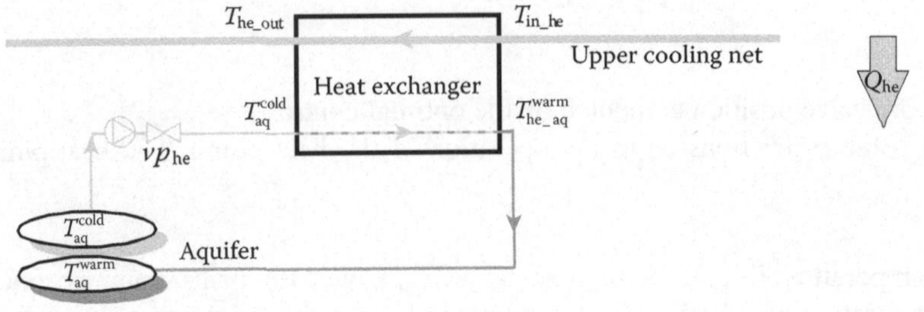

FIGURE 8.9 Configuration and energy transport heat exchanger.

$$Q_{\text{in\_uc}} = \rho_{\text{H}_2\text{O}} \cdot c_{\text{p,H}_2\text{O}} \cdot F_{\text{uc}}^{\max} \cdot T_{\text{in\_uc}} \qquad \text{(W)} \quad (8.89)$$

$$Q_{\text{uc\_out}} = \rho_{\text{H}_2\text{O}} \cdot c_{\text{p,H}_2\text{O}} \cdot F_{\text{uc}}^{\max} \cdot T_{\text{uc\_out}} \qquad \text{(W)} \quad (8.90)$$

where $\rho_{\text{H}_2\text{O}}$ (kg m^{-3}) is the density of water, $c_{\text{p,H}_2\text{O}}$ (J kg^{-1} K^{-1}) is the specific heat capacity of water, $F_{\text{uc}}^{\max} = 6 \times 10^{-3}\,\text{m}^3\,\text{s}^{-1}$ is the maximum pump flow rate of water into the upper cooling net and $T_{\text{in\_uc}}$ and $T_{\text{uc\_out}}$ (K) are the water temperatures entering and leaving the upper cooling net. The water temperature $T_{\text{in\_uc}}$ entering the upper cooling net is equal to the water temperature leaving the heat exchanger

$$T_{\text{in\_uc}} = T_{\text{he\_out}} \qquad \text{(K)} \quad (8.91)$$

The water temperature $T_{\text{uc\_out}}$ leaving the upper cooling net is given by (Van Ooteghem, 2007)

$$T_{\text{uc\_out}} = T_{\text{uc}} + \frac{Q_{\text{rad\_uc}} + Q_{\text{a\_uc}}^{\text{cons}} - Q_{\text{uc\_a}} - Q_{\text{uc\_c}} - Q_{\text{uc\_ri}} - Q_{\text{uc\_s}} - Q_{\text{uc\_sc}}}{\rho_{\text{H}_2\text{O}} \cdot c_{\text{p,H}_2\text{O}} \cdot F_{\text{uc}}^{\max}} \qquad \text{(K)} \quad (8.92)$$

where T_{uc} (K) is the upper cooling net temperature, $\rho_{\text{H}_2\text{O}}$ (kg m^{-3}) is the density of water, and $c_{\text{p,H}_2\text{O}}$ (J kg^{-1} K^{-1}) is the specific heat capacity of water.

Heat Exchanger

The water temperature $T_{\text{in\_he}}$ entering the heat exchanger is equal to the water temperature $T_{\text{uc\_out}}$ leaving the upper cooling net

$$T_{\text{in\_he}} = T_{\text{uc\_out}} \qquad \text{(K)} \quad (8.93)$$

The water temperature $T_{\text{he\_out}}$ (K) leaving the heat exchanger is the input of the upper cooling net (Equation 8.91). If the valve position of the heat exchanger $vp_{\text{he}} = 0$, then no water is led through the heat exchanger (the heat exchanger is off).

It is assumed here that a countercurrent heat exchanger is used. The energy transfer Q_{he} by the heat exchanger is defined by the energy transport due to the water flow on the aquifer side, given by

$$Q_{\text{he}} = \rho_{\text{H}_2\text{O}} \cdot c_{\text{p,H}_2\text{O}} \cdot vp_{\text{he}} \cdot F_{\text{he}}^{\max} \cdot \left(T_{\text{he\_aq}}^{\text{warm}} - T_{\text{aq}}^{\text{cold}}\right) \qquad \text{(W)} \quad (8.94)$$

where $\rho_{\text{H}_2\text{O}}$ (kg m^{-3}) is the density of water, $c_{\text{p,H}_2\text{O}}$ (J kg^{-1} K^{-1}) is the specific heat capacity of water, $F_{\text{he}}^{\max} = 5.7 \times 10^{-3}\,\text{m}^3\,\text{s}^{-1}$ is the maximum pump flow rate of water through the heat exchanger, and $vp_{\text{he}} \in [0,1]$ is the valve position of the heat exchanger (control input).

The resulting temperature $T_{\text{he\_aq}}^{\text{warm}}$ of the aquifer water heated by the heat exchanger is given by (Van Ooteghem, 2007)

$$T_{\text{he\_aq}}^{\text{warm}} = \begin{cases} \dfrac{c_{\text{he}} \cdot T_{\text{aq}}^{\text{cold}} \cdot \left(vp_{\text{he}} \cdot F_{\text{he}}^{\max} - F_{\text{uc}}^{\max}\right) + T_{\text{in\_he}} \cdot F_{\text{uc}}^{\max} \cdot \left(c_{\text{he}} - 1\right)}{c_{\text{he}} \cdot vp_{\text{he}} \cdot F_{\text{he}}^{\max} - F_{\text{uc}}^{\max}} & \text{if } c_{\text{he}} > 1 \\ T_{\text{aq}}^{\text{cold}} & \text{if } c_{\text{he}} = 1 \end{cases} \qquad \text{(K)} \quad (8.95)$$

in which $c_{he} = e^{\frac{k_{he} \cdot A_{he}}{\rho_{H2O} \cdot c_{p,H2O}} \cdot \left(\frac{1}{F_{uc}^{max}} - \frac{1}{vp_{he} \cdot F_{he}^{max}} \right)}$, where $k_{he} = 250$ W m^{-2} K^{-1} is the heat exchanger heat transfer coefficient, $A_{he} = 2500$ m^2 is the heat exchanger surface for heat transfer, and F_{uc}^{max} (m^3 s^{-1}) is the maximum pump flow rate of water into the upper cooling net.

The water temperature T_{he_out} leaving the heat exchanger is given by (Van Ooteghem, 2007)

$$T_{he\_out} = T_{in\_he} - \frac{vp_{he} \cdot F_{he}^{max} \cdot \left(T_{he\_aq}^{warm} - T_{aq}^{cold} \right)}{F_{uc}^{max}} \qquad \text{(K)} \qquad (8.96)$$

The minimum valve position vp_{he}—determined by the heat exchanger characteristics—is $vp_{he}^{min} = 0.43$. Below this value the mean temperature difference $\Delta T_{m\_he} < 0$ (Van Ooteghem, 2007; Van Kimmenade, 1986). Because the optimal control will compute the value for the valve position vp_{he}^{*} between zero and one, the valve position is scaled between the minimum valve position vp_{he}^{min} and the maximum value of one

$$vp_{he} = \begin{cases} \left(1 - vp_{he}^{min}\right) \cdot vp_{he}^{*} + vp_{he}^{min} & \text{if } vp_{he}^{*} > 0 \\ 0 & \text{if } vp_{he}^{*} = 0 \end{cases} \qquad (-) \qquad (8.97)$$

where vp_{he}^{*} is the valve position computed by the optimal control.

There are some restrictions as to the operation of the heat exchanger. The heat exchanger cannot be operated:

- If the temperature T_{in_he} is too low ($T_{in\_he} < T_{in\_he}^{min}$) the heat exchanger cannot decrease this temperature any further. The minimum temperature is equal to the temperature of the warm aquifer side $T_{in\_he}^{min} = T_{aq}^{warm}$.
- If the valve position of the heat pump vp_{he}^{*} is so low that the temperature difference $\Delta T_{m\_he} < 0$ ($vp_{he}^{*} < vp_{he}^{min}$), which would mean that heat transfer would take place in the opposite direction.
- If the resulting temperature for the warm side of the aquifer $T_{he\_aq}^{warm}$ is lower than the desired temperature T_{aq}^{warm}, because the desired temperature cannot be reached by mixing with water with temperature T_{aq}^{cold}. If this occurs, the valve position vp_{he}^{*} is decreased by 0.1, such that the water runs slower, increasing temperature $T_{he\_aq}^{warm}$. The valve position is decreased further until $T_{he\_aq}^{warm} > T_{aq}^{warm}$.
- The heat exchanger is turned off ($vp_{he} = 0$) if any of the temperature differences between the incoming water and the warm side or the outgoing water and the cold side of the heat exchanger is lower than or equal to zero, which would mean that the heat transfer would take place in the opposite direction.

8.5 MODEL OF CROP BIOPHYSICS

The Dutch solar greenhouse design aims at reducing fossil energy use in Dutch horticulture (Bot, 2001). It reduces the required heating while maintaining or increasing crop yield and quality. It is therefore beneficial if larger temperature fluctuations are allowed compared with conventional greenhouses. This may lead to temperature and humidity extremes that are beyond the range for which the current crop models are designed and tested. It is important that the crop model also gives an accurate description of the relevant crop processes for these extreme values for temperature and humidity.

In the literature many models are given for various parts of the crop growth process. To limit the on-line computational load in the optimal control computation, the model should be sufficiently small with respect to the number of differential equations. It should, however, also be sufficiently accurate. The time scale considered is also important, because a longer time scale requires a longer prediction horizon in the optimal control context.

The crop processes considered here are the rates of photosynthesis, dark respiration, and evapotranspiration. The description of the evapotranspiration process is based on the resistances for H_2O diffusion. These resistances are closely related to the resistances for CO_2 diffusion, which are important for the photosynthesis rate. From the photosynthesis and dark respiration the crop total biomass is obtained. It is assumed that the photosynthesis and respiration directly affect the biomass weight. No subdivision into vegetative and generative state or partitioning into fruit and leaves is taken into account. Temperature integration is used as a descriptive method for long-term temperature effects on crop development. It is assumed that the grower is able to set optimal control values for the greenhouse temperature and humidity and for the temperature integration such that proper crop development is ensured during all its development stages. The crop is grown on substrate, which is placed in a gutter, and covered with white plastic. It is assumed that water and nutrient supply are well controlled and not limiting crop photosynthesis and evaporation.

From the various models available in the literature for the simulation of crop and leaf photosynthesis, a new model is formed. For the new model to be suitable for optimal control purposes, it should be a complete model of sufficient accuracy over the full range of working conditions. It must describe the crop gross photosynthesis rate as a function of light intensity, CO_2 concentration, and temperature. The effect of the control and external inputs on the crop processes should be well described. Furthermore, it is favorable to have a limited number of differential equations (lower order model) to limit the online computation time.

A complete and detailed description of a new crop processes model that is suitable for optimal control purposes is given in this section. The different physical and physiological processes that together make up the model are described in different subsections. The evapotranspiration process is described in Section 8.5.1 and the crop photosynthesis and respiration are described in Section 8.5.2. A number of models from the literature are compared. The temperature integration is given in Section 8.5.3.

8.5.1 Evapotranspiration

The evapotranspiration process concerns the evaporation of water from the leaf to the greenhouse air. This process is important for the water and nutrient transport from roots to leaves and fruits. It is also important to decrease the temperature of the crop. Water is mainly evaporated through the leaf stomata. The canopy transpiration is thus a function of the resistance of the stomata and the leaf boundary layer. In the literature, these resistances are often assumed to be constant. Because we want to use the crop model for extreme temperature and humidity values, we are not in the domain where these constant resistances apply. We therefore use a model to compute the leaf resistances.

The model by Stanghellini (1987) is used for the evaporation process. This model is an adaptation of the Penman–Monteith–Rijtema method (the combination method) to determine the actual instead of the potential transpiration rate in a greenhouse. Where the Penman–Monteith–Rijtema method determines the potential transpiration rate, the Stanghellini method determines the actual transpiration rate in a greenhouse. The transpiration rate depends on light intensity, CO_2 concentration, temperature, and humidity. All relations—if not otherwise noted—are taken from Stanghellini (1987).

The canopy transpiration Φ_{H_2O,c_a} or the mass flow rate of water vapor from crop to indoor air is

$$\Phi_{H_2O,c\_a} = \max\left(A_c \cdot k_{H_2O,c\_a} \cdot \left(C_{H_2O,c}^{sat} - C_{H_2O,a}\right),\ 0\right) \qquad (kg[H_2O]\ s^{-1}) \qquad (8.98)$$

where A_c (m^2[leaf]) is surface area of the canopy, k_{H_2O,c_a} ($m\ s^{-1}$) is the mass transfer coefficient of water vapor from the crop to the indoor air, $C_{H_2O,c}^{sat}$(kg[H_2O] m^{-3}[air]) is the saturation concentration of water vapor at the temperature of the crop (see Appendix B.1) and $C_{H_2O,a}$ (kg[H_2O] m^{-3}[air]) is the concentration water vapor at the temperature of the indoor air. If $C_{H_2O,c}^{sat} \le C_{H_2O,a}$ then $\Phi_{H_2O,c\_a} =$ 0 (no evapotranspiration).

Bot (1983) describes the total resistance to diffusion of water as the boundary layer resistance in series with the cuticular resistance parallel to the stomatal resistance. From this the mass transfer coefficient k_{H_2O,c_a} from crop to indoor air is derived as

$$k_{H_2O,c\_a} = \frac{1}{R_{H_2O,b} + \dfrac{R_{cut} \cdot R_{H_2O,s}}{R_{cut} + R_{H_2O,s}}} \qquad (m\ s^{-1}) \qquad (8.99)$$

in which the leaf cuticular resistance $R_{cut} = 2000\ s\ m^{-1}$, where $R_{H_2O,s}$ ($s\ m^{-1}$) is the stomatal resistance to diffusion of water and $R_{H_2O,b}$ ($s\ m^{-1}$) is the boundary layer resistance to diffusion of water.

The stomatal resistance to diffusion of water $R_{H_2O,s}$ is described by

$$R_{H_2O,s} = R_{min} \cdot f_I \cdot f_{Tc} \cdot f_{CO_2} \cdot f_{H_2O} \qquad (s\ m^{-1}) \qquad (8.100)$$

in which the radiation dependency f_I is given by

$$f_I = \frac{\dfrac{q_{rad\_c}}{2LAI} + 4.3}{\dfrac{q_{rad\_c}}{2LAI} + 0.54} \qquad (-) \qquad (8.101)$$

the temperature dependency f_{Tc} is given by

$$f_{Tc} = \begin{cases} 1 + 0.5 \times 10^{-2} \cdot \left(T_c - T_0 - 33.6\right)^2 & \text{if } q_{rad\_c} \le 3 \\ 1 + 2.2593 \times 10^{-2} \cdot \left(T_c - T_0 - 24.512\right)^2 & \text{if } q_{rad\_c} > 3 \end{cases} \qquad (-) \qquad (8.102)$$

the CO_2 dependency f_{CO_2} is given by

$$f_{CO_2} = \begin{cases} 1 & \text{if } q_{rad\_c} \le 3 \\ 1 + 6.08 \times 10^{-7} \cdot \left(C_{CO_2,a}^{ppm} - 200\right)^2 & \text{if } q_{rad\_c} > 3 \\ 1.49 & \text{if } C_{CO_2,a}^{ppm} \ge 1100 \end{cases} \qquad (-) \qquad (8.103)$$

the humidity dependency f_{H_2O} is given by

$$f_{H_2O} = \frac{4}{\sqrt[4]{1+255\, e^{-0.5427 \Delta p^{mbar}_{H_2O,c}}}} \qquad (-) \quad (8.104)$$

and the minimum internal crop resistance R_{min} = 82.003 s m^{-1} (Jarvis's model). The term $\frac{q_{rad\_c}}{2LAI}$ (W m^{-2}[leaf]) determines the leaf shortwave radiation absorption from the heat absorbed by the canopy $q_{rad\_c} = \frac{Q_{rad\_c}}{A_s}$ (W m^{-2}[gh]) and the leaf area index LAI (m^2[leaf] m^{-2}[gh]). T_c (K) is the temperature of the crop, T_0 = 273.15 K, $C^{ppm}_{CO_2,a}$ (μmol[CO_2] mol^{-1}[air]) is the CO_2 concentration of the indoor air, and $\Delta p^{mbar}_{H_2O,c}$ (mbar) is the crop saturation deficit. All numbers in Equations 8.100 through 8.104 are model parameters, determined by Stanghellini for tomato. De Zwart (1996) also gives values for roses.

The dependencies of the stomatal resistance to diffusion of water $R_{H_2O,s}$ (s m^{-1}) are given in Figure 8.10. The radiation dependency f_I decreases from 8 to 1 for increasing values of radiation, which indicates that radiation only influences the resistance at low light intensities. The temperature dependency f_{Tc} is parabolic with a minimum at 24.5°C. The humidity dependency f_{H_2O} is constant at a maximum of 4 for vapor pressure differences above about 15 mbar and decreases at lower vapor pressure differences. This is due to stomata closure at low humidity values to prevent dehydration. The CO_2 dependency f_{CO_2} increases to 1.5 if the CO_2 concentration increases.

The boundary layer resistance to diffusion of water $R_{H_2O,b}$ is described by Monteith and Unsworth (1990) as

$$R_{H_2O,b} = Le^{\frac{2}{3}} \cdot R_{heat,b} \qquad (s\ m^{-1}) \quad (8.105)$$

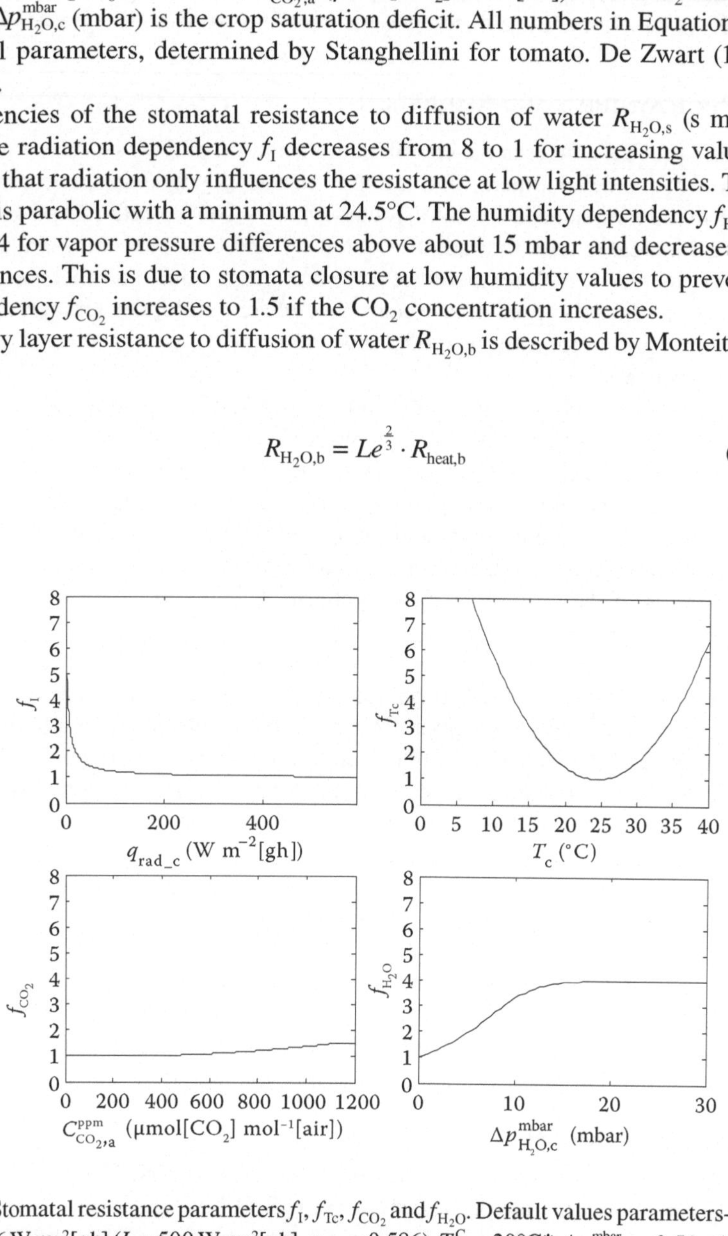

FIGURE 8.10 Stomatal resistance parameters f_I, f_{Tc}, f_{CO_2} and f_{H_2O}. Default values parameters—if not varied—are: q_{rad_c} = 293.06 W m^{-2}[gh] (I_o = 500 W m^{-2}[gh], η_{c_Is} = 0.586), T^C_c = 20°C*, $\Delta p^{mbar}_{H_2O,c}$ = 3.51 mbar (RH_a = 85%), $C^{ppm}_{CO_2,a}$ = 1000 μmol[CO_2] mol^{-1}[air] and T^C_a = 20°C*.
*T_a and T_c in (K) in computations, in (°C) here for readability.

where $R_{heat,b}$ (s m^{-1}) is the boundary layer resistance to convective heat transfer and $Le = 0.89$ (–) is the Lewis number for water vapor in air.

The boundary layer resistance to convective heat transfer $R_{heat,b}$ is given by

$$R_{heat,b} = \frac{1174\sqrt{l_f}}{\left(l_f \cdot \left|T_c - T_a\right| + 207 v_a^2\right)^{\frac{1}{4}}} \quad (\text{s m}^{-1}) \quad (8.106)$$

in which the mean leaf width $l_f = 0.035$ m and the wind speed (in the greenhouse) $v_a = 0.09$ m s^{-1}, where $\left|T_c - T_a\right|$ (K) is the temperature difference between the crop and the greenhouse air.

8.5.2 Crop Photosynthesis and Respiration

The photosynthesis process concerns the chemical assimilation of CO_2 and water to assimilates for maintenance, growth, and development. The canopy extracts CO_2 from its environment. The photosynthesis rate is mainly influenced by light intensity, CO_2 concentration, and temperature. The photosynthesis rate increases with the radiation intensity and CO_2 concentration. Furthermore the photosynthesis rate increases with temperature to a maximum value, and then decreases at higher temperatures. Because the solar greenhouse may have lower and higher temperatures than a conventional greenhouse, the photosynthesis model must describe the photosynthesis process well over a wide temperature range.

Various models are available for the simulation of crop and leaf photosynthesis. These models describe the photosynthesis process in different ways. There are two mainstream approaches to photosynthesis modeling. Leaf photosynthesis describes the photosynthesis rate of a single leaf. Crop photosynthesis describes the overall photosynthesis rate of the canopy as a whole. In principle, crop photosynthesis can be obtained from leaf photosynthesis by some form of spatial integration over the canopy. The models are compared, and from these models a new model is formed, to give an accurate description of the crop gross photosynthesis rate as a function of light intensity, CO_2 concentration, and temperature.

The new photosynthesis model is a crop photosynthesis model. For the description of the biochemical processes on a leaf level, the leaf photosynthesis model by Farquhar et al. (1980) is used because is gives the most detailed description from the models selected here. For the light interception in the crop layers and the Gaussian integration, the model by Goudriaan and Van Laar (1994) and Heuvelink (1996) is used because we need a crop and not a leaf photosynthesis model.

The photosynthesis rate can be limited by the stomatal and boundary layer resistances to CO_2 diffusion, which are a function of the resistances to H_2O diffusion. Often constant resistances to CO_2 diffusion are used. Because we are not working in the temperature and humidity ranges where these constant resistances apply, the resistances found from Section 8.5.1 by Stanghellini (1987) are used.

The new crop photosynthesis model is described here in detail. This model has been validated by Körner and Van Ooteghem (2003) and Körner et al. (2001a,b, 2002, 2003, 2009). It was found that the model showed good accordance with measured data. The resistances computed with the evaporation model resulted in better results in most cases compared to constant resistances.

8.5.2.1 Photosynthesis Model

A number of general parameters and their values are given in Table 8.11. The purpose of the model is to describe the CO_2 assimilation rate of the canopy (in kg[CO_2] m^{-2}[gh] s^{-1}) as a function of the outdoor shortwave solar radiation I_o (W m^{-2} [gh]), the CO_2 concentration $C^{ppm}_{CO_2,a}$ (μmol[CO_2] mol^{-1}[air]), the temperature of the crop T_c (K), and the relative humidity RH_a (%).

TABLE 8.11
Photosynthesis Model and General Parameters

Name	Value	Unit	Contents	Source Reference
		Constants		
R_g	8.314	$J\ mol^{-1}\ K^{-1}$	Gas constant	
ρ_{CO_2}	1.98×10^{-6}	$mg[CO_2]\ m^{-3}[CO_2]$	CO_2 density at T_0	
M_{CO_2}	44×10^{-9}	$kg[CO_2]\ \mu mol^{-1}[CO_2]$	Molar mass CO_2	
T_0	273.15	K	273.15 K = 0°C	
T_{25}	$T_0 + 25$	K	273.15 K + 25 K = 25°C	
ζ	4.59	$\mu mol[photons]\ J^{-1}$	Conversion factor, J to photons	
ρ_{Chl}	0.45	$g[Chl]\ m^{-2}[leaf]$	Superficial chlorophyll density	a
		Constants at 25°C		
$p_{O2,i}$	210	mbar	O_2 partial pressure inside stomata	a, b
K_{C25}	310	µbar	Michaelis–Menten constant Rubisco carboxylation (CO_2)	b
K_{O25}	155	mbar	Michaelis–Menten constant Rubisco oxygenation (O_2)	b
k_C	2.5	s^{-1}	Turnover number of RuP2 (carboxylase)	a
E_t	87.0	$\mu mol[CO_2]\ g^{-1}[Chl]$	Total concentration of enzyme sites	a
V_{Cmax25}	$\rho_{Chl} \cdot k_C \cdot E_t$	$\mu mol[CO_2]\ m^{-2}[leaf]\ s^{-1}$	Maximum carboxylation rate at 25°C	a
$r_{D25,L}^{\mu mol}$	1.1	$\mu mol[CO_2]\ m^{-2}[leaf]\ s^{-1}$	Dark respiration at 25°C	a
J_{max25}	$467\rho_{Chl}$	$\mu mol[e^-]\ m^{-2}[leaf]\ s^{-1}$	Maximum electron transport rate at 25°C	a
		Radiation Parameters		
slo	0.5	–	Specific leaf orientation	c
δ	0.15	–	Scattering coefficient	b, d
k_{difBL}	0.8	–	Extinction coefficient diffuse PAR and black leaves	e
k_{dif}	$k_{difBL} \cdot \sqrt{1-\delta}$	–	Extinction coefficient diffuse PAR	c
k_{dirBL}	$\dfrac{slo}{\sin\beta}$	–	Extinction coefficient direct PAR and black leaves	c
k_{dir}	$k_{difBL} \cdot \sqrt{1-\delta}$	–	Extinction coefficient direct PAR	e
τ_{dif}	$e^{-k_{dif}\ LAI}$	–	Transmittance diffuse PAR	b
τ_{dirBL}	$e^{-k_{dirBL} \cdot LAI}$	–	Transmittance direct PAR and black leaves	b
τ_{dir}	$e^{-k_{dir} \cdot LAI}$	–	Transmittance direct PAR total	b
β_{dif}	$\dfrac{1-\sqrt{1-\delta}}{1+\sqrt{1-\delta}}$	–	Reflection coefficient canopy diffuse PAR	e
β_{dir}	$\dfrac{2}{1+\dfrac{k_{difBL}}{k_{dirBL}}} \cdot \beta_{dif}$	–	Reflection coefficient canopy direct PAR	e
$I_{P,o}$	$f_{par} \cdot I_o$	$W\ m^{-2}[gh]$	PAR outside greenhouse	e
$I_{P,dif,o}$	$f_{difpar} \cdot I_{P,o}$	$W\ m^{-2}[gh]$	Diffuse PAR outside greenhouse	e
$I_{P,dir,o}$	$I_{P,o} - I_{P,dif,o}$	$W\ m^{-2}[gh]$	Direct PAR outside greenhouse	e
$I_{P,dif}$	$\tau_{difR} \cdot \tau_{sc\_Is} \cdot I_{P,dif,o}$	$W\ m^{-2}[gh]$	Diffuse PAR inside greenhouse	
$I_{P,dir}$	$\tau_{dirR} \cdot \tau_{sc\_Is} \cdot I_{P,dir,o}$	$W\ m^{-2}[gh]$	Direct PAR inside greenhouse	
I_P	$f_{par} \cdot \tau_{difR} \cdot \tau_{sc\_Is} \cdot I_o$	$W\ m^{-2}[gh]$	PAR inside greenhouse	

(continued)

TABLE 8.11 (Continued)
Photosynthesis Model and General Parameters

Name	Value	Unit	Contents	Source Reference
		Temperature Parameters, Arrhenius Function		
E_C	59356	J mol^{-1}	Activation energy K_C Rubisco carboxylation	a
E_O	35948	J mol^{-1}	Activation energy K_O Rubisco oxygenation	a
E_M	39017	J mol^{-1}	Activation energy K_M Michaelis–Menten constant	
E_{VC}	58520	J mol^{-1}	Activation energy V_{Cmax} maximum carboxylation rate	a
E_D	66405	J mol^{-1}	Activation energy r_D dark respiration rate	a
E_J	37000	J mol^{-1}	Activation energy J_{max} maximum electron transport rate	a

Source: [a] Farquhar et al. (1980); [b] Gijzen (1994); [c] Spitters (1986); [d] Heuvelink (1996); [e] Goudriaan and Van Laar (1994).

Gross Assimilation and Dark Respiration

The gross canopy assimilation rate P_g is found by multiplying the gross leaf assimilation rate $P_{g,L}$ by the leaf area index LAI

$$P_g = P_{g,L} \cdot LAI \qquad (\text{kg[CO}_2\text{] m}^{-2}\text{[gh] s}^{-1}) \quad (8.107)$$

The canopy dark respiration rate r_D is equal to

$$r_D = r_{D,L} \cdot LAI \qquad (\text{kg[CO}_2\text{] m}^{-2}\text{[gh] s}^{-1}) \quad (8.108)$$

where $r_{D,L}$ (kg[CO_2] m^{-2}[leaf] s^{-1}) is the leaf dark respiration rate.

In general, the gross leaf assimilation rate $P_{g,L}$ is determined from the negative exponential light-response curve (Goudriaan and Van Laar, 1994)

$$P_{g,L} = P_g^{max} \cdot \left(1 - e^{-\frac{\varepsilon I_A}{P_g^{max}}}\right) \qquad (\text{kg[CO}_2\text{] m}^{-2}\text{[gh] s}^{-1}) \quad (8.109)$$

where P_g^{max} (kg[CO_2]m^{-2}[leaf] s^{-1}) is the maximum gross assimilation rate, ε (kg[CO_2] J^{-1}) is the light use efficiency by photorespiration and I_A (W m^{-2}[gh]) is the absorbed radiation.

The absorbed radiation I_A depends on the position of a leaf in the canopy. It is determined by the gradual extinction of radiation with canopy depth as a whole and by the leaves being either sunlit or shaded at any single level in the canopy. Therefore the assimilation rate is computed through a three-point Gaussian integration over the crop depth. The Gaussian integration determines the canopy assimilation rate from the average assimilation rate for three layers in the canopy. Two summation counters are used: $l_1 \in \{1,2,3\}$ for the integration over the canopy depth, and $l_2 \in \{1,2,3\}$ for the correction of $I_{A,ppd}$ for the canopy depth.

The values of the relative depth X_g of the canopy and the weight factor W_g needed for the three-point Gaussian integration are

$$X_g = \left\{0.5-\sqrt{0.15}\ \ 0.5\ \ 0.5+\sqrt{0.15}\right\} = \left\{0.1127 \quad 0.5 \quad 0.8873\right\} \qquad (-) \quad (8.110)$$

$$W_g = \left\{\frac{1}{3.6} \quad \frac{1.6}{3.6} \quad \frac{1}{3.6}\right\} = \left\{0.2778 \quad 0.4444 \quad 0.2778\right\} \qquad (-) \quad (8.111)$$

Note: if LAI is higher than 3, a five-point Gaussian integration should be used for accuracy.

The leaf area index LAI_l at layer l_1—used to determine the transmittance τ_{dif} and τ_{dir}—is a function of the depth in the canopy

$$LAI_l(l_1) = LAI \cdot X_g(l_1) \qquad (\mathrm{m^2[leaf]\ m^{-2}[gh]}) \quad (8.112)$$

The gross leaf assimilation rate $P_{g,L}$ is computed from the assimilation rate of the sunlit and the shaded part with the fraction sunlit leaf area f_{SLA} (–)

$$P_{g,L} = \sum_{l_1=1}^{3} W_g(l_1)\cdot\left(f_{SLA}\cdot P_{g,sun}(l_1)+\left(1-f_{SLA}\right)\cdot P_{g,shd}(l_1)\right) \qquad (\mathrm{kg[CO_2]\ m^{-2}[leaf]\ s^{-1}}) \quad (8.113)$$

in which the fraction sunlit leaf area $f_{SLA} = \tau_{dirBL}(l_1)$. This summation moves through the crop layers from top to bottom.

The gross assimilation rates $P_{g,sun}$ of the sunlit part and $P_{g,shd}$ of the shaded part at layer l_1 are defined by

$$P_{g,sun}(l_1) = P_g^{max}\cdot\sum_{l_2=1}^{3} W_g(l_2)\cdot\left(1-e^{-\frac{\varepsilon I_{A,sun}(l_1,l_2)}{P_g^{max}}}\right) \qquad (\mathrm{kg[CO_2]\ m^{-2}[leaf]\ s^{-1}}) \quad (8.114)$$

$$P_{g,shd}(l_1) = P_g^{max}\cdot\left(1-e^{-\frac{\varepsilon I_{A,shd}(l_1)}{P_g^{max}}}\right) \qquad (\mathrm{kg[CO_2]\ m^{-2}[leaf]\ s^{-1}}) \quad (8.115)$$

The absorbed radiation $I_{A,sun}$ of the sunlit part and $I_{A,shd}$ of the shaded part of the canopy can be defined as a function of various absorbed radiation terms (Spitters, 1986; Goudriaan and Van Laar, 1994)

$$I_{A,sun}(l_1,l_2) = I_{A,shd}(l_1) + I_{A,ppd}(l_1)\cdot X_g(l_2) \qquad (\mathrm{W\ m^{-2}[leaf]}) \quad (8.116)$$

$$I_{A,shd}(l_1) = I_{A,dif}(l_1) + I_{A,tdir}(l_1) - I_{A,dir}(l_1) \qquad (\mathrm{W\ m^{-2}[leaf]}) \quad (8.117)$$

in which the diffuse flux $I_{A,dif}$, the total direct flux $I_{A,tdir}$, the direct flux $I_{A,dir}$, and the direct flux of leaves perpendicular on the direct beam $I_{A,ppd}$ are given by

$$I_{A,dif}(l_1) = (1 - \beta_{dif}) \cdot I_{P,dif} \cdot k_{dif} \cdot \tau_{dif}(l_1) \qquad (\text{W m}^{-2}[\text{leaf}]) \quad (8.118)$$

$$I_{A,tdir}(l_1) = (1 - \beta_{dir}) \cdot I_{P,dir} \cdot k_{dir} \cdot \tau_{dir}(l_1) \qquad (\text{W m}^{-2}[\text{leaf}]) \quad (8.119)$$

$$I_{A,dir}(l_1) = (1 - \delta) \cdot I_{P,dir} \cdot k_{dirBL} \cdot \tau_{dirBL}(l_1) \qquad (\text{W m}^{-2}[\text{leaf}]) \quad (8.120)$$

$$I_{A,ppd}(l_1) = \frac{1-\delta}{\sin\beta} \cdot I_{P,dir} \qquad (\text{W m}^{-2}[\text{leaf}]) \quad (8.121)$$

The summation in Equation 8.114 is needed for the sunlit leaves. The sunlit part $I_{A,sun}$ of the absorbed radiation gives an average value over all sines of incidence of the direct beam on the leaves. Because in principle any sine of incidence can occur, this part has to be integrated separately.

Photosynthesis Parameters

The light use efficiency by photorespiration ε (kg[CO_2] J^{-1}) and the maximum gross assimilation rate P_g^{max} (kg[CO_2] m^{-2}[leaf] s^{-1}) depend on the photosynthesis parameters. The photosynthesis parameters depend on the CO_2 concentration $C_{CO_2,a}^{ppm}$ (μmol[CO_2] mol^{-1}[air]) in the greenhouse and the temperature of the crop T_c (K).

The light use efficiency by photorespiration ε (Goudriaan and Van Laar, 1994) is given by

$$\varepsilon = \psi \cdot \frac{M_{CO_2}}{4} \cdot \frac{\max\left(C_{CO_2,a}^{ppm}, \Gamma\right) - \Gamma}{\max\left(C_{CO_2,a}^{ppm}, \Gamma\right) + 2\Gamma} \qquad (\text{kg}[\text{CO}_2]\ \text{J}^{-1}) \quad (8.122)$$

in which the number of electrons (e^-) per fixed CO_2 is 4, where $C_{CO_2,a}^{ppm}$ (μmol[CO_2] mol^{-1}[air]) is the CO_2 concentration in the greenhouse, Γ (μmol[CO_2] mol^{-1}[air]) is the CO_2 compensation concentration, M_{CO_2} (kg[CO_2] μmol^{-1}[CO_2]) is the molar mass of CO_2, and ψ (μmol[e^-] J^{-1}) is the conversion factor from (J) to (μmol[e^-]).

The conversion factor ψ is

$$\psi = \frac{1 - F_p}{2} \cdot \zeta \qquad (\mu\text{mol}[e^-]\ \text{J}^{-1}) \quad (8.123)$$

in which the fraction PAR (photosynthesis active radiation) absorbed by nonphotosynthetic tissues $F_p = 0.3$, the number of electrons (e^-) per absorbed photon is 2, and the conversion factor $\zeta = 4.59$ μmol[photons] J^{-1}.

The CO_2 compensation concentration Γ in the absence of dark respiration (Farquhar et al., 1980) is defined by

$$\Gamma = \frac{K_C}{2K_O} \cdot p_{O_2,i} \cdot f_{OC} \qquad (\mu\text{mol}[\text{CO}_2]\ \text{mol}^{-1}[\text{air}]) \quad (8.124)$$

in which the O_2 partial pressure inside the stomata $p_{O_2,i}$ = 210 mbar and the ratio of V_{Omax} (maximum oxygenation rate) to V_{Cmax} (maximum carboxylation rate) $f_{OC} = \frac{V_{Omax}}{V_{Cmax}} = 0.21$ (which is assumed constant). The Michaelis–Menten constants K_C for Rubisco carboxylation and K_O for Rubisco oxygenation are given by

$$K_C = K_{C25} \cdot e^{E_C \cdot \frac{T_c - T_{25}}{T_c \cdot R_g \cdot T_{25}}} \quad (\mu bar) \quad (8.125)$$

$$K_O = K_{O25} \cdot e^{E_O \cdot \frac{T_c - T_{25}}{T_c \cdot R_g \cdot T_{25}}} \quad (mbar) \quad (8.126)$$

where T_c (K) is the temperature of the crop.

The maximum gross assimilation rate P_g^{max} is determined by adding the maximum net assimilation rate and the leaf dark respiration rate

$$P_g^{max} = P_n^{max} + r_{D,L} \quad (kg[CO_2]\ m^{-2}[leaf]\ s^{-1}) \quad (8.127)$$

The leaf dark respiration rate $r_{D,L}$ is given by

$$r_{D,L} = M_{CO_2} \cdot r_{D,L}^{\mu mol} \quad (kg[CO_2]\ m^{-2}[leaf]\ s^{-1}) \quad (8.128)$$

in which the leaf dark respiration rate $r_{D,L}^{\mu mol}$ (Farquhar et al., 1980) is

$$r_{D,L}^{\mu mol} = r_{D25,L}^{\mu mol} \cdot e^{E_D \cdot \frac{T_c - T_{25}}{T_c \cdot R_g \cdot T_{25}}} \quad (\mu mol[CO_2]\ m^{-2}[leaf]\ s^{-1}) \quad (8.129)$$

The maximum (light saturated) net assimilation rate P_n^{max} is a function of the maximum net assimilation rate P_{nc} limited by CO_2, the maximum endogenous photosynthetic capacity P_{mm}, and a factor Θ for the degree of curvature (Goudriaan and Van Laar, 1994)

$$P_n^{max} = \frac{P_{mm} + P_{nc} - \sqrt{\left(P_{mm} + P_{nc}\right)^2 - 4\Theta \cdot P_{mm} \cdot P_{nc}}}{2\Theta} \quad (kg[CO_2]\ m^{-2}[leaf]\ s^{-1}) \quad (8.130)$$

in which Θ = 0.7. The rate P_n^{max} is the solution of the nonrectangular hyperbola $\Theta \cdot P_n^{max^2} - (P_{mm} + P_{nc}) \cdot P_n^{max} + P_{mm} \cdot P_{nc} = 0$. This function gives a close approximation of the negative exponential function.

The maximum endogenous photosynthetic capacity P_{mm} is defined by

$$P_{mm} = \frac{M_{CO_2}}{4} \cdot J_{max} \quad (kg[CO_2]\ m^{-2}[leaf]\ s^{-1}) \quad (8.131)$$

in which the maximum electron transport rate J_{max} (Farquhar et al., 1980; Gijzen, 1994) is given by

$$J_{max} = J_{max\,25} \cdot e^{E_J \cdot \frac{T_c - T_{25}}{T_c \cdot R_g \cdot T_{25}}} \cdot \frac{1 + e^{\frac{S \cdot T_{25} - H}{R_g \cdot T_{25}}}}{1 + e^{\frac{S \cdot T_c - H}{R_g \cdot T_c}}} \qquad (\mu mol[e^-]\ m^{-2}[leaf]\ s^{-1}) \qquad (8.132)$$

and the constants $S = 710$ J mol^{-1} K^{-1} and $H = 220000$ J mol^{-1}.

The CO_2 limited rate P_{nc} of net photosynthesis (Goudriaan and Van Laar, 1994) is defined by

$$P_{nc} = \frac{f^{ppm}}{R_{CO_2,tot}} \cdot \left(\max\left(C^{ppm}_{CO_2,a}, \Gamma \right) - \Gamma \right) \qquad (kg[CO_2]\ m^{-2}[leaf]\ s^{-1}) \qquad (8.133)$$

where Γ (μmol[CO_2] mol^{-1}[air]) is the CO_2 compensation concentration in absence of dark respiration, $C^{ppm}_{CO_2,a}$ (μmol[CO_2] mol^{-1}[air]) is the CO_2 concentration in the greenhouse, and $R_{CO_2,tot}$ (s m^{-1}) is the total resistance to CO_2 diffusion.

To convert from the unit (μmol[CO_2] mol^{-1}[air]) to (kg[CO_2] m^{-3}[air]) we use the volume of 1 mol of gas (ideal gas law)

$$V^{mol} = \frac{n \cdot R_g \cdot T}{p_{bar}} = \frac{1 \cdot 8.314 \cdot T}{101325} \qquad (m^3[gas]\ mol^{-1}[gas]) \qquad (8.134)$$

and the molar mass of CO_2 $M_{CO_2} = 44 \times 10^{-9}$ kg[CO_2] μmol^{-1}[CO_2], so we find

$$f^{ppm} = \frac{M_{CO_2}}{V^{mol}} \qquad (kg[CO_2]\ mol[air]\ m^{-3}[air]\ \mu mol^{-1}[CO_2]) \qquad (8.135)$$

This is sometimes approximated in the literature by $f^{ppm} = 1.83 \times 10^{-6}$.

The total resistance to CO_2 diffusion $R_{CO_2,tot}$ is determined by adding stomatal, boundary layer, and carboxylation resistance

$$R_{CO_2,tot} = R_{CO_2,s} + R_{CO_2,b} + R_{CO_2,c} \qquad (s\ m^{-1}) \qquad (8.136)$$

The stomatal and boundary layer resistance to CO_2 diffusion $R_{CO_2,s}$ and $R_{CO_2,b}$ are computed from the stomatal and boundary layer resistance to H_2O diffusion $R_{H_2O,s}$ (Equation 8.100), and $R_{H_2O,b}$ (Equation 8.105) from Section 8.5.1. For CO_2 these resistances are larger than for water vapor because the diffusion coefficient is lower (Monteith and Unsworth, 1990).

$$R_{CO_2,s} = 1.6 R_{H_2O,s} \qquad (s\ m^{-1}) \qquad (8.137)$$

$$R_{CO_2,b} = 1.37 R_{H_2O,b} \qquad (s\ m^{-1}) \qquad (8.138)$$

The carboxylation resistance $R_{CO_2,c}$ (Goudriaan and Van Laar, 1994; Gijzen, 1994) is given by

$$R_{CO_2,c} = \frac{K_M}{V_{Cmax}} \cdot \frac{\rho^T_{CO_2}}{M_{CO_2}} \quad (\mathrm{s\ m^{-1}}). \quad (8.139)$$

where K_M is the effective Michaelis–Menten constant for carboxylation and V_{Cmax} is the maximum carboxylation rate (Farquhar et al., 1980)

$$K_M = K_C \cdot \left(1 + \frac{p_{O_2,i}}{K_O}\right) \quad (\mu\mathrm{bar}) \quad (8.140)$$

$$V_{Cmax} = V_{Cmax25} \cdot e^{E_{VC} \cdot \frac{T_c - T_{25}}{T_c \cdot R_g \cdot T_{25}}} \quad (\mu\mathrm{mol[CO_2]\ m^{-2}[leaf]\ s^{-1}}) \quad (8.141)$$

The CO_2 density $\rho^T_{CO_2}$ at temperature T_c (K) is defined by the law for ideal gas as

$$\rho^T_{CO_2} = \rho_{CO_2} \cdot \frac{T_0}{T_c} \quad (\mathrm{mg[CO_2]\ m^{-3}[CO_2]}) \quad (8.142)$$

where ρ_{CO_2} is the CO_2 density at T_0.

8.5.3 Temperature Integration

Temperature integration is used as a descriptive method for long-term temperature effects on crop development. A descriptive method is used because—to our best knowledge—no simple accurate models for crop development exist. More elaborate models for crop development do exist. These models, however, work on a larger time scale (days, up to 10-day periods), are too detailed (many crop development stages), are too crop-specific, or are not developed for greenhouse climate but for the open field (different temperature, humidity, and CO_2 conditions).

The temperature integration concept is based on the results of horticultural research, which indicates that crop growth responds to long-term average temperatures rather than specific day and night temperature profiles (Sigrimis et al., 2000). Photosynthesis is an almost instantaneous process, while the processing of the assimilates is a slower, dynamical process. It can be assumed that the crop stores the assimilates in a carbohydrate pool (Seginer et al., 1994). The capacity of the assimilate pool is crop specific and it probably differs for each development stage. Temperature integration is a simplified approach to the same theory. The buffering capacity is not specified in this concept, but it is assumed sufficient over a period of several days (De Koning, 1988). The concept is mainly based on empirical observations.

Much research has been done on temperature integration to describe crop development (Van den Bosch, 1998; Gijzen et al., 1998; Körner and Challa, 2003; Elings et al., 2005), and it is already in use by many commercial greenhouse horticulturists. The duration of the temperature integration and the boundary values described here are based on the research by Körner and Challa (2003), who developed temperature integration rules specifically for the solar greenhouse. The underlying assumption is that crop development is determined by an average temperature, rather than the actual temperature. In addition, it is assumed that temperature deviations that occurred long ago can no longer be compensated for—as far as their influence on crop development is concerned—and should therefore not be taken into account.

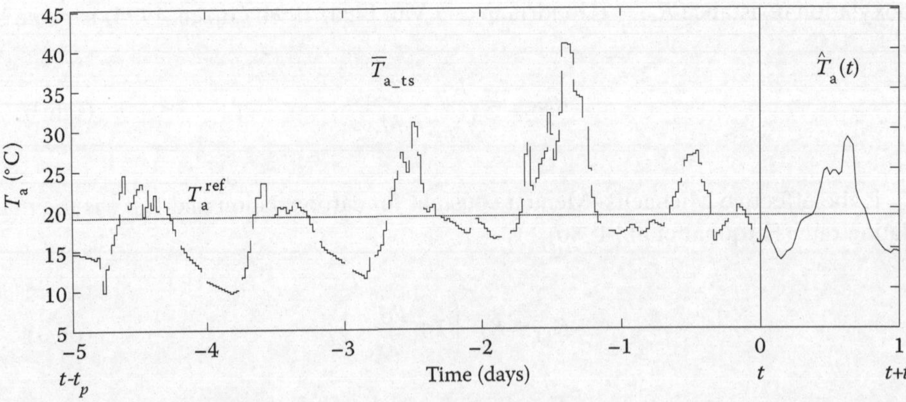

FIGURE 8.11 Temperature trajectory T_a* for temperature integral. *$T_a, \overline{T}_{a\_ts}, \hat{T}_a$, and T_a^{ref} in (K) in computations, in (°C) here for readability.

For the temperature integral used here, a time period of 6 days is considered. From these 6 days, 5 days (t_p (s)) are in the past, and 1 day (t_f (s)) is used to correct for this past.

The temperature integral is determined from the temperature of the indoor air T_a. An example is given in Figure 8.11. The average temperature of the indoor air $\overline{T}_{a\_ts}$ (dashed) is saved at every sampling interval t_s (1800 s) for the days in the past (t_p). The predicted temperature course $\hat{T}_a$ for the day in the future (t_f) is found by simulating the greenhouse-with-crop model during the next day. The temperature $T_a^{ref} = T_0 + 19$ K (solid line) is the reference temperature for the temperature integral.

The temperature integral trajectory S_T at time t is described by

$$S_T(t,\tau) = \frac{1}{n_{secs}} \cdot \int_0^{\tau} \left(\hat{T}_a(t,v) - T_a^{ref}\right) dv + S_{T0}(t) \quad \forall 0 \le \tau \le t_f \qquad \text{(K day)} \qquad (8.143)$$

in which $t_f = 1 n_{secs}$ s (1 day)* is the future horizon, where t (s) is the current time and $\hat{T}_a(t,v)$ (K) is the predicted temperature of the indoor air at time v based on information until time t. This gives a trajectory $S_T(t, \tau)$ for every time t, where τ runs from 0 to t_f (see Figure 8.12).

The initial value $S_{T0}(t)$ of the temperature integral is defined so that temperature deviations that occurred more than 5 days ago are not taken into account. The initial value of the temperature integral $S_{T0}(t)$ at time t (s) is therefore computed over the past horizon t_p (s)

$$S_{T0}(t) = \frac{1}{n_{secs}} \cdot \int_{t-tp}^{t} (T_a(v) - T_a^{ref})\, dv \qquad \text{(K day)} \qquad (8.144)$$

in which $t_p = 5 n_{secs}$ s (5 days) is the past horizon, where T_a (K) is the temperature of the indoor air.

This integral is approximated numerically by a summation, where the average temperature of the indoor air $\overline{T}_{a\_ts}$ is saved at every sampling interval t_s (1800 s)

* Because the unit of the temperature integral is (K day), time has to be converted from (s) to (day) with the number of seconds in a day $n_{secs} = 86400$ s day^{-1}.

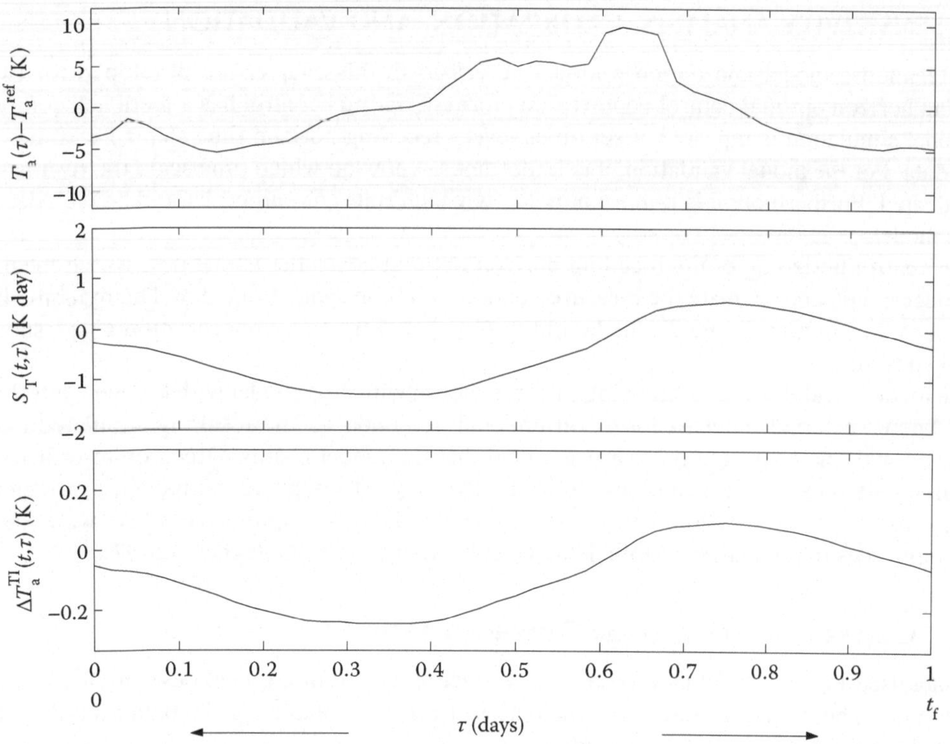

FIGURE 8.12 Temperature deviation $T_a - T_a^{ref}$, temperature integral S_T, and predicted average temperature deviation ΔT_a^{TI} at time t.

$$S_{T0}(t) = \frac{1}{n_{secs}} \cdot \sum_{k_v = \frac{t-t_p}{t_s}}^{\frac{t}{t_s}} (\overline{T}_{a\_ts}(k_v) - T_a^{ref}) \cdot t_s \qquad \text{(K day)} \quad (8.145)$$

where k_v is the discrete time step. In the example given in Figure 8.12 the initial value $S_{T0}(t) = -0.194$ K day. The initial value $S_{T0}(t)$ of the temperature integral is recomputed (and thus changed) at every time interval t_s.

From the temperature integral S_T in Equation 8.143 a predicted average temperature deviation ΔT_a^{TI} is computed, which will be used in the optimal control to ensure proper crop development.

The predicted average temperature deviation trajectory $\Delta T_a^{TI}(t,\tau)$ at time t is given by

$$\Delta T_a^{TI}(t,\tau) = \frac{S_T(t,\tau)}{\frac{t_p+\tau}{n_{secs}}} \quad \forall\, 0 \le \tau \le t_f \qquad \text{(K)} \quad (8.146)$$

This describes the average deviation between the past and forecasted temperatures T_a and $\hat{T}_a$ and the reference temperature T_a^{ref} (see Figure 8.12).

8.6 SENSITIVITY ANALYSIS, CALIBRATION, AND VALIDATION

The greenhouse model and the crop model described in this chapter are developed for use in a receding horizon optimal control context. This form of optimal control has a feedback component: the model simulation is repeated at set times (every receding horizon time step t_s) with new measured data. For the model validation, it is important to know on which time scale the results should be evaluated. Furthermore it is important to know which states or outputs should be well described by the model.

The control horizon t_f of the receding horizon optimal controller is one day, which means that the model should approximate the measured data over a time span of one day. The dynamic behavior of the temperatures T_a and T_c, the humidity $C_{H_2O,a}$, and the CO_2 concentration $C_{CO_2,a}$ should be described well.

The model is validated by simulating the greenhouse with crop model with known control inputs **u** and known external inputs **d** for a conventional greenhouse. The resulting simulated data are compared with the measured data. It was found that the model results agreed quite well with the measured data on some days, and less well on other days. To improve the model, parameter estimation was performed (Section 8.6.3). The parameters likely to improve the model were found by performing sensitivity analysis. More details can be found in Van Ooteghem (2007).

8.6.1 Conventional versus Solar Greenhouse Model

The data used for the model calibration and validation are from a greenhouse in Olsthoorn, The Netherlands. The sampling time Δt of the data is 1 min. The names of the data sets like *040323* indicate the measurement date (*yymmdd*).

8.6.1.1 Control Inputs

Some control inputs **u** are different from those defined in Table 8.2, e.g., instead of the valve positions vp_l and vp_u the water temperatures T_{in_l} and T_{in_u} (K) are measured. These differences are taken into account and corrected for. The upper heating net was not used in this greenhouse, so $vp_u = 0$.

8.6.1.2 External Inputs

All external inputs **d** used in the model are measured. From the first evaluation of the simulation results it was found that the sky temperature T_{sk} is not a measured but a computed value. It was computed with the equations given below, with a clouded fraction of the sky $c_{Tsk} = 1$. This means that it was assumed that the sky was 100% clouded (overcast sky). If this fraction is changed to 0.5 (partly cloudy sky), the computed sky temperature is 10°C lower than the "measured" sky temperature. This difference for the sky temperature was found to give a large difference in the simulation results. To correct for this, the sky temperature is computed. Because the clouded fraction of the sky c_{Tsk} is unknown, it is a parameter in the parameter estimation (Section 8.6.3).

The sky temperature T_{sk} can be computed from the temperature and the humidity of the outdoor air (Monteith and Unsworth, 1990).

$$T_{sk} = \sqrt[4]{(1-c_{Tsk}) \cdot E_{sky\_clear} \cdot T_0^4 + c_{Tsk} \cdot \left(T_0^4 - \frac{9}{\sigma} \right)} \quad \text{(K)} \quad (8.147)$$

in which the fictive emission coefficient E_{sky_clear} of the clear sky is given by

$$E_{sky\_clear} = 0.53 + 6 \times 10^{-3} \sqrt{p_{H_2O,o}} \quad (-) \quad (8.148)$$

where T_o (K) is the outdoor air temperature, $c_{T\mathrm{sk}} \in [0,1]$ (–) is the clouded fraction of the sky, σ (W m^{-2} K^{-4}) is the Stefan–Boltzmann constant and $p_{H_2O,o}$ (N m^{-2}) is the vapor pressure of the outdoor air.

8.6.1.3 States

The following states are measured, and can be compared with simulated results: temperatures of the indoor air below and above the screen, crop, and indoor side of the roof (T_a, T_{as}, T_c, T_{ri} (K)) and concentrations of CO_2 and H_2O below the screen ($C_{CO_2,a}$, $C_{H_2O,a}$ (kg m^{-3})).

8.6.2 Sensitivity Analysis

Sensitivity analysis is used to find model terms and parameters for which the model is most sensitive. This is used to find suitable model parameters to calibrate the model. It is also used to analyze sensitivity to the inputs. Parameters that are suitable for calibration are both sensitive and not well established. No adjustments are made to the well-established physical and physiological equations.

Two data sets were selected (*040323, 040925*). These data sets were expected to give good results for the sensitivity analysis, because they both show excitation of all the control and external inputs **u** and **d**. The sensitivity analysis is performed over a period of 2 days (*040322–040323*), where only the result of the second day is used for the analysis. This is done to make sure that the initial state values have little influence on the result.

Instead of changing the actual model terms and parameters themselves, they are multiplied by the parameters p, which have a nominal value of one. This introduces scaling of the model parameters, which is necessary because they are not in the same range (e.g., $c_{T\mathrm{sk}}$ and $\Phi_{CO_2}^{\max}$). Then the sensitivity to these parameters p is investigated. The results are called relative sensitivities (Bernaerts and Van Impe, 2004). The Fisher information matrix of the relative sensitivities is used to indicate which parameters are most important.

This procedure is first used to determine the sensitivity to the control inputs (vp_{CO_2}, Ap_{lsd}, Ap_{wsd}, Cl_{sc}, vp_{boil}, T_{boil}, $T_{\mathrm{in\_l}}$, $T_{\mathrm{in\_u}}$) and the external inputs (I_o, v_o, T_o, T_o^{wb}, T_{sk}, $C_{CO_2,o}$).

The most important inputs are

- *Control inputs* **u**: Ap_{lsd}, vp_{CO_2}, Cl_{sc}, T_{boil}, $T_{\mathrm{in\_l}}$, which means that Ap_{wsd}, vp_{boil}, and $T_{\mathrm{in\_u}}$ are less important. For $T_{\mathrm{in\_u}}$ it is obvious that it has no influence, because the upper net is not used.
- *External inputs* **d**: (I_o, v_o, T_o, T_o^{wb}, T_{sk}, $C_{CO_2,o}$), which means that they are all important.

For model validation and calibration it is important to have data of the sensitive control inputs **u** and the external inputs **d**. These are in general not parameters to adjust. The only exception in this case is the sky temperature T_{sk}, which is not measured but computed from measurements. Now the same procedure is used to determine the sensitivity of the 46 exchange terms in the model to determine the most important ones. The 11 most important exchange terms are:

- *Exchange terms* (Q, Φ): $Q_{\mathrm{in\_l}}$, $Q_{\mathrm{l\_out}}$, $Q_{\mathrm{rad\_c}}$, $Q_{\mathrm{ro\_o}}$, $Q_{\mathrm{ro\_sk}}$, $\Phi_{CO_2,\mathrm{a\_c}}$, $\Phi_{CO_2,\mathrm{as\_o}}$, $\Phi_{H_2O,\mathrm{as\_o}}$, $\Phi_{CO_2,\mathrm{in\_a}}$, $[Q_{\mathrm{as\_ri}}^{\mathrm{cons}}, \Phi_{H_2O,\mathrm{as\_ri}}]$, $[Q_{\mathrm{c\_a}}^{\mathrm{trans}}, \Phi_{H_2O,\mathrm{c\_a}}]$. The terms in brackets use the same parameter.

From these remaining 11 sensitive terms the underlying equations are studied in more detail to locate model parameters within these terms that are not well known. This leaves the following six parameters for calibration: $c_{T\mathrm{sk}}$, τ_{difR}, h_{win}, $F_{\mathrm{boil}}^{\max}$, $F_{\mathrm{l}}^{\max}$, and $\Phi_{CO_2}^{\max}$. These parameters are estimated to fit the model to the measured data.

8.6.3 Parameter Estimation

To calibrate the six parameters, the measured states (T_a, T_{as}, T_c, T_{ri}, $C_{CO_2,a}$, $C_{H_2O,a}$) have to be compared with the simulated states. The parameters are fitted on one day of data (*040323*). The initial state values are based on the measured values when they are available. Otherwise an estimate is made from the known values. It is assumed that the time constants are smaller than 1/3 of a day.

The following six parameter values are found from the parameter calibration: $c_{T\mathrm{sk}} = 0.6, \tau_{\mathrm{difR}} = 0.55, h_{\mathrm{win}} = 1, F_{\mathrm{boil}}^{\max} = 1.62 \times 10^{-3}, F_{\mathrm{l}}^{\max} = 1.62 \times 10^{-3}$, and $\Phi_{CO_2}^{\max} = 0.4 \times 10^{-5} A_s$. The measured and simulated states are given in Figure 8.13.

From these results it can be seen that after calibration the greenhouse with crop model gives a good description of the dynamic behavior. The temperatures T_a and T_c of the indoor air and the crop and the humidity $C_{H_2O,a}$ are well described. The CO_2 concentration $C_{CO_2,a}$ fits less well. Although the dynamics seem quite good, there is a static deviation during nighttime. Because the deviation is during the night, the problem is not that bad. The CO_2 concentration influences the total biomass B mainly through the photosynthesis rate P_g during daytime. These terms are important in the optimal control concept, so they should match well.

From the measured data it is found that the time constants of the processes are:

- 5 min: between outdoor shortwave solar radiation I_o and the temperature T_a of the indoor air below the screen
- 18 min: between CO_2 supply with valve vp_{CO_2} and the change in CO_2 concentration $C_{CO_2,a}$ of the indoor air
- 40 min: between the aperture Ap_{lsd} and Ap_{wsd} of the windows and the change in humidity $C_{H_2O,a}$

This indicates that the assumption that the time constants are smaller than $\frac{1}{3}$ of a day is valid.

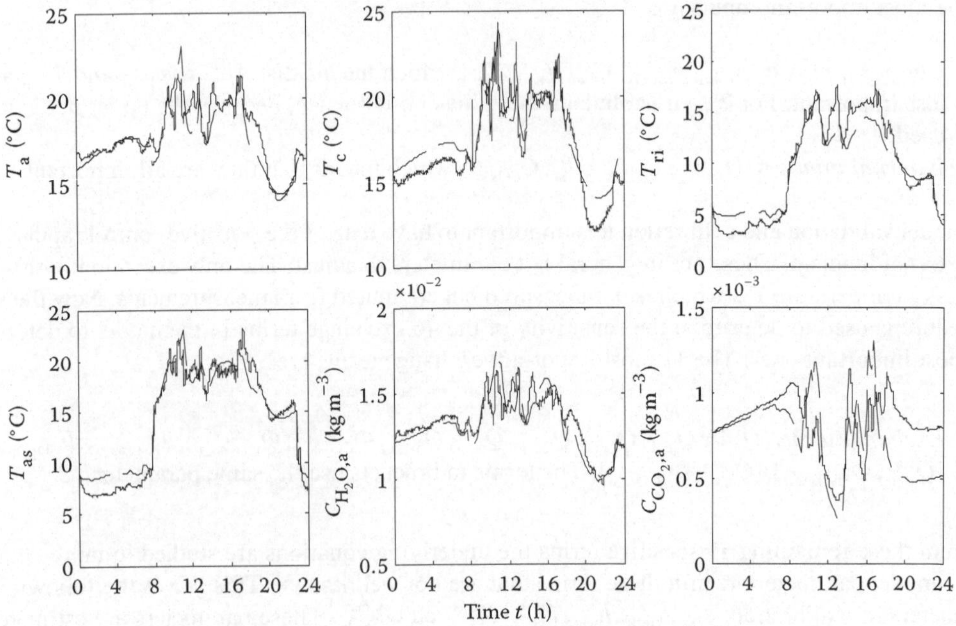

FIGURE 8.13 Estimation: measured (–) and simulated (– –) states **x**, data set *040323*.

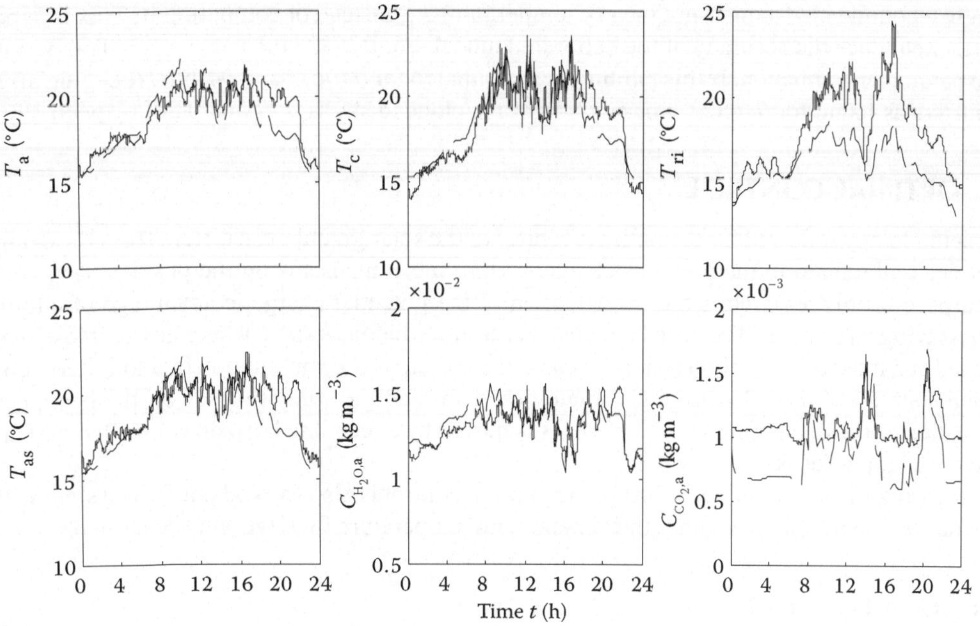

FIGURE 8.14 Validation: measured (–) and simulated (– –) states x, data set *040617*.

To verify the overall validity of the parameter values, they are subsequently used on other data sets in other seasons (*040617* and *040910*). These results are given in Figures 8.14 and 8.15. From these results, it can be seen that the simulations give a fair fit of the measurements. The deviation in the CO_2 concentration is again seen.

All calibrated parameters are assumed to have a fixed value during the day. The parameter $c_{T\mathrm{sk}}$ describes the fraction of clouded sky, which is unlikely to stay the same all day. It is therefore

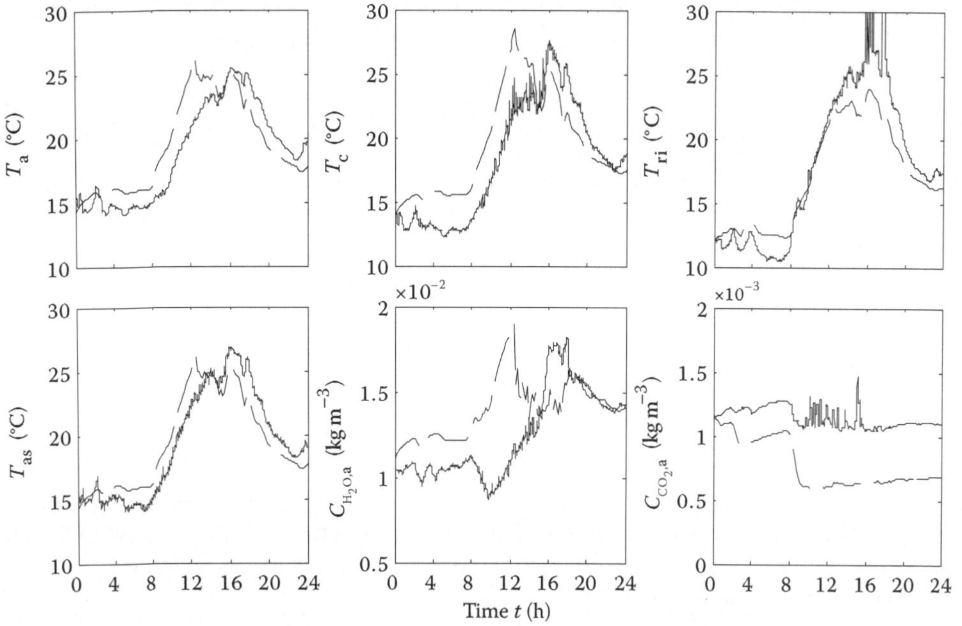

FIGURE 8.15 Validation: measured (–) and simulated (– –) states **x**, data set *040910*.

strongly recommended to measure the sky temperature T_{sk}, instead of computing it. This is expected to further enhance the accuracy of the calibrated model. For *040323* the value $c_{Tsk} = 0.60$ was found in the parameter estimation. If this parameter is estimated for *040617* and *040910* (keeping all other parameters as found for *040323*), the fraction of the clouded sky is 0.77 and 0.24, respectively.

8.7 OPTIMAL CONTROL

The methodology followed for the optimal control of the solar greenhouse differs from the examples on lettuce and tomato in the previous chapters. Here the emphasis is on the physics part, because the purpose of this research is the investigation of the potential energetic advantage of additional energy-saving equipment. The crop is modeled as a single biomass state, where any increase in mass is valued at a fixed price as soon as it is produced. So, there is no computation of a long-term costate development, as with the lettuce and tomato cases. Instead the study focuses on the best possible exploitation of the fluctuating weather. This is achieved by a receding horizon controller, which acts as a closed loop solution.

In the work of Van Ooteghem (2007), the optimal control is also worked out for a system without temperature integration. Here only the version with temperature integration is described.

8.7.1 Cost Function

Optimal control is concerned with the computation of optimal control input trajectories based on a cost function. The control solution consists of actuator trajectories (e.g., window apertures and valve positions) that result in state trajectories (e.g., temperature, humidity, and CO_2 concentration) that optimize a cost function. The aim is to minimize fossil energy consumption, while maximizing biomass and keeping temperature and relative humidity within certain bounds. In the cost function, costs are defined to penalize fossil energy consumption, to reward biomass increase and to keep temperature, humidity, temperature integral, and the aquifer energy content within bounds.

Using a state-space greenhouse-with-crop model describing the dynamic behavior of the greenhouse (Section 8.4) and the crop (Section 8.5) in time together with weather predictions (*SELyear*; Breuer and Van de Braak, 1989), the influence of control changes on greenhouse climate can be simulated. The state-space model has the general form (see also Equation 4.2)

$$\dot{\mathbf{x}} = \mathbf{f}(t, \mathbf{x}, \mathbf{u}, \mathbf{d}) \tag{8.149}$$

where t is time, $\mathbf{x} = \mathbf{x}(t) \in \mathbb{R}^{n_x}$ is the state vector, $\mathbf{u} = \mathbf{u}(t) \in \mathbb{R}^{n_u}$ is the control input vector, $\mathbf{d} = \mathbf{d}(t) \in \mathbb{R}^{n_d}$ is the external input vector, and $\mathbf{f}$ is a nonlinear vector function. The description of these variables is given in Table 8.2.

The goal is to minimize the cost function, which has the general form (see also Equation 4.3)

$$J(\mathbf{u}) = \Phi(\mathbf{x}, t_f) + \int_{t_0}^{t_f} L(\mathbf{x}, \mathbf{u}, t)\,\mathrm{d}t \qquad \text{(cost)} \quad (8.150)$$

where the terminal cost $\Phi : \mathbb{R}^{n_x+1} \rightarrow \mathbb{R}$ and the running costs $L : \mathbb{R}^{n_x+n_u+n_d+1} \rightarrow \mathbb{R}$ are differentiable a sufficient number of times with respect to their arguments. The final time t_f is set to the control horizon, which is equal to one day and therefore will not be subject to optimization.

The control inputs are constrained by

$$u_i^{\min} \le u_i(\tau) \le u_i^{\max} \quad i = 1, \ldots, n_u; t_0 \le \tau \le t_f \tag{8.151}$$

A control input trajectory $u(\tau)$ that satisfies the constraints in Equation 8.151 is called admissible. For the states there are trajectory constraints (bounds, see Equation 8.155), which are considered "soft." With these prerequisites the control problem is to find

$$\mathbf{u}^*(\tau) = \arg\min_{\mathbf{u}} J(\mathbf{u}) \tag{8.152}$$

given the expected external inputs (weather prediction) $\tilde{\mathbf{d}}(\tau)$ for $\tau \in [t_0, t_f]$, subject to the differential equations (Equation 8.149) and the control input constraints (Equation 8.151). In other words, the objective is to find admissible input trajectories $\mathbf{u}^*(\tau)$ on the time interval $\tau \in [t_0, t_f]$ such that the process given by Equation 8.149 has control and state trajectories that minimize the performance criterion (cost function value) J. The resulting control input and state trajectories are referred to as the optimal trajectories.

The values used for the weight factors c and the bounds (to be defined below) in the cost function are given in Table 8.12. Some terms are taken per square meter to enhance the portability of the cost function to other greenhouse dimensions. The weight factors indicate how important specific greenhouse conditions are; they however do not represent euros or dollars. They have to be balanced such that one penalty does not outweigh another penalty. The weight factors have been tuned based on open loop computations of single days throughout the year to make sure that they hold in different seasons.

The terminal cost Φ is determined by the yield in the form of biomass B (Equation 8.15) and the average temperature deviation ΔT_a^{TI} at the end of the control horizon t_f (determined from the temperature integral S_T in Equation 8.14)

$$\Phi(\mathbf{x}, t_f) = \underbrace{-c_B \cdot (B(t_f) - B(t_0))}_{\Phi_B(\mathbf{x}, t_f)} + \underbrace{c_{TI} \cdot \left|\Delta T_a^{TI}(t_f)\right|}_{\Phi_{TI}(\mathbf{x}, t_f)} \quad \text{(cost)} \tag{8.153}$$

TABLE 8.12
Cost Function: Weight Factors and Bounds

Symbol	Unit	x^{min}	x^{max}	$\frac{\text{cost}}{\text{day} \cdot \text{unit}}$	$J(u)$
$T_a^{C\,\dagger}$	°C	10	34	$c_T = 5$	$\int P_{Ta}\,dt$
RH_a	%	–	85	$c_{RH} = 5$	$\int P_{RHa}\,dt$
$\Delta T_{a\_TI}^{C\,\dagger}$	°C	–6	6	$c_{TI} = 25$	$\int P_{TI}\,dt$
E_{aq}	J m^{-2}[gh]	$E_{aq}^{min\,\ddagger}$	$E_{aq}^{max\,\ddagger}$	$c_{aq} = 10 \times 10^6$	$\int P_{aq}\,dt$
Q_{used}	W m^{-2}[gh]			$c_Q = 61.44$	$\int L_Q\,dt$
$C_{CO_2,a}^{ppm}$	µmol[CO_2] mol^{-1}[air]	320	1000	$c_{CO_2} = 0$	0
B	kg m^{-2}[gh]			$c_B = 76.8$	Φ_B
$T_a^{C,ref\,\dagger}$	°C	19		$c_{TI} = 25$	Φ_{TI}

† T_a, ΔT_a^{TI} and T_a^{ref} in (K) in computations, in (°C) here for readability.
‡ Aquifer energy content bounds are derived in Section 8.7.1.1.

where c_B and c_{TI} are the weight factors for biomass and temperature integral. Terminal cost Φ_B should preferably be large and negative and Φ_{TI} should be zero. These terminal costs are used as soft terminal constraints.

The running cost L contains penalties P for the loss of crop yield due to exceeding temperature, humidity and temperature integration bounds, exceeding the aquifer energy content bounds, and the cost of energy. To this end the running cost L (cost s^{-1}) is given by the sum of the penalties for temperature T_a (P_{Ta}), relative humidity RH_a (P_{RHa}), temperature integral ΔT_a^{TI} (P_{TI}), year-round aquifer energy content E_{aq} (P_{aq}), and the costs for the energy consumption Q_{used} (L_Q)

$$L(\mathbf{x},\mathbf{u},t) = P_{Ta}(\mathbf{x},\mathbf{u},t) + P_{RHa}(\mathbf{x},\mathbf{u},t) + P_{TI}(\mathbf{x},\mathbf{u},t) + P_{aq}(\mathbf{x},\mathbf{u},t) + L_Q(\mathbf{x},\mathbf{u},t) \quad \text{(cost s}^{-1}) \quad (8.154)$$

The penalties for temperature P_{Ta}, relative humidity P_{RHa}, temperature integral P_{TI}, and aquifer energy content P_{aq} are given by

$$P_x(x,u,t) = \begin{cases} c_x \cdot \left| x^{min} - x(t) \right| & x^{min} > x(t) \\ 0 & x^{min} \le x(t) \le x^{max} \\ c_x \cdot \left| x^{max} - x(t) \right| & x(t) > x^{max} \end{cases} \quad \text{(cost s}^{-1})$$

where c_x is the weight factor associated with exceeding the boundary values x^{min} and x^{max} of state x. The penalty function increases linearly in value with the deviation from the boundary values. In between the boundary values, the penalty function is zero. This is implemented as a smoothed version of this equation

$$P_x(x,u,t) = \frac{c_x}{2} \cdot \left(\sqrt{\left(x^{min} - x(t)\right)^2 + \beta} + \sqrt{\left(x^{max} - x(t)\right)^2 + \beta} - (x^{max} - x^{min}) \right) \quad \text{(cost s}^{-1}) \quad (8.155)$$

in which $\beta = 1 \times 10^{-3}$. This function is smooth around x^{min} and x^{max}. These penalties are used as soft constraints.

The cost for the energy consumption L_Q is given by

$$L_Q(\mathbf{x},\mathbf{u},t) = c_Q \cdot Q_{used} \quad \text{(cost s}^{-1}) \quad (8.156)$$

where c_Q is the weight factor for energy use and Q_{used} is the total amount of energy used per square meter greenhouse. This is defined as

$$Q_{used} = \frac{Q_{boil} + Q_{hp}}{A_s} \quad \text{(W m}^{-2}\text{[gh])} \quad (8.157)$$

where Q_{boil} (W) is the energy used by the boiler, Q_{hp} (W) is the energy used by the heat pump, and A_s (m^2[gh]) is the surface area of the soil. The energy Q_{used} is a measure for the total gas use per square meter greenhouse surface.

There is no penalty on the CO_2 concentration ($c_{CO_2} = 0$); the bounds are used for the proportional controller (see Equation 8.168).

The boundary values for temperature and temperature integral are taken from Körner (2003). Because temperature integration is used, the temperature bounds can be quite wide, as the temperature integral will keep the average temperature at its reference value T_a^{ref}.

From the penalties, running costs, and terminal costs given here, the running cost L_Q represents gas use, the penalty P_{aq} represents aquifer energy content and all other penalties (P_{Ta}, P_{RHa}, P_{TI}) and terminal costs (Φ_B, Φ_{TI}) represent terms that are important for crop growth and development.

The temperature integral is a special case. It is used as a descriptive method for long-term temperature effects on crop development, to compensate for the lack of a reliable long-term crop model. The temperature integral bounds are used to ensure proper crop development by safeguarding the system from moving into physiologically unattractive or unacceptable regions. The aims for the temperature integral are to keep the average temperature deviation ΔT_a^{TI} within the boundary values of ±6 K (see Table 8.12) and to obtain an average temperature deviation of zero at the end of the control horizon of one day, so

$$\begin{aligned} -6 \le \Delta T_a^{TI}(t,\tau) \le 6 \qquad & \forall 0 \le \tau \le t_f \\ \Delta T_a^{TI}(t_f) = 0 \qquad & \end{aligned} \qquad \text{(K)} \quad (8.158)$$

This is translated to two terms in the cost function: the terminal cost Φ_{TI}, which penalizes long-term crop processes, and the penalty $\int P_{TI}$, which penalizes short-term crop processes. These penalties both depend on the predicted average temperature deviation trajectory ΔT_a^{TI} in Equation 8.146.

The other special case is the aquifer energy content. There are long-term governmental requirements that state that the aquifer has to be approximately energy neutral year-round. Because we are only looking one day ahead with our RHOC, control this poses a bit of a problem. To incorporate this demand in the RHOC cost function, some a priori knowledge about the course of the aquifer energy content has to be used. This a priori knowledge is then translated to bounds on the aquifer energy content, which are used as constraints in the cost function. This is explained in the next section.

8.7.1.1 Derivation Bounds for Aquifer Energy Content

An aquifer is a formation of water-bearing sand material in the soil that can contain and transmit water. The aquifer has a warm and a cold side. The warm water is used by the heat pump to heat the greenhouse and the cold water is used by the heat exchanger to cool the greenhouse.

The aquifer must be approximately energy neutral year-round. This means that the amount of energy put into the aquifer must equal the amount of energy withdrawn from the aquifer. If this demand is not fulfilled the aquifer will warm up or cool down, which is unwanted. Therefore this demand must be incorporated in the cost function of the optimal control.

In the receding horizon control, the control horizon is one day. This means that the aquifer energy content cannot be directly computed for a period of one year in the optimal control procedure. A solution is found in which a year-round reference curve for the accumulated energy content of the aquifer is used. The reference curve is based on a year-round optimal control run with the grid search method (see Section 8.7.3.1). It gives an indication of what the energy content will look like. Relative to this reference curve bounds are defined which represent the freedom to deviate from this curve. These bounds for the accumulated energy content of the aquifer can then be used as optimal control bounds. The energy accumulated in the aquifer must stay between these bounds. The bounds are time-dependent because the reference curve is not a constant.

It is assumed that the aquifer has an infinite amount of warm and cold water available. The aquifer energy content bounds will limit the amount of energy that can be stored or retrieved. This indirectly corrects for the fact that the buffers are not infinite.

Government Demands

The government demands that the aquifer is energy neutral year-round. At any arbitrary reference date, the aquifer will have a specific initial energy content. Starting from that date the accumulated energy is monitored to make sure that the amount of energy stored in the aquifer is equal to the amount of energy retrieved from the aquifer over a period of one year.

If $E_{aq}(t)$ (J m^{-2}[gh]) describes the accumulated energy content of the aquifer over the period t, the government demand is

$$E_{aq}(n_{secs\_yr}) = 0 \qquad \text{(J m}^{-2}\text{[gh])} \quad (8.159)$$

in which $n_{secs\_yr} = 31536000$ s yr^{-1} is the number of seconds in a year.

The energy content E_{aq} of the aquifer, accumulated over a period t is given by

$$E_{aq}(t) = \underbrace{\frac{1}{A_s} \cdot \int_0^t Q_{he}\, dt}_{E_{he}(t)} - \underbrace{\frac{1}{A_s} \cdot \int_0^t Q_c\, dt}_{E_{hp}(t)} + E_{aq0} \qquad \text{(J m}^{-2}\text{[gh])} \quad (8.160)$$

in which E_{he} is the amount of energy extracted from the greenhouse by the heat exchanger and stored in the aquifer, E_{hp} is the amount of energy supplied to the greenhouse by the heat pump and retrieved from the aquifer, and E_{aq0} (J m^{-2}[gh]) is the initial value of the accumulated energy content of the aquifer. This is the energy that has been accumulated in the previous period, which should still be corrected for by the optimal control. At the first start of the aquifer use, $E_{aq0} = 0$ (no energy accumulated yet). The energy transport term Q_{he} (W) is the heat extracted from the greenhouse and supplied to the aquifer by the heat exchanger. The energy transport term Q_c (W) is the heat retrieved from the aquifer and supplied to the greenhouse by the heat pump.

In the receding horizon control, the control horizon is one day, which means that a demand for a year cannot be implemented directly. The actual requirement by the government is not quite as strict as defined in Equation 8.159. A grower should not deplete or warm up the aquifer too much, so the deviation of the energy content year-round should be within limits, such that it can be corrected during the next year. It is necessary to know the shape of the energy content curve to define limits relative to this curve, which can serve as bounds for the optimal control. This is the topic of the next section.

The Energy Content of the Aquifer

To achieve that the aquifer is energy neutral year-round, the optimal control needs a function that describes the bounds for the amount of energy stored in the aquifer as a function of time. To obtain a reference curve for the energy content $E^{\star}_{aq}$ year-round, the energy content is determined with the weather data from the *SELyear* (Breuer and Van de Braak, 1989) and the receding horizon optimal control with grid search as described in Section 8.7.3.1. The initial version of this reference curve has been developed by Van Dongen (2004).

The amounts of energy E_{he} stored in the aquifer and E_{hp} retrieved by the heat pump are given in Figure 8.16. The heat exchanger is used in spring and summer (May through August). This results in the energy curve for E_{he} stored by the heat exchanger shown in Figure 8.16a, which has a clear S shape. The heat pump is used intensively in fall and winter to heat the greenhouse, and slightly less in spring and summer to reduce the relative humidity in the greenhouse. This results in the energy curve for E_{hp} retrieved by the heat pump shown in Figure 8.16b, which is almost linear with time with a slight S shape.

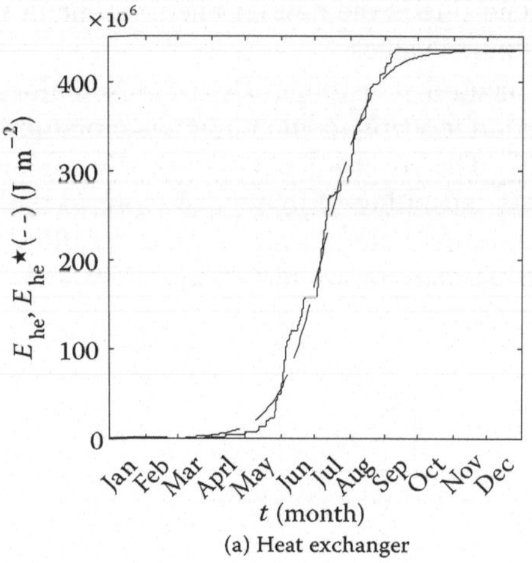

FIGURE 8.16 Energy extraction and supply for the aquifer with *SELyear*, computed (—) and estimated (– –) curves.

From the initial run it was found that more energy is retrieved from the aquifer (E_{hp}) than stored in the aquifer (E_{he}). The aim is to get an energy content reference curve $E_{aq}^{\star}$ that is equal to zero at the end of the year. This means that the energy curves E_{hp} and E_{he} should have the same value at the end of the year. The energy E_{he} is therefore scaled to match E_{hp}. It is assumed that the optimal control can increase the amount of energy E_{he} supplied to the aquifer by the heat exchanger to match the amount of energy E_{hp} retrieved from the aquifer by using the heat exchanger more intensively.

To find a function for the amount of energy E_{aq}, functions are estimated for E_{he} and E_{hp}. The energy E_{he} stored by the heat exchanger is approximated by an S-shaped curve. The energy E_{hp} retrieved by the heat pump is approximated by a linear function in combination with an S-shaped curve. This gives

$$E_{he}^{\star}(t) = \frac{0.5\max(E_{he})}{1+e^{-p_{he,a}\cdot\left(\frac{t}{n_{secs\_yr}}-p_{he,b}\right)}} \qquad (\mathrm{J\ m^{-2}[gh]}) \quad (8.161)$$

$$E_{hp}^{\star}(t) = \frac{\max(E_{he})-p_{hp,c}}{1+e^{-p_{hp,a}\cdot\left(\frac{t}{n_{secs\_yr}}-p_{hp,b}\right)}} + p_{hp,c}\cdot\frac{t}{n_{secs\_yr}} \qquad (\mathrm{J\ m^{-2}[gh]}) \quad (8.162)$$

in which $\max(E_{he}) = 435 \times 10^6$ J m^{-2}[gh] and the fraction of the year is $\frac{t}{n_{secs\_yr}}$, where t (s) is time, n_{secs_yr} (s yr^{-1}) is the number of seconds per year and p_{he} and p_{hp} are the parameters for the heat exchanger and the heat pump curve. Parameter calibration gives the following values: $p_{he,a} = 18$, $p_{he,b} = 0.52$, $p_{hp,a} = 9$, $p_{hp,b} = 0.50$, and $p_{hp,c} = 610 \times 10^6$. The estimated curves are given in Figure 8.16 as dashed lines.

The estimated function for the amount of energy $E_{aq}^{\star}$ stored in the aquifer as a function of time is given by

$$E_{ap}^{\star}(t) = E_{he}^{\star}(t) - E_{hp}^{\star}(t) \qquad (\mathrm{J\ m^{-2}[gh]}) \quad (8.163)$$

The result is shown in Figure 8.17, where the dashed line is the aquifer energy content reference curve $E^{\star}_{aq}$ and the line represents the originally computed values.

The aquifer does not have to be exactly energy neutral year-round. Bounds are defined around the aquifer energy content reference curve. These aquifer energy content reference bounds are relative to the aquifer energy content reference curve $E^{\star}_{aq}$. The government does not allow net heat storage in the aquifer year-round. The energy content is thus allowed to deviate more to the negative side than to the positive side. Furthermore the bounds are wider during summer to allow for more deviation in the period that energy is harvested. The bounds are set to E^{min}_{aq} and E^{max}_{aq}

$$E^{min}_{aq}(t) = E^{\star}_{aq}(t) - \left(200 \times 10^6 \sin\left(\frac{t}{n_{secs\_yr}} \cdot \pi\right) + 100 \times 10^6\right) \quad (\mathrm{J\ m^{-2}[gh]}) \quad (8.164)$$

$$E^{max}_{aq}(t) = E^{\star}_{aq}(t) + \left(200 \times 10^6 \sin\left(\frac{t}{n_{secs\_yr}} \cdot \pi\right) + 75 \times 10^6\right) \quad (\mathrm{J\ m^{-2}[gh]}) \quad (8.165)$$

where $\frac{t}{n_{secs\_yr}}$ is the fraction of the year, t (s) is time, and n_{secs_yr} (s yr^{-1}) is the number of seconds per year.

The resulting demand for the optimal control is to keep the aquifer energy content E_{aq} between these bounds

$$E^{min}_{aq} \leq E_{aq}(t) \leq E^{max}_{aq} \quad (\mathrm{J\ m^{-2}[gh]}) \quad (8.166)$$

These bounds are shown in Figure 8.17 as dashed lines.

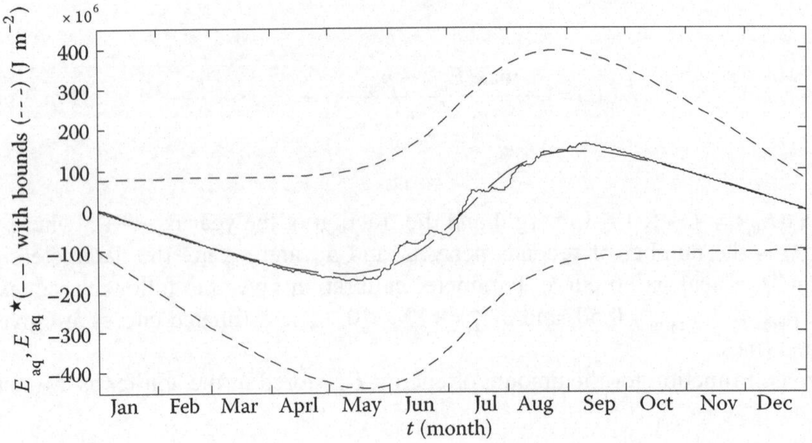

FIGURE 8.17 Aquifer energy content with *SELyear*, computed (—) and estimated curve (– –) and bounds (- - -).

The impact of the penalty $\int P_{aq}$ for the aquifer energy content has been tested by running the solar greenhouse optimization for a week in summer and winter. In summer, the initial aquifer energy content E_{aq} was set to a value above its upper bound E_{aq}^{max}. This prevented the use of the heat exchanger; the greenhouse was cooled by opening the windows. In winter, the aquifer energy content E_{aq} was set to a value below its lower bound E_{aq}^{min}. This prevented the use of the heat pump; the greenhouse was heated by the boiler. The difference between the bound and the initial value was chosen small enough to see that when the energy content was between the bounds again, the heat exchanger c.q. heat pump were used again. This indicates that the penalty is working correctly.

8.7.2 Receding Horizon Optimal Control

The knowledge about the greenhouse and crop has been incorporated in a dynamic model which has been described in Sections 8.4 and 8.5. Optimal control uses this model and furthermore requires that the control objectives—such as maximizing crop growth and minimizing gas use—are quantified and made explicit in the cost function. For maximizing crop growth the biomass increase must be maximized, while the temperature, the temperature integral, and the relative humidity are kept within bounds to obtain good development conditions and to decrease the risk for diseases and fungi. These bounds are translated to penalties, which are used as soft constraints.

The receding horizon optimal control (RHOC) concept is used to compute the control inputs u (actuator trajectories) (see Section 4.3.1). A conjugate gradient method in combination with a line search method is used to improve the search direction (see Section 3.3.1).

In the weather three time scales can be distinguished: the fast variation of the weather on a minute scale, the slow variation of the weather on a daily scale, and the seasonal variation on a monthly scale. Also in the crop growth process different time scales are found: the crop has a day–night rhythm in which the assimilates that have been formed during the day are converted to biomass during the night. Furthermore, it has several development stages, from initial growth from a seed to the formation of leaves, buds, flowers, and fruit up until fruit maturity.

The crop dry weight is a slow state, which depends heavily on the greenhouse temperature, which is a fast state. As the temperature in the solar greenhouse is expected to fluctuate a lot due to the aims of the optimal control, it does not make sense to incorporate the time scale decomposition for the crop dry weight. The same holds for the temperature integral and the loading and unloading of the aquifer heat and cold content, which also strongly depend on the greenhouse temperature.

The time-scale decomposition described in Section 5.2.5 is therefore not used in this research. It is assumed that the increase of crop biomass is instantaneous. We want to make maximum use of all small momentary variations in the weather to maximize the solar radiation absorbed by the crop (thus producing biomass) and to optimize the temperature in the greenhouse with minimum use of heating input (gas). The offline computation of the slow dynamics, which would be based on averaged weather, is therefore omitted.

A feasibility study of the optimal control of the solar greenhouse is performed in this section. Unfortunately the solar greenhouse only exists on paper, so the results of the feasibility study are entirely based on simulations. In these simulations, we tried to mimic reality as closely as possible. Year-round simulations have been performed with a receding horizon optimal controller to control the greenhouse, where the actual weather is different from the forecasted weather that is used for the on line optimal control computations (as in reality). This long period will provide insight concerning the use of the heat pump, boiler, and heat exchanger in the different seasons of the year.

The initial guess of the optimal control input $\mathbf{u}(t)$ is a very important step. If the guess of the optimal control is poor the algorithm needs more iterations (computation time) to find the solution or it may not find a solution at all. Also depending on the control guess the algorithm may find a local solution. To make sure that the optimal control starts with a good initial guess a grid search method is used to find good control input trajectories (see Section 8.7.3.1).

General conclusions with discussion on the optimal control are given in Section 8.7.8.

Time Intervals RHOC

The receding horizon optimal controller uses a number of time intervals, which are defined as

- $t_f = 86400$ s (one day) *control horizon*: the time interval over which the optimal control input trajectories are computed
- $t_{s,u} = 1800$ s (30 min) *time interval u**: the time interval over which the computed optimal control inputs u^* are kept piecewise constant
- $t_s = 1800$ s (30 min) *sampling interval RHOC*: the time interval between the RHOC computations (time shift)
- $t_{s,RK} = 60$ s (1 min) *integration time interval Runge-Kutta*: the time interval for the Runge-Kutta integration

This means that $\frac{t_f}{t_{s,u}} = 48$ values are determined for each control input at each receding horizon time step. For the sampling interval t_s a relatively large value is chosen, because the values of the weather conditions in our case are hourly values (*SELyear*). In a setup where the actual weather conditions are measured, the sampling interval t_s should be as small as the sampling interval of the weather observations (e.g., 2 min) (Van Willigenburg et al., 2000). The small integration time interval used in the Runge-Kutta integration ensures that faster dynamics are correctly incorporated in the computed results. Smaller time intervals (t_s, $t_{s,u}$) or a longer control horizon (t_f) will result in a longer computation time.

8.7.3 Control Inputs

In the first tests of the optimal control all control inputs were optimized by the optimal control. In a number of computations, the optimal control results would yield control inputs where heating and cooling were used at the same time. In view of the cost function, as it has been defined, this was unexpected. Evaluation of these results led to the conclusion that the optimal control got stuck in a local minimum. The same would hold for the window aperture: opening the lee- or windward-side windows makes no difference in the results (but it does require extra computations). Therefore these control inputs are coupled into a combined control input, which is optimized by the optimal control.

A number of control inputs are computed by the optimal control, while other control inputs are determined directly from other control inputs, external inputs or states:

- Computed by the optimal control (see Figure 8.18a): the valve positions for heating and cooling with the boiler (vp_l, vp_u), heat pump (vp^*_{hp}), and heat exchanger (vp^*_{he}), the window apertures (Ap_{lsd}, Ap_{wsd}) and the option ventilation with heat recovery (op_{vhr})
- Determined directly from other control inputs, external inputs, or states (see Figure 8.18b): the valve position for CO_2 supply (vp_{CO_2}) and the thermal screen closure (Cl_{sc})

In Figure 8.2 this difference is denoted by dotted boxes (set by optimal control) and dash-dotted boxes (directly derived from other inputs).

Control Inputs Set by Optimal Control

The control inputs computed by the optimal control are combined into two control inputs:

- The combined heating valve position vp_h [–1,2], which determines the valve positions vp_l, vp_u, vp^*_{hp}, and vp^*_{he}, as well as the option ventilation with heat recovery op_{vhr}
- The combined window aperture Ap_{csd} [0,2], which determines the window apertures Ap_{lsd} and Ap_{wsd}

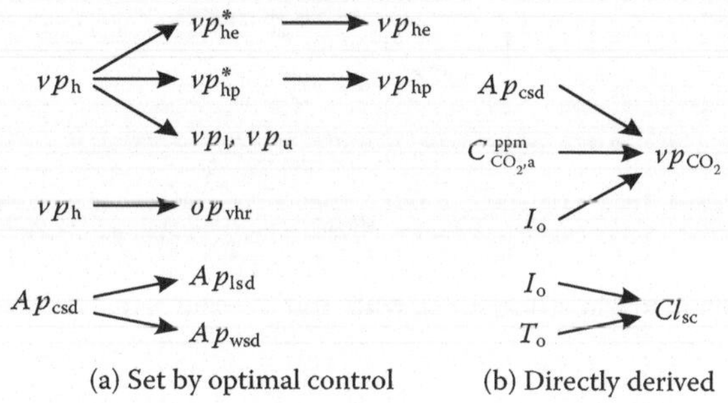

FIGURE 8.18 Relations control inputs.

These combined optimal control inputs (vp_h, Ap_{csd}) are computed by the receding horizon optimal controller.

The relations between the combined control inputs computed by the optimal control and the control inputs used by the model are shown in Figure 8.18a. The actual control inputs vp_{hp} and vp_{he} used by the model are derived from the computed control inputs vp^*_{hp} and vp^*_{he} with Equations 8.88 and 8.97.

For heating/cooling, the combined heating valve position vp_h [–1,2] is used. It is subdivided into the valve positions for heat exchanger vp^*_{he}, heat pump vp^*_{hp}, lower net vp_l, and upper net vp_u (see Figure 8.19). The idea of this subdivision is that heating and cooling at the same time makes no sense, so this should be ruled out. When heating is needed, it should preferably be done at the lowest cost. Therefore the first choice is heating with the heat pump and the second choice is heating with the boiler if the heat pump cannot supply enough heat.

When ventilation is needed to decrease humidity, but not to decrease temperature (as at times of heat demand), ventilation with heat recovery should be used. This is determined by the use of heat pump or boiler ($vp_h > 0$). When ventilation with heat recovery is used, 90% of the sensible heat is recovered. Its value is either 0 (false) or 1 (true).

For ventilation the combined window aperture Ap_{csd} [0,2] is used. It is subdivided into the lee-side Ap_{lsd} and windward-side Ap_{wsd} window aperture (see Figure 8.20). To prevent the wind from blowing through the greenhouse first the lee-side windows are opened, and if more ventilation is needed, the windward-side windows are opened.

Control Inputs Directly Derived from Other Inputs

The control inputs for CO_2 supply (vp_{CO_2}) and thermal screen closure Cl_{sc} are determined directly from other control inputs, external inputs, or states. The relations are shown in Figure 8.18b.

In the solar greenhouse CO_2 supply is independent from boiler operation. The valve position for CO_2 supply vp_{CO_2} is controlled with a proportional controller. The idea is that CO_2 supply is only

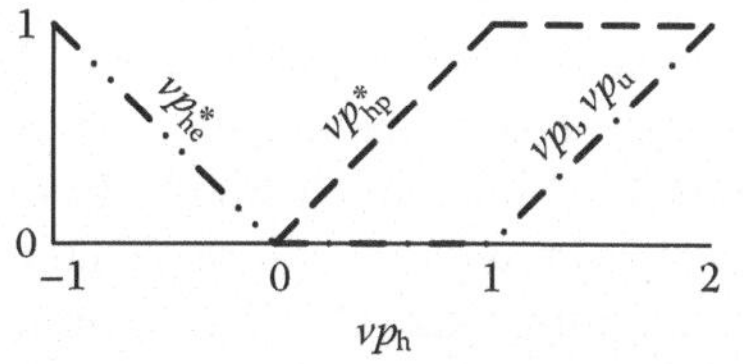

FIGURE 8.19 Combined heating valve position vp_h, with vp^*_{he} (– ·· –), vp^*_{hp} (– –), and {vp_l, vp_u} (–·–).

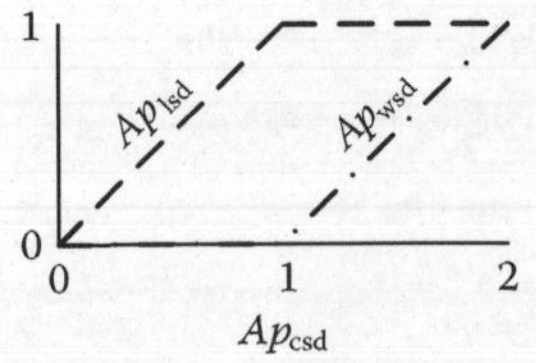

FIGURE 8.20 Combined window aperture Ap_{csd}, with Ap_{lsd} (—) and Ap_{wsd} (− · −).

needed during daytime, when there is radiation, because then it is needed for photosynthesis. If the windows are opened, CO_2 is ventilated out, so it makes sense to restrict the CO_2 supply depending on the window aperture. CO_2 supply is only needed when the CO_2 concentration in the greenhouse is below its maximum value. The CO_2 setpoint $C_{CO_2,a}^{ppm,sp}$ (μmol[CO_2] mol^{-1}[air]) is determined directly based on the combined window aperture Ap_{csd} and the incoming short-wave radiation I_o

$$C_{CO_2,a}^{ppm,sp}(t) = \begin{cases} C_{CO_2,a}^{ppm,max} - \dfrac{Ap_{csd}}{4}.(C_{CO_2,a}^{ppm,max} - C_{CO_2,a}^{ppm,min}) & I_o > 0 \\ 0 & I_o = 0 \end{cases} \quad (\mu mol[CO_2]\ mol^{-1}[air]) \quad (8.167)$$

$$vp_{CO_2}(t) = 0.01(C_{CO_2,a}^{ppm,sp}(t) - C_{CO_2,a}^{ppm}(t)) \quad [0,1] \quad (8.168)$$

in which $C_{CO_2,a}^{ppm,min} = 320$ and $C_{CO_2,a}^{ppm,max} = 1000$ (μmol[CO_2] mol^{-1}[air]) as given in Table 8.12. This valve position is constrained to the range [0,1]. The valve position for CO_2 supply vp_{CO_2} is still partly set by the optimal control, because it depends on Ap_{csd}.

With this controller a setpoint $C_{CO_2,a}^{ppm,sp}$ of 1000 μmol[CO_2] mol^{-1}[air] is used when the windows are fully closed ($Ap_{csd} = 0$), and a setpoint of 660 μmol[CO_2] mol^{-1}[air] when the windows are fully opened ($Ap_{csd} = 2$). This setting was chosen because according to Nederhoff (1994) a setpoint of twofold the outdoor concentration (of about 320 μmol[CO_2] mol^{-1}[air]) already has a large positive effect on the photosynthesis rate.

The thermal screen is used to decrease heat loss during cold periods with little solar radiation. The screen is opened and closed in about 3 min, which is much smaller than the time interval $t_{s,u}$ for the control inputs of 30 min. The "rules" for the control are quite straightforward (see Section 8.4.4.1), similar to those used in greenhouse horticulture. The thermal screen closure Cl_{sc} is determined directly from the screen condition $c_{sc} \in \{0,1\}$. This screen condition is a discrete switch, which can be seen as an external input d, which only depends on the outdoor shortwave solar radiation I_o and the temperature T_o of the outdoor air. The value of the screen closure Cl_{sc} in the optimal control is 0 (open) or 0.97 (closed, with a 3% crack opening to carry off moisture).

8.7.3.1 Initial Guess Control Inputs

Control input trajectories **u*** (Equation 8.152) that minimize the cost function value J have to be found. Only two control inputs are set by the optimal control (Section 8.7.3). Each control input consists of 48 values (Section 8.7.2), so at each receding horizon time step 96 values have to be computed. This can be done with several optimization methods. In this research, a conjugate gradient algorithm is used. This is the first algorithm for the indirect method described in Section 4.3.1. The optimization is repeated with the sampling interval t_s. The optimization starts with initial values $\mathbf{u}_0$ for the control input trajectories and changes these values until the minimum cost function value J is found. By default, the optimization for the next interval is started with the values for the control inputs found in the previous optimization, from which the first value is omitted, and the last value is equal to the last but one.

Because the model used is highly nonlinear, the search procedure is likely to find a local minimum instead of the global minimum when the search is started from ill-chosen initial values. Therefore a good initial guess for the control input trajectories $\mathbf{u}_0$ is needed. The procedure suggested here is partly based on a priori knowledge of the system, and partly on common sense. It is called the grid search method, and was first described in Van Ooteghem et al. (2003). An example of the procedure is given in Section 8.7.3.3.

At first the initial values for the control input trajectories vp_{h} [−1,2] and Ap_{csd} [0,2] are kept constant. This means that during the whole control horizon t_{f} (one day), the same values are used for each control input. With two (constant) control inputs, it is easy to imagine a grid spanned over all possible control input values. The control space is discretized by restricting the possible values of the control inputs vp_{h} and Ap_{csd} to {−1 −0.5 0 0.5 1 1.5 2} and {0 0.5 1 1.5 2}, respectively. With weather predictions for the next day (external inputs **d**), the influence of the control (control inputs **u**) on the greenhouse climate (states **x**) and the cost function value J can be simulated. If the cost function values $J(\mathbf{u})$ are plotted against the control values vp_{h} and Ap_{csd}, a surface is formed. The control input combination $\mathbf{u}_0$ with the lowest cost function value $J_{\min}$ is chosen. This is a good first guess for the control input values.

Because the control horizon t_{f} is one day, it may not always be desirable that the initial guesses for the control values are constant during this whole day. Therefore, so-called state dependent control input bounds are introduced, to rule out control values that make no sense based on knowledge of the system.

8.7.3.2 State-Dependent Control Input Bounds

Based on a priori knowledge of the system, bounds are set on the initial guess for the control inputs to push the optimal control solutions into the correct direction. These bounds are based on the initial states **x** for the time interval $t_{\mathrm{s,u}}$ (30 min) for the control inputs. From these states, the values of the indoor air temperature T_{a} and the relative humidity of the indoor air RH_{a} (based on the H_2O concentration of the indoor air $C_{\mathrm{H_2O,a}}$) are used to determine the input bounds. The minimum and maximum values for T_{a} and RH_{a} are the boundary values given in Table 8.12. For the combined window aperture Ap_{csd} also the screen condition c_{sc} is used, which depends solely on the external inputs d (Equation 8.48).

The control input bounds on the combined heating valve position vp_{h} [−1,2] are defined by

$$\begin{array}{ll} vp_{\mathrm{h}}^{\min}(t)=0 & T_{\mathrm{a}}<T_{\mathrm{a}}^{\min} \\ vp_{\mathrm{h}}^{\max}(t)=1 & T_{\mathrm{a}}^{\mathrm{ref}}<T_{\mathrm{a}}\le T_{\mathrm{a}}^{\max} \\ vp_{\mathrm{h}}^{\max}(t)=0 & T_{\mathrm{a}}^{\max}<T_{\mathrm{a}} \end{array} \tag{8.169}$$

This can be interpreted as

- No cooling with the heat exchanger if temperature T_{a} is below its lower bound $T_{\mathrm{a}}^{\min}$
- No heating with the boiler if temperature T_{a} is above the reference temperature $T_{\mathrm{a}}^{\mathrm{ref}}$
- No heating with the heat pump or the boiler if temperature T_{a} is above its upper bound $T_{\mathrm{a}}^{\max}$

The control input bound on the combined window aperture Ap_{csd} [0,2] is defined by

$$\begin{array}{ll} Ap_{\mathrm{csd}}^{\max}(t)=1 & T_{\mathrm{a}}<T_{\mathrm{a}}^{\mathrm{ref}} \text{ and } RH_{\mathrm{a}}<0.9RH_{\mathrm{a}}^{\max} \\ Ap_{\mathrm{csd}}^{\max}(t)=0.1 & c_{\mathrm{sc}}=1 \end{array} \tag{8.170}$$

This can be interpreted as:

- Less ventilation if temperature T_a is below the reference temperature T_a^{ref} for the temperature integral and relative humidity RH_a is below its upper bound. A 10% safety margin is used for the upper bound of the relative humidity, because it can increase very fast and the time interval $t_{s,u}$ for the control inputs is relatively large (30 min).
- Much less ventilation when the screen is closed (as is done in greenhouse horticulture practice). The influence of the climate above the screen on the climate below the screen is small if the screen is closed (only 3% crack opening; see Section 8.4.4.1). Furthermore, this prevents a sudden drop in temperature or humidity when the screen is opened.

8.7.3.3 Example Grid Search

The control inputs u_{grid}^* are determined with the grid search method with state-dependent control input bounds (see Sections 8.7.3.1 and 8.7.3.2). In Figure 8.21, the resulting cost function values are given for an open loop computation for a winter and a summer day, where the best control input combinations found are denoted with a star (★):

$$\text{winter:} \quad vp_h = 1.5 \quad \text{and} \quad Ap_{csd} = 0.5$$
$$\text{summer:} \quad vp_h = -0.5 \quad \text{and} \quad Ap_{csd} = 2.0$$

which yields the cost function values $J_{grid} = 10.26$ for the winter day and $J_{grid} = -20.79$ for the summer day.

The following can be observed for the winter day:

- The combined heating valve position $vp_h = 1.5$ corresponds to the valve positions $vp_{hp}^* = 1$, $vp_l = vp_u = 0.5$ and $vp_{he}^* = 0$. This means that the greenhouse is heated with the heat pump, and additional heat is supplied by the boiler. The heat exchanger is off.
- The combined window aperture $Ap_{csd} = 0.5$ corresponds to the window apertures $Ap_{lsd} = 0.5$ and $Ap_{wsd} = 0$. This means that the lee-side window is partly opened. Ventilation with heat recovery is used ($op_{vhr} = 1$), because the greenhouse is heated ($vp_h > 0$), which means there is less heat loss due to this ventilation compared to normal ventilation.

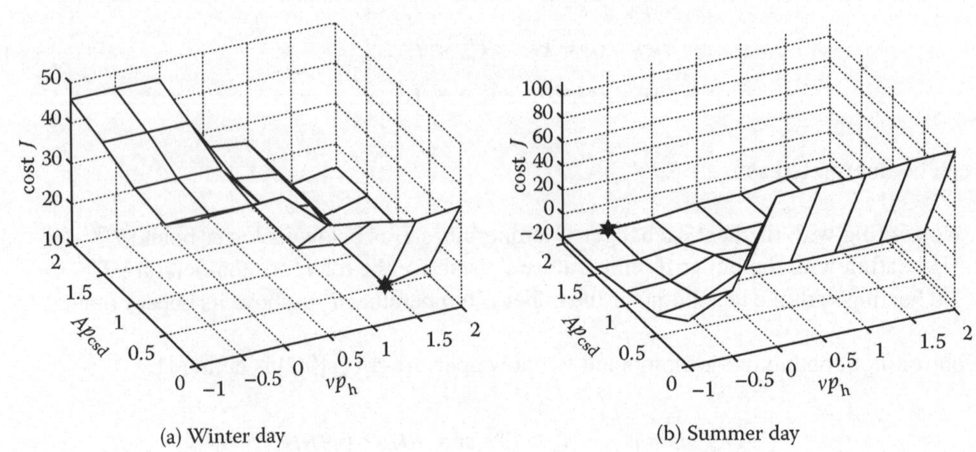

FIGURE 8.21 Cost function values J_{grid} (grid of 5 × 7 values).

From the responses (not shown here), it is found that the boiler is only used when the temperature of the indoor air is below its reference temperature ($T_a \leq T_a^{ref}$). Because the screens are closed ($c_{sc} = 1$) until 9:30 and from 16:30, the lee-side window aperture is limited to $Ap_{lsd} = 0.1$ for most of the time. This is the result of the state-dependent control input bounds.

For the summer day we find:

- The combined heating valve position $vp_h = -0.5$ corresponds to the valve positions $vp^*_{hp} = 0$, $vp_l = vp_u = 0$ and $vp^*_{he} = 0.5$. This means that the greenhouse is cooled with the heat exchanger. The heat pump and the boiler are off.
- The combined window aperture $Ap_{csd} = 2$ corresponds to the window apertures $Ap_{lsd} = 1$ and $Ap_{wsd} = 1$. This means that the windows are fully opened on both sides. Ventilation with heat recovery is not used ($op_{vhr} = 0$), because the greenhouse is cooled ($vp_h \leq 0$).

From the responses (not shown here), it is found that the heat exchanger is only used when the temperature of the indoor air is above its minimum value ($T_a \geq T_a^{min}$). This is the result of the state, dependent control input bounds.

8.7.4 External Inputs: The Weather Predictions

For the current weather conditions and the weather predictions, the *SELyear* weather data is used (Breuer and Van de Braak, 1989). The *SELyear* data consists of Dutch climate data on selected months (Jan. 1971, Feb. 1973, etc.) that are fairly representative for the Dutch climate. The *SELyear* weather data contains hourly values for I_o, v_o, T_o, T_o^{wb}, and T_{sk}. The relative humidity RH_o of the outdoor air is determined from the temperatures of the outdoor air T_o (dry bulb) and T_o^{wb} (wet bulb) (see Appendix B). For the CO_2 concentration of the outdoor air no value is given in the *SELyear* data, so it is assumed that $C^{ppm}_{CO_2,o} = 320$ μmol[CO_2] mol^{-1}[air] ($C_{CO_2,o} = 585.6 \times 10^{-6}$ kg[CO_2] m^{-3}[air]). This data is shown in Figure 8.22.

For the weather prediction a so-called lazy-man weather prediction is used. Tap et al. (1996) used this method to predict the weather conditions during the next hour. They assumed that the weather conditions during the next hour were the same as the weather conditions during the past hour.

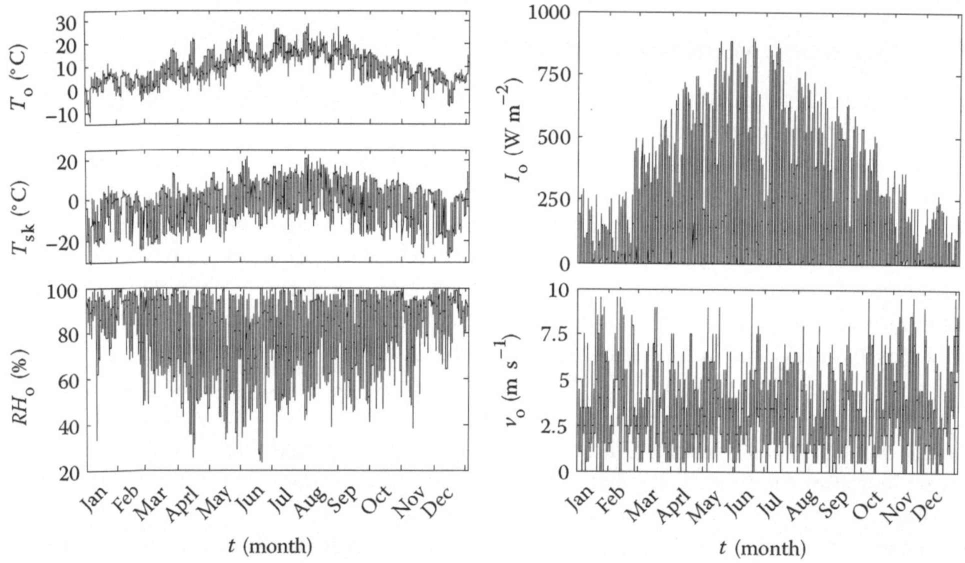

FIGURE 8.22 *SELyear* weather data *d*.

Now a weather prediction for one day is needed, because our control horizon t_f is one day. Assuming that the predicted weather conditions $\tilde{\mathbf{d}}(t,\tau)$ on the current day at time t are equal to the weather conditions $\mathbf{d}(t, t - t_f + \tau)$ of the previous day would be too crude an assumption, therefore a small correction is made. The weather conditions $\mathbf{d}(t, t - t_f + \tau)$ of the previous day are adjusted to match the current weather conditions $\mathbf{d}(t, t_0)$, where $\tau \in [t_0, t_f]$.

$$\tilde{\mathbf{d}}(t,\tau) = \mathbf{d}(t, t - t_f + \tau) + (\mathbf{d}(t, t_0) - \mathbf{d}(t, t - t_f + t_0)) \quad \forall \tau \in [t_0, t_f] \tag{8.171}$$

i.e., if the current outdoor temperature $T_o^C(t,t_0) = 15°C$ and the temperature at the same time one day earlier was $T_o^C(t, t - t_f + t_0) = 10°C$, then the correction (offset) for the whole temperature trajectory of the previous day $T_o^C(t, t - t_f + \tau)$ is 5°C.

Equation 8.171 is used for the weather conditions: v_o, T_o, T_o^{wb}, T_{sk}, and $C_{CO_2,o}$. The wind speed v_o is set to zero if Equation 8.171 gives a negative value. For the outdoor shortwave solar radiation I_o, the value of the previous day is used without correction, because the correction would lead to incorrect radiation profiles.

In the receding horizon concept, this adjustment is made at every sampling interval $t_s = 30$ min for the receding horizon controller to obtain a correction for the weather prediction. The weather conditions (hourly values) are therefore interpolated by linear interpolation.

Unless it is otherwise stated, the weather $\mathbf{d}$ is not equal to the weather prediction $\tilde{\mathbf{d}}$ in the year-round computations.

8.7.5 Initial Values States

In the closed loop computation with receding horizon optimal control (RHOC), the control input trajectories are determined again for every sampling interval t_s for the receding horizon controller based on the initial state values $\mathbf{x}_0$ and the expected external inputs $\tilde{\mathbf{d}}$ (the weather prediction, see Section 8.7.4). The initial state values $\mathbf{x}_0$ for the next time interval are determined with the greenhouse-with-crop model, where instead of the weather prediction $\tilde{\mathbf{d}}$, the actual weather $\mathbf{d}$ is used. Because these external inputs are different from the ones on which the computation was first based ($\mathbf{d} \neq \tilde{\mathbf{d}}$), this will cause the state values $\mathbf{x}$ to deviate from the expected state values.

8.7.6 Optimization Method: Gradient Search

The initial guess for the control inputs is found with the grid search method as described in Section 8.7.3.1 with the state-dependent control input bounds given in Section 8.7.3.2. This results in discrete values for the control inputs vp_h and Ap_{csd} ({−1 −0.5 0 0.5 1 1.5 2} and {0 0.5 1 1.5 2}, respectively). Each control input trajectory consists of $\frac{t_f}{t_{s,u}} = 48$ values, so a total of 96 values has to be determined by the optimal control. This method is used for the computation of the control input values at every full hour. The optimal control input trajectories $\mathbf{u}^*_{grid}$ correspond to the minimum cost function value J_{grid} found with the grid search method. This grid search method is a (rather rough) global minimization method. The gradient search method uses the conjugate gradient algorithm as described by Pagurek and Woodside (1968). The optimal control input trajectories $\mathbf{u}^*_{grad}$ correspond to the minimum cost function value J_{grad} found with the gradient search method. Because the model used in this research is highly nonlinear, this nonlinear iterative conjugate gradient method cannot guarantee that the global minimum is found. A wisely chosen starting point for the control input values $\mathbf{u}^*$ increases the probability that the global minimum for the cost function value J is found. The results $\mathbf{u}^*_{grid}$ of the grid search method are therefore used as an initial guess for the control input values. Resetting the algorithm from time to time will further increase this probability. By default the control input results of the previous optimization (shifted over the sampling time t_s) are used as the initial guess for the next time interval. At every full hour, the control inputs $\mathbf{u}^*_{grid}$ are determined again, and whenever the cost function value J_{grid} is lower than J_{grad}, these control inputs are used as the new initial guess, thus reinitializing the gradient search. This gradient search method is a local minimization method. The combination of the gradient search method with the grid search method (reset) is meant to give less local minima results.

The year-round RHOC computation has first been performed with the grid search method (Van Ooteghem et al., 2004, 2005). This was done to get an idea of the year-round values with a fast computation (about 8 hours). These results were also used to determine the aquifer energy content curve

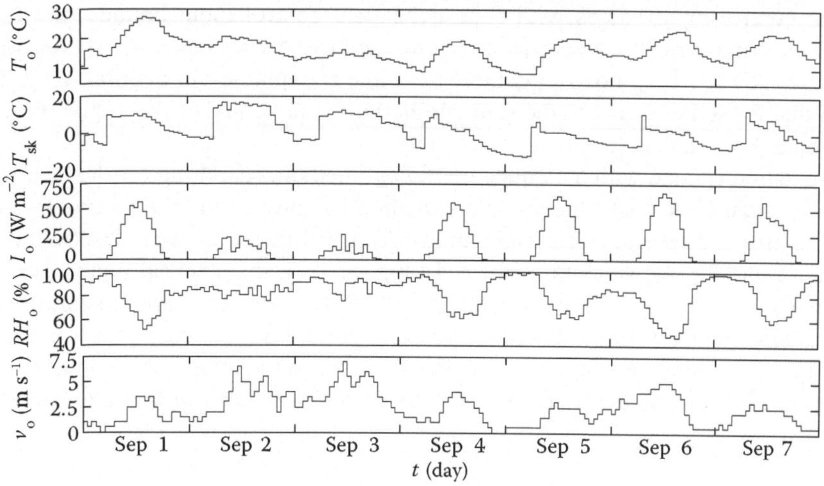

(a) External inputs **d**

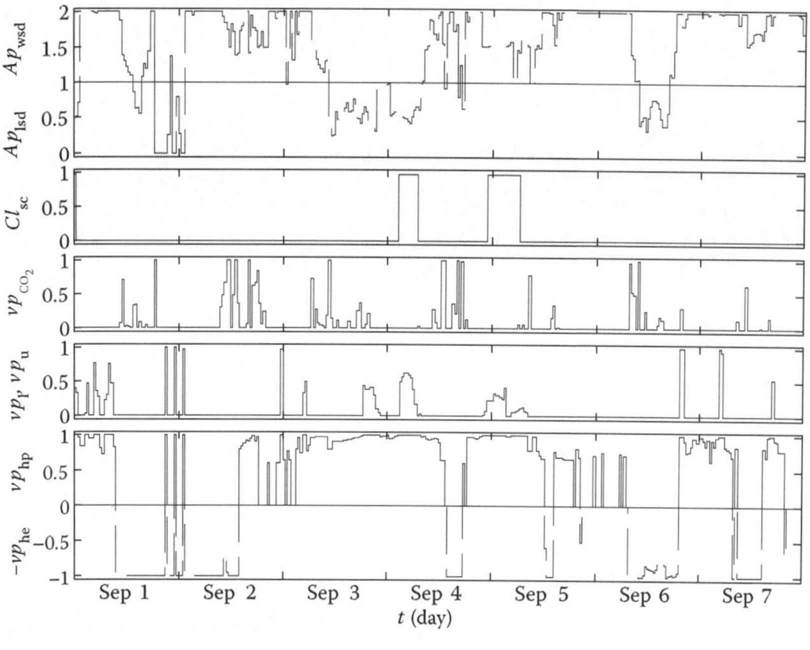

(b) Control inputs $\mathbf{u}^*_{grad}$

FIGURE 8.23 Computation RHOC with gradient search, one week: inputs **d** and $\mathbf{u}^*_{grad}$.

(Section 8.7.1.1). Then the gradient search method was applied (Section 8.7.7; Van Ooteghem et al., 2006), which was more time consuming (about 8 days).

It was found that the grid search method with state-dependent input bounds already gives good results. The gradient search method further improves these results, as it has more freedom in the control input values.

8.7.7 Results RHOC with Gradient Search

Instead of the grid search method, the "real" optimal control is now used. Starting from the initial guess for the control input values with state-dependent control input bounds (the results of the grid search), the control input trajectories are computed by solving the optimal control problem in Equation 8.152. The receding horizon control principle is explained in Section 4.3.1. The weather prediction $\tilde{d}$ that is used to compute the state predictions is not equal to the actual weather d that is used to compute the actual state values.

The optimization is done with a conjugate gradient method, searching for the best possible control inputs minimizing the cost function. This method has proven to be effective for conventional greenhouse control and several other applications (Van Willigenburg et al., 2000). In Figures 8.23 and 8.24, the results for one week are shown. In Figure 8.23, the external inputs (weather data) **d** and the optimal control inputs **u*** are shown. Figure 8.24 shows the resulting greenhouse conditions, energy use, crop growth, and costs. There are some warmer and some colder days in this week. The relative humidity outdoors is average to high (above 50%) and the temperature outdoors varies from 10°C at night to 27°C during daytime. On the colder days more heating is needed and less ventila-

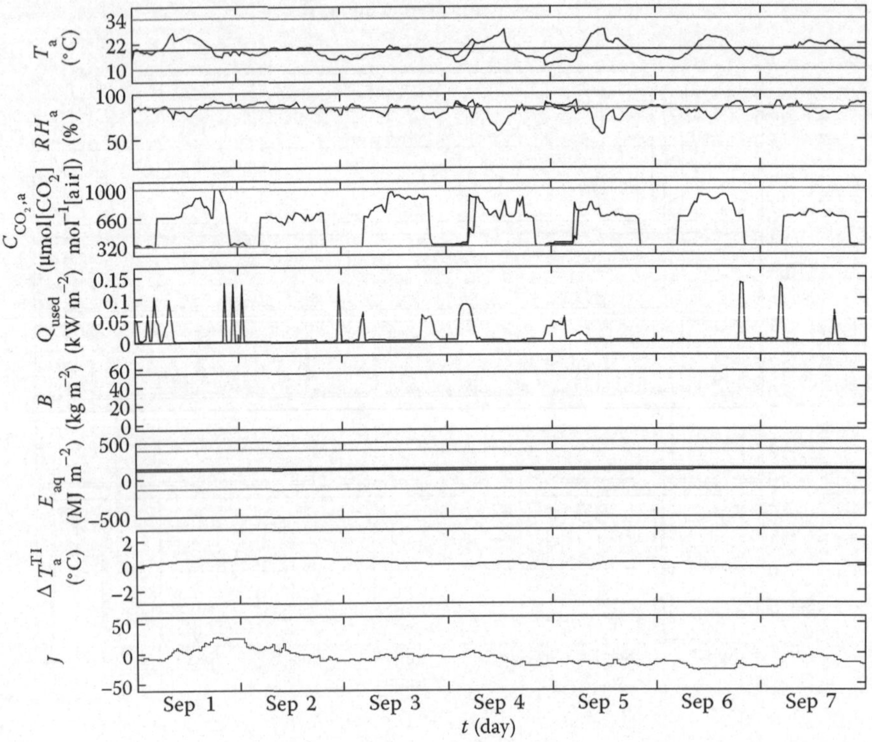

FIGURE 8.24 Computation RHOC with gradient search, one week: outputs y^*_{grad} and cost function values J_{grad}.

tion (with heat recovery) is used. On the warm days the greenhouse is cooled and there is more ventilation (without heat recovery). During the cold nights (days 4 and 5) the screens are closed.

In the year-round simulations with the receding horizon optimal controller, the actual weather is different from the forecasted weather that is used for the online optimal control computations. From these year-round computations the use of the heat pump, boiler, and heat exchanger are evaluated in the different seasons of the year. The year-round results are therefore split up per season:

- Winter (Dec. 21–Mar. 19)
- Spring (Mar. 20–Jun. 20)
- Summer (Jun. 21–Sep. 22)
- Fall (Sep. 23–Dec. 20)

The results of the RHOC computations with gradient search for fall, winter, spring, and summer are given in the Figures 8.25, 8.26, 8.27, and 8.28. These figures show control inputs and the corresponding outputs.

Because it is difficult to compare the numerical values for the control inputs and the outputs from the figures, average values and ranges are given in Tables 8.13a and 8.13b. In Table 8.13c, the actual obtained resource use and yield are given. All values are all taken over the whole season.

The heat pump (vp_{hp}) is used year-round, either to increase temperature or to decrease humidity. In fall and winter, the boiler (vp_l, vp_u) is used to supply additional heat to the greenhouse, because the capacity of the heat pump is limited, while in spring and summer it is used less often. The thermal screen (Cl_{sc}) is closed depending on the outdoor temperature and radiation. In fall and winter, the thermal screen (Cl_{sc}) is closed almost every night, and sometimes stays closed during daytime if outdoor temperature and radiation are low. In the second half of spring and in summer, the thermal screen rarely closes because outdoor temperature and radiation increase. The heat exchanger (vp_{he}) is frequently used to decrease the temperature in spring and summer but seldom in fall and winter. The windows (Ap_{lsd}, Ap_{wsd}) are mainly opened to decrease humidity, because the temperature can be decreased with the heat exchanger. When ventilation is used in fall and winter, it is mainly ventilation with heat recovery (op_{vhr}, indicated by dashed lines for Ap_{lsd} and Ap_{wsd}). Normal ventilation is only used at times of heat surplus, so when the heat exchanger is used. In spring and summer, ventilation with heat recovery (op_{vhr}) is used less often, so more normal ventilation is used. In summer, the windows are also used to decrease temperature. The CO_2 supply (vp_{CO_2}) is used whenever there is radiation. When the windows are opened, the CO_2 supply is limited. In the second half of spring and in summer, the uptake by the crop of CO_2 leads to low CO_2 concentrations in the greenhouse. The temperature T_a seldom exceeds its bounds. At times of high radiation, temperature is allowed to rise, because this yields a higher biomass increase. The average temperature deviation ΔT_a^{TI} over six days is small. It never reaches its bounds of ±6°C; the maximum deviation is 2°C. The reference temperature $T_a^{C,ref}$ of 19°C is well met in all seasons. The average temperatures are 19.22°C (fall), 19.11°C (winter), 19.10°C (spring), and 19.29°C (summer). The relative humidity RH_a exceeds its bound quite frequently, although the optimal control is doing everything it can (heating, ventilating) to decrease it. The main biomass increase ΔB is found in spring and summer, which is obvious due to higher radiation in these seasons. The main gas use is found in fall and winter due to the low outdoor radiation and temperature.

8.7.7.1 A Priori versus A Posteriori Results

The costs of the year-round computation with RHOC for the solar greenhouse are evaluated from Table 8.14. Tables 8.14a and 8.14b present the running costs, penalties, terminal costs, and cost function values.

The values reported in Table 8.14 are the averages over the whole season of the costs integrated over a day (t_f) evaluated at each half-hour (t_s). There are two ways of presenting the costs, such as running costs, penalties, terminal costs, and cost function values. The so-called a priori costs

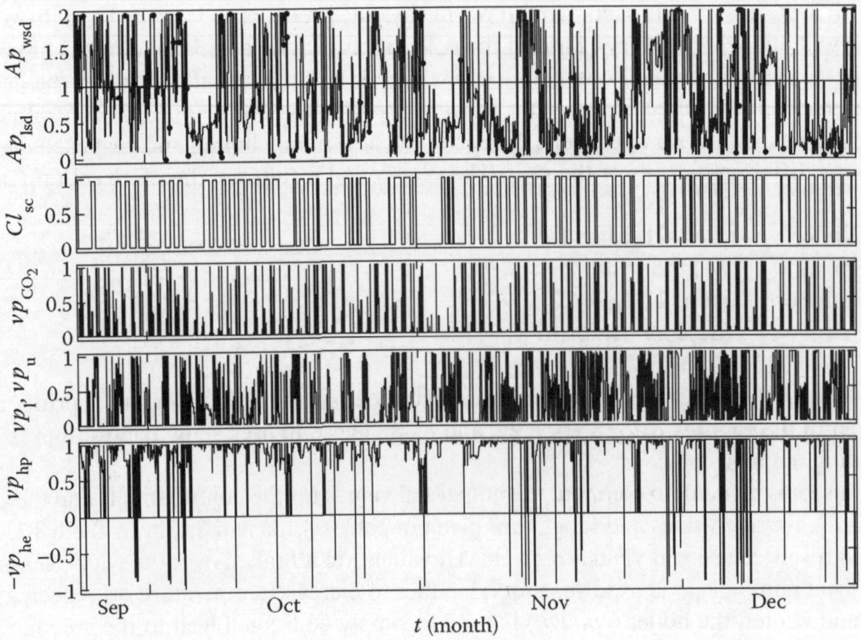

(a) Control inputs $\mathbf{u}^*_{\text{grad}}$

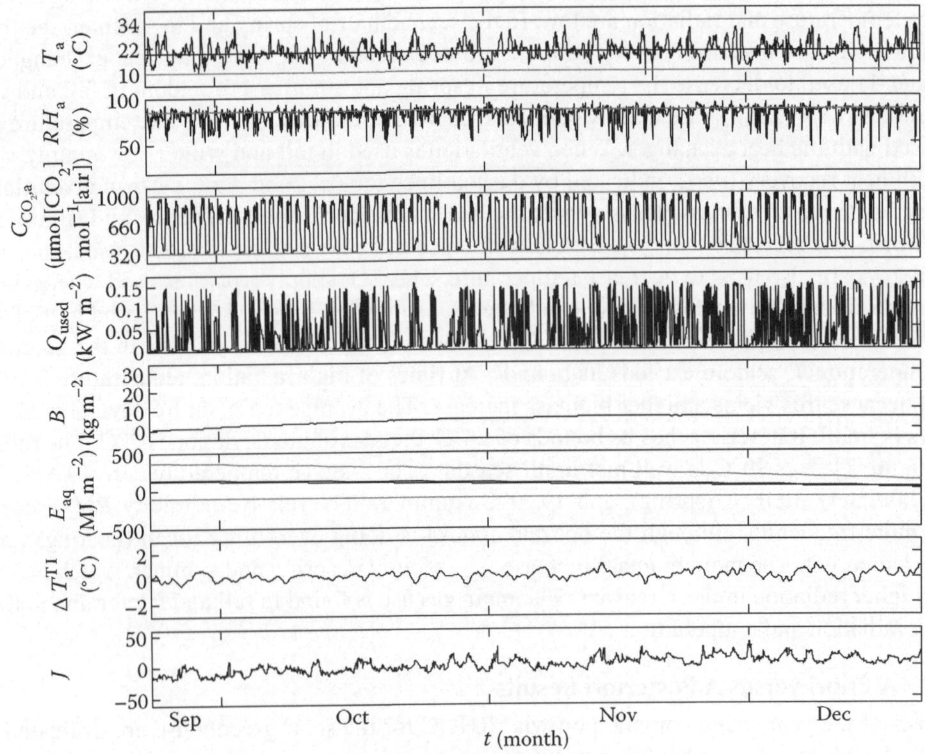

(b) Outputs $\mathbf{y}^*_{\text{grad}}$ and cost fuction values J_{grad}

FIGURE 8.25 Computation RHOC with gradient search, fall.

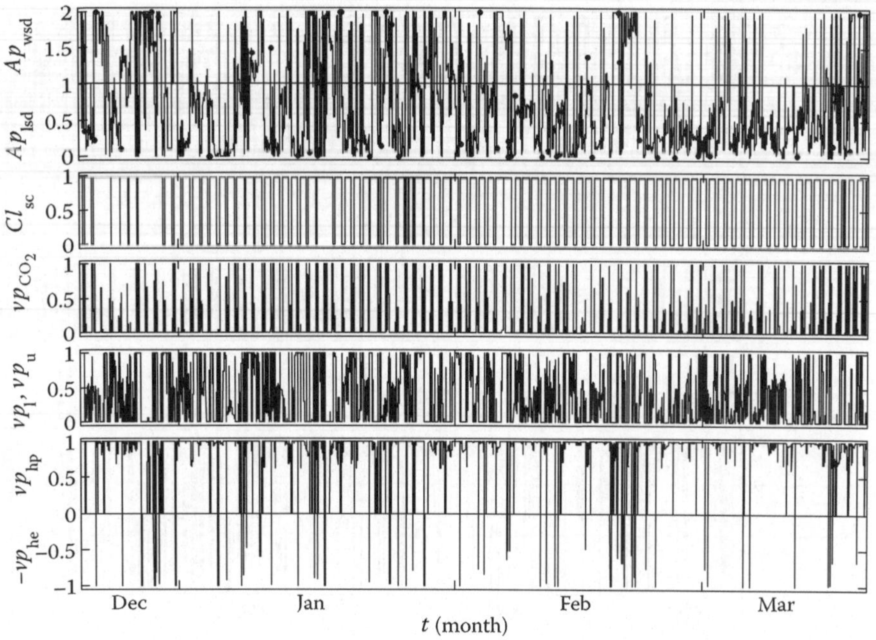

(a) Control inputs $\mathbf{u}^*_{grad}$

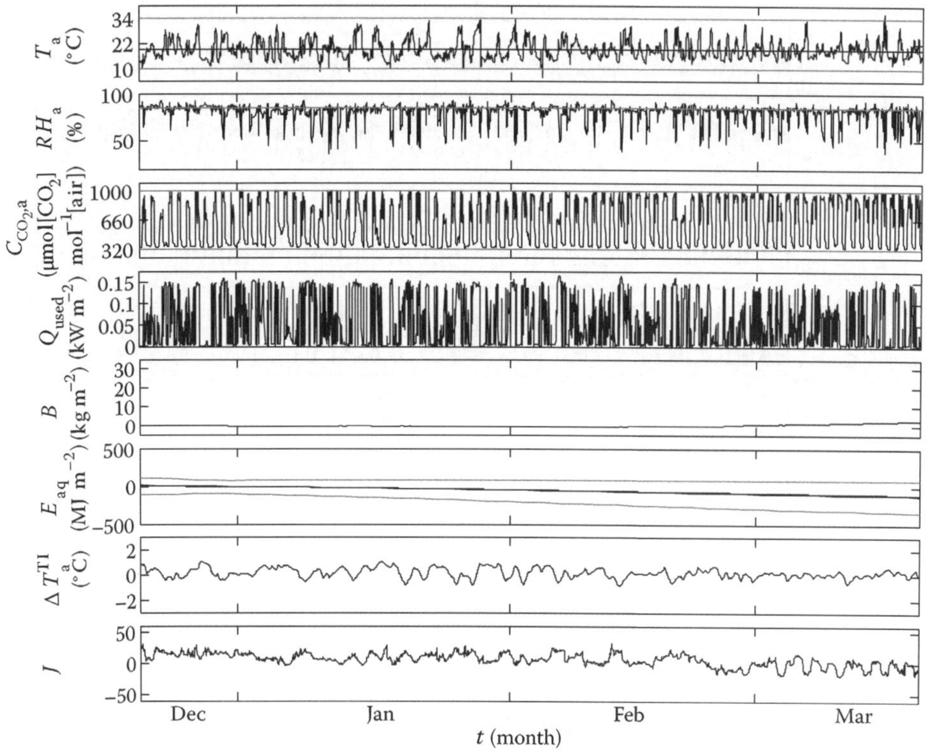

(b) Outputs $\mathbf{y}^*_{grad}$ and cost function values J_{grad}

FIGURE 8.26 Computation RHOC with gradient search, winter.

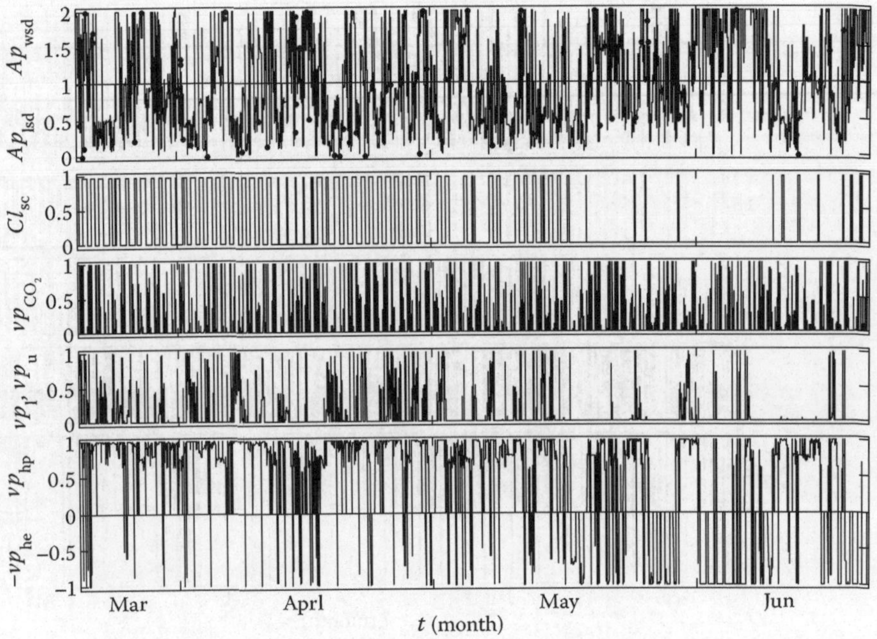

(a) Control inputs $\mathbf{u}^*_{grad}$

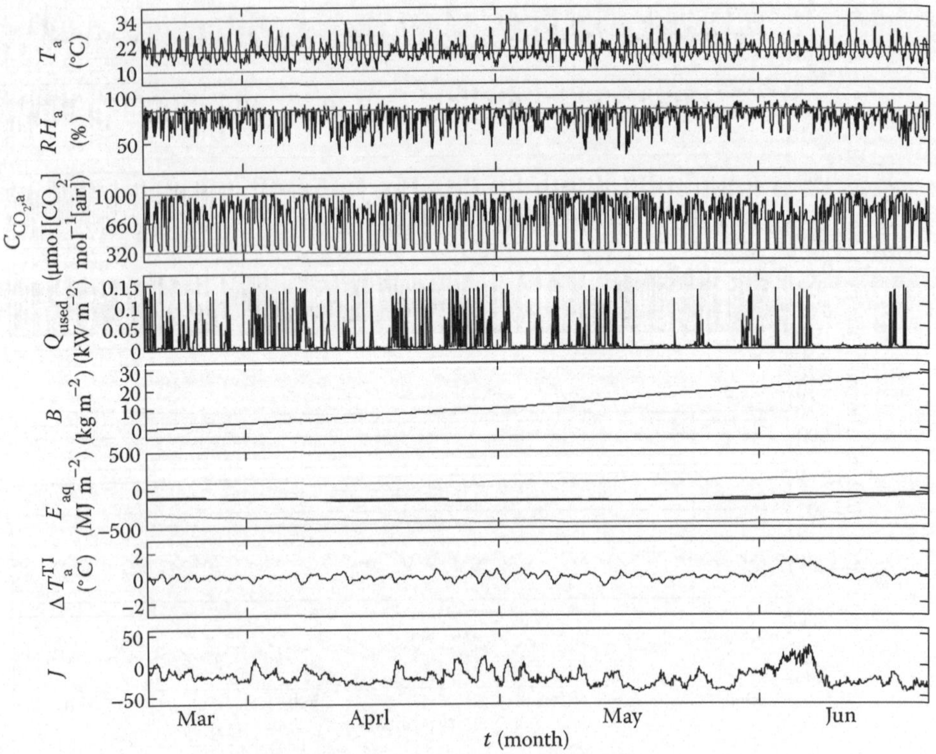

(b) Outputs $\mathbf{y}^*_{grad}$ and cost function values J_{grad}

FIGURE 8.27 Computation RHOC with gradient search, spring.

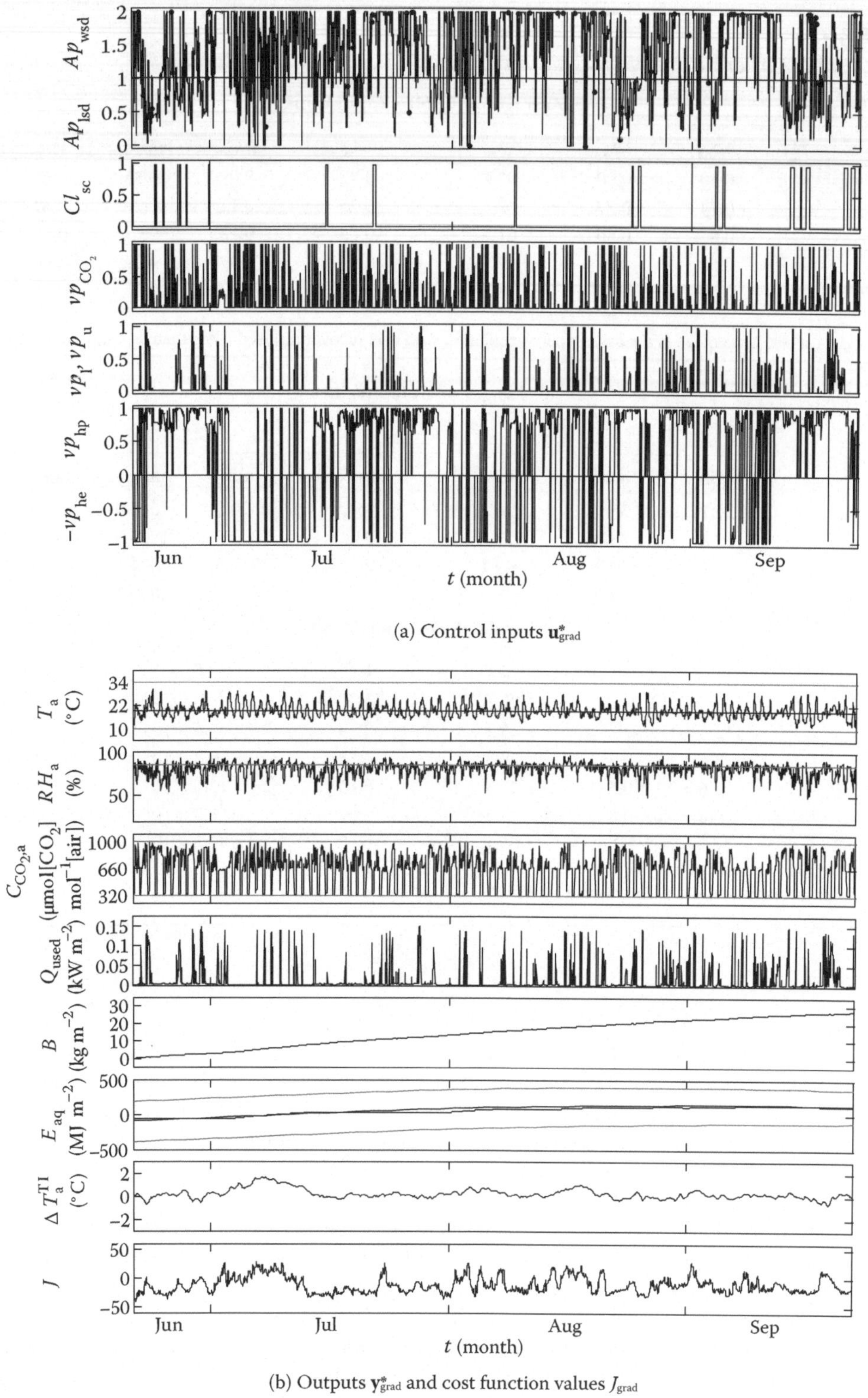

FIGURE 8.28 Computation RHOC with gradient search, summer.

TABLE 8.13
RHOC Solar Greenhouse with Gradient Search, Averages, and Ranges

(a) Averages of Control Input Values

	Ap_{lsd}	Ap_{wsd}	op_{vhr} †	Cl_{sc}	vp_{CO_2}	vp_l, vp_u	vp_{hp}	vp_{he}
Winter	0.56	0.20	92%	0.72	0.09	0.35	0.89	0.02
Spring	0.74	0.27	69%	0.25	0.11	0.09	0.62	0.15
Summer	0.87	0.50	61%	0.05	0.12	0.07	0.54	0.28
Fall	0.62	0.23	92%	0.51	0.09	0.29	0.88	0.01
Year-round	0.70	0.30	78%	0.37	0.10	0.20	0.73	0.12

† op_{vhr} is given as the percentage of the cases where ventilation with heat recovery is used when there is ventilation ($Ap_{csd} \neq 0$).

(b) Ranges of Output Values

	T_a		RH_a		$C^{ppm}_{CO_2,a}$		Q_{used}	ΔB
	min	max	min	max	min	max	max	max
Winter	5.7	35.9	36.3	96.6	320.0	1001.0	165.9	3.8
Spring	10.2	38.4	37.4	94.4	320.5	1000.8	150.4	28.9
Summer	11.4	30.3	49.1	96.7	320.6	1024.1	148.6	27.5
Fall	4.9	32.6	39.4	93.3	320.5	1000.7	158.4	5.9

(c) Results, A Posteriori Values

	F_{gas} ($m^3\ m^{-2}$)	ΔB (kg m^{-2})	$\Phi^{max}_{CO_2}$ (kg m^{-2})	F_{as_o} ($m^3\ m^{-2}$)	Q_{as_o} (W m^{-2})
Winter	12.1 = 11.3 + 0.8	3.2	32.9	9.6×10^4	12.8×10^7
Spring	3.4 = 2.9 + 0.5	28.8	42.8	9.3×10^4	15.4×10^7
Summer	2.6 = 2.1 + 0.5	27.5	50.3	13.5×10^4	9.4×10^7
Fall	9.8 = 9.0 + 0.8	5.6	34.2	8.2×10^4	9.9×10^7
Year-round	27.9 = 25.3 + 2.6	65.1	160.2	40.6×10^4	47.5×10^7

Note: F_{gas} is given as total gas use = gas use by boiler + gas use by heat pump.

(Table 8.14a) are the values obtained during the optimization using the predicted weather $\tilde{\mathbf{d}}$. Once the controls $\mathbf{u}^*$ are computed, they are applied to the greenhouse using the actual weather $\mathbf{d}$, which in this case was the *SELyear* weather data. Because the actual weather deviates from the predicted weather, the actually attained temperatures, humidities, and other states deviate from what was expected. From these actually attained state values the costs are computed afterward, which gives the a posteriori costs (Table 8.14b).

The reason for doing this is to see in what respect the realized costs deviate from the ones expected during the optimization. This is important if the a priori results are going to be used as a prediction of the costs for a presentation tool for the grower.

The deviation between the a priori values (Table 8.14a) and the a posteriori values (Table 8.14b) is partly due to the fact that the a priori results are open-loop results whereas the a posteriori results are closed-loop (feedback) results, and partly due to the difference between the actual weather $\mathbf{d}$ and the weather predictions $\tilde{\mathbf{d}}$. The main deviations are found in the running cost $\int L_Q$ and in the terminal cost Φ_{TI}, while for the other costs the deviations are smaller. The a posteriori running costs $\int L_Q$ for energy use are much higher in fall and winter, which indicates that more energy is used than

TABLE 8.14
Costs RHOC Solar Greenhouse with Gradient Search

(a) Average Costs, A Priori Values

	$\int P_{Ta}$	$\int P_{RHa}$	$\int L_Q$	Φ_B	Φ_{TI}	J
Winter	0.01	3.20	5.90	−3.25	0.83	6.69
Spring	0.00	2.04	1.79	−24.26	2.71	−17.72
Summer	0.00	2.64	1.32	−22.65	7.69	−10.99
Fall	0.00	2.84	4.58	−5.49	0.83	2.76
Year-round	0.00	2.67	3.35	−14.14	3.08	−5.04

Note: $\int P_{TI} = 0$ and $\int P_{aq} = 0$.

(b) Average Costs, A Posteriori Values

	$\int P_{Ta}$	$\int P_{RHa}$	$\int L_Q$	Φ_B	Φ_{TI}	J
Winter	0.01	2.72	8.36	−2.52	7.65	16.23
Spring	0.01	3.45	2.21	−23.60	7.28	−10.65
Summer	0.00	5.73	1.69	−22.23	10.44	−4.37
Fall	0.05	3.21	6.76	−4.81	7.44	12.65
Year-round	0.02	3.80	4.69	−13.52	8.22	3.21

Note: $\int P_{TI} = 0$ and $\int P_{aq} = 0$.

initially expected. The a posteriori terminal cost Φ_{TI} for the temperature integral is much higher in all seasons, which indicates that the realized average temperature over a period of 6 days is different from what was initially expected. This was likely to happen, because Φ_{TI} is a soft terminal constraint: any deviation of the average temperature from the target value $T_a^{C,ref} = 19°C$ at time t_f (the end of the receding horizon) is penalized. When the horizon is shifted, this terminal constraint is no longer imposed in the cost function, because it is shifted in time along with the receding horizon. It is therefore unlikely that it will be maintained in the receding horizon approach. These deviations in Φ_{TI} strongly affect the cost function value J, which is higher in all seasons.

It is, of course, of interest to know the consequences of these deviations. First, it should be stated that in general Φ_{TI} will never be equal to zero. To maintain $\Phi_{TI} = 0$ at all times would mean either to keep the temperature constant, or to have a fixed periodic symmetrical pattern. Because the temperature integral is meant as a primitive means to ensure proper crop development and crop quality, deviations might result in less development or quality, but the extent of this is unknown. This holds for any control with temperature integration.

From the a posteriori costs it can be seen that

- The temperature penalty is always low, which indicates that the temperature bounds are seldom exceeded.
- The relative humidity bound is exceeded more frequently in summer.
- The running cost for energy (and thus the energy use) is much higher in fall and winter than in spring and summer, because in these seasons the greenhouse needs to be heated.
- The terminal cost for the biomass increase is low in fall and winter and high in spring and summer. This shows that the main crop growth is found in the seasons with high radiation, irrespective of temperature and humidity conditions in the greenhouse.
- The terminal cost for the temperature integral—indicating how well the average temperature equals the reference temperature $T_a^{C,ref}$ (19°C) over a period of 6 days—is higher in

summer compared with the other seasons. The average temperature over the whole season is, however, close to the reference temperature T_a^{ref}.

- The aquifer energy content E_{aq} stays within its bounds all year.
- The resulting cost function value J is lower in spring and summer than in fall and winter. This is mainly due to the lower energy use and the higher biomass increase.

To investigate the influence of the weather prediction on the a priori and a posteriori results, a year-round computation is performed where the weather prediction $\tilde{\mathbf{d}}$ is equal to the actual weather $\mathbf{d}$. These results are indicated with "gradient*" in Table 8.15.

The deviation between the a priori and a posteriori costs for the gradient* search are entirely due to the fact that the a priori costs are open-loop results, and the a posteriori costs are closed-loop (feedback) results. Although the gradient* version uses the actual weather for both the prediction and the actual results, the a priori and a posteriori costs differ. In the receding horizon principle, the actual states will be equal to the predicted states when there is no disturbance for the open-loop (a priori) results. However, when the horizon is shifted, new weather information becomes available, which may cause the optimal control to change the control inputs, leading to different closed-loop (a posteriori) results.

Comparing the a posteriori and the a priori costs in Table 8.15, it is found that:

- The influence of the weather prediction is the highest on the relative humidity penalty $\int P_{RHa}$ and the terminal cost Φ_{TI} for the temperature integral.
- The a posteriori terminal cost Φ_{TI} for the temperature integral is much higher in all seasons. This is because it is a soft terminal constraint, which will always deviate from the target value ($T_a^{C,ref} = 19°C$) when the receding horizon is shifted. As expected, the deviations with actual weather (gradient*) are smaller than with predicted weather (gradient).

TABLE 8.15
Differences between A Priori and A Posteriori Results

	$\int P_{RHa}$		$\int L_Q$		Φ_B		Φ_{TI}		J	
Winter										
Gradient	3.20	2.72	5.90	8.36	−3.25	−2.52	0.83	7.65	6.69	16.23
Gradient*	3.20	1.95	5.33	8.25	−3.57	−2.76	1.33	6.47	6.31	13.93
Spring										
Gradient	2.04	3.45	1.79	2.21	−24.26	−23.60	2.71	7.28	−17.72	−10.65
Gradient*	2.71	2.10	1.15	1.63	−24.46	−24.31	1.35	3.58	−19.25	−17.01
Summer										
Gradient	2.64	5.73	1.32	1.69	−22.65	−22.23	7.69	10.44	−10.99	−4.37
Gradient*	2.66	2.54	0.86	1.22	−22.56	−22.57	5.41	6.64	−13.63	−12.18
Fall										
Gradient	2.84	3.21	4.58	6.76	−5.49	−4.81	0.83	7.44	2.76	12.65
Gradient*	2.57	1.72	4.35	6.29	−5.34	−4.88	1.19	5.28	2.77	8.41
Year-Round										
Gradient	2.67	3.80	3.35	4.69	−14.14	−13.52	3.08	8.22	−5.04	3.21
Gradient*	2.78	2.08	2.88	4.28	−14.21	−13.87	2.35	5.49	−6.20	−2.02

Note: Black = a priori values; gray = a posteriori values; gradient* = gradient search, in which $\tilde{\mathbf{d}} = \mathbf{d}$; $\int P_{TI} = 0$, $\int P_{Ta}$, and $\int P_{aq}$ are very small.

- All a posteriori costs are higher, except for the relative humidity penalty $\int P_{RHa}$. The relative humidity penalty is much higher a posteriori than a priori in summer for the gradient search, while it is smaller for the gradient* search.
- For the gradient search with and without weather prediction, the a posteriori running costs for energy use $\int L_Q$ are much higher in winter and fall, causing a higher year-round energy use than initially anticipated.

The weather prediction is found to influence all results. Little deviation is seen in the terminal cost Φ_B for biomass increase, because the biomass increase ΔB mainly responds to solar radiation I_o. In the weather prediction (in this study), the solar radiation is merely shifted over a day, so that the prediction and the actual weather have practically the same radiation sum year-round. Nevertheless, with a perfect weather prediction ($\tilde{\mathbf{d}} = \mathbf{d}$) the results are better. A real weather forecast will probably lead to larger deviations in the biomass increase than the weather prediction used here.

The difference between the a priori costs of the version with and without weather prediction ($J = -5.04$ vs. -6.20) is due to the difference between the weather prediction and the actual weather. This difference is found to be quite small compared with the influence of the terminal constraint on the temperature integral.

The difference between the a priori and the a posteriori costs of the gradient* search ($J = -6.20$ vs. -2.02) is entirely due to the receding horizon concept, because ($\tilde{\mathbf{d}} = \mathbf{d}$). The only cost that can change much when the horizon is shifted is the soft terminal constraint of the temperature integral. The optimal control inputs are adjusted when the horizon is shifted to minimize the terminal cost of the temperature integral.

8.7.7.2 Influence of the Separate Solar Greenhouse Elements

The solar greenhouse has the following enhancements compared with a nonsolar greenhouse:

- Heat pump, heat exchanger, and aquifer
- Ventilation with heat recovery
- CO_2 separate from boiler operation
- Zigzag cover
- Thermal screen

The extra devices that are evaluated here are the heat pump, heat exchanger and aquifer, ventilation with heat recovery, and CO_2 supply independent of boiler operation. These are called the solar greenhouse elements.

The closest to a real comparison between the solar greenhouse and a nonsolar greenhouse can be seen when comparing the solar greenhouse with all solar greenhouse elements with a greenhouse without all these features. The nonsolar greenhouse does have a zigzag cover and a thermal screen. Both the solar and the nonsolar greenhouse are controlled by optimal control to obtain a fair comparison.

We found that the gas use F_{gas} is decreased by 52%, the biomass increase ΔB is higher (139%), the CO_2 use $\Phi^{max}_{CO_2}$ is much higher (352%), and more ventilation is used F_{as_o} (118%) with much less energy loss Q_{as_o} (32%). This shows that it is possible to obtain a higher biomass increase (39% more) with a much lower gas use (52% less). These benefits compensate for the higher use of CO_2 (252% more).

The solar greenhouse uses 27.9 $m^3\ m^{-2}$ gas, 160.2 kg m^{-2} CO_2, and it produces 65.1 kg m^{-2} biomass per year. The nonsolar greenhouse with CO_2 from the boiler uses 57.9 $m^3\ m^{-2}$ gas, 45.5 kg m^{-2} CO_2, and it produces 46.7 kg m^{-2} biomass per year.

Additional computations were done to distinguish the influence of the separate solar greenhouse elements (Van Ooteghem, 2007). It was found that

- The use of the heat pump, heat exchanger, and aquifer decreases the gas use by 23%. This is due to the use of the heat pump which uses less gas ($COP \approx 5$).
- Ventilation with heat recovery decreases the gas use by 26%. This is due to the decrease of the energy loss through ventilation by 59%.
- The use of CO_2 supply independent of boiler operation leads to a much higher CO_2 use $\Phi_{CO_2}^{max}$ (289%), which leads to a higher biomass increase ΔB (137%).

From this we can conclude that the main gas use reduction is due to the ventilation with heat recovery.

8.7.8 Conclusions and Discussion

It is found that receding horizon optimal control of the solar greenhouse can be used, even with a nonlinear, complex model.

The Solar Greenhouse: Does It Use Less Gas?

Yes, it does! A gas use reduction of 52% can be accomplished. Furthermore the total biomass weight is increased by 39%, which is partly due to the possibility to use (252% more) free CO_2. It is found that the main gas use reduction is due to the ventilation with heat recovery.

These values are based on a comparison between a nonsolar greenhouse where the CO_2 supply depends on boiler operation with the solar greenhouse, including all its enhancements. It should be noted that, unlike common practice, this nonsolar greenhouse is controlled by optimal control as well. Because the same requirements were put on maintaining humidity and temperature integral, it is likely that this greenhouse uses more gas and gives a better yield than greenhouses in practice, which are controlled by classical controllers. This means that the yield improvement expected from the solar greenhouse with respect to common practice is even higher than presented here, at the expense of somewhat less gas use reduction. In all cases, the constraints for crop development and crop quality are maintained far better than in current practice.

A Priori versus A Posteriori Results

The a priori results are open-loop results, and the a posteriori results are closed-loop (feedback) results. The open-loop results will always differ from the closed-loop results. The results in Section 8.7.7.1 indicate that part of the open-loop (a priori) results are not achieved in the actual closed-loop (a posteriori) results, even when the actual weather is equal to the predicted weather ($\tilde{\mathbf{d}} = \mathbf{d}$). When the a priori results are used in a presentation tool for the grower, these results should be carefully interpreted because they may be delusive.

The deviations between the a priori and a posteriori results of the costs $\int L$ are mainly due to the difference between the predicted weather $\tilde{\mathbf{d}}$ and the actual weather $\mathbf{d}$, which stresses the importance of accurate weather predictions. It must be noted that the "lazy man weather prediction" as used in this chapter is not a very accurate weather prediction. The deviations between the forecasted and the actual weather are probably smaller in practice, which will lead to smaller differences between the a priori and a posteriori results. Better weather predictions are available from meteorological institutes. These should be used in a real implementation of this optimal control.

The deviations in the terminal costs Φ are mainly the result of the open loop results not being achieved due to shift of the horizon. This is clearly found in the terminal cost Φ_{TI}, which is a soft terminal constraint: any deviation from the target value at the end of the receding horizon is penalized. The terminal constraint will no longer be maintained when the receding horizon is shifted. Because the difference in the costs is quite large, it is important to further investigate this effect.

The same principle holds for the terminal cost Φ_B, but because the biomass increase ΔB mainly responds to solar radiation I_o, which is shifted over a day in the weather prediction $\tilde{\mathbf{d}}$, the deviation is small.

Toward a Presentation Tool

A good quality of optimal control is that it allows freedom to individual growers to make their own judgment, and to adjust the weights in the cost function according to their entrepreneurship and experience. An important advantage over classical greenhouse climate control is that these weight factors have a clear and evident meaning. Balancing the weight factors in the cost function, however, is a delicate matter. In the end, the choice of the weight factors depends on what the grower thinks is important. If the grower has to set these weight factors it is important that a presentation tool is available that shows the results of these settings in the long run. With the optimal control approach it is possible to show such results. It would be possible to advise the grower on these settings based on weather predictions and a reference temperature for the temperature integral. This is an interesting subject for further research.

REFERENCES

Bernaerts, K., and J.F. Van Impe. 2004. Data-driven approaches to the modelling of bioprocesses. *Transactions of the Institute of Measurement and Control* 26 (5): 349–372.

Bot, G.P.A. 1983. *Greenhouse Climate: From Physical Processes to a Dynamic Model.* Ph.D. dissertation, Wageningen Agricultural University, Wageningen, The Netherlands, 240 pp.

Bot, G.P.A. 2001. Developments in indoor sustainable plant production with emphasis on energy saving. *Computers and Electronics in Agriculture* 30: 151–165.

Breuer, J.J.G., and N.J. Van de Braak. 1989. Reference year for Dutch greenhouses. *Acta Horticulturae* 248: 101–108.

De Jong, T. 1990. *Natural Ventilation of Large Multi-span Greenhouses.* Ph.D. dissertation, Wageningen Agricultural University, Wageningen, The Netherlands.

De Koning, A.N.M. 1988. More efficient use of base load heating with a temperature integrating control programme; effect on development, growth and production of tomato. *Acta Horticulturae* 229: 233–237.

De Zwart, H.F. 1996. *Analyzing Energy-Saving Options in Greenhouse Cultivation Using a Simulation Model.* Ph.D. dissertation, Wageningen Agricultural University, Wageningen, The Netherlands, 236 pp.

Defant, A., and F. Defant. 1958. *Physikalische Dynamik der Atmosphäre.* Akademische Verlagsgesellschaft, Frankfurt, Germany (in German).

Elings, A., F.L.K. Kempkes, R.C. Kaarsemaker, M.N.A. Ruijs, N.J. Van de Braak, and T.A. Dueck. 2005. The energy balance and energy-saving measures in greenhouse tomato cultivation. *Acta Horticulturae (ISHS)* 691: 67–74.

Farquhar, G.D., S. von Caemmerer, and J. A. Berry. 1980. A biochemical model of photosynthetic CO_2 assimilation in leaves of C_3 species. *Planta* 149: 78–90.

Ferrel, W. 1885. Annual report of the chief signal officer of the army to the secretary of war for the year 1886. Appendix 24, Washington DC.

Gijzen, H. 1994. Ontwikkeling van een simulatiemodel voor transpiratie en wateropname en van een integraal gewasmodel. AB-DLO, Report 18, Wageningen, The Netherlands (in Dutch).

Gijzen, H., E. Heuvelink, H. Challa, E. Dayan, L.F.M. Marcelis, S. Cohen, and M. Fuchs. 1998. Hortisim: a model for greenhouse crops and greenhouse climate. *Acta Horticulturae* 456: 441–450.

Goudriaan, J., and H.H. Van Laar. 1994. *Modelling Potential Crop Growth Process*, vol. 2. Kluwer Academic Publishers, Dordrecht, The Netherlands, 238 pp.

Heesen, L. 1997. *Definitie, gevoeligheidsanalyse en evaluatie van een dynamisch model van het kas-gewasproductieproces.* M.Sc. thesis, Wageningen Agricultural University, Wageningen, The Netherlands (in Dutch).

Heuvelink, E. 1996. *Tomato Growth and Yield: Quantitative Analysis and Synthesis.* Ph.D. thesis, Wageningen Agricultural University, Wageningen, The Netherlands, 326 pp.

Körner, O. 2003. *Crop Based Climate Regimes for Energy Saving in Greenhouse Cultivation.* Ph.D. dissertation, Wageningen University, Wageningen, The Netherlands.

Körner, O., and H. Challa. 2003. Design for an improved temperature integration concept in greenhouse cultivation. *Computers and Electronics in Agriculture* 39 (1): 39–59.

Körner, O., H. Challa, and R.J.C. Van Ooteghem. 2001a. Modelling temperature effects on crop photosynthesis at high radiation in a solar greenhouse. In *Joint Meeting on Modeling for the 21st Century: Agronomic and Greenhouse Crop Models. ISHS Symposium.*

Körner, O., H. Challa, and R.J.C. Van Ooteghem. 2001b. Modelling temperature effects on crop photosynthesis at high radiation in a solar greenhouse. In *Workshop Modelling for the 21st Century: Agronomic and Greenhouse Crop Models. Joined Meeting of International Society of Horticultural Science (ISHS) and Biological Systems Simulation Group (BSSG).* Beltsville, MD, March 25–29.

Körner, O., H. Challa, and R.J.C. Van Ooteghem. 2002. Modelling temperature effects on crop photosynthesis at high radiation in a solar greenhouse. *Acta Horticulturae* 593: 137–144.

Körner, O., H. Challa, and R.J.C. Van Ooteghem. 2003. Modelling temperature effects on crop photosynthesis. In *Crop Based Climate Regimes for Energy Saving in Greenhouse Cultivation.* Ph.D. dissertation, Wageningen University, Wageningen, The Netherlands, pp. 29–36.

Körner, O., Q. Niu, and E. Heuvelink. 2009. Quantification of temperature, CO_2 and light effects on crop photosynthesis as a basis for model based-greenhouse climate control. *Journal of Horticultural Science and Biotechnology* 84 (2): 233–239.

Körner, O., and R.J.C. Van Ooteghem. 2003. Simulating crop gross photosynthesis at high temperatures. In *Crop Based Climate Regimes for Energy Saving in Greenhouse Cultivation.* Ph.D. dissertation, Wageningen University, Wageningen, The Netherlands, pp. 75–90.

List, R.J. (ed). 1966. *Smithsonian Meteorological Tables*, 6th rev. ed. Washington, DC,

Monteith, J.L., and M.H. Unsworth. 1990. *Principles of Environmental Physics*, 2nd ed. Edward Arnold, London, 291 pp.

Nederhoff, E.M. 1994. *Effects of CO_2 Concentration on Photosynthesis, Transpiration and Production of Greenhouse Fruit Vegetable Crops.* Ph.D. dissertation, Wageningen Agricultural University, Wageningen, The Netherlands, 213 pp.

Pagurek, B., and C.M. Woodside. 1968. The conjugate gradient method for optimal control problems with bounded control variables. *Automatica* 4: 337–349.

Press, W.H., B.P. Flannery, S.A. Teukolsky, and W.T. Vetterling. 1986. *Numerical Recipes*, 2nd ed. Cambridge University Press, Cambridge, UK, 291 pp.

Seginer, I., C. Gary, and M. Tchamitchian. 1994. Optimal temperature regimes for a greenhouse crop with a carbohydrate pool: a modelling study. *Scientia Horticulturae* 60: 55–80.

Sigrimis, N., A. Anastasiou, and N. Rerras. 2000. Energy saving in greenhouses using temperature integration: a simulation survey. *Computers and Electronics in Agriculture* 26 (1): 321–341.

Sparrow, E.M., and R.D. Cess. 1970. *Radiation Heat Transfer*, rev. ed. Brooks/Cole Publishers, Belmont, CA, 340 pp.

Spitters, C.J.T. 1986. Separating the diffuse and direct component of global radiation and its implications for modeling canopy photosynthesis. Part II. Calculation of canopy photosynthesis. *Agricultural and Forest Meteorology* 38: 231–242.

Stanghellini, C. 1987. *Transpiration of Greenhouse Crops—An Aid to Climate Management.* Ph.D. dissertation, Wageningen Agricultural University, Wageningen, The Netherlands, 150 pp.

Stoffers, J.A. 1989. Tuinbouwtechnische aspecten van de druppelprofilering bij kasverwarmingsbuis. Internal report, IMAG-DLO, Wageningen, The Netherlands, 24 pp. (in Dutch).

Tap, R.F. 2000. *Economics-Based Optimal Control of Greenhouse Tomato Crop Production.* Ph.D. dissertation, Wageningen Agricultural University, Wageningen, The Netherlands.

Tap, R.F., L.G. Van Willigenburg, and G. Van Straten. 1996. Receding horizon optimal control of greenhouse climate using the lazy man weather prediction. In *Proceedings of the 13th IFAC World Congress, San Francisco, CA, USA.* June 30–July 5, paper 4a-013.

Van den Bosch, J.A.M. 1998. Opportunities and bottlenecks for model applications in practice. *Acta Horticulturae* 456: 505–508.

Van Dongen, B.N.J. 2004. *Zonnekas tussen theorie en praktijk.* B.Sc. thesis, (MRS034), Wageningen University, Wageningen, The Netherlands, 21 pp.

Van Henten, E.J. 1994. *Greenhouse Climate Management: An Optimal Control Approach.* Ph.D. dissertation, Wageningen Agricultural University, Wageningen, The Netherlands, 329 pp.

Van Kimmenade, A.J.M. 1986. *Warmteleer voor technici.* Educaboek, Culemborg, The Netherlands, 576 pp.

Van Ooteghem, R.J.C. 2007. *Optimal Control Design for a Solar Greenhouse.* Ph.D. dissertation, Wageningen University, Wageningen, The Netherlands, 304 pp.

Van Ooteghem, R.J.C., J.D. Stigter, L.G. Van Willigenburg, and G. Van Straten. 2003. Optimal control of a solar greenhouse. In *Proceedings of the European Control Conference (ECC) 2003,* University of Cambridge, Cambridge, UK. September 1–4.

Van Ooteghem, R.J. C., J.D. Stigter, L.G. Van Willigenburg, and G. Van Straten. 2004. Receding horizon optimal control of a solar greenhouse. In *GreenSys2004, International Symposium on Sustainable Greenhouse Systems*, G. Van Straten, G.P.A. Bot, W.T.M. Van Meurs, and L.M.F. Marcelis, eds. Catholic University Leuven, Leuven, Belgium, September 12–16.
Van Ooteghem, R.J.C., L.G. Van Willigenburg, and G. Van Straten. 2005. Receding horizon optimal control of a solar greenhouse. *Acta Horticulturae* (*ISHS*) 691: 797–806.
Van Ooteghem, R.J.C., L.G. Van Willigenburg, and G. Van Straten. 2006. Receding horizon optimal control of a solar greenhouse. In 25th *Benelux Meeting on Systems and Control*, B. De Jager and G. Meinsma, eds, Heeze, The Netherlands, March 13–15: 90.
Van Strien, A. 1988. *Klimaat in een enkel- en dubbeldekskas—metingen en simulaties.* M.Sc. thesis, Wageningen Agricultural University, Wageningen, The Netherlands, 2 dl (in Dutch).
Van Willigenburg, L.G., E.J. Van Henten, and W.T.M. Van Meurs. 2000. Three time-scale receding horizon optimal control in a greenhouse with a heat storage tank. In *Proceedings of the Agricontrol 2000 Conference*. Wageningen, The Netherlands.

APPENDICES

A. Solar Radiation Parameters

Often the measured data only holds the outdoor shortwave solar radiation I_o (W m^{-2}[gh]). The fractions PAR f_{par} and diffuse PAR f_{difpar} (–) in the outdoor shortwave solar radiation, and other terms like the transmittance τ_{dirR} of the roof for direct radiation have to be computed. These parameters depend on the position of the sun in relation to the location of the greenhouse (Goudriaan and Van Laar, 1994) and on time.

The location of our greenhouse is given by its latitude $\lambda_{gh} = 52°$ and longitude $\varphi_{gh} = 4.2°$. With the day number day_{NR} [1365] and the hour of the day *hour* [0,23], the solar parameters, such as the position of the sun (azimuth α_{sun} and elevation β_{sun}), the sine of the solar elevation $\sin\beta$ and the solar constant $solar_C$ can be determined.

A.1 Solar Parameters

The declination of the sun δ_{sun} with respect to the equator is given by

$$\delta_{sun} = -\arcsin\left(\sin\left(2\pi\cdot\frac{23.45}{360}\right)\cdot\cos\left(2\pi\cdot\frac{day_{NR}+10}{365}\right)\right) \quad \text{(rad)} \quad (8.172)$$

where day_{NR} [1365] is the day number. The angle of 23.45° is the tilt of the Earth axis with regard to the plane in which the Earth moves around the sun.

The elevation of sun β_{sun} is the angle between the direction of the sun and the horizon described by

$$\beta_{sun} = \arcsin(\sin\beta) \quad \text{(rad)} \quad (8.173)$$

in which the sine of solar elevation $\sin\beta$ is given by

$$\sin\beta = \sin(\lambda_{gh})\cdot\sin(\delta_{sun}) + \cos(\lambda_{gh})\cdot\cos(\delta_{sun})\cdot\cos\left(2\pi\cdot\frac{SOL_{hr}-12}{24}\right) \quad (–) \quad (8.174)$$

the time of day (solar time) SOL_{hr} with time correction for Middle European Time (MET) is

$$SOL_{hr} = hour - \left(1 - \frac{\varphi_{gh}}{15}\right) \quad \text{(h)} \quad (8.175)$$

and *hour* is the hour of the day [0,23]. The Earth rotates 360° every 24 hours, which gives the term 15° h^{-1}.

The azimuth of the sun α_{sun} is the angle between the direction of the sun and the south (in which east is negative and west is positive) described by

$$\alpha_{sun} = \begin{cases} \arccos(\cos\alpha) & \text{if } SOL_{hr} > 12 \\ -\arccos(\cos\alpha) & \text{if } SOL_{hr} \leq 12 \end{cases} \quad \text{(rad)} \quad (8.176)$$

in which the cosine of the azimuth cos α is given by

$$\cos\alpha = \frac{\sin(\lambda_{gh}) \cdot \sin\beta - \sin(\delta_{sun})}{\cos(\lambda_{gh}) \cdot \cos(\beta_{sun})} \quad -1 \leq \cos\alpha \leq 1 \quad (-) \quad (8.177)$$

The solar constant $solar_C$ is the solar radiation received at the outer layer of the Earth's atmosphere. It is described by

$$solar_C = 1367\left(1 + 0.033\cos\left(2\pi \cdot \frac{day_{NR}}{365}\right)\right) \quad (\text{W m}^{-2}) \quad (8.178)$$

which gives the atmospheric transmission τ_{atm}

$$\tau_{atm} = \frac{I_o}{solar_C \cdot \sin\beta} \quad (-) \quad (8.179)$$

A parameter sun_{up} is defined, to verify if the sun is up or down, where 1 denotes true and 0 denotes false. The sine of solar elevation sin β is used to indicate if the sun is up: if $\sin\beta > 0$, then the sun is up. A small margin (10^{-3}) is used to prevent numerical problems in the computation.

$$sun_{up} = \begin{cases} 1 & \text{if } \sin\beta > 10^{-3} \\ 0 & \text{if } \sin\beta \leq 10^{-3} \end{cases} \quad (0,1) \quad (8.180)$$

With the solar parameters azimuth α_{sun}, elevation β_{sun}, sine of elevation sin β and atmospheric transmission τ_{atm}, the following parameters can be computed: the fraction diffuse radiation f_{dif} in the outdoor shortwave solar radiation, the fraction PAR radiation f_{par} in the outdoor shortwave solar radiation, the fraction diffuse radiation f_{difpar} in the PAR radiation, and the transmittances τ_{difR} and τ_{dirR} of the roof for diffuse and direct PAR radiation.

A.2 Radiation Parameters

The solar radiation parameters f_{dif}, f_{difpar}, and f_{par} are determined according to Gijzen (1994).

The fraction diffuse radiation f_{dif} in outdoor shortwave solar radiation is given by

$$f_{dif} = \begin{cases} \max(f_{dif1}, f_{dif2}) & \text{if } sun_{up} = 1 \\ 1 & \text{if } sun_{up} = 0 \end{cases} \quad (-) \quad (8.181)$$

in which

$$f_{dif1} = \begin{cases} 1 & \text{if } \tau_{atm} \leq p_{db} \\ 1 - p_{da} \cdot (\tau_{atm} - p_{db})^2 & \text{if } p_{db} < \tau_{atm} \leq p_{dc} \\ 1 - p_{da} \cdot \left((\tau_{atm} - p_{db})^2 - (\tau_{atm} - p_{dc})^2\right) & \text{if } \tau_{atm} > p_{dc} \end{cases} \quad (-) \quad (8.182)$$

$$f_{\text{dif2}} = p_{\text{dd}} + (1 - p_{\text{dd}}) \cdot \left(1 - e^{-\frac{0.1}{\sin\beta}}\right) \qquad (-) \quad (8.183)$$

in which the parameter values are: p_{da} = 6.4, p_{db} = 0.22, p_{dc} = 0.35, and p_{dd} = 0.15 (parameters for De Bilt, The Netherlands).

The fraction diffuse f_{difpar} in PAR radiation is given by

$$f_{\text{difpar}} = \begin{cases} \min(f_{\text{dif}} \cdot (1 + 0.35 f_{\text{clear}}), 1) & \text{if } sun_{\text{up}} = 1 \\ 1 & \text{if } sun_{\text{up}} = 0 \end{cases} \qquad (-) \quad (8.184)$$

in which the apparent fraction clear f_{clear} is given by

$$f_{\text{clear}} = \begin{cases} 0 & \text{if } \tau_{\text{atm}} < 0.3 \\ 2(\tau_{\text{atm}} - 0.3) & \text{if } 0.3 \le \tau_{\text{atm}} \le 0.8 \\ 1 & \text{if } \tau_{\text{atm}} > 0.8 \end{cases} \qquad (-) \quad (8.185)$$

The fraction PAR f_{par} in outdoor shortwave solar radiation is given by

$$f_{\text{par}} = \begin{cases} \max\left(\dfrac{p_{\text{pa}} - p_{\text{pe}} \cdot e^{\frac{p_{\text{pf}}}{\sin\beta}} \cdot \left(1 - e^{-p_{\text{pb}} \cdot \tau_{\text{atm}}{}^{p_{\text{pc}}}}\right)}{\zeta}, 0 \right) & \text{if } sun_{\text{up}} = 1 \\ 0 & \text{if } sun_{\text{up}} = 0 \end{cases} \qquad (-) \quad (8.186)$$

in which the parameter values are p_{pa} = 2.9, p_{pb} = 4.9, p_{pc} = 0.51, p_{pe} = 0.84, and p_{pf} = 0.033 (parameters for Wageningen and Assen, The Netherlands) and the conversion factor ζ = 4.59 μmol[photons] J^{-1}.

The transmittance τ_{difR} of the roof for diffuse PAR radiation is equal to

$$\tau_{\text{difR}} = \begin{cases} 0.78 & \text{solar gh. (De Zwart, 1996)} \\ 0.55 & \text{conv. gh. (parameter estimation Section 8.6.3)} \end{cases} \qquad (-) \quad (8.187)$$

The transmittance τ_{dirR} of the roof for direct PAR radiation is determined from transmissivity tables by De Zwart (1996) for single glass, double glass, and hortiplus glass. The tables contain values for the transmittance depending on the azimuth α_{sun} and the elevation β_{sun} of the sun. The azimuth and elevation both range from 0 to $\frac{\pi}{2}$ (0° to 90°).

Because interpolation in these tables—depending on the current position of the sun—is time consuming, in this research the values from the tables have been approximated by functions. They have been determined by fitting an equation for t_{dirR} as a function of α_{sun} and β_{sun} on the values from the table. In the functions found, the azimuth was found to have little influence on the correctness of the fit. The transmittance t_{dirR} of the roof for direct PAR radiation is then given by

$$\tau_{dirR} = \begin{cases} 0.85\left(1-e^{-0.083\cdot\frac{360}{2\pi}\cdot\beta_{sun}}\right) & \text{single glass} \\ 0.82\left(1-e^{-0.066\cdot\frac{360}{2\pi}\cdot\beta_{sun}}\right) & \text{double glass} \\ 0.76\left(1-e^{-0.083\cdot\frac{360}{2\pi}\cdot\beta_{sun}}\right) & \text{hortiplus glass} \\ 0 & \text{if } sun_{up}=0 \end{cases} \quad (-) \quad (8.188)$$

For the zigzag roof used in this research, it is assumed that the transmittance of the roof for direct PAR radiation τ_{dirR} is as high as with a single glass roof.

B. Humidity Parameters

The humidity of the air is related to the saturation water vapor pressure, which depends on temperature. The relations between humidity, saturation deficit, and relative humidity and temperature are given in the next sections.

B.1 Saturation Pressure and Concentration

The saturation deficit between object x and air is computed by

$$\Delta p_{H_2O,x} = p^{sat}_{H_2O,x} - p_{H_2O,a} \quad (\text{N m}^{-2}) \quad (8.189)$$

where $p^{sat}_{H_2O,x}$ (N m^{-2}) is the saturation water vapor pressure at object temperature T_x and $p_{H_2O,a}$ (N m^{-2}) is the water vapor pressure at the temperature T_a of the indoor air.

The saturation vapor pressure $p^{sat}_{H_2O,x}$ (N m^{-2}) at a specific temperature T_x (K) is computed with the Magnus-Tetens equation (Defant and Defant, 1958)

$$p^{sat}_{H_2O,x} = c_{s1}\cdot e^{\frac{c_{s2}\cdot(T_x-T_0)}{c_{s3}+(T_x-T_0)}} \quad (\text{N m}^{-2}) \quad (8.190)$$

in which the correction factor from temperature in Kelvin (K) to Celsius (°C) T_0 = 273.15 K, where c_{s1}, c_{s2}, and c_{s3} are the saturation pressure coefficients. For the pressure in (mbar), divide the pressure in (N m^{-2}) by 100.

The values of the saturation pressure coefficients (Smithsonian Meteorological Tables, 1966) depend on the temperature T_x of object x, which determines the phase condition of the water vapor (water ($T_x \geq T_0$) or ice ($T_x < T_0$))

$$\forall T_x \geq T_0 = \begin{cases} c_{s1} = 610.780 \\ c_{s2} = 17.08085 \\ c_{s3} = 234.175 \end{cases} \quad \forall T_x < T_0 = \begin{cases} c_{s1} = 610.714 \\ c_{s2} = 22.44294 \\ c_{s3} = 272.440 \end{cases} \tag{8.191}$$

The saturation concentration of water vapor $C_{H_2O,x}^{sat}$ at a specific temperature T_x (K) is computed from the saturation vapor pressure $p_{H_2O,x}^{sat}$ at temperature T_x using the law for ideal gas

$$C_{H_2O,x}^{sat} = \frac{p_{H_2O,x}^{sat} \cdot M_{H_2O}}{R_g \cdot T_x} \qquad (\text{kg[H}_2\text{O] m}^{-3}) \tag{8.192}$$

in which $M_{H_2O} = 18 \times 10^{-3}$ kg mol^{-1} is the molar mass of water and $R_g = 8.314$ J mol^{-1} K^{-1} is the gas constant.

The water vapor pressure $p_{H_2O,x}$ at temperature T_x is computed by

$$p_{H_2O,x} = p_{H_2O,x}^{sat} \cdot \frac{C_{H_2O,x}}{C_{H_2O,x}^{sat}} \qquad (\text{N m}^{-2}) \tag{8.193}$$

where $p_{H_2O,x}^{sat}$ (N m^{-2}) is the saturation water vapor pressure at temperature T_x, $C_{H_2O,x}^{sat}$ (kg[H$_2$O] m^{-3}) is the saturation concentration water vapor at temperature T_x, and $C_{H_2O,x}$(kg[H$_2$O] m^{-3}) is the water concentration of object x.

B.2 Relative Humidity

If the dry bulb temperature T_x and the wet bulb temperature T_x^{wb} are known, the relative humidity RH_x can be computed from these temperatures

$$RH_x = \begin{cases} 100 \cdot \dfrac{p_{H_2O,x}}{p_{H_2O,x}^{sat}} & \text{if } p_{H_2O,x} > 0 \text{ and } p_{H_2O,x}^{sat} > 0 \\ 0 & \text{if } p_{H_2O,x} \leq 0 \text{ or } p_{H_2O,x} \leq 0 \end{cases} \qquad (\%) \tag{8.194}$$

where $p_{H_2O,x}$ (N m^{-2}) is the water vapor pressure at dry bulb temperature T_x and $p_{H_2O,x}^{sat}$ (N m^{-2}) is the saturation water vapor pressure at dry bulb temperature T_x.

The water vapor pressure $p_{H_2O,x}$ at dry bulb temperature T_x is given by the psychrometric equation

$$p_{H_2O,x} = p_{H_2O,x}^{sat,wb} - p_{bar} \cdot A_{psy} \cdot \left(T_x - T_x^{wb}\right) \qquad (\text{N m}^{-2}) \tag{8.195}$$

in which the atmospheric pressure $p_{bar} = 101325$ N m^{-2}, where $p_{H_2O,x}^{sat,wb}$ (N m^{-2}) is the saturation water vapor pressure at wet bulb temperature T_x^{wb}. The psychrometric coefficient A_{psy} (Ferrel, 1885) is given by

$$A_{psy} = 0.00066\left(1 + 0.00115\left(T_x^{wb} - T_0\right)\right) \qquad (\text{K}^{-1}) \tag{8.196}$$

in which the correction factor from temperature in Kelvin (K) to Celsius (°C) $T_0 = 273.15$ K.

B.3 Dewpoint Temperature

The dewpoint temperature indicates the crop temperature at which water would condense on the crop surface. The difference between the crop temperature T_c and the dewpoint temperature T_d can therefore be used to indicate crop wetness. The dewpoint temperature is given by

$$T_d = T_0 + \frac{c_{s3} \cdot \log\left(\frac{p_{H_2O,a}}{c_{s1}}\right)}{c_{s2} - \log\left(\frac{p_{H_2O,a}}{c_{s1}}\right)} \quad \text{(K)} \quad (8.197)$$

where $p_{H_2O,a}$ (N m^{-2}) is the water vapor pressure at indoor air temperature T_a and the saturation pressure coefficients are given in Equation 8.191.

9 Developments, Open Issues, and Perspectives

9.1 INTRODUCTION

The purpose of this last chapter is to discuss developments, open issues, and perspectives for greenhouse cultivation control. The development of control methodology cannot be detached from the developments in the industry as a whole. Therefore, we start with a brief overview of some recent developments in the greenhouse industry, without having the pretension to be complete. We then try to abstract the requirements that most likely will be posed to the control systems to come. Control system designers will face a number of challenges to meet the demands of future greenhouse systems. We argue that the framework of optimal control theory as elucidated in this book is versatile and generic enough to meet these challenges. However, despite its favorable features, real application is still very limited. We try to explore possible reasons why this might be the case. We distinguish between issues regarding the methodology itself and the human factor as elements of reluctant acceptance in the market. We then discuss for both categories remedies and potential solutions. This is a fantastic playing field for researchers and practitioners alike, and we hope to offer a stimulus to the scientific community and the industry for continued work and progress in this exciting field.

9.2 DEVELOPMENTS IN THE GREENHOUSE INDUSTRY AND CONSEQUENCES FOR CONTROL

9.2.1 Recent Advances in the Greenhouse Industry

Except in Chapter 8, the greenhouse systems we studied had a structure that is relatively simple. In practice, there are many enhancements to the basic structure, depending on local factors. For instance, in the past few decades in The Netherlands, which is home to the world's leading greenhouse industry for temperate climate zones, the greenhouse equipment has become more and more involved in response to, on the one hand, the favorable natural conditions such as the availability of cheap natural gas and, on the other hand, the need to suppress the costs and to reduce environmental impact such as the use of natural gas and the emission of CO_2. The development of the technology that has occurred and its implication for control can be sketched as follows:

1. Use of flue gas as CO_2 source. This means that there is a coupling between heat generation and CO_2 availability, thus reducing the degrees of freedom for control.
2. Use of a short-term heat buffer. This step restores the number of degrees of freedom because it allows the decoupling of heat generation from CO_2 generation. In this concept, natural gas is burned to produce CO_2, and the heat is stored for use in periods when there is heat demand. The control system needs to provide strategies for filling and emptying the buffer.
3. Screening. Greenhouses get energy screens, which, however, take away light during the day. A screen, on the other hand, increases the degrees of freedom for energy savings. The screen is yet another control variable, and its operation needs to be incorporated in the control.

4. Artificial lighting. To prolong day length and to control the production, artificial light is widely used. This again increases the number of degrees of freedom for control but requires the incorporation in the control system. There are side effects such as heat generation and hindrance to the public that need to be dealt with as well.
5. Cogeneration of power and heat. Power units are used to generate electricity while burning gas for CO_2 and heat generation. The heat is stored in a short-term buffer for later use, part of the electricity is used for artificial lighting, and the remainder is sold to the market. In this way, the grower becomes an electricity producer, and with the current fluid energy market, the management of the cogeneration unit constitutes a new control problem.
6. Closed greenhouse with long-term heat storage. Although still in its infancy, some systems have been realized where the greenhouse is kept closed. Heat surplus in summer is harvested by heat exchangers and is stored in aquifers under the greenhouse, whereas in winter the heat is recovered using a heat pump. This is similar to the solar greenhouse described in Chapter 8. The heat exchanger also removes surplus moisture from the air by condensation. The number of degrees of freedom for control is larger in such systems because the heat exchanger can be viewed as an air conditioner, thus giving more free control over moisture content. On the other hand, the interactions between the various components of the system increase and the abundance of equipment now introduces the additional problem of what device to use in the most economical way to meet the climate demands.

Although the examples above are typical for the Dutch situation, one can see similar adjustments of greenhouse design in other regions. For instance, in semiarid regions, more emphasis is put on cooling, using evaporator systems or air conditioners, and on water preservation.

The advances in greenhouse structure designs as sketched earlier are accompanied by developments in sensors and the use of information. For example, many growers have installed weighing gullets to keep track of crop evapotranspiration. Also, leaf photosynthesis sensors can be found occasionally. However, the proper incorporation of such sensor information in monitoring and control is still a challenge. The use of weather forecasts retrieved from Internet services is also widespread and is currently used in energy management overlays of the standard greenhouse control computer.

Apart from these technical innovations, there are also developments in the area of economy and marketing that affect greenhouse operations. In The Netherlands, a clear tendency can be observed that nurseries become bigger. Greenhouses from 5 to 30 ha are a reality today. This poses questions of risk. If a control system fails, the damage may be horrendous. Another trend is branding of products, which implies strict requirements to the constancy of the produce. Thus, in view of the variable outside conditions, reproducibility and being in control at all times are important issues.

The developments sketched earlier are the daily practice at the time of writing this book, some of them already for years. It is certain that increasing demands for energy savings, CO_2 emission reduction, water and nutrient recovery, and market evolutions will act as drivers for further innovations that will become reality in the (near) future. The next section gives a number of examples of the ongoing new greenhouse design work, which will pose additional challenges to the control community.

9.2.2 Future Developments in the Greenhouse Industry

There is no doubt that the greenhouse industry will show further innovative progress. To outline the motivation for these developments, we quote Giacomelli et al. (2008): "Innovations in greenhouse engineering are technical developments which help evolve the state-of-the-art in CEA (Controlled Environment Agriculture). They occur in response to the operational demands on the system, and to strategic changes in expectations of the production system. Influential operational factors include availability of labor, cost for energy, logistics of transport, etc. Influential strategic factors result from broader, regional issues such as environmental impact, product safety and consistency, and consumer demand. These are industry-wide concerns that have the effect of changing the production system in the long term. Global issues are becoming more influential on greenhouse production sustainability, and include less tangible issues such as social acceptance, political stability, quality of life benefits, and environmental stewardship. These offer much more complex challenges and are generally beyond the realm of engineering. However global issues do affect greenhouse engineering innovation." This quote clearly indicates both the role of technology and the other factors that play a part. Each new solution will also constitute a challenge to the operational control, and although it is tempting to apply proven methods to these new concepts—it is likely that this is going to be the first choice in practice—we must realize that this is most likely not the best solution. As illustrated by the difficulties encountered in current complex greenhouses, such as the problem of choosing the best deployment of available alternative devices, as mentioned before, new control solutions may be needed from scratch.

Let us dwell a little on a number of interesting new developments that are emerging today. This is certainly not a complete list but serves the purpose of illustrating the options.

9.2.2.1 Innovations Motivated by Sustainability: Energy and CO_2

For long, cutting down on energy consumption has been a motivation for better designs. Bakker et al. (2008) summarized a number of innovative technologies for energy saving, both from a North-Western European perspective and from a Mediterranean perspective. Novel cover materials, energy screens, and energy conserving heat exchange solutions are mentioned for cooler climates and heat-repellent cover materials, shading screens, and direct evaporative cooling methods for hotter climates. They point to the significant role of humidity control on energy use and summarize new control concepts that explicitly take energy savings into account, such as temperature integration and, indeed, optimal control. Energy savings can be obtained with a thermal screen, but dehumidification is required to prevent risk of fungal diseases. Campen, Kempkes, and Bot (2009) described a novel system to inject cold dry outdoor air instead of the usual dehumidification by transport of above screen dryer air to below via the crack usually left open for this purpose. The latter has disadvantages because it tends to cause horizontal temperature gradients with negative crop production effects. They also describe a control strategy for the air injection, illustrating once again the role of control in interaction with innovative designs.

In regions with a suitable underground, the use of aquifers to store thermal energy will become more important. The solar greenhouse concept discussed in Chapter 8 is an example of this and was at the same time an illustration on how optimal control can help operate such complex systems. Other accounts on thermal control with aquifer water are given in an experiment of a chili (*Capsicum*)-producing greenhouse by Sethi and Sharma (2007). The performance is expressed in terms of the ability to maintain, on average, higher temperatures than ambient in winter and lower than ambient in summer, while reducing the amplitude of daily temperature fluctuations. Sethi and Sharma (2008) provided a comprehensive survey and evaluation of a wide variety of heating technologies for worldwide agricultural greenhouse applications. Seen from a control point of view, some of these methods, such as north wall storage as practiced intensively in China (Figure 9.1), are increasing the heat capacity of the system, with a profound effect on the dynamics of the system. Others, like storage in phase change material, will introduce an additional piece of equipment. This not only increases the number of degrees of freedom for control but also introduces an additional decision problem.

The concept of a completely closed greenhouse is another innovation that was triggered by the need for energy conservation. Because higher CO_2 concentrations can be maintained, the expectation is that the higher investment costs can be earned back from higher crop yield. Completely closed greenhouses could also reduce the use of pesticides, meaning a minimization of effects on the environment and a better product safety. De Zwart (2008) presented an overall energy analysis

FIGURE 9.1 North wall heat buffering of typical Chinese greenhouses. (Courtesy of Prof. Luo Weihong, Nanjing University.)

of (semi)closed greenhouses. Heuvelink et al. (2008) reported experimental studies showing that yield increase can indeed be realized. Similar experimental studies are under way (e.g., see Hoes et al., 2008). Experience with the application of traditional greenhouse climate controllers to the closed greenhouse has taught that the control task is not straightforward.

The closed greenhouse even has the potential to produce energy, albeit in the form of low-valued hot water. The thought that a greenhouse is after all a solar collector has provoked other ideas to produce energy. One such innovation under study is to reduce the heat load by a special coating and a curved roof, such that the near-infrared component of the light is shielded off from entering the greenhouse and is reflected and concentrated at photovoltaic cells to produce electricity (Figure 9.2) (Sonneveld et al., 2010). Apart from the obvious control problem of maintaining the best orientation toward the sun during the day, there is a host of other problems to be solved in relation to the trade-off between the best crop value and the production of electricity.

A somewhat different way of energy savings is obtained by replacing conventional assimilation lighting by LED lighting (Morrow, 2008). Because LEDs have the potential to influence crop morphology and plant composition via the light spectrum and are easy to dim, their deployment will introduce issues of controlling the light color and intensity and of balancing assimilation light against marginal value. In addition, there may be side effects, such as the effects on the humidity balance and its control, which are still waiting to be evaluated.

9.2.2.2 Innovations Motivated by Sustainability: Water

Greenhouse operation in hot climates is not possible without cooling. The most used method for cooling of greenhouses is evaporative cooling. Such methods require water, which is often scarce

FIGURE 9.2 Electricity-producing greenhouse. (Courtesy of the Business Unit Glass, Plant Research International.)

in those regions. The desire for water savings is therefore an important driver for innovations, in addition to desires to levy disadvantages of common systems. For instance, in standard pad and fan technology, there are undesirable horizontal temperature gradients. These can be avoided by using fogging. As an example, Perdigones et al. (2008) described the research of cooling by fogging in combination with a screen and showed that fogging above the screen gives similar temperature effects as below, but with better humidity conditions in the crop compartment. In contrast to standard fixed on–off control, a temperature-dependent control using pulse width modulation contributed to substantial water savings.

Because the greenhouse is in fact a solar collector, it could potentially be used for desalination. In fact, greenhouses have been advocated as suitable methods to alleviate water shortage problems (Van Kooten, Heuvelink, and Stanghellini, 2008). Novel greenhouse designs have been proposed and tested to combine cooling needs, water savings, and desalination or treatment of gray water, for example, the Watergy greenhouse (Zaragoza and Buchholz, 2008; Zaragoza et al., 2007) (Figure 9.3) and the Seawater greenhouse (Davies and Paton, 2005). The main challenge in these approaches is to find suitable solutions for condensation of evaporated water (e.g., see Dawoud et al., 2006). The degree of integration of such systems, combining water recovery and plant production, poses new challenges to control. An attempt to use the integrative power of optimal control for integration of functions in the Watergy greenhouse has been described by Speetjens, Stigter, and Van Straten (2009) and Speetjens (2008).

FIGURE 9.3 Watergy greenhouse under construction (Almería, Spain). In the middle, the heat exchanger tower, harvesting heat during the day and releasing heat during the night. (Courtesy of Gerrit van Straten.)

9.2.2.3 Innovations Motivated Mainly by Consumer Demands

Some of the innovations discussed above also lead to better control over crop quality, as in the closed greenhouse, or better service to local markets by prolonged production periods enabled by suitable cooling. In addition, innovations have been invoked by problems encountered with product quality.

An important factor in product quality is homogeneity. In standard greenhouses, usually considerable spatial gradients exist, leading to heterogeneity of product and loss of economic value. Therefore, novel concepts have been developed aiming at reducing the spatial gradients. This begins with computational fluid dynamics (CFD) studies and with monitoring. A trend is to install wireless sensor networks in greenhouses, sometimes in conjunction with the GPRS and the Internet (e.g., see Sun et al., 2006). The development of hardware and protocols has been described in a number of studies, particularly in China (e.g., see Yu et al., 2009). The new technology allows more measurements at the crop level to gather information from a large number of subsystems (Park et al., 2009; Van Tuijl, Van Os, and Van Henten 2008). This poses a new challenge to control system design because it is not clear how the multiple sensor information should be used, let alone how the control system should be enhanced to avoid or exploit spatial gradients.

Other developments in sensors and monitoring also have an effect on the control design. Some growers have installed sensors for photosynthesis, leaf temperature, and stem thickness of individual plants, but it is an issue of research how such information can be used to control the crop as a whole (Ehret et al., 2001). Evapotranspiration sensors, such as measurement gullets, give data on local evapotranspiration, but they too need to be interpreted and integrated into the control system.

To conclude this section, it can be said that the consequences of the exciting new developments in the greenhouse industry, as described above, for control system design and development can be summarized as changes in degrees of freedom, fusion of data and information, and shifts in control goal paradigms. Unthoughtful application of common standard methods will not bring the expected results and might easily lead to disappointments. In many cases, a fundamental rethinking of the control system is mandatory.

9.3 PREREQUISITES FOR FUTURE CONTROL SYSTEMS

It is likely that the development of control systems for greenhouses will follow two major lines. On the one hand, one will see a further development along the lines of setpoint controllers. This will start from "quick and dirty" implementation of simple on–off or PI-like controllers in areas where the greenhouse industry is coming up toward more elaborate systems, including expert system overlays, in more mature markets. Although PI control does not require high skills on the part of the developer, this is not true for the more elaborate solutions. On the user side, there is a learning curve to know how to choose the settings to achieve a decent result. These solutions will be suboptimal, and adopting them means that suboptimality is accepted in exchange for perceived implementation advantages and believed ease of use. Experience in the last decades of the twentieth century has taught that wishes by users expanding over time create systems that are the opposite of easy-to-use systems and will require considerable operator training, without ever providing true optimality. Nevertheless, we expect that these systems will continue to exist.

The other path is optimal control in one form or another. If one really wishes to go for the best possible economic result, there is no other way. The implementation of these systems requires considerable skills of the developer and system designer. On the other hand, for the user, the use of the system is relatively easy and will probably require only a short training period. In the sequel, we will explore with which requirements an optimal control system has to comply.

9.3.1 Demands of the Future

In view of the discussion above, it can be envisaged that a successful control system

1. Provides economic optimality based on scientific knowledge while backing off from risks with high sequential damage
2. Enables a user-defined trade-off between acceptance of suboptimality in exchange for a larger risk margin
3. Is robust against unexpected changes and avoids risk of complete harvest loss
4. Is able to deal with the increased interaction of system components; in particular, it provides optimal selection among available alternative utilities at all times
5. Creates trust with the grower by providing transparency for the reasons for its actions and by allowing the grower to assess the effect of decisions he might take to overrule the recommendations of the system in case of unexpected or unacceptable behavior
6. Is able to communicate with the outside world for exchange and updates of models and procedures and to retrieve important data on the production process for economic analysis and to allow production tracing
7. Is able to adapt flexibly to new economic conditions
8. Has flexibility to adapt to changes invoked by the user, such as different crop types, modifications in configuration, and installation of equipment and utilities
9. Is able to handle and integrate data and information from different sensors and others sources
10. Has a modular software structure so that tasks and responsibilities in designing parts of the system can be distributed over a team of designers

11. Has the capability to integrate component vendors' local control intelligence (e.g., boiler, heat pump, irrigation, and fertigation) with the overall system without jeopardizing the optimality of the system performance as a whole
12. Is easy to install and maintain

Items 1–4 provide the base top level functionality. Items 5–7 are additional user functionalities, whereas items 10–12 represent implementation functionalities that are important for the developer and vendor of the control system. Items 8 and 9 are important for the end user as well as for the supplier.

9.3.2 How Does Optimal Control Fit In?

The prerequisites for optimal control are as follows: (1) the availability of relevant and adequate models; (2) as much information about future external inputs as possible; (3) the proper formulation of a goal function, including constraints; (4) a solution method to solve the open-loop problem over a prolonged time horizon; and (5) a methodology to provide feedback in response to deviations while accepting information from the open-loop solution. The theory outlined in this book is generic and can therefore, in principle, be applied to any new design and to any expansion of the demands and requirements set to the system.

In view of the requirements set out in Section 9.3.1, it is clear that the very nature of optimal control is to offer the functionalities in items 1 and 2. Other system solutions can, at best, provide only approximate solutions. Additional work is needed to provide suboptimal solutions in exchange for robustness and risk avoidance, but the optimal control framework offers an excellent starting point for this.

The work needed regarding safeguarding to complete production loss (item 3), that is, alarms and abnormal situation handling, is not different from what has to be done in current traditional climate computers.

Because optimal control is a multivariable control solution, it provides an automatic solution to the problems related to interaction and the optimal deployment of utilities (item 4). This is a very large advantage of optimal control. To provide trust to growers (item 5), work has to be done in the form of presentation tools, but as optimal control is inherently based on models, forecasting and simulation features are offered without difficulty, which is not the case in traditional control solutions.

Connectivity (item 6) is already a fact nowadays and does not have more problems than in conventional systems. Moreover, connection to the Internet offers superb chances to keep user systems up to date with the latest scientific findings.

On the issue of adaptability to new economic conditions (item 7), optimal control as developed here is at its best. Any change in market prices or utility costs can be accommodated immediately, thus leading immediately and automatically to adapted control strategies, which is far from what is possible with traditional control solutions. Moreover, new demands can be expressed as expansions of the goal function, without any need to adjust the control algorithms themselves. We believe that this versatility of optimal control makes it preeminently suitable to meet the challenges of the future.

Regarding flexibility (item 8), the requirements for the optimal controller are not different from those of more traditional controllers. Obviously, in view of flexibility to structural change, automation of this task would be beneficial to all control solutions and therefore also to optimal control. New designs require the development of models for the new elements in the system in a suitable form. However, this has to be done only for the first nursery that adopts a new cultivation system, as due to the property that models are the best way to condense existing knowledge, the solutions obtained can be used over and over again. In traditional systems changes in crop type will necessitate the use of another cultivation blueprint, in optimal control it will entail downloading of the appropriate cultivation model. A bank of suitable models is needed, but one can start with only three or four fundamentally different models that cover the majority of crop types, for example, single-harvest vegetable crop, continuous fruit crops, flowers, and pot plants. Changes in crop are reflected by the choice of the model and the loading of a crop-specific parameter file.

In the area of sensor fusion (item 9), the systems theory underlying optimal control is providing a very natural framework for this, lacking in more traditional approaches. This pertains to model-based state and parameter estimators, where the optimal control framework provides clear guidelines on what is needed.

On the implementation side, responding to configuration changes (second part of item 8) is not really different from today's situation. Configuration changes usually need the help of the vendor or extension service and are facilitated by a modular setup (item 10). As the optimal control system consists of the components listed (prerequisites 1–5) in the beginning of this section, this framework offers a natural way to break down the various tasks, whereas the methodology gives a clear indication of what is the information that goes into each system component and what needs to come out, as shown in the schemes presented in Chapters 2–5. Hence, in principle, the interconnection is reduced to a technical software design problem.

The point of integration of vendor components (item 11) is similar to the situation in conventional systems. If local loops can be considered as fast enough, there is no problem in the optimal control framework of handling setpoints for local controllers as control inputs of the system as a whole to be provided by the optimal control system. Some pieces of equipment have so-called built-in intelligence. Examples are the filtering of sensor data or the implementation of low-level intelligent controllers. Although this is useful in many applications, it is not always desirable in the frame of optimal control. Therefore, equipment vendors should be encouraged to make provisions that allow direct access to the raw sensor data or, in the case of local controllers, to allow direct steering of the actuators. Such provisions are particularly important in a research environment, but also in commercial applications, because they will enable the optimal control system to take over in case such action serves overall economy. The costs for such provisions, when envisaged right from the beginning, are usually very low and hence can hardly be a reason not to do it.

If the system is set up in a modular way, it will greatly simplify installation and maintenance (item 12), although work must be done to design test beds and diagnosis tools. Perhaps the way this has been done in the car industry today, with plug-in diagnosis tools, can serve as an inspiring example.

The discussion above shows that optimal control can meet the requirements of the future. Although we strongly believe in the power of the methodology, it would not be justified to say that there are no more problems to be addressed. These are the challenges to scientists and engineers working in the field. The issues and potential solutions will be analyzed to some degree in the sections below.

9.4 CHALLENGES FOR SCIENCE AND TECHNOLOGY

In this section, we discuss the major themes, improvement ideas, and open issues that deserve further research or development. We follow more or less the same modular line as before.

9.4.1 Sensors and Monitoring

9.4.1.1 External Input Information

The incorporation of weather data collected locally is already common practice in most greenhouse areas. Interaction with the control system is commonplace, for instance, to close the windows when it rains, to irrigate based on solar radiation, and to interfere with the ventilation controllers. Increasingly, weather forecasts are being used, in particular, to serve as input to daily heat storage–buffering strategies. In the context of optimal control, more intensive use is made of knowledge of current and expected external signals. There are opportunities to further improve on this. To compute the uncertainty in slow costates, more work is needed to use a collection of real weather sets rather than just one single nominal set. In view of the weather uncertainty, the expected annual profit is a stochastic variable. This aspect has been studied to investigate yield variation (e.g., see Cooman and Schrevens, 2007; Schrevens, Bojacá, and Cooman, 2006) but has not yet been fully explored in the context of optimal control. Online, the use of ensemble forecasts to compute the extremes of actual control actions is another interesting area. It has been shown that local forecasts can be improved considerably by incorporating weather sensor data from neighboring nurseries (Doeswijk and Keesman, 2005) The implementation work required to incorporate weather and weather forecasts into optimal control systems is mainly a matter of information and communication technology.

9.4.1.2 Feedback from the Crop

From the point of view of the crop, the optimal control as developed in this book is still open loop; that is, it completely relies on the correctness of the predictions of crop models. Hence, what is urgently needed is feedback from the crop. An obvious improvement would be if the marginal total biomass increase could be measured. In principle, this could be accomplished with load cells, but the water household of plants considerably complicates the matter (e.g., see Helmer, Ehret, and Bittman, 2005). On a somewhat faster timescale, monitoring of photosynthesis rate might also be used as an indirect measure of crop biomass increase.

Instead of focusing on the rate of change of biomass, it is another avenue to estimate the current biomass states from sensor information from individual plants. At the plant level, monitoring of sap flow, leaf thickness, leaf photosynthesis, and leaf evapotranspiration can be done (Ehret et al., 2001). Also, there has been some experimentation going on with electrical signals produced by plants (Huang et al., 2010; Wang et al., 2009). Information from several plants can be collected on the basis of sensor networks (e.g., see Van Tuijl, Van Os, and Van Henten, 2008). A different avenue is to approach the crop status at the gene level. An example is the search for quantitative trait loci (QTLs) that link to crop "vigor," as indicated by characteristics such as shoot length and biomass, leaf area, number of roots, root biomass, partitioning coefficients, and growth rates (Cairns et al., 2009; Yan et al., 2007). So far, these applications are mainly directed to selection of species, but it is conceivable that such methods will eventually be useful on the operational level as well, although online application is still a long way to go.

9.4.1.3 Sensor Fusion; Soft Sensors

All this monitor information in itself is not enough to provide appropriate crop feedback to the grower or to the optimal control system. There are two additional challenges: sensor fusion and translating individual plant information to information at the canopy level. Sensor fusion deals with the problem of how to merge data from various sources with varying reliability, measurement frequency, and so forth into useful information. The optimal controller, in the end, will need reliable crop state information. In the optimal control context, both the fusion and the interpretation problem can be tackled by model-based observers or state estimators. Such model-based reconstruction of rates or states is often denoted as "soft sensor" technology. As crop models are needed anyway in the optimal control approach, this avenue is attractive and feasible. Plant sensors must, however, become more reliable and robust before this will find widespread application, and much more work is needed to develop these soft sensors for use in practice.

9.4.2 Physical Modeling

9.4.2.1 Lumped Physical Models

Lumped models for greenhouse physics have reached a mature state. These models are in state–space form by nature and can easily be accommodated in optimal control.

9.4.2.2 Moisture and Condensation Prediction

The proper modeling of humidity in the greenhouse in the presence of varying crop evapotranspiration and the possibility of condensation has made great progress but is still a challenge, especially in hot climates. A way to circumvent the problem is to make model parameters adaptive (see Section 9.4.4.3).

9.4.2.3 Spatial Distribution

Lumped models are not able to deal with issues of spatial heterogeneity, which nevertheless is important in greenhouses. The major approach to deal with spatial distributions is based on CFD methods, but this approach is not suitable in direct form for optimal control. There is a clear need, in general, for control methods that can be applied to spatially distributed systems and flow-dominated systems. An intermediate way can be to use the CFD outcome to build a coarse compartmental model in state–space form for use in optimal control. It is likely that better spatial control will also require a wider spectrum of actuators or at least a distributed control of the actuators present (window sections, separate CO_2 and air circulation hoses, and separate controllable heating sections). Although most approaches focus on creating larger homogeneity, another avenue would be to exploit the spatial distribution. In the vertical direction, gradients are not necessarily bad and can be exploited, under circumstances, for energy savings. Horizontal gradients may be exploited in the overall logistics of greenhouse operations, that is, by moving pot plants around on automatic racks. This is a whole new area of research.

9.4.3 Crop Models

9.4.3.1 State–Space Form, Hybrid Models, and Time–Variable Structure

On the crop side, the situation is less favorable than on the physical modeling side. Fairly detailed crop models exist for major crops, such as tomato (TOMGRO, for references to various versions of this model, see Cooman and Schrevens, 2007; or TOMSIM, Heuvelink, 1999), sweet pepper (Marcelis et al., 2006), and lettuce. Several agronomic crop models contain constructs with varying timescales, where the photosynthetic production is collected over a day and then is distributed over the various organs only once a day. This does not comply with the state–space form required, although it can relatively easily be remedied at the expense of more computation time. In addition, agronomic models that describe various development stages are hybrid models, where the structure changes when new organs are formed. These are not in the general state–space form, although the various stages can be described by state–space models. The remaining issue is then to handle the switching moments or, in more elaborate models, the age distributions of the various organs. It should be noted that there are no fundamental barriers against using variable structure models in an optimal control framework, as was demonstrated in an experiment on the control of sweet pepper fruit set (Buwalda et al., 2006; Van Henten et al., 2006).

9.4.3.2 Crop Model Process Details

In optimal control, regions of temperature are found where crop yield is favorable as compared with energy costs. To obtain trustworthy results, it is important that the temperature dependencies in the model are as accurate as possible. Let us analyze this point a bit further. Suppose that a linear relationship is used for crop growth with a positive slope. In moderate climate zones, the optimal control will on the one hand try to increase the temperature as much as possible to boost yield but on the other hand try to decrease the temperature to save energy. The exact position of the resulting temperature depends on the price ratios, but the counteracting forces will reduce chances that temperature reaches a range where the linear relationship will no longer hold. However, in hot climates, where costs are associated with cooling, it is beneficial to the optimal control to increase the temperature as much as possible, thus making it almost sure that the model will be stretched out of its linear validity range, almost surely leading to erroneous control. This underlines the need to include as accurate relationships as possible. Mathematical simplicity must not be used as an excuse to avoid a little more elaborate expressions.

In chemical engineering, it is usual to model the dependency of a reaction rate by an Arrhenius relation. As Thornley and France (2007) pointed out, this is hardly a proper way of modeling in aggregated biological models. In general, there is a temperature range where crop growth and development proceed at maximum rate, and deviation below or above leads to diminishing returns. One would think that, in principle, such emerging behavior should be the result of underlying reactions that all follow the exponential laws of chemistry, where some processes act as propagators and others as inhibitors, such that it results in the overall bell-shaped temperature dependency as observed in practice. The emerging behavior can be viewed as the outcome of millions of years of evolution, where plants have been adapted to the natural temperature variations. In current crop modeling, the degree of detail on the cellular level is not large enough to "explain" these emerging properties. This is not really a problem for optimal control because a smooth empirical relationship can be used without difficulty. The price that has to be paid is that calibration per cultivar is necessary. If, for a specific crop, temperature relations in the extreme ranges are not known, it is time to start comprehensive experiments to elucidate these.

An element in crop modeling that requires more attention for use in optimal control is the issue of how growth and maintenance respiration need to be modeled. The basic equations for build-up of assimilates (nonstructural biomass W_B) and structural biomass (W_V) in the case where the assimilate buffer is always plenty are given by

$$\frac{dW_B}{dt} = P - (1+\theta)G - R \tag{9.1}$$

and

$$\frac{dW_V}{dt} = G \tag{9.2}$$

so that for the sake of the analysis here, the total biomass is expressed as

$$\frac{dW}{dt} = P - \theta G - R \tag{9.3}$$

The rates P, θG, and R are the photosynthesis, growth respiration, and maintenance respiration rates, respectively, with θ expressing the surplus amount of assimilates needed to produce one unit of structural biomass. Further detailing this with constitutive relations gives

$$P = P_{max} f\left\{C_{CO_2}, I^{PAR}\right\} f_m\{W\} \tag{9.4}$$

where $f\{C_{CO_2}, I^{PAR}\}$ represents the attenuation of the maximum growth rate P_{max} as a function of CO_2 and PAR, and $f_m\{W\}$ is some "maturity" function, expressing the changing interception of light by the canopy when the crop matures. Whatever the exact form of $f_m\{W\}$, it has the property that $f_m\{W\} \propto W$ when W is small, whereas $\lim_{W \to \infty} f_m\{W\} = 1$.

In Chapters 6 and 7, respiration terms were set proportional to the biomass, that is,

$$G, R \propto W \tag{9.5}$$

Hence, when the crop is young, in view of the properties of $f_m\{W\}$, the rate of change of the biomass is proportional to the biomass, and we have exponential growth. With the closure of the canopy, the term $f_m\{W\}$ approaches 1 so that photosynthesis per unit area reaches its maximum, whereas the respiration terms increase with biomass and hence with age. Therefore, this model leads to the leveling off of biomass with plant age as expressed by, for instance, logistic growth curves and similar empirical descriptions of biomass evolution over time.

However, in his development of models for a lettuce crop, Seginer (2003) argued that it is more appropriate to set the maintenance and also the growth rate terms proportional to $f_m\{W\}$, with proportionality factors that may depend on environmental variables, but not on crop biomass, that is,

$$G, R \propto f_m\{W\} \tag{9.6}$$

This conjecture is supported by reasoning and some experimental results that the growing parts take more part in the cell processes than the mature parts of the leaves. With this assumption, there is again exponential growth when the leaves are young, but now the biomass evolution moves asymptotically to linear growth later on so that overall an expolinear curve results. This model does not encapsulate the leveling off during senescence but is perhaps more realistic in the mid range of crop age.

The relevance for optimal control is that, with Equation 9.6, a model results, where the relative net growth can be approximated by a product of two terms, both not depending on the greenhouse climate conditions and the crop state (see Equation 5.45). This construct allows for interesting analyses of crop behavior (e.g., see Seginer, 2004). More importantly from the point of view of optimal control, it was shown to lead to considerable simplification of the optimal control problem because it allows for a transformed costate that is roughly constant throughout the season (Seginer and Ioslovich, 1998). Moreover, only one slow costate is enough.

In view of the desire to make optimal crop cultivation as widely applicable as possible, it is desirable to have a description that combines the elements of both approaches. Clearly, the simplification that allows the development of analytical control strategies will then only hold over a particular crop range. The problem of knowing when the model is valid can be overcome by using a more comprehensive model in conjunction with numerical procedures.

Another matter of uncertainty in crop models refers to the temperature dependency of terms that determine the partitioning over leaves and fruits in generative crops. The parameters of the temperature functions can be chosen in such a way that high temperatures favor more fruit growth or the other way around. It is clear that for the optimal control, it is of paramount importance to have the correct relationships for the crop at hand, and the same is true for model calculations, for instance, the computations made by Heuvelink et al. (2008), to evaluate the potential benefits of a closed greenhouse.

9.4.3.3 Crop Development

The relatively simple models used in this book are not able to encapsulate development issues, such as internodal length, leaf thickness, fruit set and fruit number, and flower abortion. There are two fundamentally different approaches to tackle this.

The first is to use more elaborate models, like TOMGRO or TOMSIM. If these models can be cast in state–space form, there is no fundamental problem in using these models in an optimal control setting. Gradient methods can be used if nondifferentiable functions, like growth rate–temperature tables, are smoothed. Also, special attention is needed when there is a transition in structure, for instance, going from the stage with leaves only to a stage with fruits. In methods that do not use gradients, there is no real problem when the number of states varies. It will, however, not be easy to obtain and use costate information, and it may be worthwhile to investigate simplified approaches such as direct use of observations by the grower.

The second approach is to circumvent the problem by formulating constraints, as discussed later.

9.4.3.4 Expansion of the Operational Range

As discussed above, in the details of crop modeling, still quite a number of issues must be solved. Moreover, models developed for current circumstances may not hold in the more extreme circumstances that can be expected in novel greenhouses with optimal control. It should be noted that it will probably be very hard to distinguish between various alternate models on the basis of collected biomass samples over time in view of the large variance of biomass measurements in comparative tests. Once more, it will be necessary to experiment over wider operational ranges than currently customary and to support experiments by model computations right from the start, for instance, to mitigate the unavoidable effects of spatial heterogeneity (e.g., see Bojacá, Gil, and Cooman, 2009). Accurate models are crucial to prevent the optimal control from dwelling in regions that are optimal only on paper.

Although further development of crop modeling is in fact the "royal" way, it is not necessary to wait until all issues have been resolved. In the meantime, operational constraints can be used to limit the playground for optimal control. This is discussed further in Section 9.4.5.

9.4.3.5 Stress and Vulnerability Models

In the frame of optimal control, not much has been done so far to attempt to incorporate stress modeling, although this would not be essentially different. Examples are the models and the optimal control strategies developed to describe nitrate shortage effects on lettuce and nitrate in lettuce (De Graaf, 2006; Linker, Seginer, and Buwalda, 2004; Seginer, 2003).

Conceptually more difficult is the prediction of diseases and pests as a function of climate variables because there is a stochastic aspect here. Even if the circumstances are favorable for crop health attacks, this does not mean that the attack is actually occurring. One could make efforts to predict unfavorable conditions, such as condensation on fruits or leaves, and then set constraints to back off from these regions. This would be an improvement over fixed conservative bounds for humidity.

9.4.3.6 Crop Quality

Product quality in terms of morphology (pot plants, cut flowers), taste (tomato), or composition (e.g., nitrate in lettuce) is another issue of considerable interest. When it becomes possible to express the effect of environmental conditions on quality attributes in the form of models, quality and its economic value can be incorporated in the goal function. A review on the use of NIR to detect ripeness has been given by Nicolaï et al. (2007). A start has been made to link agronomic crop models to models that describe morphology (L-system models). An entrance to this so-called functional–structural plant modeling field was studied by Hanan and Prusinkiewicz (2008). Other current developments go in the direction of relating measurable crop genetic properties to the phenotype (for gene expression in iris flowers during flowering, see Van Doorn et al., 2003), although much work is still needed to clarify how the greenhouse climate enters the game. There is no doubt that we will see significant developments in the genetic and "omics" arena, which eventually can be incorporated in control.

9.4.4 Modeling Methodology

9.4.4.1 Model Identification, Calibration, and Sensitivity

Modeling has two components: incorporating accumulated knowledge from the past and confronting the model with current data. The first takes the form of established relations and handbook sets of parameters. The second involves calibration or, in a wider sense, model system identification from data, which ideally can reveal alternate model structures as well, thus serving as hypothesis testing. However, it is well known from systems theory that a particular input–output behavior might be represented by alternative state–space representations. The biophysical models that arise from collecting past knowledge, in general, do not represent minimal representations and can be assumed to be overparameterized. As long as reliable parameter information is available from separate experiments, this is not a problem. It can even be seen as an advantage because the model may have predictive power over a wider range than covered by current input–output data, as long as the conditions under which the original parameters have been collected remain equally valid. Where input–output models lose their meaning when stretched beyond the validity range, physics-based state–space models have the potential to remain valid over a wider range. Nevertheless, the simplifications made when deriving the model may jeopardize the results in a different setting. It is remarkable that there are few really convincing validation and cross-validation studies in greenhouses, with proper analysis of residual error, and a clear indication of estimated parameter uncertainty. It seems that this is generally considered as a past station, but the truth is that calibrations that are valid in a particular season fail in other seasons, and when a model like TOMGRO, for instance, is transferred to another region, it has to be recalibrated for reasons that are still unclear.

Sensitivity analysis is a first step to elucidate the role of various parameters (e.g., see Cooman and Schrevens, 2007; Linker, Seginer, and Buwalda, 2004; Schrevens, Bojacá, and Cooman, 2006; Van Henten and Van Straten, 1994; Van Straten et al., 1999). In the optimal control context, it is not so much the sensitivity of the model that counts, but rather the effect of model parameter sensitivity and uncertainty on the ultimate control solution, as pointed out by Van Henten (2003). In this area, more work can be rewarding because it may indicate which aspects of model inaccuracy are most important to tackle.

Calibration methodology is another issue of considerable interest. Because there are many parameters to calibrate in a combined physics and crop model, an immediate question to the model developer is which parameters to select for calibration. In Chapters 7 and 8, methodologies on the basis of the Fisher information matrix were indicated. A more in-depth treatment in the frame of greenhouse optimal control has been presented by Ioslovich, Gutman, and Seginer (2004). The methodology can also be used for the design of informative experiments (Ioslovich and Gutman, 2007).

9.4.4.2 Model Reduction

Increased computational power has allowed the admittance of more and more complex models. However, comprehensive models will tend to be relatively slow, which in the context of dynamic optimization is undesirable. Hence, there is a case for model reduction, with the purpose to catch only the essentials and to increase the confidence in the remaining parameters. The tomato model case in Chapter 7 was already an example of model reduction, albeit on the basis of heuristic arguments. More formal methods are available on the basis of local linearization and singular value decomposition. In a study on tomato, Ioslovich and Gutman (2008) presented a stage-oriented process, where the optimal control problem associated with a reduced model of similar type as in Equations 9.1 through 9.4 and Equation 9.6 is solved and a corresponding set of costate variables are used for the optimization of the TOMGRO trajectory. The reduced model had two state variables and five parameters in contrast to seventy-one state variables and fifty parameters in the TOMGRO model. It was shown that the parameters for the simplified model could successfully be identified from the TOMGRO data. In the study of Seginer (1997), successful work has been quoted to mimic the TOMGRO behavior by an artificial neural net. In addition, there also seems room to reduce the space of relevant external inputs. The article also discusses the possibility to mimic the optimal control itself with an artificial neural net.

Common to these solutions is the use of an elaborate model as a kind of virtual reality against which reduced models are calibrated. Obviously, calibration on the basis of real data instead of virtual data would be more direct. However, the advantage of using detailed models over data is that they allow "virtual" experiments with different excitations, which in a real greenhouse would be time consuming and expensive. Care should be taken to make sure that the validity domain of the original as well as the reduced version match each other and is wide enough to capture the operation range of the optimally operated greenhouse. In fact, as pointed out by Seginer (1997), the success of, for instance, an artificial neural net to describe the observed system's behavior can be viewed as a benchmark for the quality of the physical model.

9.4.4.3 Parameter Variability and Adaptation

As models are simplifications by design, model parameters can be seen as expressions for underlying unmodeled subprocesses. The basic assumption is that the variability in these subprocesses can be ignored with respect to the overall dynamics. Experience with greenhouse and crop models shows that in elaborate physical models this assumption holds more than for crop models. However, even then, there is carryover from crop uncertainty to the physics because the latter is strongly influenced by biophysical processes like evapotranspiration and photosynthesis. Moreover, there are variations due to gradual as well as sudden structural changes in the system dynamics, such as crop development, pollution of the cover, whitening, and partial failure of equipment. Hence, overall, it can be expected that model parameters will vary over time and that information to predict changes automatically is not available. This calls for online parameter adaptation, such as, for instance, the study of Speetjens, Stigter, and Van Straten (2009), where evapotranspiration and water balance parameters were estimated in an adaptive way. Issues related to application of adaptive models in control are discussed in Section 9.4.7.2.

9.4.5 Goal Function

9.4.5.1 Formulation of Goal Function

Proper formulation of an economic goal function is not trivial. Fluctuating market prices, contracting, and energy policies all need to be incorporated. In the evaluation of optimal control solutions, models and assumptions for these aspects are needed, and so far, little is known about the effects of uncertainties in economic model parameters on the results of optimal control or the control policy itself. Despite these, the fact that optimal control requires the explicit formulation of a goal function is a definite advantage. Moreover, it allows rapid and automatic adaptation of the control actions to new economic parameters. Market prices are known to vary over a season; see, for instance, the analysis by Van Henten (1994) for lettuce. In any case, the grower is the entrepreneur and should make the final decisions regarding his expectations of the market. It should be noted that the framework also offers a method to make a choice out of possible indifferent alternate solutions arising from multiple-objective optimization because, in the end, there are reasons for a grower to prefer one solution above another. Once known, this kind of decision making can be incorporated within a single goal function, with parameters that express the grower's preferences.

9.4.5.2 Constraints and Penalties

In fact, setting environmental constraints can be seen as a poor man's answer to the lack of adequate models. Environmental constraints on the system states should ideally not be necessary if the models would be able to predict detrimental effects of extremes. Researchers have attempted to relax the restrictive effect of fixed constraints by replacing them by time variant versions or by the temperature integral concept. The developmental aspects of the temperature integral and temperature and humidity regimes in the context of optimal control have been investigated among others by Körner (2003). Apart from having predefined constraints, the grower may like to overrule them when the visually observed development is not as expected, but it is clear that this limits the room for optimization. Constraints will also be used to back off from regions that are considered to be a risk.

9.4.5.3 Risk

An issue that definitely requires more research efforts is the question of how risk could be included in the goal function. Currently, risk is avoided by setting bounds on the states and by handling constraint violation by penalties. Improvements can be expected if better stress and vulnerability models become available, as discussed before. The degree to which backing off from constraints is selected will still be somewhat arbitrary.

9.4.5.4 Stochastic Variability

In the design and evaluation stages of optimal controllers, the fact that the weather is a stochastic variable makes the ultimate goal function value a stochastic variable itself. Although uncertainty propagation has been investigated in models, much less information is available on the expected distribution of the goal function and, in addition, on the slow costates that in fact also become stochastic variables. The uncertainty in goal function can be viewed as the normal agricultural variability that cannot be avoided, but the costate uncertainty will have an effect in the online receding horizon controller. It is therefore essential to asses these uncertainties. Results presented by Van Henten (1994) suggest that the variability of the slow costate with weather is not large. Seginer and Ioslovich (1998) performed a transformation such that the costate remains constant over the season, although the analysis is limited because it is based on stationary periodic weather. Nevertheless, there is good hope that nominal slow costates are good stand-ins for the actual values. In any case, it would always be possible to solve the slow subproblem from time to time on the basis of the currently available historical weather information to improve the estimates. More research in this direction would be welcome.

9.4.6 Offline: Dynamic Optimization Methods

Although dynamic optimization is a well-matured field in science (see Chapter 3), still attention is required for implementation in online controllers. In particular, it is necessary to have tests on whether the computed optimum is not a local optimum. Another solution is to use optimization methods directed at maximizing chances to find a global optimum, such as genetic algorithms (e.g., see López Cruz, 2002). It also occurs that there are nondecisive patterns that may lead to unduly long searches. For instance, imagine that heat loss and other temperature effects were purely linear with temperature, then all temperature trajectories with the same integral over time would yield the same goal function value. However, in the setup proposed in this book, this problem is less severe, as the major outcome of the dynamic optimization is the trajectory of the slow-state variables. Therefore, techniques to check the reasonability of the trajectories found would suffice.

In the examples worked out in this book, aquifer storage buffers have not been associated yet with slow costates. This is an issue for further research, although we do not foresee fundamental difficulties.

Computationally, dynamic optimization methods have difficulty with high-frequency components of the external signals. Because of the timescale decomposition, smoothing is required, and as the computation is done with the smooth nominal trajectory, this is not a real problem at the start of a season. To keep the slow costates up to date, the computation is repeated from time to time with actual external signal trajectories that are known until then. In that case, a choice must be made of what the smoothing filter for the real data should be. There is no experience with this yet.

Reduction of computation time is always welcome, but as the dynamic optimization has to be done at the start of the cultivation only and perhaps a few times during the cultivation, it is no problem to spend, for instance, a day to achieve the optimal slow costates. It is even conceivable in standard greenhouses with standard equipment to have them available for a specific climate zone in advance via the Internet or dedicated for each individual grower on the basis of his equipment and his goal function. However, speed of computation might be an issue in investigating the stochastic nature of the slow costates discussed earlier because this requires many repetitions of the optimization. Model reduction (e.g., see Ioslovich and Gutman, 2008) and approximate error propagation methods could be a solution here.

9.4.7 Online Control

9.4.7.1 Receding Horizon Optimal Control

9.4.7.1.1 MPC Issues in General

As receding horizon optimal control (Chapter 4) is, in fact, a special form of MPC, some of the issues related to MPC are also playing a part in RHOC. In a review on a number of architectures for control of large-scale systems with MPC, Scattolini (2009) presented the following points, listed here with our comments pertaining to the greenhouse cultivation problem.

- Need for new algorithms with guaranteed properties. Empirical experience needs to be supported by theoretical analysis. In the case of greenhouse cultivation, stability is probably not a big issue. However, theoretical support for the question on how to increase performance in the presence of structural and external uncertainties is much needed.
- Selection of the control structure. Which disturbances can be handled at the local level, under what circumstances is it feasible to use pseudostatic approximations, and what information needs to be transferred between components? In this book, we have offered a structure via the transmission of slow costates to the online control, but there is still the issue of multiple timescales and time variability of the slow costates themselves, which may have an effect on the hierarchical decompositions that are possible and which could reduce the system interconnectivity and complexity.
- Reconfigurable control structures and hybrid systems. The need for more insight on this point was already discussed in the previous sections. It is also related to temporal unavailability of sensors or actuators due to maintenance and to component failures.
- Optimization algorithms. Important criteria are computational speed and the avoidance of local minima.
- (Distributed) State estimators. The presence of reliable sensors for most physical variables of interest has diverted the interest in state estimators for greenhouses. However, the tendency to cope with spatial distributions puts state estimators back on the agenda. In addition, there is an urgent need for estimators for the crop states.
- System partitioning. Although the theory of optimal control as developed in this book may seem to make partitioning less urgent, the modern greenhouses have several subsystems, such as the boiler/power generator unit, the heat exchangers and heat pumps, the cooling systems, and the short- and long-term buffering. The solution of the optimal control problem becomes simpler if dynamic couplings can be reduced. Moreover, model reduction would be beneficial to simplifying the optimization task. Scattolini (2009) suggested pathways, some of which might be relevant to the greenhouse industry as well, such as, for instance, relative gain array methods.
- Synchronization and communication protocols. With the arrival of "intelligent" sensors and sensor networks, the distribution of multiple actuators and pieces of equipment, and field bus concepts, the standard assumption of synchronous communication cannot be made anymore. Although the processes in greenhouses seem to be slow enough not to bother too much, the issue has to be investigated carefully because performance and stability can deteriorate considerably under delayed communication. Solutions have been offered in the general literature (e.g., see Van Willigenburg and De Koning, 1995). In addition, in any practical application, decisions must be made on sampling intervals, control intervals, handling of events, and so on (e.g., see Van Henten and Bontsema, 2008).

9.4.7.1.2 Implementation Aspects

The pragmatic approach to have the receding horizon controller as a front end that computes the setpoints for the existing classical controllers is a solution that, although suboptimal, can help system vendors to step over the doorstep toward more advanced control. A successful implementation of a receding horizon controller used this way for cold storage of potatoes was discussed by Lukasse et al. (2009). The authors gave an interesting discussion on issues that are encountered when implementing receding horizon optimal control. In particular, they point to the shortsightedness of the RHOC if the prediction horizon is chosen too short. Paraphrasing the observation of the authors, we state that in general the prediction horizon should be larger than the longest time constant of the system. This time constant can be evaluated as the reciprocal of the eigenvalues of the linearized system. In contrast to the cold storage application, in greenhouses the longest timescale is associated to the crop, and for this very reason, the timescale decomposition as explained in Chapter 5 has been developed. If, however, one would be prepared to assume blueprint temperatures and humidity settings, then the relevant timescale is related to the greenhouse dynamics. The assessment of the most appropriate horizon can then be based on this timescale and tested for the nonlinear model by simulation, in the same way as done by Lukasse et al. (2009).

9.4.7.1.3 Computational Speed

In real implementations in greenhouses, performing the online optimization during the control interval of, typically, five minutes is not a big issue, but if the controller has to be tested in simulation, the computation time of a single optimization must be short. The total time needed for a year-round receding horizon simulation with computation time θ_c seconds per optimization and control interval T_s (s) becomes $\theta_c(365 \times 24 \times 3600)/T_s$, meaning that, for instance, with $\theta_c = 60$ s and $T_s = 300$ s, a year-round simulation will take over two months. Thus, repeated online optimization can be time consuming, as was experienced by Tap (2000), who was not able to perform a year-round computation at the time (see also Chapter 7). To be feasible, it is therefore imperative that θ_c/T_s is as small as possible, preferably <1/10000, to have results within an hour. With the advent of faster computers and more efficient search methods, such simulations, however, become increasingly feasible, as shown by Van Ooteghem (2007) (Chapter 8).

Tackling of the computational burden can be done in various ways. One is to use model reduction, as discussed before. The other is to use approximate optimization methods, such as subspace optimization, to approach the true solution more rapidly.

The appearance of fast parallel computing devices on the market, such as field programmable gate arrays (FPGA) (e.g., see Castañeda-Miranda et al., 2006), might bring about a complete change of online optimization possibilities, of which the exploration is yet to begin.

9.4.7.1.4 Stability, Robustness, and Reliability

Experience with receding horizon control in greenhouses thus far suggests that the procedure is robust. Instability because of modeling errors has not been observed. The worst that can happen is that the control solutions are not optimal when the model predictions are wrong. The possibility to adapt the model when persistent bias is observed has, however, not yet been explored (see Section 9.4.7.2).

An issue that needs to be investigated further is what will happen in case of failure of a sensor or other components in the system. This requires, on one hand, automatic fault detection for triggering alarms, but this is not different from standard control. The analysis on how to act is somewhat more involved in the optimal control framework. One possibility is to use fallback mechanisms that keep the system within constraints at all times because this becomes a more important target during failure than optimization.

9.4.7.2 Adaptive Receding Horizon Optimal Control

Several attempts have been reported to develop adaptive controllers for greenhouses (Arvanitis, Paraskevopoulos, and Vernardos, 2000; Rodríguez et al., 2008; Speetjens, 2008; Young et al., 1987).

However, the development of an adaptive optimal control scheme directed to achieve an economic optimal control strategy and online control scheme in the spirit of this book with learning or self-tuning properties is a challenge for the future.

Parameter adaptation could also be used to gradually improve on the models and to move gradually from robust traditional but conservative control toward optimal control. By actively probing the system to provide sufficient excitation, the model parameters are updated to yield more and more accurate predictions. It is known that identification in the loop can lead to instability, so the investigation of stability will need to be part of such research. In contrast to more or less steady industrial applications, the natural excitation already offered by the sun and wind conditions may already be sufficient, but this needs to be investigated. A self-learning optimal active adaptive controller would be a big advantage for implementation and maintenance.

9.4.7.3 Tracking Necessary Conditions for Optimality

An alternative to RHOC for online control is tracking the necessary conditions for optimality (Srinivasan et al., 2003), also denoted by the abbreviation NCO. This is based on the idea that whenever the control is not on a constraint, the condition,

$$\partial H / \partial \mathbf{u} = \mathbf{0} \tag{9.7}$$

must be fulfilled.

The Hamiltonian for the fast subproblem follows from Equation 5.44 and is given by

$$H(\mathbf{u}) = L\left(\mathbf{x}^{s*}, \mathbf{x}^{f}, \mathbf{u}, \mathbf{d}, \mathbf{p}\right) + \boldsymbol{\lambda}^{s*T} \mathbf{f}^{s}\left(\mathbf{x}^{s*}, \mathbf{x}^{f}, \mathbf{u}, \mathbf{d}, \mathbf{p}\right) + \boldsymbol{\lambda}^{fT} \mathbf{f}^{f}\left(\mathbf{x}^{s*}, \mathbf{x}^{f}, \mathbf{u}, \mathbf{d}, \mathbf{p}\right) \tag{9.8}$$

The use of Equation 9.8 to derive the control **u** directly from Equation 9.7 in closed form is not possible as the fast costate $\boldsymbol{\lambda}^{f}$ must be known, which requires the solution of a two-point boundary value problem (see Chapters 3 and 4). In a study on nitrate in lettuce, De Graaf (2006) tried to circumvent the problem in the following way. He starts with a physical model that in the terms of Chapter 7 has a form similar to

$$K_{m} \frac{dT_{m}}{dt} = q_{o\_m}^{rad} - q_{m\_a}^{cond} - q_{m\_a}^{trans} \tag{9.9}$$

$$K_{a} \frac{dT_{a}}{dt} = q_{m\_a}^{cond} - q_{a\_o}^{vent} - q_{a\_o}^{cond} + u_{q} \tag{9.10}$$

$$\frac{V_{a}}{A_{s}} \frac{dC_{H_2O}}{dt} = E_{c} - \varphi_{H_2O,a\_o}^{vent} \tag{9.11}$$

$$\frac{V_{a}}{A_{s}} \frac{dC_{CO_2}}{dt} = -\eta_{CO_2/dw}(P - R) - \varphi_{CO_2,a\_o}^{vent} + u_{CO_2} \tag{9.12}$$

where T_m is the temperature (K) of a virtual material component consisting of greenhouse materials, plant containers, subsoil, and crop; T_a is the air temperature (K); C_{H_2O} is the water vapor concentration (kg[H_2O] m^{-3}); and C_{CO_2} is the CO_2 concentration (kg[CO_2] m^{-3}). K_m is the heat capacity of the virtual material components (J m^{-2}[gh] K^{-1}), $K_a = \rho_a V_a c_{pa}$ is the heat capacity of the air (J m^{-2}[gh] K^{-1}), and V_a/A_s is the air volume per unit greenhouse projected soil area (m). The right-hand side terms consist of flows per unit greenhouse area. The meaning of the remaining symbols is as follows:

$q_{o\_m}^{rad}$	Solar radiation absorbed by the virtual materials	(W m^{-2}[gh])
$q_{m\_a}^{cond}$	Sensible heat transfer from material to air	(W m^{-2}[gh])
$q_{m\_a}^{trans}$	Latent heat associated with evapotranspiration	(W m^{-2}[gh])
$q_{a\_o}^{vent}$	Sensible heat exchange with outside by ventilation	(W m^{-2}[gh])
$q_{a\_o}^{cond}$	Sensible heat exchange with outside by conduction	(W m^{-2}[gh])
u_q	Sensible heat input from the heating pipe system	(W m^{-2}[gh])
$E_c = q_{m\_a}^{trans}/\Lambda$	Evapotranspiration rate of the canopy	(kg[H_2O] m^{-2}[gh] s^{-1})
Λ	Heat of evaporation of water	(J kg^{-1})
$\varphi_{H_2O,a\_o}^{vent}$	Loss rate of water due to ventilation	(kg[H_2O] m^{-2}[gh] s^{-1})
P, R	Photosynthesis and respiration rate, respectively	(kg[dw] m^{-2}[gh] s^{-1})
$\eta_{CO_2/dw}$	Ratio CO_2 per kg dry weight	(kg[CO_2] kg^{-1}[dw])
$\varphi_{CO_2,a\_o}^{vent}$	Loss rate of CO_2 due to ventilation	(kg[CO_2] m^{-2}[gh] s^{-1})
u_{CO_2}	Supply rate of CO_2	(kg[CO_2] m^{-2}[gh] s^{-1})

Next, the problem of needing to know the fast costates is avoided by assuming that the air temperature, humidity, and CO_2 are in pseudo-steady state at all times and by including the costate for the virtual material component in the list of slow costates, similar to how it is done for the crop biomass. In doing so, the optimal slow states and costates can be computed for nominal smooth external inputs in advance by dynamic optimization as usual (see Chapter 3). Once available, the online control problem can be solved without optimization by applying the necessary condition Equation 9.7. This is possible as Equations 9.10 through 9.12 degenerate to

$$0 = q_{m\_a}^{cond} - q_{a\_o}^{vent} - q_{a\_o}^{cond} + u_q \tag{9.13}$$

$$0 = E_c - \varphi_{H_2O,a\_o}^{vent} \tag{9.14}$$

$$0 = -\eta_{CO_2/dw}(P - R) - \varphi_{CO_2,a\_o}^{vent} + u_{CO_2} \tag{9.15}$$

which constitutes an algebraic relation between $\mathbf{u}$, $\mathbf{d}$, $\mathbf{x}^{s*}$, and $\mathbf{x}^f$, whereas the $\boldsymbol{\lambda}^f$ in Equation 9.8 is disappearing as $\mathbf{f}^f = \mathbf{0}$ by virtue of Equations 9.13 through 9.15.

The main remaining problem with this approach is that Equation 9.7 does not hold when the control is on a constraint. It is therefore necessary to find out when that is the case. The author developed a rather complicated decision tree that is based on the analysis of a number of optimal control patterns to decide on periods when the control is at a constraint. For instance, during the night in spring, summer, and autumn, the ventilation is always at its lower bound, whereas it is between bounds in autumn, and so forth. In addition, the quality of the approach depends on the effects on the optimal result of the assumptions in Equations 9.13 through 9.15. Unfortunately, these have not been investigated. Another issue is whether the greenhouse air temperature is really a slow variable. Some difficulties were encountered in finding the true optimum for the offline optimization problem. This may be associated with the rather high frequencies that need to be maintained in the external input sequence to obtain realistic daily patterns of the air temperature. Despite all this, the idea to obtain a closed control law is certainly appealing.

9.4.7.4 Self-Optimizing Control

In the decomposition of the control problem proposed in this book, the linking pin variables are the slow costates. Online, a receding horizon control problem must be solved with these costates as inputs. In Section 9.4.7.3, a solution was discussed that tried to avoid the online optimization. Another interesting idea originating from control of chemical plants was described by Skogestad (2000a, 2000b, 2004). This approach is coined "self-optimizing" control. The basic idea is as follows. After the computation of the nominal optimal solution by dynamic optimization (our "slow problem"), one looks for combinations of variables with the following properties: (1) they are weakly dependent on the external disturbances, (2) they are easily deducible from measurements, and (3) they are sensitive to the controls. If these conditions are fulfilled, one could implement a feedback controller to keep these variables near their nominally optimal path, thus ensuring that the online solution responds to the external disturbances and nevertheless remains close to the optimum. In greenhouse cultivation, the straightforward approach of using the nominal state trajectories as set points clearly does not fulfill requirement item 1.

On the other hand, the slow costates were identified as relatively insensitive to the disturbances, so they do comply with condition item 1. It has been proposed by Seginer (2008) to give the grower direct access to the slow costates. As these represent the marginal values of biomass components, retardation or boosting of growth or development processes may be achieved by changing the precalculated costate variable for the variable of interest. This could also be the mechanism to correct visually observed deviations from the envisaged cultivation path. This requires user intervention but it may be a step toward truly self-optimizing control. For the latter, the slow costates are not directly suitable because they cannot be measured and controlled directly. Nevertheless, there may be combinations of measurable variables, reliably computable variables, and measured external disturbances that, when controlled, keep the system near its optimum. It would therefore be an interesting project to investigate whether such variables exist in the greenhouse cultivation systems. In view of the relatively large disturbances, perhaps a combination of the ideas of self-optimizing control and NCO is an option to go for.

9.4.8 User Interaction

A black box without explanation of why a certain control action is taken is not likely to be accepted by growers. Hence, there is a need to have one or more of the following features:

- A knowledge base attached to the controller that explains its actions to the user (cf. SERRISTE, Tchamitchian et al. 2006; albeit not in optimal control framework).
- A simulation tool to show the grower the effects of user adjustable parameters, such as constraint conditions and market prices.
- A method to translate optimal control results into transparent rules that finally are implemented using the existing infrastructure. In this case, the optimal controller is used only offline to design such a system. It will be suboptimal because the control is hindered by the dynamics of existing devices, but it can be acceptable in practice.

Especially a simulation tool that can tell growers how much backing off from constraints costs and what the consequences will be of changing economic parameters is of paramount importance to allow the grower to play his role as entrepreneur. Such a presentation tool can be developed relatively easily because all ingredients are already available. It is also conceivable to use offline models—perhaps offered via the Internet—for this purpose. Internet tools for design simulations are available (Fitz-Rodríguez et al., 2010), guidance of online operation is just beginning (e.g., see Buwalda et al., 2008), but, as far as we know, remote tools to support online optimal control are not yet available.

9.4.9 Information and Communication Technology

Software and information and communication technology offer increasing opportunities to solve, to circumvent, or to alleviate some of the methodological and technical issues listed in the previous section. Table 9.1 summarizes the potential role of new information and communication technologies in bringing the goal of optimal greenhouse cultivation control closer.

9.5 SHOWSTOPPERS FOR OPTIMAL CONTROL

Despite the need for additional work as listed in Section 9.4, it was argued and demonstrated in this book that optimal control is already possible today. Yet, if optimal control is so fabulous, why do we not see it applied in current practice? This is the topic addressed in this section.

The issue was addressed before by Challa and Van Straten (1991) and Van Straten (1999). To our knowledge, no research has been done regarding factors that prevent radical innovations in the greenhouse controller industry. Hence, the discussion below is somewhat speculative. The analysis is partly based on discussions with climate computer vendors and growers and partly on intuition and experience.

The discussion is split into hard factors related to the methodology itself and soft factors regarding the psychology of acceptance of new technology.

9.5.1 Limitations in State of the Art

Although there are definitely a number of points regarding the methodology that need further scientific attention (Section 9.4), we do not believe that the limitations in current state of the art are the real showstoppers for the introduction of optimal control. However, what is frequently put forward in discussions on optimal control is what will be the benefit gained in the end. Although typically savings in energy of 10–20% have been reported, lack of clear indications on the benefits that can be obtained under various economic and climatological conditions may have been a factor in precluding penetration of optimal systems in the market. Hence, the "what do I gain" question is definitely an important question to address. Yet, this must be placed in perspective: if it were true that more time needs to be invested in optimal control, then the extra investment could be balanced against the gained benefit. We suspect, however, that there is hardly a big difference in time investment

TABLE 9.1
Role of New Information and Communication Technologies and Challenges for Control

Aspect or Issue	Technology	Control Challenges
Feedback from the crop	Individual crop sensors	Data interpretation to canopy level
Spatial heterogeneity	Wireless sensor networks	Data fusion Spatial models Distributed actuation
Model maintenance	Web-based models	Automatic model update
Computational burden	Model reduction	Approximate suboptimal control; near-optimal control
Downward compatibility (old computer hardware)	Model and control reduction	Approximate suboptimal control; near optimal control
	Parallel hardware	Vectorizing control solutions
User support	Web-based model-based advice	Policies to be translated to goal or costate
Effect of user settings and goal choices	Online simulation	Model-based forecast presentation tool
Control system maintenance	Web-based control	Asynchronous control
Economic uncertainty	Web-based price forecast	Price forecast models

between developing advanced traditional control systems and optimal control systems, but there is no objective information on whether this conjecture is true or not. If the difference is marginal, then the investment argument is no reason not to go for optimal control. However, there can be other factors, which we analyze next.

9.5.2 The Human Factor: The Grower

Over time, growers got used to the current way of operation and have learned how to use the computer settings to accommodate their needs. Although a method that would allow them to directly convey their needs to the system would be much easier to use, they are so much used to the current systems that it will not be easy to make a change. It is even conceivable that many of them do not even perceive the need for change.

Yet, it is a fact that growers with the same crop and the same climate computer have large differences in yield. It is only possible to speculate on possible reasons. We cannot say that the handling of the control system is the one and only reason. In the ideal case, the differences are caused by factors that reflect the entrepreneurship of a grower: (1) ability to organize and execute appropriate cultivation methods, such as watering, pruning, and leaf picking; (2) skills in handling abnormal situations; and (3) thoughtful economics-based selection of settings of the climate computer. Of these, the first two have no relation to the climate computer. Thoughtful computer settings can be different among growers because growers may differ in their willingness to accept risk in exchange for potential higher income.

In practice, the situation is less ideal. There are various paradigms among growers on how to choose climate computer settings. Two growers can have remarkable differences in physical knowledge and insight, depending on their level of education, their beliefs on how the system works, and their experience. For instance, it is not common knowledge that moisture control is more important than temperature control to achieve the lowest energy costs. In addition, there are differences among growers regarding their attitude toward computer technology and their willingness and interest to spend time behind the computer screen.

In view of these differences in skills and background, it would be very valuable to have a system that assists the grower in making the correct decisions on the basis of the ultimate goal. This is precisely what model-based optimal control can do. Should such a system be on the market, it will require an adjustment period, but if the grower still has the feel to be provided by information on the state of his crop and that he is the person in control, the benefits will soon become apparent, and rapid adoption can be expected. This is somewhat comparable with the original resistance of pilots against the fly-by-wire technology or, to be closer to the agro sector, of farmers against the milking robot, but once pioneers saw the benefit, others followed quickly.

The current dilemma is that growers will only have the opportunity to start using optimal control methodology if this solution is offered on the market, whereas, on the other hand, vendors are hesitating to adopt the methodology if they believe that growers will not buy it. Therefore, in this way, there is a deadlock and a transition problem. In addition, it has to be said that early attempts to introduce model-based control in The Netherlands failed and caused an aversion against anything related to models for a long time.

9.5.3 Human Factor: The Control Engineer

The realization of the optimal methodology in commercial greenhouse automation equipment requires a fairly high level of expertise. In practice, this solution—if it is considered at all—will be judged against the efforts needed for the design of other more traditional control solutions that are perhaps more familiar to control engineers. By education, control engineers are trained to think in desired response speeds and disturbance rejection sensitivity. In addition, the natural tendency is to think of solutions in "sequence" of required complexity. The first choice is classical PI and PID feedback control, which is appropriate if simple transfer function models are easily obtained. Gain scheduling is used as a practical way to handle the major nonlinearities. Next, when parameters appear uncertain and time varying, either adaptive control or robust control comes into the picture. Large external measurable inputs and state and controller gains are more easily handled by model-predictive control methods. Quadratic optimal output feedback designs together with state observers are less customary, probably because of the relatively poor robustness properties. In any case, what is common in all these solutions is the thinking in terms of response speeds, disturbance rejection, and stability arguments. Optimality in resource use or costs is considered to be a matter of the higher level.

Another perhaps more important factor is that in the greenhouse industry the lead in design of control and automation is not with control engineers but with people with a background in physics. It is tempting and indeed interesting to base the design on quasi-steady state balance equations, where the greenhouse's actual state is characterized by notions like heat demand, CO_2 demand, and (de)humidification demand or ventilation demand. In itself, there is nothing wrong with this kind of thinking because it is a first step to provide control on the basis of feed-forward compensation. However, closure of the balance is possible only if one is prepared to assume an ideal setpoint level for temperature, CO_2, and moisture content. When designers implemented these strategies together with low-level controllers to compensate for small disturbances, it soon became clear that fixed setpoints were not what users needed. One responded to this by introducing night–day patterns and by allowing deviations under high solar radiation input. Next it was found that these are not enough to meet the needs for energy savings. The temperature integral concept came into play, and the industry's answer was to create yet another layer on top of the existing feed-forward/feedback layer. With every new demand, new layers were created. In this way, in many modern climate computers, one can see stacks of solutions, for instance, (1) for temperature integration; (2) energy management to decide on the best deployment of utilities for CO_2, heat, and electricity supply based on demand for CO_2, electric energy, and heat computed from balances; and (3) control of electricity production for the public net by cogeneration on the basis of an agreed power delivery profile provided for the next day by the power company on the basis of current electricity prices.

This approach, while leading to the commercial solutions on the market today, leads to ever more complex systems, with increasing demands to the user who may easily lose track of the many options offered. And perhaps more importantly, the layered approach creates internal conflicts regarding priorities, and we suspect that the final solution is much less optimal than suggested in the glossy selling brochures.

It is interesting to observe that often objections are made against economic optimal control as described in this book on the arguments that (1) much knowledge needs to be available about the process and (2) it requires a high level of expertise. If we compare this with the efforts needed with the control solutions listed above, it can hardly be maintained that these are disadvantages of the optimal control methodology. If a control engineer is able to design an MPC, he would, in principle, have the intellectual capacity to design an economic optimizing integrating controller as well.

Yet, a factor that may well be precluding the adoption of optimal control as a method is that an individual engineer may not have received his boss's support to rethink the complete system. Much effort has been put into components that already exist, and it is understandable that a company requires upward compatibility. There may be a belief that optimal controllers are more difficult to design in teamwork. In modern design, teamwork between developers and software and hardware engineers is necessary. The ability to split a task into several subtasks is necessary on the company level because it makes it easier to hold someone accountable for specific components and allows step-by-step expansion and development of the control system. There might be a perception that optimal controllers are less easy to set up in modular form because, after all, all parts are components of the solution. However, the breakdown in models, optimization, weather input, goal presentation, and online control as outlined earlier, shows that a modular setup is equally possible with optimal controllers. Yet, it must not be denied that there may be a transition problem in the process of rethinking the design strategy.

9.5.4 Leveling the Barriers

The barriers for truly optimal control discussed above can be categorized in three major themes. The first is education, training, and sharing of experience. We hope that this book will contribute to lowering the threshold. The second is the opinion that control is playing a marginal role in the economy of the nursery. If one reckons that a commercial tomato grower in The Netherlands has an annual turnover of about 100 €/m^2 and that roughly 20% of this is for energy, savings of 10% in energy—which are easy to achieve with advanced control—will increase the margin by 2 €. With an original margin in the same order of magnitude, the profit doubles, and this will make the difference between being in business or not. The third limiting factor is the problem of transition from current practice to desired practice. It is clear that the deadlock between suppliers and growers needs to be broken. Will we see the challenge being taken up by the well-established greenhouse climate computer suppliers, or will innovation be brought about by new players who can start from scratch? There is a growing awareness of the need for a new direction, as testified by the following address to an International Society for Horticultural Science meeting by a representative of the greenhouse climate computer industry:

> "A new generation of climate, irrigation and nutrition control will employ crop sensors and models. Feed-forward controllers anticipate the effects of disturbances on the greenhouse climate and take corrective action before they are allowed to occur. Greenhouse models can be used to predict the effects of disturbances. By using a crop model to estimate the benefits to the crop and a greenhouse model to estimate the costs, optimum setpoints can be generated. The reliability of model-based control is significantly enhanced when feedback on the crop's status and growth rate are added. For this purpose, crop sensors need to be developed. Sensor data combined with intelligent algorithms, collectively called 'soft sensors', represent a promising way of obtaining additional information on the growth process. Crop monitoring can also be used as an early warning system (by comparing sensor measurements with reference data) and so help limit the consequences of human error or technical failure. Optimal controllers use a model-based economic assessment to determine the optimum values for various processes and resource input levels. Optimal control will first be introduced as decision support systems at crop process level" (De Koning, 2006).

This quote may perhaps be seen as an indication that the barriers are indeed gradually being leveled.

9.6 CONCLUSIONS AND PERSPECTIVES

In view of the many advantages of greenhouse cultivation over open air cultivation, it can be expected that the greenhouse industry will show substantial growth, especially in upcoming economies. At the same time, the greenhouse industry in its current form is criticized for its excessive use of energy and water. Therefore, it is clear that sustainability will be imperative to be acceptable to society. Innovative solutions will be found that exploit the fact that the greenhouse is, after all, a solar collector. All this will lead to more complex designs. It is very obvious that the associated control problem cannot be solved in a satisfactory manner with a collection of single-loop controllers. The problems that require solutions are

- The proper handling of interactions, complexity, and nonlinear behavior
- A high degree of automation
- Transparency of the automated systems to the user (not hundreds of settings)
- Proper balancing of multiple objectives (energy, crop yield, water)
- Optimality of the result to ensure maximum sustainability and profit

The optimal control framework described and developed in this book offers an elegant, science-based, and feasible solution to these issues. Beyond the greenhouse as entity, there are a number of other interesting perspectives that deserve to be mentioned. In view of the need for sustainability, solutions will be sought beyond the boundaries of the traditional greenhouse industry. For instance, integrating greenhouse production with other farming activities, such as animal breeding, is likely to be developed further. These innovations once again lead to more interaction and interdependency between system components, and the complexity of proper control will grow. The optimal control framework will be attractive to study control policies for the operation of such integrated facilities. It has not generally been noticed that the optimal control framework can also be beneficial to the design stage of greenhouse systems. This is because in the framework, the sensitivity of the predicted economic operation to design parameters is easily investigated. Because economics is part of the game right from the beginning, return on investments under various scenarios is easily evaluated. In this frame, dynamics becomes part of the design, and the problems invoked by the classical separation between the system design stage and the control design stage will be remedied.

Another perspective is that the integrated design and control solutions offered by the methodology of optimal control for greenhouse cultivation may have a spin-off to other sectors. The application to cold storage warehouses has already been mentioned (Lukasse et al., 2009). Other challenging areas are developments toward bioproduction in urban environments (e.g., see Nelkin and Caplow, 2008) or for manned space missions (Albright et al., 2001). The foreseeable end of the fossil fuel and the global warming problem have triggered interest in producing biomaterials and biofuels from algae. Because algae are microscopic plants, production with solar radiation as input has a number of similarities with greenhouse production, and we expect that it will be a rewarding area of research to apply and adapt the methodology to this new, exciting field.

Central to the control solutions offered in this book are the incorporation of crop dynamics and the exploitation of the opportunities offered by the weather and in particular the sun. Only science-based optimal control can achieve this. Despite scientific and technological challenges, there is sufficient basis to start implementing these methodologies in daily practice and to gain experience on the fly. In this way will the control system become not only a instrument to the grower but also an instrument to reach sustainability, which is a *conditio sine qua non* for the vitality of crop production in protected environments.

REFERENCES

Albright, L.D., R.S. Gates, K.G. Arvanitis, and A.E. Drysdale. 2001. Environmental control for plants on earth and in space. *IEEE Control Systems Magazine* 21 (5): 28–47.

Arvanitis, K.G., P.N. Paraskevopoulos, and A.A. Vernardos. 2000. Multirate adaptive temperature control of greenhouses. *Computers and Electronics in Agriculture* 26 (3): 303–320.

Bakker, J.C., S.R. Adams, T. Boulard, and J.I. Montero. 2008. Innovative technologies for an efficient use of energy. *Acta Horticulturae* 801: 49–62.

Bojacá, C.R., R. Gil, and A. Cooman. 2009. Use of geostatistical and crop growth modelling to assess the variability of greenhouse tomato yield caused by spatial temperature variations. *Computers and Electronics in Agriculture* 65 (2): 219–227.

Buwalda, F., G.J. Swinkels, F. De Zwart, J. Kipp, F. Kempkes, T. Van Gastel, C. Burema, and H. Van Bokhoven. 2008. Exchange of knowledge between research and horticultural practice through the Internet. *Acta Horticulturae* 801: 531–538.

Buwalda, F., E.J. Van Henten, A. De Gelder, J. Bontsema, and J. Hemming. 2006. Toward an optimal control strategy for sweet pepper cultivation: I. A dynamic crop model. *Acta Horticulturae* 718: 367–374.

Cairns, J.E., O.S. Namuco, R. Torres, F.A. Simborio, B. Courtois, G.A. Aquino, and D.E. Johnson. 2009. Investigating early vigour in upland rice (*Oryza sativa* L.): Part II. Identification of QTLs controlling early vigour under greenhouse and field conditions. *Field Crops Research* 113 (3): 207–217.

Campen, J.B., F.L.K. Kempkes, and G.P.A. Bot. 2009. Mechanically controlled moisture removal from greenhouses. *Biosystems Engineering* 102 (4): 424–432.

Castañeda-Miranda, R., Jr E. Ventura-Ramos, R. del Rocío Peniche-Vera, and G. Herrera-Ruiz. 2006. Fuzzy greenhouse climate control system based on a field programmable gate array. *Biosystems Engineering* 94 (2): 165–177.

Challa, H., and G. Van Straten. 1991. Reflections about optimal climate control in greenhouse cultivation. In *Mathematical and Control Applications in Agriculture and Horticulture*, ed. H. Hashimoto and W. Day. Oxford: Pergamon Press.

Cooman, A., and E. Schrevens. 2007. Sensitivity of the TOMGRO model to solar radiation intensity, air temperature and carbon dioxide concentration. *Biosystems Engineering* 96 (2): 249–255.

Davies, P.A., and C. Paton. 2005. The seawater greenhouse in the United Arab Emirates: Thermal modelling and evaluation of design options. *Desalination* 173 (2): 103–111.

Dawoud, B., Y.H. Zurigat, B. Klitzing, T. Aldoss, and G. Theodoridis. 2006. On the possible techniques to cool the condenser of seawater greenhouses. *Desalination* 195 (1–3): 119–140.

De Graaf, S.C. 2006. *Low Nitrate Lettuce Cultivation in Greenhouses—Optimal Control in the Presence of Measurable Disturbances*. Wageningen: Wageningen University.

De Koning, A.N.M. 2006. Models and sensors in greenhouse control. *Acta Horticulturae* 718: 175–182.

De Zwart, H.F. 2008. Overall energy analysis of (semi) closed greenhouses. *Acta Horticulturae* 801: 811–817.

Doeswijk, T.G., and K.J. Keesman. 2005. Adaptive weather forecasting using local meteorological information. *Biosystems Engineering* 91 (4): 421–431.

Ehret, D.L., A. Lau, S. Bittman, W. Lin, and T. Shelford. 2001. Automated monitoring of greenhouse crops. *Agronomie* 21 (4): 403–414.

Fitz-Rodríguez, E., C. Kubota, G.A. Giacomelli, M.E. Tignor, S.B. Wilson, and M. McMahon. 2010. Dynamic modeling and simulation of greenhouse environments under several scenarios: A Web-based application. *Computers and Electronics in Agriculture* 70 (1): 105–116.

Giacomelli, G., N. Castilla, E. Van Henten, D. Mears, and S. Sase. 2008. Innovation in greenhouse engineering. *Acta Horticulturae* 801: 75–88.

Hanan, J., and P. Prusinkiewicz. 2008. Foreword: Studying plants with functional–structural models. *Functional Plant Biology* 35 (10): i–iii.

Helmer, T., D.L. Ehret, and S. Bittman. 2005. CropAssist, an automated system for direct measurement of greenhouse tomato growth and water use. *Computers and Electronics in Agriculture* 48 (3): 198–215.

Heuvelink, E. 1999. Evaluation of a dynamic simulation model for tomato crop growth and development. *Annals of Botany* 83 (4): 413–422.

Heuvelink, E., M. Bakker, L.F.M. Marcelis, and M. Raaphorst. 2008. Climate and yield in a closed greenhouse. *Acta Horticulturae* 801: 1083–1092.

Hoes, H., J. Desmedt, K. Goen, and L. Wittemans. 2008. The GESKAS project, closed greenhouse as energy source and optimal growing environment. *Acta Horticulturae* 801: 1355–1362.

Huang, L., Z.-Y. Wang, L.-L. Zhao, D. Zhao, C. Wang, Z. Xu, R.-F. Hou, and X.-J. Qiao. 2010. Electrical signal measurement in plants using blind source separation with independent component analysis. *Computers and Electronics in Agriculture* 71 Supplement 1: 554–559.

Ioslovich, I., and P.O. Gutman. 2007. Evaluation of experiments for estimation of dynamical crop model parameters. *Bulletin of Mathematical Biology* 69 (5): 1603–1614.

Ioslovich, I., and P.O. Gutman. 2008. Fitting the MBM-A model of plant growth to the data of TOMGRO: Implication for greenhouse optimal control. *Acta Horticulturae.* 801: 515–521.

Ioslovich, I., P.O. Gutman, and I. Seginer. 2004. Dominant parameter selection in the marginally identifiable case. *Mathematics and Computers in Simulation* 65 (1–2): 127–136.

Körner, O. 2003. *Crop Based Climate Regimes for Energy Saving in Greenhouse Cultivation*. Wageningen: Wageningen University.

Linker, R., I. Seginer, and F. Buwalda. 2004. Description and calibration of a dynamic model for lettuce grown in a nitrate-limiting environment. *Mathematical and Computer Modelling* 40 (9–10): 1009–1024.

López Cruz, I.L. 2002. *Efficient Evolutionary Algorithms for Optimal Control, Systems & Control*. Wageningen: Wageningen University.

Lukasse, L., J. Van Maldegem, E. Dierkes, A.-J. Van der Voort, J. De Kramer-Cuppen, and G. Van der Kolk. 2009. Optimal control of indoor climate in agricultural storage facilities for potatoes and onions. *Control Engineering Practice* 17 (9): 1044–1052.

Marcelis, L.F.M., A. Elings, J.A. Dieleman, P.H.B. De Visser, E. Brajeul, M.J. Bakker, and E. Heuvelink. 2006. Modelling dry matter production and partitioning in sweet pepper. *Acta Horticulturae* 718: 121–128.

Morrow, R.C. 2008. LED lighting in horticulture. *HortScience* 43 (7): 1947–1950.

Nelkin, J., and T. Caplow. 2008. Sustainable controlled environment agriculture for urban areas. *Acta Horticulturae.*

Nicolaï, B.M., K. Beullens, E. Bobelyn, A. Peirs, W. Saeys, K.I. Theron, and J. Lammertyn. 2007. Nondestructive measurement of fruit and vegetable quality by means of NIR spectroscopy: A review. *Postharvest Biology and Technology* 46 (2): 99–118.

Park, D.H., B.J. Kang, K.R. Cho, C.S. Shin, S.E. Cho, J.W. Park, and W.M. Yang. 2009. A study on greenhouse automatic control system based on wireless sensor network. *Wireless Personal Communications*: 1–14.

Perdigones, A., J.L. García, A. Romero, A. Rodríguez, L. Luna, C. Raposo, and S. De la Plaza. 2008. Cooling strategies for greenhouses in summer: Control of fogging by pulse width modulation. *Biosystems Engineering* 99 (4): 573–586.

Rodríguez, F., J.L. Guzmán, M. Berenguel, and M.R. Arahal. 2008. Adaptive hierarchical control of greenhouse crop production. *International Journal of Adaptive Control and Signal Processing* 22 (2): 180–197.

Scattolini, R. 2009. Architectures for distributed and hierarchical model predictive control—A review. *Journal of Process Control* 19 (5): 723–731.

Schrevens, E., C. Bojacá, and A. Cooman. 2006. Uncertainty of predicted greenhouse tomato production: Mechanistic versus stochastic models? *Acta Horticulturae* 718: 139–146.

Seginer, I. 1997. Some artificial neural network applications to greenhouse environmental control. *Computers and Electronics in Agriculture* 18 (2–3): 167–186.

Seginer, I. 2003. A dynamic model for nitrogen-stressed lettuce. *Annals of Botany* 91 (6): 623–635.

Seginer, I. 2004. Equilibrium and balanced growth of a vegetative crop. *Annals of Botany* 93 (2): 127–139.

Seginer, I. 2008. Co-state variables as strategic set-points for environmental control of greenhouses. *Acta Horticulturae* 797: 69–74.

Seginer, I., and I. Ioslovich. 1998. Seasonal optimization of the greenhouse environment for a simple two-stage crop growth model. *Journal of Agricultural Engineering Research* 70 (2): 145–155.

Sethi, V.P., and S.K. Sharma. 2007. Experimental and economic study of a greenhouse thermal control system using aquifer water. *Energy Conversion and Management* 48 (1): 306–319.

Sethi, V.P., and S.K. Sharma. 2008. Survey and evaluation of heating technologies for worldwide agricultural greenhouse applications. *Solar Energy* 82 (9): 832–859.

Skogestad, S. 2000a. Plantwide control: The search for the self-optimizing control structure. *Journal of Process Control* 10 (5): 487–507.

Skogestad, S. 2000b. Self-optimizing control: The missing link between steady-state optimization and control. *Computers & Chemical Engineering* 24: 569–575.

Skogestad, S. 2004. Near-optimal operation by self-optimizing control: From process control to marathon running and business systems. *Computers & Chemical Engineering* 29 (1): 127.

Sonneveld, P.J., G.L.A.M. Swinkels, G.P.A. Bot, and G. Flamand. 2010. Feasibility study for combining cooling and high grade energy production in a solar greenhouse. *Biosystems Engineering* 105 (1): 51–58.

Speetjens, S.L. 2008. *Towards Model Based Adaptive Control for the Watergy Greenhouse*. Wageningen: Wageningen University.

Speetjens, S.L., J.D. Stigter, and G. Van Straten. 2009. Towards an adaptive model for greenhouse control. *Computers and Electronics in Agriculture* 67 (1–2): 1–8.

Srinivasan, B., D. Bonvin, E. Visser, and S. Palanki. 2003. Dynamic optimization of batch processes: II. Role of measurements in handling uncertainty. *Computers & Chemical Engineering* 27 (1): 27–44.

Sun, Z., H. Cao, H. Li, K. Du, Y. Wang, X. Su, T. Cai, S. Liu, and J. Chu. 2006. GPRS and WEB based data acquisition system for greenhouse environment. *Nongye Gongcheng Xuebao/Transactions of the Chinese Society of Agricultural Engineering* 22 (6): 131–134.

Tap, R.F. 2000. *Economics-Based Optimal Control of Greenhouse Tomato Crop Production, Systems and Control*. Wageningen, The Netherlands: Wageningen Agricultural University.

Tchamitchian, M., R. Martin-Clouaire, J. Lagier, B. Jeannequin, and S. Mercier. 2006. SERRISTE: A daily set point determination software for glasshouse tomato production. *Computers and Electronics in Agriculture* 50: 25–47.

Thornley, J.H.M., and J. France. 2007. *Mathematical Models in Agriculture*. 2nd ed. Wallingford, UK: CABI International.

Van Doorn, W.G., P.A. Balk, A.M. Van Houwelingen, F.A. Hoeberichts, R.D. Hall, O. Vorst, C. Van Der School, and M.F. Van Wordragen. 2003. Gene expression during anthesis and senescence in iris flowers. *Plant Molecular Biology* 53 (6): 845–863.

Van Henten, E.J. 1994. *Greenhouse Climate Management: An Optimal Control Approach*. Wageningen: Agricultural University Wageningen.

Van Henten, E.J. 2003. Sensitivity analysis of an optimal control problem in greenhouse climate management. *Biosystems Engineering* 85 (3): 355–364.

Van Henten, E.J., and J. Bontsema. 2008. Open loop optimal temperature control in greenhouses: Choosing the length of the sample interval in a control parameterization solution. *Acta Horticulturae* 801 (pt. 1): 629–635.

Van Henten, E.J., F. Buwalda, H.F. De Zwart, A. De Gelder, J. Hemming, and J. Bontsema. 2006. Toward an optimal control strategy for sweet pepper cultivation: II. Optimization of the yield pattern and energy efficiency. *Acta Horticulturae* 718: 391–398.

Van Henten, E.J., and G. Van Straten. 1994. Sensitivity analysis of a dynamic growth model of lettuce. *Journal of Agricultural Engineering Research* 59 (1): 19–31.

Van Kooten, O., E. Heuvelink, and C. Stanghellini. 2008. New developments in greenhouse technology can mitigate the water shortage problem of the 21st century. *Acta Horticulturae* 767: 45–52.

Van Ooteghem, R.J.C. 2007. *Optimal Control Design for a Solar Greenhouse, Systems and Control.* Wageningen: Wageningen University.

Van Straten, G. 1999. Acceptance of optimal operation and control methods for greenhouse cultivation. *Annual Reviews in Control* 23: 83–90.

Van Straten, G., I. Lopez Cruz, I. Seginer, and F. Buwalda. 1999. Calibration and sensitivity analysis of a dynamic model for control of nitrate in lettuce. *Acta Horticulturae* 507: 149–156.

Van Tuijl, B., E. Van Os, and E. Van Henten. 2008. Wireless sensor networks: State of the art and future perspective. *Acta Horticulturae* 801: 547–554.

Van Willigenburg, L.G., and W.L. De Koning. 1995. Derivation and computation of the digital LQG regulator and tracker in the case of asynchronous and aperiodic sampling. *Control Theory and Advanced Technology* 10 (4, pt. 5): 2083–2098.

Wang, Z.-Y., Q. Leng, L. Huang, L.-L. Zhao, Z.-L. Xu, R.-F. Hou, and C. Wang. 2009. Monitoring system for electrical signals in plants in the greenhouse and its applications. *Biosystems Engineering* 103 (1): 1–11.

Yan, Z., P.B. Visser, T. Hendriks, T.W. Prins, P. Stam, and O. Dolstra. 2007. QTL analysis of variation for vigour in rose. *Euphytica* 154 (1–2): 53–62.

Young, P., M.A. Behzadi, C.L. Wang, and A. Chotai. 1987. Direct digital and adaptive control by input–output state variable feedback pole assignment. *International Journal of Control* 46 (6): 1867–1881.

Yu, C., Y. Cui, L. Zhang, and S. Yang. 2009. ZigBee wireless sensor network in environmental monitoring applications. In *Proceedings of the 5th International Conference on Wireless Communications, Networking and Mobile Computing, WiCOM 2009, Beijing, IEEE.*

Zaragoza, G., and M. Buchholz. 2008. Closed greenhouses for semi-arid climates: Critical discussion following the results of the Watergy prototype. *Acta Horticulturae* 797: 37–42.

Zaragoza, G., M. Buchholz, P. Jochum, and J. Perez-Parra. 2007. Watergy project: Towards a rational use of water in greenhouse agriculture and sustainable architecture. *Desalination* 211 (1–3): 296–303.

Index

D

E

F

G

H

T